Manufacturing techniques for polymer matrix composites (PMCs)

Related titles:

Failure mechanisms in polymer matrix composites: Criteria, testing and industrial applications
(ISBN 978-1-84569-750-1)
Polymer matrix composites are increasingly replacing traditional materials, such as metals, for applications in the aerospace, automotive and marine industries. This important book explores the main types of composite failure and examines their implications in specific applications. Part I discusses various failure mechanisms, including manufacturing defects, and addresses a variety of loading forms, such as impact and the implications for structural integrity. Testing techniques and modelling methods for predicting potential failure in composites are also reviewed. Part II investigates the effects of polymer-matrix composite failure in a range of industries and looks at recycling issues and environmental factors affecting the use of composite materials.

Polymer matrix composites and technology
(ISBN 978-0-85709-221-2)
Given such properties as low density and high strength, polymer matrix composites have become a widely used material in the aerospace and other industries. *Polymer matrix composites and technology* provides a helpful overview of these materials, their processing and performance. After an introductory chapter, Part I reviews the main reinforcement and matrix materials used as well as the nature of the interface between them. Part II discusses forming and molding technologies for polymer matrix composites. The final part of the book covers key aspects of performance, including tensile, compression, shear and bending properties as well as impact, fatigue and creep behaviour.

Creep and fatigue in polymer matrix composites
(ISBN 978-1-84569-656-6)
Creep and fatigue in polymer matrix composites reviews ways of modelling creep and fatigue in polymer matrix composites with the aim of predicting and preventing failure. The first part of the book reviews the modelling of viscoelastic and viscoplastic behaviour as a way of predicting performance and service life. Part II discusses techniques for modelling creep rupture and failure. The final part of the book discusses ways of testing and predicting long-term creep and fatigue in polymer matrix composites.

Details of these and other Woodhead Publishing materials books can be obtained by:

- visiting our web site at www.woodheadpublishing.com
- contacting Customer Services (e-mail: sales@woodheadpublishing.com; fax: +44 (0) 1223 832819; tel.: +44 (0) 1223 499140 ext. 130; address: Woodhead Publishing Limited, 80 High Street, Sawston, Cambridge CB22 3HJ, UK)
- contacting our US office (e-mail: usmarketing@woodheadpublishing.com; tel. (215) 928 9112; address: Woodhead Publishing, 1518 Walnut Street, Suite 1100, Philadelphia, PA 19102-3406, USA)

If you would like e-versions of our content, please visit our online platform: www.woodheadpublishingonline.com. Please recommend it to your librarian so that everyone in your institution can benefit from the wealth of content on the site.

Manufacturing techniques for polymer matrix composites (PMCs)

Edited By
Suresh G. Advani and Kuang-Ting Hsiao

WOODHEAD PUBLISHING

Oxford Cambridge Philadelphia New Delhi

Published by Woodhead Publishing Limited,
80 High Street, Sawston, Cambridge CB22 3HJ, UK
www.woodheadpublishing.com
www.woodheadpublishingonline.com

Woodhead Publishing, 1518 Walnut Street, Suite 1100, Philadelphia,
PA 19102-3406, USA

Woodhead Publishing India Private Limited, G-2, Vardaan House, 7/28 Ansari Road,
Daryaganj, New Delhi – 110002, India
www.woodheadpublishingindia.com

First published 2012, Woodhead Publishing Limited

British Library Cataloguing in Publication Data
A catalogue record for this book is available from the British Library.

Library of Congress Control Number: 2012938538

ISBN 978-0-85709-067-6 (print)
ISBN 978-0-85709-625-8 (online)

The publisher's policy is to use permanent paper from mills that operate a sustainable forestry policy, and which has been manufactured from pulp which is processed using acid-free and elemental chlorine-free practices. Furthermore, the publisher ensures that the text paper and cover board used have met acceptable environmental accreditation standards.

Typeset by Newgen Publishing and Data Services
Printed by TJ International Ltd, Padstow, Cornwall, UK

Contents

Contributor contact details *xi*

1 Introduction to composites and manufacturing processes 1
S. G. Advani, University of Delaware, USA and K.-T. Hsiao, University of South Alabama, USA

1.1 Processing of polymer matrix composites 1
1.2 Focus and scope of this book 3
1.3 References 11

Part I Manufacturing of polymer matrix composites (PMCs): short fiber and nanomaterial based processing 13

2 Injection molding in polymer matrix composites 15
S.-J. Liu, Chang Gung University, Taiwan

2.1 Introduction to injection molding 15
2.2 Molding compounds 21
2.3 Characterization and prediction of fiber orientation 34
2.4 Molding defects 37
2.5 Conclusions and future trends 41
2.6 Acknowledgment 42
2.7 References 42

3 Compression molding in polymer matrix composites 47
C. H. Park, University of Le Havre, France and W. I. Lee, Seoul National University, Korea

3.1 Introduction 47
3.2 Molding materials 53
3.3 Process defects and remedies 61
3.4 Recent developments in press design and process optimization 67
3.5 Modeling and simulation 69
3.6 Future trends 90

3.7 Conclusions 92
3.8 References 92

4 Processing of polymer nanocomposites 95
J. Gou, J. Zhuge and F. Liang, University of Central Florida, USA

4.1 Introduction 95
4.2 Process description 96
4.3 Methods to improve process 107
4.4 References 115

Part II Manufacturing of polymer matrix composites (PMCs): thermoplastic based processing 121

5 Sheet forming in polymer matrix composites 123
T. Creasy, Texas A&M University, USA

5.1 Introduction: key objectives 123
5.2 Process description 124
5.3 Matrix flow and fibre deformation 134
5.4 Changes to the process to improve product quality 136
5.5 Future trends 136
5.6 Sources of further information and advice 137
5.7 References 137

6 Fabric thermostamping in polymer matrix composites 139
J. A. Sherwood, K. A. Fetfatsidis and J. L. Gorczyca, University of Massachusetts Lowell, USA and L. Berger, General Motors Research and Development Center, USA

6.1 Introduction 139
6.2 Process description 142
6.3 Material characterization 147
6.4 Modeling 159
6.5 Methods of improving the process to improve product quality 169
6.6 Future trends 175
6.7 Sources of further information and advice 175
6.8 Acknowledgments 178
6.9 References 178

7 Filament winding process in thermoplastics 182
J. MACK, Institut für Verbundwerkstoffe GmbH, Germany and R. SCHLEDJEWSKI, Montanuniversitaet Leoben, Austria

7.1 Introduction 182
7.2 Winding basics 184
7.3 Winding process 186
7.4 Simulation tools 199
7.5 Component quality 202
7.6 Future trends 203
7.7 Sources of further information and advice 204
7.8 References 205

8 Continuous fiber reinforced profiles in polymer matrix composites 209
P. MITSCHANG and M. CHRISTMANN, Institut für Verbundwerkstoffe GmbH, Germany

8.1 Introduction 209
8.2 Pultrusion 210
8.3 Continuous compression molding 225
8.4 Preferred application areas for the pultrusion process and continuous compression molding process 238
8.5 References 239

Part III Manufacturing of polymer matrix composites (PMCs): thermoset based processing 243

9 Resin transfer molding (RTM) in polymer matrix composites 245
E. M. SOZER, Koc University, Turkey and P. SIMACEK and S. G. ADVANI, University of Delaware, USA

9.1 Introduction 245
9.2 Resin transfer molding (RTM) process steps 252
9.3 Fibers, fabrics and preform manufacturing 253
9.4 Resin system 259
9.5 RTM mold 261
9.6 Resin injection equipment 267
9.7 Issues that influence manufacturing with RTM 270
9.8 The need for process modeling 277
9.9 Resin flow models for RTM 278

9.10 Heat transfer and cure model 292
9.11 Numerical simulation of resin flow 295
9.12 Process control 300
9.13 Conclusions and future trends 303
9.14 References 303

10 Vacuum assisted resin transfer molding (VARTM) in polymer matrix composites 310
K.-T. Hsiao, University of South Alabama, USA and
D. Heider, University of Delaware Newark, USA

10.1 Vacuum assisted resin transfer molding (VARTM) processing 310
10.2 Fundamentals of VARTM 315
10.3 Defects and challenges of VARTM 323
10.4 Recent advances in VARTM 335
10.5 Conclusion and future trends in VARTM 337
10.6 Membrane-based infusion processing 339
10.7 Membrane evaluation 340
10.8 Process and material property improvement 342
10.9 Summary of membrane-based infusion processing 344
10.10 References 345

11 Compression resin transfer moulding (CRTM) in polymer matrix composites 348
S. Bickerton and P. A. Kelly, The University of Auckland, New Zealand

11.1 Introduction 348
11.2 Process description 351
11.3 Material properties and characterisation 355
11.4 Modelling and analysis of the CRTM-1 process 357
11.5 Modelling and analysis of the CRTM-2 process 364
11.6 Optimisation of CRTM 370
11.7 Future trends 375
11.8 References 376

12 The pultrusion process in polymer matrix composites 381
S. C. Joshi, Nanyang Technological University, Singapore

12.1 Introduction 381
12.2 Process description 384
12.3 Improvements in pultrusion 391
12.4 Innovation, industrial and future trends 403

12.5 Acknowledgement 408
12.6 References 408

13 Autoclave processing for composites 414
P. Hubert, McGill University, Canada and G. Fernlund and A. Poursartip, The University of British Columbia, Canada

13.1 Introduction 414
13.2 Autoclave processing model 417
13.3 Process development 425
13.4 Conclusions and future trends 432
13.5 References 432

14 Out-of-autoclave curing process in polymer matrix composites 435
J. Schlimbach and A. Ogale, Schlimb@ch Business Solutions, Germany

14.1 Introduction 435
14.2 Reasons for using the out-of-autoclave (OoA) process 435
14.3 Strategies 438
14.4 Technical description of different OoA processes 439
14.5 Process comparison and classification 472
14.6 Future trends 475
14.7 Sources of further information and advice 479
14.8 References 479

Index 481

Contributor contact details

(* = main contact)

Editors and Chapter 1

Suresh G. Advani*
Department of Mechanical Engineering and Center for Composite Materials
University of Delaware
Newark, DE 19716
USA

E-mail: advani@udel.edu

Kuang-Ting Hsiao
University of South Alabama
Department of Mechanical Engineering
150 Jaguar Drive
SHEC 3130
Mobile, AL 36688-0002
USA

E-mail: KTHsiao@SouthAlabama.edu

Chapter 2

Shih-Jung Liu
Polymer Rheology and Processing Lab
Department of Mechanical Engineering
Chang Gung University
Tao-Yuan 333
Taiwan

E-mail: shihjung@mail.cgu.edu.tw

Chapter 3

Chung Hae Park
Laboratoire d'Ondes et Milieux Complexes
UMR 6294 CNRS
University of Le Havre
53 rue de Prony
BP540 76058 Le Havre
France

Professor Woo Il Lee*
Department of Mechanical & Aerospace Engineering
Seoul National University
Seoul 151–742
Korea

E-mail: wilee@snu.ac.kr

Chapter 4

Jihua Gou*, Jinfeng Zhuge and Fei Liang
Composite Materials and Structures Laboratory
Department of Mechanical, Materials and Aerospace Engineering
University of Central Florida
4000 Central Florida Blvd
Orlando, FL 32816
USA

E-mail: jihua.gou@ucf.edu

Chapter 5

Terry Creasy
Department of Mechanical Engineering
Texas A&M University
200 Engineering/Physics Building
College Station
TX 77843-3123
USA

E-mail: tcreasy@tamu.edu

Chapter 6

James A. Sherwood*, Konstantine A. Fetfatsidis and Jennifer L. Gorczyca
Department of Mechanical Engineering
University of Massachusetts Lowell
1 University Avenue
Lowell, MA 01854
USA

E-mail: james_sherwood@uml.edu

Libby Berger
General Motors Research and Development Center
Warren, MI
USA

Chapter 7

Jens Mack
Institut für Verbundwerkstoffe GmbH
Erwin-Schrödinger-Straße Gebäude 58
67663 Kaiserslautern
Germany

Ralf Schledjewski*
Chair in Processing of Composites
Montanuniversitaet Leoben
Otto-Gloeckel Strasse 2
8700 Leoben
Austria

E-mail: Ralf.Schledjewski@unileoben.ac.at

Chapter 8

Peter Mitschang* and Marcel Christmann
Institut für Verbundwerkstoffe GmbH
Erwin-Schrödinger-Straße Gebäude 58
67663 Kaiserslautern
Germany

E-mail: peter.mitschang@ivw.uni-kl.de

Chapter 9

E. Murat Sozer
Koc University
Mechanical Engineering Department
Rumelifeneri Yolu
Sariyer
Istanbul 34450
Turkey

E-mail: msozer@ku.edu.tr

Pavel Simacek
Department of Mechanical Engineering and Center for Composite Materials
University of Delaware
Newark, DE 19711
USA

E-mail: psimacek@udel.edu

Suresh G. Advani*
Department of Mechanical Engineering and Center for Composite Materials
University of Delaware
Newark, DE 19711
USA

E-mail: advani@udel.edu

Chapter 10

Kuang-Ting Hsiao*
University of South Alabama
Department of Mechanical Engineering
150 Jaguar Drive
SHEC 3130
Mobile, AL 36688-0002
USA

E-mail: KTHsiao@SouthAlabama.edu

Dirk Heider
Center for Composite Materials
University of Delaware
Newark, DE 19716
USA

E-mail: heider@udel.edu

Chapter 11

Simon Bickerton*
Centre for Advanced Composite Materials
Department of Mechanical Engineering
The University of Auckland
Private Bag 92019
Auckland 1020
New Zealand

E-mail: s.bickerton@auckland.ac.nz

Piaras A. Kelly
Centre for Advanced Composite Materials
Department of Engineering Science
The University of Auckland
Private Bag 92019
Auckland 1020
New Zealand

E-mail: pa.kelly@auckland.ac.nz

Chapter 12

Sunil C. Joshi
Division of Aerospace Engineering
School of Mechanical and Aerospace Engineering
Nanyang Technological University
N3.1-b2c-03, Yunnan Garden Campus
Nanyang Avenue
Singapore – 639 798

E-mail: mscjoshi@ntu.edu.sg

Chapter 13

Pascal Hubert*
Department of Mechanical Engineering
McGill University
817 Sherbrooke Street West
Montreal, QC
H3A 0C3
Canada

E-mail: pascal.hubert@mcgill.ca

Göran Fernlund and Anoush Poursartip
Department of Materials
The University of British Columbia
6350 Stores Road
Vancouver BC
V6T 1Z4
Canada

Chapter 14

Jens Schlimbach* and Amol Ogale
Schlimb@ch Business Solutions
Säulenstraße 12
82008 Unterhaching
Germany

E-mail: jens@schlimbach.com

1 Introduction to composites and manufacturing processes

S. G. ADVANI, University of Delaware, USA and
K.-T. HSIAO, University of South Alabama, USA

Abstract: This chapter introduces the range of manufacturing processes for polymer matrix composites discussed in detail in the book *Manufacturing techniques for polymer matrix composites (PMCs)*. It discusses short-fiber composite molding processes, such as injection and compression molding, as well as the processing of polymer nanocomposites. The chapter also highlights high-viscosity thermoplastics-based processes such as sheet forming, thermostamping and filament winding. Finally, the chapter considers processes for thermoset matrices such as resin transfer molding, pultrusion, autoclave and out-of-autoclave processing.

Key words: polymer matrix composites, short-fiber composite molding processes, polymer nanocomposites, thermoplastics, thermosets.

1.1 Processing of polymer matrix composites

Polymer matrix composites consist of particles or fibers embedded in polymer matrices. The particles or/and fibers are introduced to enhance selected properties of the composite. The particles on the micron scale usually provide better thermal stability and toughness whereas particles at the nanoscale such as nanoclays can improve physical and mechanical properties. Long aspect ratio particles such as whiskers and short fibers or continuous fibers are added to reinforce physical and mechanical properties. Continuous fiber reinforcements usually provide the highest improvement in mechanical properties such as stiffness and strength and can be introduced in various forms such as continuous random mat, woven fabric, and stitched fabric, unidirectional or bidirectional fabric. The polymer matrices addressed in this book are thermosets and thermoplastics. The fiber material is usually glass, carbon or aramid. Often these materials are introduced in a precursor form in the manufacturing process. The precursor form usually mixes the fibers and resin in the form of a pellet (short fibers embedded in a solid matrix) or prepregs (continuous fibers with resin attached to them in terms of a powder or pre-impregnated partially cured resin). For further reading on types of reinforcements and various forms available, we refer the interested reader to chapters in other textbooks on the topic.[1–4]

Although the properties of the reinforcing phase have a significant influence on the final properties of the composite, the quality of the starting material form and the processing route selected to manufacture the composite can impact the final properties and performance of the composite. The goal in composite manufacturing is to produce components with the desired properties that combine the best properties of the fibers (or particles) and the resin while masking or minimizing their weaknesses. Nowhere is this truer than in composite manufacturing where one makes the properties while manufacturing the part. This is true for short fiber composites as well as continuous fiber composites. For short fiber systems, the processing history and material properties can influence the final fiber orientation and distribution density, which is guided by the flow, and can result in non-homogeneous properties in the final component. For continuous fiber systems, the challenge during making the composite is to fill all the empty regions between the fibers with resin so there is good matrix–fiber interfacial bonding, which is essential to prevent delamination and avoid crack initiation under applied loads.

The composite manufacturing industry has always recognized the importance of initial material forms and processing for part performance and over the last three decades there have been (i) improvements in material chemistry such as reduction in the viscosity of resin, delayed curing of the resin, or fabrics that are easily conformable, (ii) improvements in existing composite manufacturing processes or introduced new processes and (iii) the industry has resorted to more science-based processing than the trial and error prevalent manufacturing approach to mitigate the cost of manufacturing and make composite materials more competitive with their metal counterparts. Hence over the years many reference books have been written to summarize the advancement of the science by researchers and practitioners in the field.[5–8] In the last two decades, modeling and simulation textbooks, tools, and courses have been introduced to provide science-based understanding of these complex material interactive behaviors as they combine to form a composite with superior properties.[1–8] As a result, polymer composites are continuing to replace their metal counterparts in many industries from defense to infrastructure due to their lower weight (which usually translates into energy savings) and corrosion resistance (which is attractive for durability and lifetime cost savings). This book is a collection of chapters by various experts in selected composite manufacturing processes. The general approach followed for each chapter is for the authors to (i) describe the process, its benefits and challenges, (ii) provide a window into our current level of understanding of the material behaviors and processes and (iii) introduce the tools for improving the manufacturing of composites.

There are many different manufacturing processes that have evolved to fabricate polymer matrix composites. These processes were modified or developed to address various needs such as (i) new fiber or matrix systems, (ii) new and improved initial precursor material forms, (iii) composite part geometrical constraints, (iv) cost-effectiveness, (v) multi-functionality of the part, (vii) enhancement of a specific physical, electric or mechanical property and (viii) defect constraints. The composite industry continues to engage in the development of new manufacturing processes that will allow them to fabricate composite parts with higher quality, reduced cost, and embrace emerging opportunities such as new reinforcement systems and new polymer matrix systems. In addition to the development of new processes, the adoption and the utilization of the fundamental science in the manufacturing processes have accelerated the improvement of the manufacturing technologies. Scientific approaches along with analytical modeling and numerical simulations are better utilized today to predict the physics and reaction during the manufacturing processes before investing in the prototype manufacturing and testing. The variability from one part to the next can also be reduced by improving material forms and introducing processing methods that can produce defect free parts with desired properties, despite the undesired but inevitable variations in the materials or process conditions during fabrication. The coupling of the composite structural design with the manufacturing and assembly process is a potential future trend for the composite industry to accelerate the growth of composite materials in various industries by continuing to improve composites manufacturing in addition to developing new materials at lower costs.

1.2 Focus and scope of this book

This book aims at presenting the audience with state-of-the-art yet comprehensive knowledge of selected commonly used or emerging composite manufacturing processes. Each chapter, written by an expert of that process, first describes the process and provides physics insights, and then introduces the corresponding models and scientific understanding. The common issues and various approaches to address are outlined. Each chapter endeavors to present state-of-the-art knowledge of the manufacturing process and provides editorial discussion on the future outlook for that process.

The manufacturing processes included in this book are grouped into three parts based on their processing characteristics or physics. Part I: Short fiber and nanoparticle based processing, focuses on the two commonly used polymer processes used for composite manufacturing: injection molding and compression molding. Chapter 4 reviews the recent developments in processing of nanoparticle based composites. Note that the polymer matrix nanocomposites processing techniques introduced in this chapter could also be combined with other mainstream polymer matrix composite manufacturing processes with slight modification to make hybrid composites. Hence this chapter also offers a prelude to the remaining 12 chapters as a digest to important composite manufacturing processes when nanoparticles are included in the composite systems. Part II discusses thermoplastic based processing, which due to the high viscosity of the resin and the ability to melt and re-solidify make composite sheet forming, fabric thermostamping, thermoplastic filament winding and continuous fiber reinforced thermoplastic profile manufacturing processes very attractive for automation and large volume production. Part III of the book addresses the most commonly used processes with the easy to process thermoset matrix. There are six chapters in this part. The processes discussed are resin transfer molding, vacuum assisted resin transfer molding, compression resin transfer molding, pultrusion, autoclave processing and the newly popular and cost-effective out-of-autoclave processing.

Chapter 2 is a comprehensive overview of the injection molding process for short fiber reinforced composites. This process is capable of high production volume, relatively low fiber volume fraction (compared with the other type of composites) and good part finish. The precursor materials used in this process are pellets usually containing thermoplastic resins with embedded chopped short fibers. A screw-based extruder is used to melt and mix the molten resin with fibers and pump the suspension into a mold to form the net shape part. The physics of the process such as the rheology of the suspension and the evolution of fiber orientation during the molding process is described in this chapter. This chapter also reviews different variations of the injection molding process that have emerged in the last couple of decades. The reasons for the common defects such as jetting, fiber exposure, weld lines, shrinkage and warpage are provided along with various strategies based on experience, scientific understanding and modeling, and simulation tools that can be adopted to avoid them. The future trends to evolve the injection mold process toward employing and maintaining long fibers in thermoplastic composites are discussed so that one can manufacture parts with this process for load-bearing applications.

Chapter 3 discusses the other most commonly adopted polymer process: the compression molding process for composites containing chopped fibers. Although this process can be used with thermoplastics as well as thermoset matrix for short fibers the main emphasis has always been the use of thermosets due to their ease in processing and low viscosity. This process is suitable for mid-high production volume, shell-like part geometry, and can provide good part finish. This chapter reviews the compositions and characteristics of several precursor materials employed such as sheet molding compound (SMC), bulk molding compound (BMC), thick molding compound (TMC), directionally reinforced molding compound (XMC) and glass mat thermoplastics (GMT). This precursor material is cut into layers and placed inside the mold. As the compression molding press lowers the platen to close the mold, the mold squeezes the resin and the fibers displacing the resin to fill the regions in between the fibers devoid of resin. The placement of the charge can dictate the final part quality and performance. The modeling and simulation of the process can help in tailoring material properties and help avoid knit line formation. Other common defects including the surface defects and internal defects are explained and their remedies are discussed. To better understand and simulate this process, the modeling approaches regarding the resin cure, material flow, fiber motion and orientation are briefly described in this chapter. They show how process modeling is a useful tool to help in the selection and placement of the material and how the process design can improve the production speed and yield.

No state-of-the-art review book of polymer composite manufacturing is complete without addressing the recent trend to include nanoparticles in the mix to improve composite properties. Hence we have included a review chapter that addresses the current trends. Chapter 4 discusses the processing of polymer nanocomposites. Polymer matrix nanocomposites consist of polymer filled or reinforced by nanoparticles. Commonly investigated nanoparticles include single-walled carbon nanotubes (SWNTs), multi-walled carbon nanotubes (MWNTs), carbon nanofibers (CNFs), grapheme sheets (GS), montmorillonite (MMT), nanoclay and polyhedral oligomeric silsesquioxanes (POSS). The nanoparticles are added into the polymer matrix to improve the mechanical, thermal or electrical properties of the composites. Due to the small size of nanoparticles and the Van der Walls force among nanoparticles, the uniform dispersion of nanoparticles in the polymer matrix is an important issue. This chapter reviews several methods such as the solution processing, *in situ* polymerization, and the melting process for dispersing nanoparticles in different types of polymer matrices. Current developments in nanoparticle dispersion, alignment and functionalization are reviewed. Methods for further forming the nanoparticle-based paper are also discussed. In the future, more improvements in nanoparticle dispersion, polymer compatibility, nanoparticle–polymer interfacial bonding, and nanoparticle alignment control are important to exploit the full potential of polymer matrix nanocomposites. As nanoparticles could also be used in the composites fabricated by other manufacturing processes, this chapter could be studied with the other chapters when nanoparticles need to be included in the composites manufactured via other mainstream manufacturing processes.

Chapters 5, 6, 7 and 8 specifically address issues associated with high viscosity thermoplastic processing. Chapters 5 and 6 are based on the stamping operation usually encountered in metal forming. Chapter 5 features the composite sheet forming process for thermoplastic matrix composites, which was attractive as one could save initial investment by modifying metal stamping equipment. In this process, a thermoplastic prelaminate (a flat panel consisting of a thermoplastic matrix and continuous or discontinuous long fibers) is shaped into a composite part by (i) softening the thermoplastic matrix with heat, (ii) using pressure to press the softened prelaminate against the surface of the mold to form the part geometry and (iii) cooling the mold to consolidate the composite part. The mold tools can be either dies or diaphragms. During the forming process of a non-flat part, the fibers in the prelaminate could deform, move, stretch or break due to the resin flow, inextensibility of the material in the fiber direction and the interaction with the mold tool surface. Hence, the challenge in this process is to form the part without introducing wrinkles, instabilities and fiber slippage. If one tries to deep draw, it is difficult to maintain the uniform thickness due to the anisotropy introduced by the presence of the fibers. Hence the layup of the thermoplastic sheets with the imbedded fibers plays an important role in the formability in addition to the heating and pressurizing rate. Good understanding of this anisotropy on the forming process is the key to successful part manufacturing with sheet forming in addition to understanding the limits of formability with certain material forms. Three sheet forming methods including the matched die method, the rubber pad method and the diaphragm method are discussed. The future trends of this process include the improved mechanical properties by using new stretch-broken fibers (or SBXF) reinforcement system and the shortened cycle time by use of advanced thermoplastic matrices with higher thermal conductivities for more efficient heat transfer.

Chapter 6 deals with fabric thermostamping for manufacturing continuous fiber reinforced composites. A stack of continuous fiber reinforced prepregs are heated and stamped to form a composite structure. The polymer matrix could be either a thermosetting resin or a thermoplastic resin. The major challenges are managing the fabric deformation, which includes the fiber displacement, shear and deformation, during the stamping process. Many factors such as the friction, the fiber yarn structure, the fabric pattern, the fiber volume fraction, the resin lubrication effect, the temperature control and the mold geometry, etc., all can affect the fabric deformation during the stamping. This chapter reviews the experiment characterization techniques and the modeling approach to predict the fabric deformation during the stamping process. As this process exhibits promising potential for mass producing continuous fiber reinforced composites, further improvements in the physical understanding, the experimental characterization and numerical simulations for both fabric deformation and composite properties are considered important for the future development of this technology.

Chapter 7 focuses on the thermoplastic filament winding process for continuous fiber reinforced composites. Thermoset filament winding has been around for over four decades and is routinely used to make axisymmetric structures. In the last couple of decades, researchers have tried to translate that expertise to manufacture thermoplastic filament wound parts. The main advantage is that one can make thick composite tanks and cylinders that can withstand high pressures. By using the thermoplastic matrix, the one-step *in situ* consolidation process and composite part welding are possible. In addition, this process further reduces the thermal residual stresses by an order of magnitude. During the process, the fibers are continuously fed and impregnated with the molten matrix and are wound on a mandrel with a winding head. Other precursor materials used are thermoplastic tapes containing unidirectional fibers and as the tape is wound over the mandrel or a head, local heat and pressure can be applied to consolidate the incoming tape to the substrate below. With good design of the winding path, the heating and the consolidation process, high performance composite parts such as pressure vessels and pipes can be produced. This chapter discusses the process steps, the building elements and different winding technologies. The future outlook such robotized winding systems, self-reinforced polymers, process monitoring and inline process simulation tools can be used to make composites of consistent quality.

Two processing methods that imbed and shape continuous fibers are discussed in Chapter 8 'Continuous fiber reinforced profiles in polymer matrix composites'. The first process discussed is the well-established pultrusion process. In this process reinforcing fibers along with thermoplastic resin which may be present in the powder form, fiber form or pre-impregnated form are pulled through a die to form the profile of the die. The die is half-heated and half-cooled to achieve the resin melting and impregnation, forming, consolidation and solidification. The fiber weight fraction by the pultrusion process can be as high as 85% when continuous unidirectional fibers are used. The steps, key elements and varieties of techniques used in this pultrusion process are explained and discussed in this chapter. The second method for creating complex composite profiles is a new process called the continuous compression molding (CCM) process, which utilizes several pressing tools connected in series to decouple the functions required to form, consolidate and solidify the profile. Complex part geometries, complex fiber reinforcement systems, and even the structures with embedded form cores can be realized by the CCM process. The process techniques and the materials and tooling consideration are outlined in this chapter.

Chapters 9 through 11 all belong to a class of processes called liquid composite molding (LCM). In this class of processes, one usually transfers the resin from a reservoir into a network of fibers. The fiber network is formed from any type of reinforcement from random mat to woven, stitched, non-crimp and bidirectional fabrics. This class of processes is very attractive as one can make net shape parts from composites using continuous fibers and have good control over the orientation of the fibers so one can tailor the properties as required by the design. Usually thermoset matrices are used due to their low viscosity, which is necessary to have the resin penetrate and fill spaces in between the fibers.

The first process in this class that was introduced is called resin transfer molding (RTM) and is discussed in detail in Chapter 9. RTM consists of a mold cavity in the shape of the final part. Fabric or fiber preforms are placed in this cavity and the mold cavity is closed with upper and lower rigid mold platen to completely seal the cavity. Resin is introduced into the part through one or more gates using positive pressure. The goal is to fill all the empty spaces between the fibers with resin. Over the last two decades this process has evolved where it can be easily used to make composite structures for load-bearing applications. The advances in this process have mainly been due to the research effort in developing the science base of this process and creating numerical simulation of the flow in the mold. Issues have been identified and addressed. Optimization and control routines have been developed to improve consistency in this process in manufacturing parts. This chapter discusses all the issues and provides a basic understanding of the physics of the process and highlights the current state-of-the-art along with what one can expect in the future.

Vacuum-assisted resin transfer molding (VARTM) evolved as the manufacturers became more cost conscious and wanted to develop larger scale parts such as sides of a train or truck walls, boat decks and hulls and wind blades. Chapter 10 provides a systematic review about this process along with some of the variations including addition of a semi-permeable membrane to reduce void content. The membrane processing is relatively a new process in which one side of the part is covered by a special membrane which can make high quality parts as it allows air to escape but keeps the resin intact thus mimicking the idea of having a vent over the entire part. Currently such membranes are very expensive and cannot be used in a cost-effective way when manufacturing large structures. The VARTM process utilizes the resin infusion fundamental of the RTM process does; however, the VARTM uses a vacuum to draw the resin into the fiber preform and uses a vacuum bag to seal the fiber preform instead of a rigid mold. Hence in this process, the thickness of the preform varies as a function of the resin pressure into the mold and must be addressed. Also as the maximum pressure one can pull the resin into the mold is 1 atm, the fill time can become much higher than the time one has for the resin to gel. Hence variations of this process have sprouted – the most common among them being SCRIMP,[9–11] which add a high permeability flow distribution media ply on top of the preform before enveloping the entire assembly inside a vacuum bag. Thus the resin can first flood the top surface of the preform and flow in the thickness direction dramatically reducing the fill time. This chapter introduces the key physics, modeling concepts and approaches to defect control. Several useful modeling tools including flow, curing and dimensional accuracy control for VARTM process are discussed in the chapter. The approach to infuse large structures with resin loaded with nanoparticles is also discussed. The membrane-based infusion processing, which is derived from VARTM by adding the semi-permeable membrane (permeable to gas and impermeable to liquid), is introduced. Its potential improvements such as resin saving and continuous degassing are discussed.

In Chapter 11, the compression resin transfer molding (CRTM) process is reviewed. The process takes advantage of both the good dimensional control and complex part manufacturing of RTM and reduction in fill time obtained by the use of distribution media in VARTM. In this process, the preform is placed in a mold cavity, the top rigid mold platen is lowered to the extent where there is a gap between the top of the preform and the top mold half. The measured amount of resin is injected into this gap. Next, the injection gate is closed and the mold top is lowered to close it to the desired thickness, pushing the resin into regions which are devoid of resin from the resin rich areas. It targets to produce a composite part with the quality close to a RTM part and with a shorter cycle time. However, due to the *in situ* mold compression requirement, the CRTM may be less competitive than the RTM and the VARTM in terms of the part complexity and size of FRP. The important processing issues of CRTM are due to the preform compression and the related physics in terms of both the solid fibers and the liquid resin flow. This chapter discusses the processing issues, presents the modeling approach and explores the design optimization for the CRTM process. The CRTM could be further improved with the ongoing research activities in process modeling, better understanding of preform and the flow during the compression process, selection of suitable preform materials and study of utilizing semi-rigid mold tooling.

There are other variations of the RTM, VARTM and CRTM process which belong to the class of LCM processes such as RTM Light in which a caul plate is introduced on top of the bag, resin infusion under flexible tooling (RIFT), resin film infusion (RFI) and vacuum injection preform relaxation (VIPR). In addition many of the same processes are known by different acronyms. It is useful to note that as the science base of RTM improved and led to better understanding of the variables and parameters that influence the flow and dry spot creation in RTM, it led to approaches and creative ideas to improve the process and cut down the cost. This learning continues today to evolve new modifications to the LCM family of processes. For example in VIPR one can now introduce flow control effectively to modify the flow front and create a part that is better than the SCRIMP process and is of consistent quality.[12,13]

Although LCM processes have been popular in manufacturing parts that have some complexity, the lack of automation and intensive labor costs have precluded this class of processes for medium to high volume production. However, other processes such as pultrusion can be used for high volume when the part needs to be made with a certain profile with a reasonable length. Chapter 8 focuses on advancement in this profiling with pultrusion and a new process called continuous compression molding mainly as applied to thermoplastics. Chapter 12 reviews the pultrusion process for composites manufactured with thermoset resins. Thermoset pultrusion is a high volume automated process for manufacturing composite parts with a constant cross-sectional geometry and continuous fiber reinforcement. Fiber reinforced polymer parts with high fiber volume fractions and well-controlled fiber alignment can be obtained by this process. The fibers are pulled through a resin bath (or through a resin injection system) for resin impregnation and into a heated die with the desired profile for curing to form the part. The solid FRP extruded from the die is cut into the desired length for further distribution and shipment. The components such as the reinforcement dispenser, resin impregnator, forming guides, die and temperature controller, puller and clamp system and cut-off saw are reviewed. This chapter also provides the models for curing and thermal control, thermal shrinkage and matrix shrinkage, and material property changes. Although the earlier versions of these models have been published in other texts,[2,6,8] the authors here have relaxed certain assumptions to improve the predictions. The experimental methods and numerical simulations for process development are discussed. The future trends in terms of both technology and marketing and many variant techniques of this process are highlighted.

Chapter 13 presents autoclave processing, which is a well-known and almost a gold-standard method for manufacturing high performance composite parts with complex shapes. However, the high manufacturing cost and the part size limitation due to the requirement of a large and expensive autoclave are well-known disadvantages compared with the other processes such as LCM. During this process, the partially cured resin pre-impregnated fiber reinforcement, namely the 'prepreg', is placed on a single-side mold surface and covered with bleeder, breather and vacuum bag. The entire assembly is placed into an autoclave that can provide well-controlled high pressure and temperature. The vacuum tube is connected to the vacuum bagged breather to remove the air or gas and to draw the excess resin into the bleeder. To produce a part with good control in terms of fiber volume fraction, void fraction, residual stresses and part dimension fidelity, a careful plan about how to change and control the vacuum, temperature and autoclave pressure is critical. Due to the complexity of this process and the high performance and the high development cost, process simulations have proven to be very useful tools to (i) reduce the time spent by a part in the autoclave and (ii) produce parts with minimum voids. Many chapters on the autoclave processing have been written in the past;[2,6,8] however this chapter focuses more on the tool and part interaction to address dimensional stability in addition to the reduction of voids and optimal strategy for controlling the pressure and temperature cycle.

Autoclave processing can produce good parts, and process modeling and simulations help, but the main hurdle to the process of making parts with the autoclave is the cost of the autoclave in addition to all the plumbing and safety requirements needed to operate one. One aspect that was made clear by modeling of this process is that one does not necessarily need very high positive pressure if the goal is to have fiber reinforced parts that are within 50–60% fiber volume fraction. This gave birth to the idea of eliminating the autoclave and consolidating and curing the thermoset composite part by other means. This new class of processing approach has been termed out-of-autoclave (OoA) processing. Chapter 14 provides a systematic survey of various OoA curing techniques and the related processes. This chapter highlights the technical issues and economic reasons along with the trend to develop the various OoA processes. The most recent development regarding the vacuum bag only OoA (VBO-OoA) process and its related industrial activities are highlighted in this chapter. In addition to the VBO-OoA process, other types of OoA curing techniques and processes are presented, including the use of fluid heated/cooled mold tools on autoclave prepreg materials (such as the Quickstep™ process), electromagnetic heating induced curing (such as induction heating, infrared heating, microwave heating and radio-frequency curing), LCM (such as RTM, VARTM, etc.), *in situ* polymerization and E-beam curing. The chapter is also very positive about the future development of various OoA processes.

Although each chapter in the book focuses on a selected process, an effort is made to draw similarities and highlight differences between processes where possible. Our hope is that by recognizing the useful and functional features of each process, the reader will be encouraged to combine different processing features into a new manufacturing process for their application. In fact, it is our expectation that many useful polymer matrix composite manufacturing techniques will be developed based on our current understanding of the composite processing techniques and will further facilitate and accelerate the use of polymer matrix composites in many future applications.

1.3 References

1. B. T. Astrom, *Manufacturing of Polymer Composites*. London: Chapman & Hall, 1997.
2. S. G. Advani and E. M. Sozer, *Process Modeling in Composites Manufacturing*, New York: Marcel Dekker (2002), 2nd edn, Boca Raton, FL: CRC Press, 2010.
3. A. B. Strong, *Fundamentals of Composite Manufacturing, Materials, Methods and Applications*. Dearborn, MI: Society of Manufacturing Engineers, 2008.
4. S. V. Hoa, *Principles of the Manufacturing of the Composite Materials*. Lancaster, PA: DesTech Publications, 2009.

5. T. G. Gutowski (ed.), *Advanced Composites Manufacturing*. New York: John Wiley, 1997.
6. S. G. Advani (ed.), *Flow and Rheology in Composites Manufacturing*. Amsterdam: Elsevier, 1994.
7. T. D. Papathanasiou and D. C. Guell (eds.), *Flow-Induced Alignment in Composite Materials*. Cambridge: Woodhead Publishing Ltd., 1997.
8. R. S. Dave and A. C. Loos (eds.), *Processing of Composites*. Cincinnati, OH: Hanser Gardner Publications, 1999.
9. W. H. Seemann, Plastic transfer molding techniques for the production of fiber reinforced plastic structures. US Patent No. 4902,215; 1990.
10. W. H. Seemann, Plastic transfer molding apparatus for the production of fiber reinforced plastic structures. US Patent No. 5052,906; 1991.
11. W. H. Seemann, Unitary vacuum bag for forming fiber reinforced composite articles. US Patent No. 5316,462; 1991.
12. J. B. Alms, J. L. Glancey and S. G. Advani, 'Mechanical properties of composite structures fabricated with the vacuum induced preform relaxation process', *Composite Structures*, **92** (2010), 2811–2816.
13. J. B. Alms, S. G. Advani and J. L. Glancey, 'Liquid Composite Molding control methodologies using Vacuum Induced Preform Relaxation', *Composites Part A – Applied Science and Manufacturing*, **42**(1), 57–65. DOI: 10.1016/j.compositesa.2010.10.002. January 2011.

Part I

Manufacturing of polymer matrix composites (PMCs): short fiber and nanomaterial based processing

2

Injection molding in polymer matrix composites

S.-J. LIU, Chang Gung University, Taiwan

Abstract: This chapter is intended to provide some general introduction regarding the injection molding process and the molding of fiber filled polymers. Some new injection molding techniques are introduced. Rheological properties of compounds are given. Characterization and prediction of fiber orientation in molded parts are also discussed. Commonly used fiber filled polymer compound and their properties are addressed. Molding faults and possible corrections are provided. Finally the future trend for injection molding is discussed.

Key words: injection molding, processes and variants, mechanical properties, fiber orientation, molding defects and corrections.

2.1 Introduction to injection molding

Injection molding[1] is one of the most important polymer processing methods for producing plastic and plastic composite parts. Injection molding of glass fiber reinforced composites[2] has the capability of producing near net shape articles having exceptional physical and mechanical properties. The process typically employs a reciprocating single-screw extrusion machine, as shown schematically in Fig. 2.1. The machine is used for transporting, melting and pressurizing the fiber filled polymeric materials, which is fed into the machine in granular form. The polymer melts within the barrel by heat conduction through the barrel wall and via the dissipation of heat generated within the sheared polymer melt. During plastication, the melt accumulates in front of the screw, which is driven back against an adjustable pressure within the hydraulic system until a desired shot size (melt volume) is achieved. This is followed by injection where the screw pushes forward to force the polymer melt through a runner system and into the relatively cold empty cavity of the already closed mold. In order to compensate for any shrinkage caused by the cooling of the melt within the cavity, the melt in front of the screw is held under pressure so as to force more materials into the cavity. When the gate into the mold freezes, no more material can be supplied through the gate and the product cools down further without compensation for shrinkage. The mold temperature is regulated by water that circulates through channels to keep the mold cavity walls at a temperature

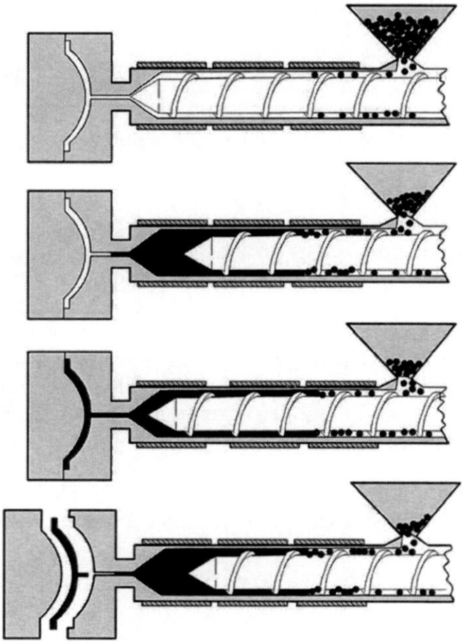

2.1 Schematic of the injection molding process.

between room temperature and the glass transition temperature (for amorphous polymers) or the melt temperature (for semi-crystalline polymers) of the polymeric materials. When the product is cooled to a state of sufficient rigidity, which in most cases occur when all regions of the part have cooled down to below glass transition temperature or melt temperature of the polymer, the mold opens and the product is ejected.[3,4]

Injection molding is a cyclical process. Cycle times may range from 10 to 100 s and are controlled by the cooling time of the melted plastic. Close tolerances on small intricate parts is possible with injection molding. Typically there is very little post-production work required because the parts usually have a very finished look upon ejection. All scrap may be reground to be reused to reduce the waste. Furthermore, full automation is also possible with injection molding. Injection molding produces various fiber filled parts in high volumes for applications in the aerospace industry, automotive industry (air intake manifolds, rocker covers, cooling modules, etc.), electrical/electronic industries (connectors), medical and dental products (components for blood analyzer equipment, heart pump parts, orthopedic devices, EKG and oxygen parts), model shops, toys and hobbies, household appliances (washing machine cylinders), etc. As injection molding is a complex process, a number of factors can affect the molded product quality. A good understanding of the process can assist the product and mold designer to avoid molding defects.

2.1.1 Co-injection (sandwich) molding

Co-injection molding,[1] also named sandwich molding, is a process that creates a skin and core material arrangement in a molded part. The process requires two injection/processing units that inject two melt streams into a centrally located nozzle. The arrangement of the two units on the machine is more or less independent of the mold technology but is governed essentially by the geometry of the molded part. The units can be arranged either parallel or perpendicular to each other. The molding machine controls the injection units to achieve a skin–core–skin flow sequence through the manifold located at the end of the barrels into the mold. The skin material is injected first into the mold cavity, followed by a core material. As the skin material flows into the cavity, the material next to the cavity walls contacts the walls and freezes. When the core material is filled, it displaces the skin material in the center of the channel by pushing the skin ahead. The two melts contact with each other but do not flow into one another during the filling process. Figure 2.2 shows schematically the sandwich injection molding process.

There are many factors in favor of using co-injected structures. The use of the two-material co-injection molding process allows several functions to be integrated in one part to satisfy stringent specifications. It also improves quality and minimizes costs, for example by reducing assembly work. The parts can be produced economically by co-injection molding in one step. Only one mold and thus only one machine is needed to manufacture the moldings. The two- or multi-material products can find applications for everyday use in food packaging, toys, components for electrical equipment or other engineering components such as rubber bearings, rollers and sealing elements or various kinds of damping elements and housings with integrally molded seals. Furthermore, it is possible to obtain or improve certain performance properties such as functional, tactile or design features. Composites of rigid and flexible materials open up a wide range of new possibilities. These composites have a variety of functions, with the flexible component (e.g., virgin polymer) exhibiting the typical springy and elastic properties to provide resilience and a cushioning effect, or non-slip characteristics for good grip, while the rigid component (e.g., fiber reinforced polymer) contributes its strength and stiffness to prevent distortion where loads are applied.

2.1.2 Fluid-assisted injection molding

Fluid-assisted injection molding technologies, including gas-assisted[5–11] and water-assisted[12–19] injection moldings, have received extensive attention in recent years, due to their light weight, relatively lower resin cost per part, faster cycle time and their flexibility in the design and manufacture of plastic parts. Fluid-assisted injection molding is similar to co-injection molding, with the core materials being replaced by either gas or water. In the process, the mold cavity is partially filled with the polymer melt followed by the injection of gas or water into the core of the polymer melt. A schematic diagram of the fluid-assisted injection molding is illustrated in Fig. 2.3a. Fluid-assisted injection molding can produce parts incorporating both thick and thin sections with less shrinkage and warpage and a better surface finish. The fluid-assisted injection molding process can also enable greater freedom of design, material savings, weight reduction and cost savings in tooling and press capacity requirements. Applications of fluid-assisted injection molding include clothes hangers, hammer handles, chair armrests, car bumpers, back mirror housings, golf club shafts, window frames, ladders, collapsible shipping containers, motorcycle rear handles, etc., to name only a few.[20]

Different from co-injection molding which requires a special molding machine with two plastication (injection) units, fluid-assisted injection molding has usually been treated as an enhancement of the injection molding process rather than a dedicated stand-alone technology. Conventional injection molding machines can be modified with 'add-on' gas or water control units.[11] The injection unit can be either an independently operated

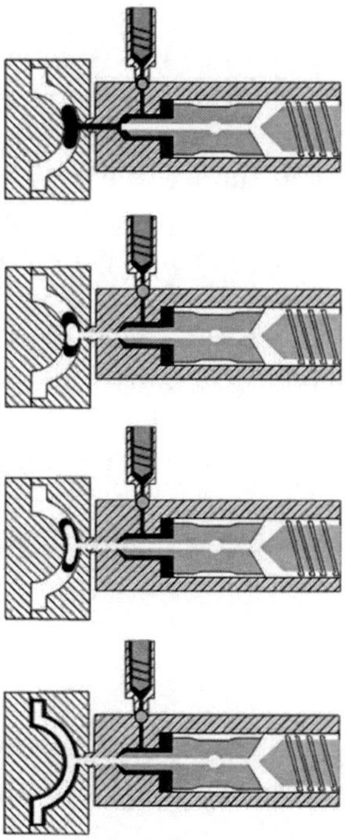

2.2 The sandwich injection molding process.

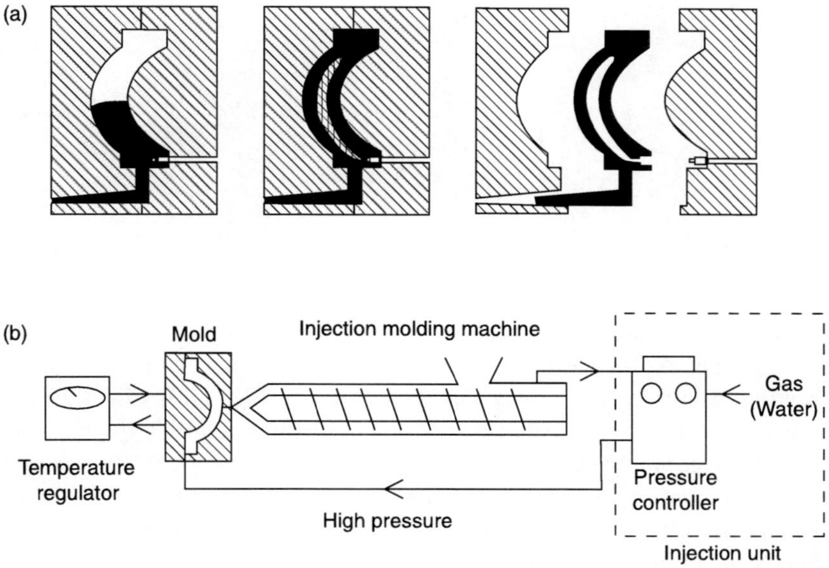

2.3 Schematic of (a) the fluid-assisted injection molding process, (b) the setup for the process.

unit or be integrated into the injection molding machine. For the machine-integrated type, the unit is fully integrated into the machine design and is not an add-on portable unit. A fluid control panel is permanently mounted to the machine injection base and all lines are hard piped between valves and the gas (water) nozzle manifolds. An individual gas (water) pressure regulator and a flow control are provided for each shot sequence. Each part can have its own independent gas (water) and flow settings. Independent controls optimize part quality and improve ease of setup and adjustment. For the independent injection type, on the other hand, the system has its own hydraulic and control system, and is able to operate independently of the molding machines. The unit is usually a mobile one and can be easily adapted to all molding machines due to its self-sufficiency. Gas or water can be injected through the machine nozzle or by direct injection into the cavity. Injectors are either stationary or are inserted into the melt using hydraulics (depending on the molded part geometry). Figure 2.3b shows schematically the setup of the independent injection type fluid-assisted injection molding.

2.1.3 In-mold decoration (IMD) injection molding

In-mold decoration (IMD) injection molding[21] is a process of over-molding on decorated thermoplastic film or applied with an overlay of melted

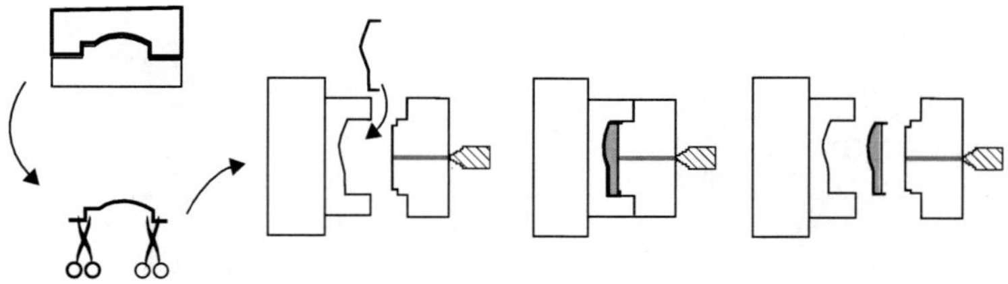

2.4 Schematic of the IMD injection molding process.

thermoplastic material. Before the process, a decoration film (e.g., PET film) can be first thermoformed.[22,23] The film is then placed into an injection molding cavity in such a way that its printed surface faces directly to the flow of melt front. After the close of the mold, hot polymer (e.g., glass fiber filled PET composites)[24,25] melt is injected into the cavity. When the melts come into contact with the film, heat is transferred to the film, which causes the film to adhere to the molten polymer. Once the component has been ejected from the mold, the graphic pattern is transferred onto the surface of the cooled component. Figure 2.4 shows schematically the IMD injection molding process. IMD offers many design flexibility and productivity advantages versus other decoration methods done after molding.[26–29] These benefits include design flexibility; multiple colors, effects, and textures with a single operation; long-lasting graphics manufacturing productivity; and systems cost reductions. IMD can achieve different colors, effects and textures that are complete when the part comes out of the mold. When any of these factors need to change, there is no need to re-tool or change resin color. Simply altering the film means that the appearance or texture of molded parts can be dramatically changed.

2.1.4 Microcellular injection molding of filled polymers

Microcellular injection molding of filled polymers has received significant attention due to their beneficial properties such as light weight, thermal and acoustic insulation, and improved energy-absorption performance on impact.[30–34] The process was first commercialized by Trexel.[35] The cell size of microcellular foam ranges from 1 to 100 microns, while the cell density is greater than 10^9 cells/cm^3. This provides improved mechanical and thermal properties such as high strength/weight ratio, enhanced toughness and fatigue life of molded parts. Microcellular foams with appropriate cell size exhibit 4–5 times higher impact strength than their unformed counterparts.[36] In a microcellular injection molding process, 'supercritical' nitrogen (N_2) or carbon dioxide (CO_2) is first injected into the barrel and dissolved into polymer melt to create a single-phase polymer–gas solution. The whole foaming process consists of three major steps: polymer/gas solution formation, cell nucleation and cell growth. Microcellular parts are the results of a thermodynamic instability in a polymer/gas solution to create a huge number of nuclei in the polymer matrix.

The viscosity and the glass transition temperature of the polymer melt in a microcellular injection molding process can be significantly reduced due to the fact that the gas fills the interstitial sites between polymer molecules. This technique can therefore produce plastic composite parts with microcellular structures using lower injection pressure, shorter cycle time and less material. It also eliminates the need for a packing stage and improves the dimensional stability of the molded parts. In addition, microcells also greatly enhance the part toughness of many polymeric materials by acting as crack arrestors. Nevertheless, the improvement of the mechanical properties can only be achieved when the cell size is smaller than some critical size which is material dependent. Microcellular injection molding of nanocomposites also exhibits the potential to produce a component that has excellent physical and mechanical properties at a significantly reduced part weight and material cost.[37]

2.2 Molding compounds

The manufacturing of thermoplastic parts by injection molding is a two-step process. The first step consists of mixing together, by compounding in an extruder, a neat thermoplastic resin with either particulate (including talc, $CaCO_3$, TiO_2, etc.),[38] chopped carbon or glass filaments (and other additives if needed),[39,40] natural fibers (e.g., wood fibers,[34,41] wheat straw fibers,[42] etc.), or other polymers with high modulus such as liquid crystal polymer (LCP)[43,44] in order to obtain the reinforced thermoplastic pellets. These reinforced thermoplastic pellets are then injection molded to obtain various components, including those of complex design which may incorporate a high degree of function integration.

The compounding process to manufacture particulate or fiber reinforced thermoplastic pellets is essentially the same as that for un-reinforced or filled thermoplastic compounds. The fiber content in fiber filled thermoplastics is generally between 20 and 50 wt%, and the fiber length and fiber diameter are around 500 and 15 μm, respectively. Twin-screw extruders[45] are generally used for higher throughput and thorough mixing. For glass fiber filled materials, chopped strands are introduced into the melt through a side feeder positioned as far down the twin-screw extruder as possible, in order to achieve the highest residual fiber length in the compound.

The main difference with un-reinforced or filled thermoplastic compounding is the treatment of the screw to limit its wear. The screw profile will also be optimized for glass filament thermoplastic compounding, more particularly in terms of the mixing forces after the introduction of the chopped strands. The barrel temperature is increased by 10–30°C over that for un-reinforced thermoplastic compounding because of higher melt viscosity. Table 2.1 lists the processing properties of some commercial injection molding compounds, while Table 2.2 shows the mechanical properties of injection molded plaques molded from these compounds.[46] Among all these composites, the short glass fiber filled thermoplastics are the most widely used materials in injection molding.

Theoretically, all injection moldable polymeric materials (including short glass fiber reinforced thermoplastic composites) can be successfully molded with the fluid-assisted injection molding technology. Nevertheless, to successfully mold a part, the plastic melt must possess a certain level of hot melt strength to prevent fluid 'blow through', that is, the gas or water, instead of pushing the polymer melt forward, penetrates through the high viscosity melt. On the other hand, glass fiber reinforced composites are more difficult to mold by water-assisted injection molding technology. This is due to the fact that as long as the water enters into the cavity, it begins to cool the polymer melt and increases the viscosity of the melt significantly, especially for glass fiber reinforced materials. It then becomes more difficult for the water to penetrate into the core of the parts for successful moldings. But this is generally not a problem. Various thermoplastic composites including 20% and 30% short glass fiber filled polypropylene,[15,16,47–49] 20% and 30% glass fiber reinforced Nylon-6[17,19,49] and 20% glass fiber filled PBT composites[18] have been reported successfully molded by the water-assisted injection molding technique.

2.2.1 Rheological properties of molding compound

The viscosity of a fluid represents its resistance to flow. While the viscosity of Newtonian fluid such as water is a constant, the viscosity of polymeric materials, either polymer melt, solution or fiber filled compounds, exhibits shear thinning behavior.[50] The incorporation of short fibers into thermoplastic polymer melts increases their viscosity without changing the basic rheological character-shear rate dependency.[51] As shown schematically in Fig. 2.5, this shear thinning behavior is mainly due to alignments and disentanglement of the long polymer chains during shearing. One simple way to fit and model the viscosity of fiber filled polymer composites is the Power-low expression,

$$\tau = m\dot{\gamma}^{n} \quad [2.1]$$

or

$$\eta = m\dot{\gamma}^{n-1} \quad [2.2]$$

Table 2.1 Typical processing properties of commercial molding compounds

	Melt flow index (g/10 min)	Melting temp., °C	Injection temp., °C	Molding pressure, MPa	Linear mold shrinkage	Coefficient of thermal expansion 10–6/°C	Thermal conductivity (10–4 cal/s °C)	Specific gravity	Water absorption %, 24 h
Testing method (ASTM)	D1238				D955	D696	C177	D792	D570
Acrylonitrile butadiene styrene (ABS) transparent	0.6–28		218–248	35.3–124	0.003–0.007	120		1.05–1.09	0.3
ABS 30% Gfr	1–8		221–276	48–138	0.001–0.005	28–50	5.1–5.5	1.16–1.45	0.12–1.3
Poly(methyl methacrylate)	0.5–30		204–260	41–138	0.001–0.01	54–115	4.4–5.3	1.11–1.19	0.2–0.41
Methyl methacrylate-styrene copolymer			204–243		0.0005–0.006			1.03–1.15	0.1–0.35
Polyfluoroalkoxy (PFA) fluoro-plastics	1.9–25	300–310	248–398	55–138	0.001–0.01	14–220	4.4–13	2–2.24	0.01–0.04
Fluorinated ethylene propylene	1–24	240–276	343	69–138		99	4.8	2.14–2.17	0.01
Polyaryletherketone			398–426	69–124		39.6		1.3	
Polyaryletherketone 30% Gfr			398–426	69–124	0.002–0.005	30.7		1.45–1.53	0.07–0.1
Polyether ether ketone (PEEK)	4–45	334–374	365–426	14–17	0.0005–0.022	45–54	5.8–6	1.26–1.61	0.1–0.5
PEEK 30% Gfr	3–16	334–343	360–430	14–17	0.002–0.005	21–22	10.2–10.3	1.45–1.66	0.1–0.15

(*Continued*)

Table 2.1 Continued

	Melt flow index (g/10 min)	Melting temp., °C	Injection temp., °C	Molding pressure, MPa	Linear mold shrinkage	Coefficient of thermal expansion 10–6/°C	Thermal conductivity (10–4 cal/s °C)	Specific gravity	Water absorption %, 24 h
PEEK 30% Cfr	3–34	334–340	382–426	14–17	0.0005–0.005	0–30	21.9	1.4–1.5	0.06–0.15
Liquid crystal polymer (LCP)		280	287–387	7–150	0–0.004	–5–6	21–717	1.37–1.96	0.1
LCP 30% Cfr		280	287–315	14–150	0.0005–0.002	–2–30		1.48–1.55	0.035–0.1
LCP 50% Mf		327	290–360	7–138	0.003	120		1.89	
Polyamide(PA)6 30% long Gfr		225	243–298	69–151		14		1.36–1.37	1–1.1
PA6 Gfr		210–225	240–300		0.001–0.007			1.2–1.62	0.8–1.3
PA6		212–220	237–298	3.4–138	0.009–0.02	70–190	4.1–5.8	1.06–1.2	0.7–1.96
PA66 30% long Gfr			260–304	69–138	0.001–0.003	25		1.36–1.42	0.6–1
PA66		237–264	248–304	3.5–138	0.012–0.03	80–140	5.8	1.07–1.14	0.7–1.4
PA66 15–33% Gfr		248–260	260–304	7–138	0.002–0.01	19–30	5.3–8.2	1.32–1.56	0.4–0.8
Polyimide-imide			343–371		0.0014–0.0085	25–38	6.2–12.8	1.4–1.59	0.27–0.35
Polyimide-imide 30% Gfr			343–371		0.001–0.0025	9–27	8.6	1.51–1.62	0.24–0.3
Polybutylene terephthalate (PBT) unfilled	2.2–75	107–250	232–271	27.5–138	0.0015–1.5	16–133	4–6	1.07–1.7	0–0.1
PBT 30% Gfr	1.5–42	220–248	232–282	55–138	0.01–0.017	11–57	4–8	1.43–1.73	0.007–0.2
Polyethylene terephthalate (PET) 15% Gfr		243–254	265–310	3.5–82	0.003–0.005	22–35		1.33–1.58	0.1
PET	2–40	182–310	254–310	3.5–138	0.0015–0.025	16–1400	3.3–7	1.2–1.9	0.07–0.25

Low density polyethylene (LDPE)	0.25–35	110–160	148–232	34.5–69	0.002–0.05	120–200	7–11.5	0.4–2.04	0.01–0.1
High density polyethylene (HDPE)	0.25–50	132	176–260			150–155		0.96–0.97	
Ultra high molecular weight-PE	5–12	132–143			0.006	20–250	9.6–10.1	0.92–1.45	0–0.07
HDPE 30% Gfr	2–4.5		193–232	7–138	0.002–0.006	50–70	7–8.5	1.08–1.2	0.01–0.06
Polymide	6		390–420	76–241	0.005–0.012	5–100	4.2–11	1.31–1.95	0.08–0.9
Polymide 30% Gfr			390–420	69–206	0.003–0.006	17	8.3	1.54–1.63	0.09–0.23
Polyethylene sulfide	30–167	275–285	293–337	34.5–138	0.0001–0.02	10–55	6.8–10	1.34–1.81	0.01–0.05
Polypropylene (PP)	0.25–100	147–230	93–287	7–138	0.002–0.04	37–100	0.3–5.4	0.8–1.25	0.008–0.04
PP 10–40% talc	0.6–35	171–193	176–287	7–138	0.0008–0.02	40–80	3.8–7.8	0.96–1.32	0.02–0.06
PP 10–30 Gfr	1.5–20	171–193	190–300	7–138	0.0002–0.04	30–60	4.5–8.6	0.94–1.48	0.01–0.06
High impact polystyrene	2.9–8			69–138	0.004–0.008	172–182	1.03		
Styrene acrylonitrile (SAN)	6.5–35		190–275	34.5–138	0.002–0.007	52–68		1.07	0.03–0.225
SAN 20 Gfr	2–18		220–275	69–138	0.001–0.05	30–41	6.5–6.8	1.15–1.23	0.2–0.3
Styrene maleic anhydride (SMA)	0.3–5.1		205–285	69–117	0.001–0.007	35–93		1.06–1.54	0.1–0.3
Polystyrene	1.4–38	180–250	176–280	34–275		35–75		0.9–1.2	0.03–0.14
Polyurethane	20				0.0014–0.013		0.9–6.2	0.09–1.28	0.3–2.9
Polysulfone	3–25.5	351	240–400	69–138	0.0001–0.01	11.7–56	5.8–9.2	1.23–1.6	0.1–0.7
Polysulfone 20% Gfr			340–400	69–138	0.002–0.005	31		1.37–1.48	0.2–0.5

Source: From reference 46.
Notes: Gfr, Glass fiber filled; Cfr, carbon fiber filled; Mf, mineral filled.

Table 2.2 Typical mechanical properties of injection molded plaques

	Tensile strength at break, GPa	Elongation at break, %	Compressive strength, MPa	Flexural strength, MPa	Tensile modulus, MPa	Compressive modulus, MPa	Flexural modulus, 103 MPa, 23°C
Testing method (ASTM)	D638b	D638b	D695	D790	D638b	D695	D790
Acrylonitrile butadiene styrene (ABS) transparent	29.6–38	27–50		54–78	1.74-2.17		1.65–2.37
ABS 30% Gfr	36.5–72.4	1–8.5	69–117	96–172	6.17–9.65		4.61–10.34
Poly(methyl methacrylate)	36.5–72.4	0.2–25	55–117	60–131	1.37–3.99	1.66	1.26–3.51
Methyl methacrylate-styrene copolymer		10–70		68–88	1.72–2.2		1.57–3.03
Polyfluoroalkoxy (PFA) fluoro-plastics	24.8–30	290–450			39–70		
Fluorinated ethylene Propylene	20–30	290–400	15		50–90		
Polyaryletherketone			117	146			
Polyaryletherketone 30% Gfr	18	2.3–3	138.5	186–251	11.72–12.41		9.65–10.34
Polyether ether ketone (PEEK)	76–93	1.4–60	117–138	121–173	2.75–19.37	3.1–4.48	3.64–4.09
PEEK 30% Gfr	154–178	1.5–3	179–215	193–269	5.17–12.41	3.44–10.34	9.10–10.3
PEEK 30% Cfr	179–233	1–2	144–240	215–351	11.99–31.02	3.78–13.78	17.92–21.37
Liquid crystal polymer (LCP)	186	0.4–5.8	142	131–244.7	9.79–38.26	11.72	8.75–20
LCP 30% Cfr	172–241	0.6–1.12	124–237		26.2–37.23	17.23–33.09	22.75–29
LCP 50% Mf	110	2			19.99		17.23
Polyamide(PA)6 30% long Gfr		D:2.6–4	222	122–290	10.34		8.54–9.03
PA6 Gfr	131–201	D:1.3–5.5		158–227	6.19–6.88		2.6–10.34
PA6	25–85	D:8–365		7–179	0.45–3.20		

PA66 30% long Gfr	110–228	D:2.4–4	215–235	243–278	9.28–10.34		
PA66	44.81–82	D:3–93	34.5	65.5–89	2.15–3.37		1.58–3.1
PA66 15–33% Gfr	124–134	D:2.5–7	170	179–225	6.48–8.27		6.89–8.84
Polyimide-imide		6–15	124–179	138–194	3.03–6.57	2.55–6.59	4.96–7.30
Polyimide-imide 30% Gfr		7	253–264	150–248	6.20–10.75	4.13–7.92	11.72
Polybutylene terephthalate (PBT) unfilled	18–62	1–305	63–109	48–109	1.51–3.72	2.58–2.96	3.03
PBT 30% Gfr	87–142	1.5–5	85–202	134–221	6.48–13.10	4.58–7.58	3.79–9.37
Polyethylene terephthalate (PET) 15% Gfr	67–110	3–4			4.27		1.55–6.13
PET	23.4–79	2–340	103	60–124	2.20–14.47	2.89	2.48–2.62
Low density polyethylene (LDPE)	7.6–15.1		4.48	7.5–18.6			
High density polyethylene (HDPE)	9.6–40	1–1000		48	1.03–1.34		0.80–1.80
Ultra high molecular weight-PE	19–52	50–330	25	26–48	0.52–1.03		160
HDPE 30% Gfr	44.8	39457	24–49	26–76	0.2–0.58		25–70
Polymide		1.5–11	76–120	96–137	2.41–3.10	2.13	370–580
Polymide 30% Gfr		2–4	88–188	207–289	10.34–11.03		1200–1400
Polyethylene sulfide	38–124	0.45–10.5	41–125	124–158	2.24–14.18	2.82–8.96	300–2453
Polypropylene (PP)	17.2–49.5	2–605	41–47	24–56	0.06–2.48		120–380
PP 10–40% talc	19–31	2–40	34.5–52	20.6–58	1.99–4.68		170–670
PP 10–30 Gfr	44.8–89	1–10	44.8–82	49–144	1.99–10.34		180–900
High impact polystyrene		55			1.99–2.17		270–310
Styrene acrylonitrile (SAN)	63	2.3–3	110–115	83–127.5	3.44		450–560
SAN 20 Gfr	103	1.5–6	103–131	117–157	4.89–8.96		867–1107

(*Continued*)

Table 2.2 Continued

	Tensile strength at break, GPa	Elongation at break, %	Compressive strength, MPa	Flexural strength, MPa	Tensile modulus, MPa	Compressive modulus, MPa	Flexural modulus, 103 MPa, 23°C
Styrene maleic anhydride (SMA)	25–96.5	1.6–20	186–206	55–156.5	1.93–6.89		310–1010
Polystyrene	17–56	1–85	0.89	0.27–1.03	1.6–3.5		1.1–3.6
Polyurethane	2800–8910	200–850	3.5–4.8	24.8–62	6.23	0.93–2.86	
Polysulfone	6500–24000	1–77	82–186	81–173	2.41–12.06	2.58	333–394
Polysulfone 20% Gfr	16680	2.2–3.5		138–153	5.17–6.48		825–1060

Source: From reference 46.
Notes: Gfr, glass fiber filled; Cfr, carbon fiber filled; Mf, mineral filled.

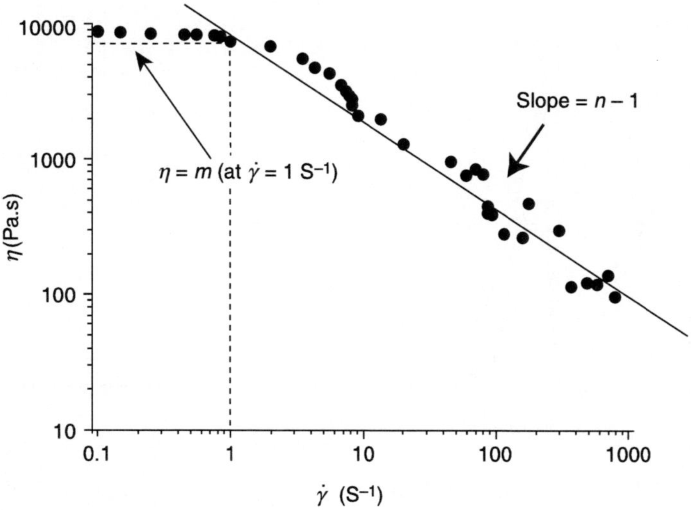

2.5 **Shear thinning behavior of fiber filled compounds.**

where m is the measure of consistency (the larger the m, the more viscous the compound) and n indicates the degree of non-Newtonian behavior (n is equal to one for Newtonian fluid while n is less than one for shear thinning fluid, for example glass fiber filled compound). The consistency coefficient m is sensitive to temperature and can be described by the following expression,

$$m = m_0 e^{-b(T-T_0)} \quad [2.3]$$

where m_0 is the consistency index at a reference temperature T_0.

2.2.2 Molding of glass fiber filled materials

The injection molding process for glass filament reinforced pellets is identical to that for un-reinforced or filled thermoplastics. The main differences will be in the treatment of the screw to limit its wear, and a 10–30°C higher barrel temperature compared to un-reinforced thermoplastic compounds because of higher melt viscosity.

Fiber breakage occurs during the whole injection molding process, starting from at the solid–melt interface in the melting zone, and continuing in the nozzle, gate and cavity. Tremblay *et al.*[52] showed that the plastication and flow through the nozzle accounts for 55%, the gate for 31% and the cavity for 13% of the fiber length reduction in injection molded glass fiber reinforced polystyrene. Kamal *et al.*[53] proposed that the mean glass fiber

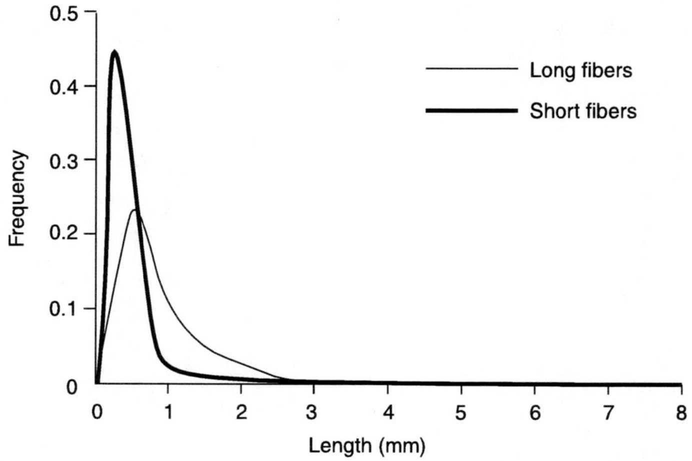

2.6 Fiber length distribution of injection molded 30 wt% short and long glass fiber reinforced polypropylene. (Reproduced based on the data of reference 54.)

length decreased from 0.71 mm in the resins to 0.27 mm in injection molded fiber reinforced polypropylene parts. Vincent[54] compared the fiber length distribution in injection molded 30% short and long glass fibers filled parts. The initial fiber lengths of short and long fibers were 0.56 and 12 mm, while after molding their lengths reduced to 0.41 and 0.87 mm, respectively. The length reduction is more severe for the initially long fibers. Figure 2.6 shows the fiber length distribution of injection molded 30 wt% short and long glass fiber reinforced polypropylene.[54] One can reduce the injection speed to minimize the fiber breakage. In addition to fiber breakage, the molding process also alters the homogeneity of the fibers in molded parts, which in turn influences their mechanical properties. Kamal *et al.*[53] proposed that for 10 wt% glass fiber filled polypropylene the fiber concentration tends to be higher at the core (11.5%) than near the surface (10%). Others reported that for long glass fiber reinforced polyamide the fiber concentration at the core and near the surface were 58 and 40 wt% respectively. [55] Additionally there exists a thin layer near the surface with sharply reduced fiber concentration.[56]

To successfully mold a part, the 'moldability' diagram (Fig. 2.7) based on key processing factors has been used to describe the molding of a specific system,[57] which is determined by the material, the mold and processing parameters. For example, to test the moldability of a new material, the polymer in question is forced under standard conditions into a spiral runner-type mold. The degree of filling of the spiral mold by the polymer gives an indication of its moldability. The moldability of an injection molding process is usually defined based on the molding area on the plane of key processing

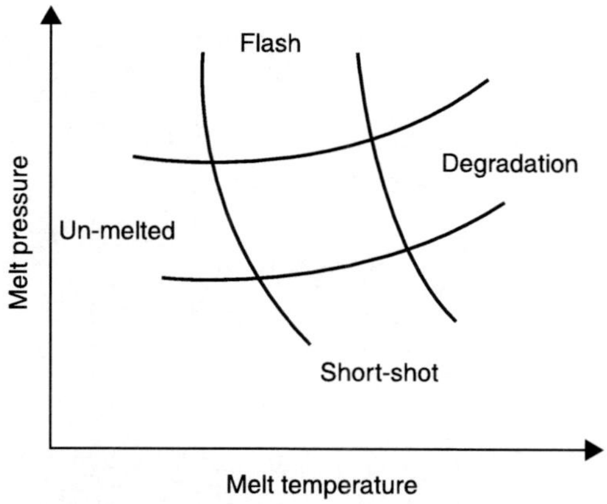

2.7 Moldability diagram for the injection molding process. (From reference 57.)

parameters 'melt filling pressure' and 'melt temperature'. Moldability diagrams can be employed to evaluate new mold designs. Molds can be designed with different melt delivery and mold cooling systems. To compare the effectiveness of these mold designs, the moldability diagram can be defined for each molding system. A system with larger molding area implies that the molding operation can be successfully carried out over a wide range of conditions. Therefore, the process is more stable. Even a different, perhaps less expensive, molding machine can be used. The characterization of moldability defined by the molding area on the plane of processing parameters can serve as the first guideline for process optimization.

2.2.3 Fiber orientation

With an injection molding machine, one cycle is characterized by the phases of filling, packing and cooling. During the filling stage of injection molding, the non-Newtonian and compressible polymer melt is injected to fill the empty cold cavity. Polymer melt elements are subjected to both shear and elongational stresses during this process. Shear stresses are generally predominant. The net effect of shear and elongational stresses is the orientation of polymer chains in the direction of the applied stress. In injection molded parts, two regions of distinct fiber orientations are usually observed: the thin skin and the core. In the skin region near the wall, all fibers are oriented parallel to the flow direction, while at the core, the fibers are oriented substantially perpendicular to the flow direction,[58–61] as shown schematically in Fig. 2.8a. Furthermore, a high injection speed generally gives alignment of fibers transverse to the flow direction, while for very low speed the fibers align parallel the flow.

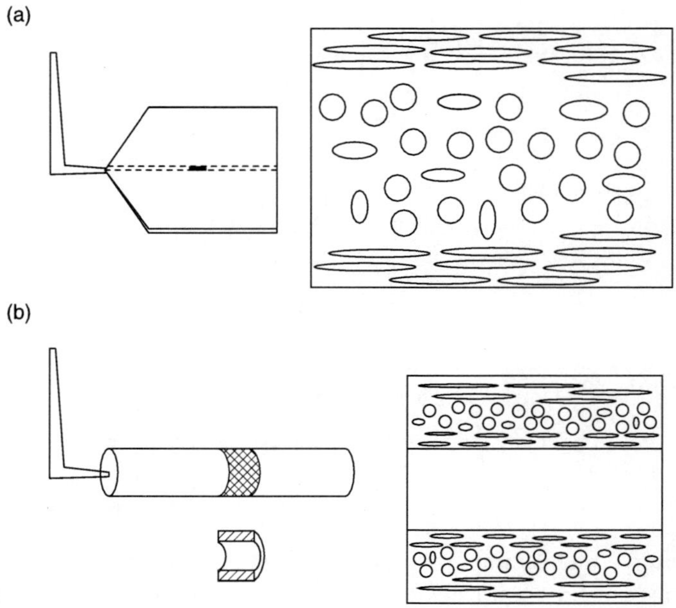

2.8 Fiber orientation in (a) injection molded parts, (b) fluid-assisted injection molded parts.

Compared to conventional injection molding, fluid-assisted injection molding technology is different in the way the mold is filled. In the fluid-assisted injection molding process, the mold cavity is partially filled with the polymer melt followed by the injection of gas or water into the core of the polymer melt. The novel filling process influences the orientation of fibers and matrix in a part significantly.[16,18] The fiber orientation in fluid-assisted injection molded parts can be approximately divided into three zones, as shown schematically in Fig. 2.8b. In the zone near the mold-side surface where the shear is more severe during the mold filling, fibers are principally parallel. For the zone near the fluid-side surface, the shear is smaller and the velocity vector greater. In this case, the fiber tends to be positioned more transversally in the direction of injection. At the core, the fibers tend to be oriented more randomly. Generally speaking, the glass fibers near the mold-side surface of molded parts orient mostly in the flow direction, and orient substantially perpendicular to the flow direction with increasing distance from the mold-side surface.

2.2.4 Long glass fiber reinforced injection molding

Long glass fiber reinforced plastic materials have received increasing attention mainly due to their superior mechanical performances

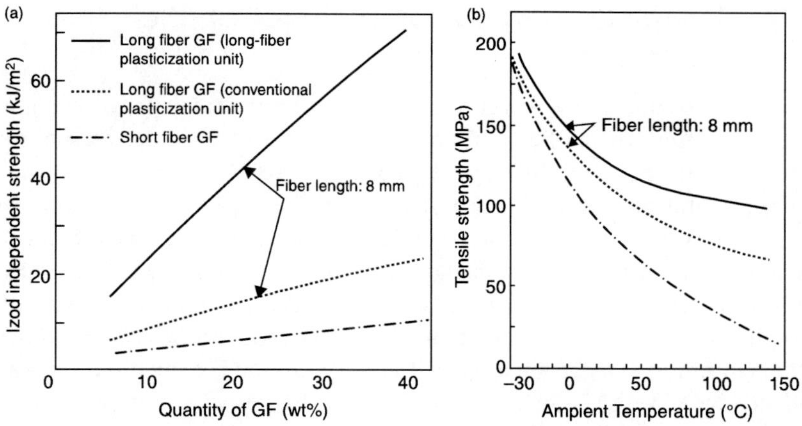

2.9 (a) Izod impact strength and (b) tensile strength of long glass fiber reinforced parts. (Reproduced based on the data of reference 65, p. 130.)

which make them as good substitutes for engineering plastics or metals.[63] As an example, polypropylene resins reinforced with 60–70 wt % of 6–8 mm long glass fibers have been widely used in the automotive industry.[64] Furthermore, Fig. 2.9 shows that the mechanical strengths of long glass fiber reinforced materials are superior to those of short glass fiber filled materials.[65] The preparation of long glass fiber compounds usually employs a single-screw extruder to mix resin and additives and then feed the melt to a second extruder where the glass is blended in. The fibers are preheated before entering the extruder to open the bundles and ensure good wet-out. The second extruder's screw should be designed to minimize glass breakage.

Conventional injection molding machines are widely used to process short glass fiber filled resins, which contain glass fiber of approximately 0.5 mm in length. However, when long glass fiber reinforced resins are molded with these machines, molded parts exhibit low mechanical properties, mainly due to the breakage of the fibers during the plastication process inside the barrel. Special machines that have optimized screw and nozzle geometries should be employed for molding these materials. Figure 2.10 shows a comparison of the fiber length distribution for the long glass fiber reinforced resins molded by conventional plastication unit and specially designed long-fiber plastication unit of injection molding machines. The measured mechanical strength results in Fig. 2.10 also suggest that the mechanical performance of injection molded long glass fiber reinforced parts can reduce significantly if conventional injection molding machines are used.[69]

2.3 Characterization and prediction of fiber orientation

As the mechanical properties of fiber filled composites depend strongly on the orientation of the fibers, there is a need to correlate the flow of polymer melt during filling and the resultant fiber orientation. To measure the fiber orientation in injection molded parts, three methods can be employed, namely by observing a polished cross-section, by employing a confocal laser scanning microscopy, or by using X-ray microtomography.[66] The observation of a polished cross-section of molded parts is probably the most widely employed method, mainly due to its simplicity and effectiveness. Cylindrical fibers orientating a certain angle with respect to the cutting plane appear as ellipses (see Fig. 2.11). As shown in Fig. 2.11, the measurement of the axes of the ellipses a and b leads to the orientation angles θ and ϕ.

The orientation of fibers can be represented by the second order orientation tensor defined as,[61]

$$a_{ij} = \int_P p_i p_j \psi(P)\mathrm{d}P \qquad [2.4]$$

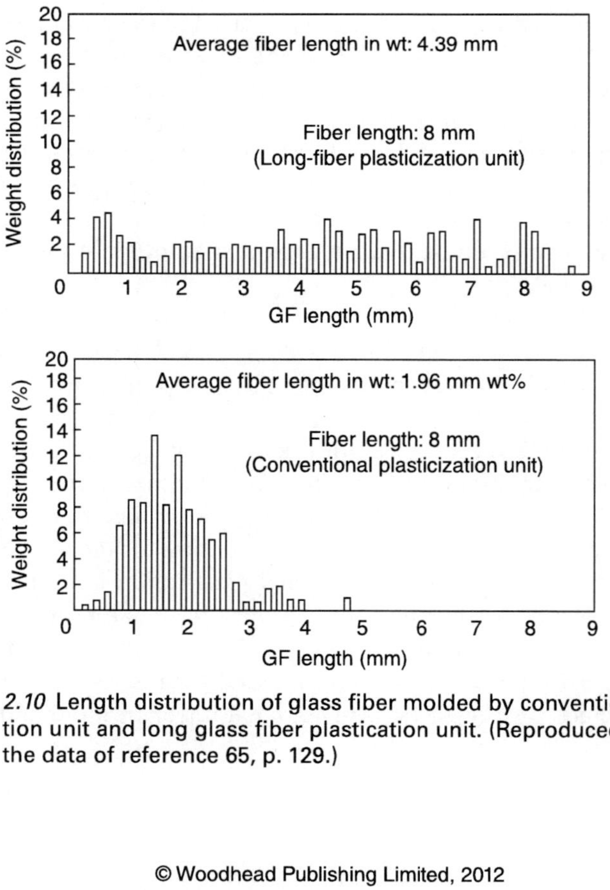

2.10 Length distribution of glass fiber molded by conventional plastication unit and long glass fiber plastication unit. (Reproduced based on the data of reference 65, p. 129.)

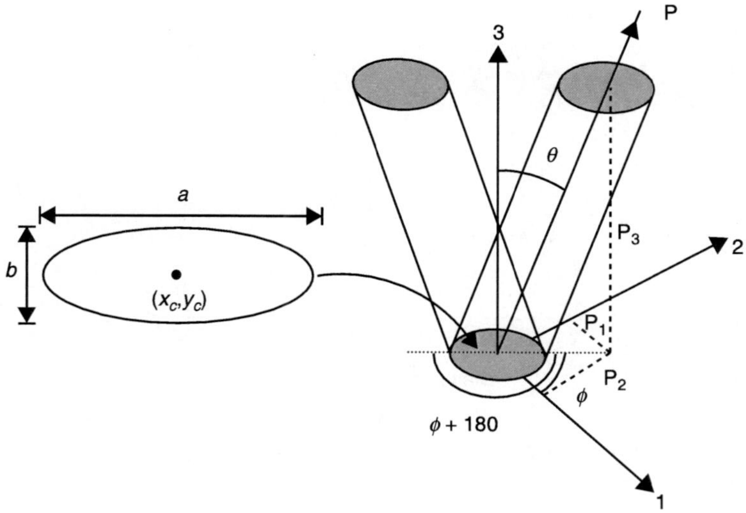

2.11 Definition of the orientation angles (θ, φ) of a fiber.

where P is unit vector parallel to the fiber, with components p_i, which are related to the angles θ and ϕ. $\psi(P)$ is the orientation distribution function and $\psi(\theta, \phi)\sin\theta d\theta d\phi$ is the probability of finding a fiber between angles θ and $\theta + d\theta$ and ϕ and $\phi + d\phi$. The orientation tensor is a symmetric one (i.e., $a_{11} + a_{22} + a_{33} = 1$ or $a_{xx} + a_{yy} + a_{zz} = 1$), and the diagonal components represent the strength of orientation in the three principal directions (i.e., x, y and z directions). While $a_{33} = 0$ represents that the fibers are perpendicular to direction 3, $a_{33} = 1$ indicates that fibers are parallel to that direction. A random orientation of fibers will lead to the diagonal components of 1/3. Vincent[67] studied the fiber orientation in injection molded short glass filled polyarylamide and found that the values of a_{zz} (z is the thickness direction) are negligible (very close to 0). This indicates that fibers are mainly parallel to the flow direction (x direction). Figure 2.12 shows the fiber orientation distribution in injection molded plates of 1.7 and 5 mm in thickness.[67] A high a_{xx} value represents that fibers are highly oriented in parallel to the x direction, while a low value indicates that fiber are oriented in perpendicular to the x direction. Obviously the thinner the plate is, the higher the probability the fibers orient in the flow direction, especially at the core region.

To model and predict the fiber orientation in injection molded parts, Folgar and Tucker[68] modified the theory proposed by Jeffery[69] by introducing a diffusion term as follows:

$$\frac{da_2}{dt} = \Omega a_2 - a_2\Omega + \lambda(\dot{\varepsilon}a_2 + a_2\dot{\varepsilon} - 2\dot{\varepsilon} : a_4) + 2C_I\dot{\varepsilon}(I - 3a_2) \qquad [2.5]$$

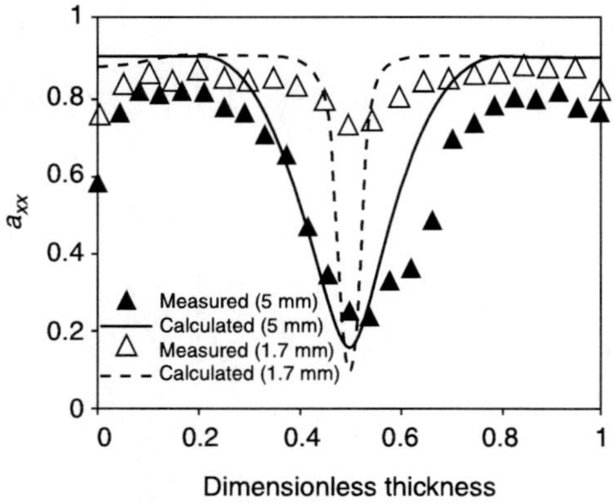

2.12 Fiber orientation (a_{xx}) in injection molded plates of different thicknesses. (Reproduced based on the data of reference 67.)

where $\lambda = (r^2 - 1)/(r^2 + 1)$, and r is the fiber aspect ratio. Ω and $\dot{\varepsilon}$ are the vorticity and rate of strain tensors, while C_I is the interaction coefficient. a_4 is the fourth order orientation tensor and is defined as $a_{ijkl} = \int_P p_i p_j p_k p_l \psi(P) \mathrm{d}P$. Figure 2.12 also shows the predicted fiber orientation[67] and the results suggest that the calculated results are in reasonable agreement with the experimental data.

Although the Folgar–Tucker model has proved useful for predicting the fiber orientation in short fiber reinforced thermoplastic composite, it is less accurate for long fiber filled materials. Long fiber thermoplastics develop similar fiber orientation pattern qualitatively to that of short fiber composites, but they exhibit quantitatively lower flow-direction alignment than short fiber ones during the injection molding process. Fan *et al.*[70] and Phan-Thien *et al.*[71] developed, based on the Folgar–Tucker theory, an anisotropic rotary diffusion model in which the scalar interaction coefficient C_I in Equation [2.5] was replaced by a second order tensor C. Furthermore, Wang *et al.*[72] introduced the so-called reduced-strain closure (RSC) model in order to obtain better agreement between experimental data and predicted results as well as to better model the orientation of long fibers in injection molded composite. The model can be given by the following kinetic theory,

$$\frac{\mathrm{d}\psi}{\mathrm{d}t} = -\nabla_s \cdot [\psi \dot{p}^h - C_I \dot{\varepsilon} \nabla_s \psi + w_i (e_i e_i \cdot p - e_i e_i : p_i p_j p_k)] \qquad [2.6]$$

where $\dot{p}^h$ is Jeffery's equation for particle motion and e_i are eigenvectors of the orientation tensor. The term with the summation represents an

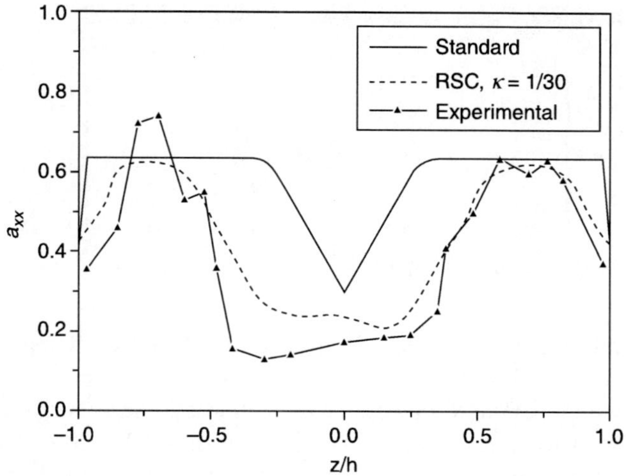

2.13 Long-fiber orientation (a_{xx}) in injection molded plates: a comparison of experimental data and various prediction models. (Reproduced based on the data of reference 73.)

orientation-dependent probability flux. The scalar coefficient w_i is calculated as,

$$w_i = -\frac{5(I-\kappa)}{4\pi}[\xi(\lambda_i D : e_i e_i - e_i e_i : a_{ij} : D) + C_I \dot{\varepsilon}(I - 3\lambda_i)] \qquad [2.7]$$

where λ_i are the eigenvalues of the orientation tensor a_{ij}. Phelps and Tucker[73] compared the accuracy of the RSC model and the Folgar–Tucker model against their experimental works on the orientation characterization of injection molded long fiber filled composites. Their results showed that the RSC model predicts better the fiber orientation (Fig. 2.13). In addition, while the Folgar–Tucker model predicts a narrow core region, the RSC model predicts a wide core which is consistent with the experimental data.

2.4 Molding defects

Here we introduce some commonly seen molding defects in injection molded short glass fiber reinforced parts. Solutions for other defects not described in this chapter can be found in Rubin (1972).[4]

2.4.1 Jetting

As a polymer melt is injected from a small size gate or runner into a large size cavity, the phenomenon of 'jetting' or irregular flow can be observed.[3]

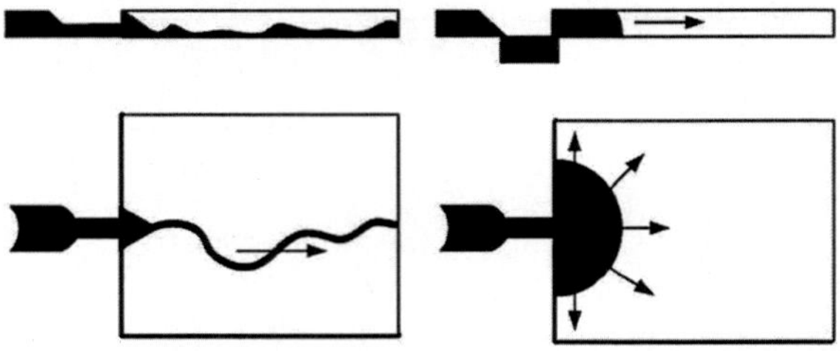

2.14 The use of a tab gate to reduce the jettings.

The melt emerging from the gate forms a jet that rapidly advances until it is stopped by the mold wall. It has been experimentally observed[74] that jetting can occur whenever the dimension of the fluid stream is smaller than the smallest dimension in the plane perpendicular to the flow. It is thus related both to the gate size and to the degree of extrudate swelling of the melt, rather than to the level of that axial momentum. Filled polymers, which swell less than unfilled melts, exhibit jetting at lower filling rates. Jetting occurs when polymer melt is pushed at a high velocity through restrictive areas, such as the nozzle, runner or gate, into open, thicker areas, without forming contact with the mold wall. The buckled, snake-like jetting stream (Fig. 2.14) causes contact points to form between the folds of melt in the jet, creating small-scale 'welds'. Jetting may lead to part weakness, surface blemishes and a multiplicity of internal defects. One can direct the melt against a metal surface to eliminate jetting by repositioning the gate or use an overlap or a submarine gate. An alternative is to use a tab or fan gate to slow down the melt with a gradually divergent flow area. This reduces the melt shear stress and shear rate. One can also employ an optimized ram-speed profile so that melt-front velocity is initially slow when the melt passes through the gate, then increases once a dispersed flow is achieved. These all will help eliminate the occurrence of jetting in injection molded composite parts.

2.4.2 Fiber exposure

During the melt filling process, as soon as the hot polymer melt contacts the mold wall it tends to cool. This is followed the packing stage of the process. The high-pressure melt pushes the composite against the mold wall to minimize the shrinkage of the materials including the polymer matrix and the reinforcing fibers. However, if the packing pressure is not high enough, polymeric material shrinks more than glass fibers when the materials cool down. This leaves the fibers exposed to the parts surface and leads to the fiber exposure phenomenon,[75] which in turn influences the surface quality

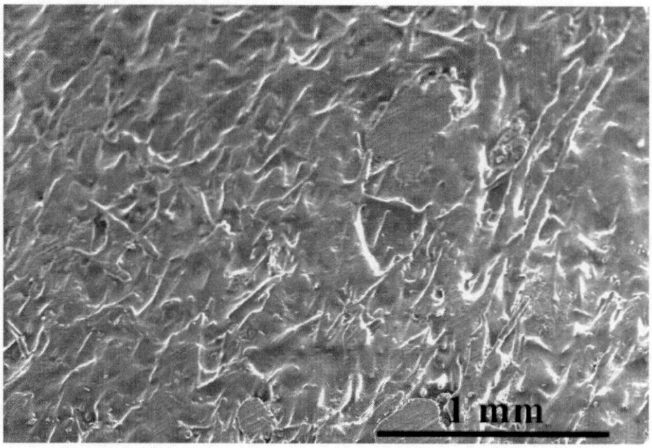

2.15 Microscopically the fiber exposure phenomenon in molded composite parts.

of molded composites. Surface roughness also occurs in jetting areas. As soon as the jetted melt contacts the mold wall it tends to cool. The polymer shrinks much more than the glass fiber and leaves the fiber exposed to the parts surface, as can be seen in Fig. 2.15. Parts roughness may thus form at the jetting and irregular flow filled surfaces. On the other hand, fiber roughness may also occur in thick sections of injection molded composites, mainly due to the insufficient packing pressure in these regions.

When designing a part, one should avoid the occurrence of jetting or irregular flows (e.g., by adopting a smaller size-variation geometry design) during filling to eliminate the surface roughness. One can also improve the surface quality of an injection molded composites by: increasing the melt and mold temperature, increasing the melt filling speed, and increasing the filling pressure as well as the packing pressure.

2.4.3 Weldlines

Where melted material flows together during molding to form a visible line or lines on a finished part that may cause weakening or breaking of the component. Weldlines form because of the presence of cores, pins and multiple gates that divide the polymer melt during flow in the mold. They also form when jetting occurs because of an abrupt change in geometry along the flow path, such as near a gate. It has been observed that the presence of weldlines significantly reduces the strength of injection molded parts.[76–81] Savadori *et al.*[82] studied the weldline strength of inorganic particle filled polypropylene (PP), and proposed that the weldline strength depends on the content

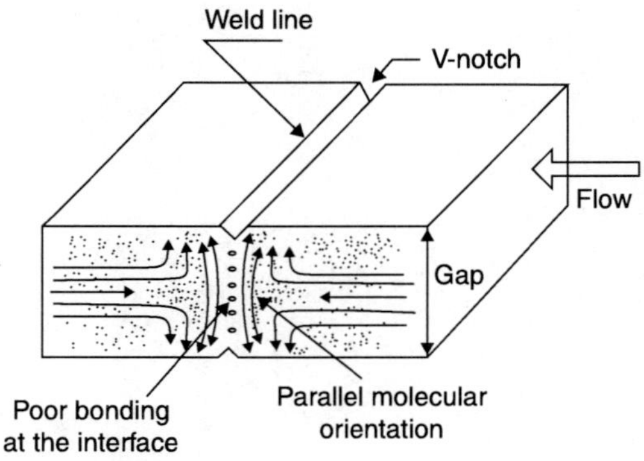

2.16 Schematic of the formation of injection molded weldlines. (Reproduced from reference 79.)

and the shape factor of the filler. Fisa and his co-workers[83,84] investigated the weldline zone of injection molded glass filled PP composites. They found that the weldline is a zone between 2 and 8 mm wide extending throughout the thickness in which the fibers are oriented almost perfectly in a plane parallel to the weldline. In addition, they also found that the weldline strength of molded composites decreased with increased fiber content. Vaxman *et al.*[85] investigated the weldlines in short fiber reinforced thermoplastic materials. They proposed that the orientation of the fibrous reinforcement parallel to the weld interface caused a significant reduction of the tensile strength compared to the weld-free products. Nadkarni and Ayodhya[86] also studied the weldline strength of injection molded composites, and found that the presence of glass fibers significantly reduced the strength as well as the strain to failure of the polymeric composites.

Weldline strength of injection molded thermoplastic composites in general decreases with the content of reinforcing fibers in the composites. As shown schematically in Fig. 2.16, the weakness of the weld area is mainly attributed to the lack of fiber flow across the weldline, which makes the material in the weldline region act as if it is not reinforced. The orientation of the fibrous reinforcement parallel to the weldline caused a significant reduction of the tensile strength compared to the weld-free products. Whereas the total elimination of weldlines is not always possible, their negative influence on part performance and appearance can be minimized, by increasing the melt and mold temperature, increasing the melt filling and packing pressures or by adopting a smaller size obstacle.[87] Increasing the melt filling speed can also increase the weldline strength of the composite materials. However, if the injection speed is too high, melt flow will be too fast and result in reinforcing fiber alignment transverse to the weld surface. It then results in a decrease in weldline strength.

2.4.4 Warpage

The warpage of an injection molded plastic product can be seen as primarily due to non-uniform differential shrinkages within the product, which leads to the development of uneven residual stresses within the product. During the cooling process, both the stiffness and the transient temperature of the polymeric materials drive the formation of the thermally induced residual stress. Part warpage is directly related to residual stress,[88–90] which results from locally varying strain fields that occur during a solidification process of the material. Such strain gradients are caused by non-uniform thermo-mechanical properties and temperature variations inside the material. The strain gradients across thickness can result from asymmetrical cooling (which is the case of rotationally molded part cooled from one side of its surface), or inappropriate processing conditions. Such strain gradient can lead to an uneven residual stress build-up during a manufacturing process, which induces a bending moment of the part. This bending moment tends to warp the part so that the unbalanced residual stress can be balanced. For fiber reinforced materials, the short fibers become aligned during the mould filling process, leading to orientation-dependent material properties. This anisotropy exerts a strong influence on the component behavior as well as on warpage and shrinkage. Despite the fibers increasing the stiffness of the parts and reducing shrinkage, they may also induce unbalanced shrinkage which in turn results in part warpage. One should try to balance the flow field during filling and the temperature distribution during cooling in the injection molding process, so as to minimize warpage of injection molded glass fiber reinforced parts.

2.5 Conclusions and future trends

Injection molding has become one of the most important methods for processing short and long glass fiber reinforced thermoplastics, and is expected to continue to grow annually for more applications in novel fields such as optics, electricals and electronics, aerospace, biomedicine and communications. New fillers such as benign filler or 'eco' fillers made from recycled paper, structural lignocellulosic fibers from corncobs and high cellulose fiber from wheat straw that are sustainable or recyclable will continue to be developed. Meanwhile, other process variants such as micro-injection molding of micro- and nanocomponents, high speed injection molding of ultra-thin composite parts, new foaming process that produces parts free of sink marks, low in warpage and with a high quality surface, injection molding of long-fiber-reinforced thermoplastics that exhibit good creep resistance and impact performance, and injection molding of conductive thermoplastics that provide effective EMI shielding in engineering applications, etc., will continue to be explored and developed. The injection molding technology will play a very important role in contributing to the growth and success of the plastic composites industry.

2.6 Acknowledgment

The author thanks Ms Chao-Ying Hsiao for the drawing and preparation of the figures in this chapter.

2.7 References

1. M. Kamal, A. I. Isayev and S. J. Liu, *Injection molding-technology and fundamental*. Munich: Hanser, 2009.
2. D. Hull and T. W. Clyne, *An introduction to composite materials*, 2nd edn. Cambridge Solid State Science Series, Cambridge University Press, 1996.
3. Z. Tadmor and C. G. Gogos, *Principles of polymer processing*. New York: John Wiley, 1979.
4. I. I. Rubin, *Injection molding: Theory and practice*. New York: John Wiley, 1972.
5. K. C. Rusch, Gas-assisted injection molding – a new technology is commercialized, *Plastics Engineering*, July, 35–38 (1989).
6. S. Shah and D. Hlavaty, Gas injection molding of an automotive structural part, *Plastics Engineering*, October, 21–25 (1991).
7. S. C. Chen, K. F. Hsu and K. S. Hsu, Simulation of the melt front advancement in injection molded plate with a rib of semicircular cross section, *Numerical Heat Transfer, Part A*, **28**, 121–129 (1995).
8. S. Y. Yang, F. Z. Huang and W. N. Liau, A study of rib geometry for gas-assisted injection molding, *Polymer Engineering and Science*, **36**, 2824–2831 (1996).
9. S. J. Liu and K. H. Chang, ., An empirical study of gas penetration in full-shot gas-assisted injection molded parts, *Proceedings of the Institute of Mechanical Engineering Part B: Journal of Engineering Manufacturing*, **216**, 1549–1559 (2002).
10. H. Potente and M. Hansen, The gas-assisted injection molding process, *International Polymer Processing*, **8**, 345–351 (1993).
11. S. J. Liu, Gas assisted injection molding. In M. Kamal, A. I. Isayev and S. J. Liu (eds.), *Injection molding: Technology and fundamental*. Munich: Hanser Publishers, 2009, pp. 195–222.
12. M. Knights, Water injection molding makes hollow parts faster, lighter, *Plastic Technology*, **April**, 42–63 (2002).
13. M. Knights, Water injection – its all coming together, *Plastic Technology*, **September**, 54–61 (2005).
14. W. Michaeli, T. Juntgen and A. Brunswick, WIT-an route to series production: first industrial application of the water injection technique, *Kunststoff Plastics*, Europe, **91**, 37–42 (2001).
15. S. J. Liu and Y. S. Chen, Water assisted injection molding of thermoplastic materials: effects of processing parameters, *Polymer Engineering Science*, **43**, 1806–1817 (2003).
16. S. J. Liu and Y. S. Chen, The manufacture of thermoplastic composites by water assisted injection molding technology, *Composites Part A: Applied Science and Manufacture*, **35**, 171–180 (2004).

17. S. J. Liu and C. C. Shih, Water assisted injection molding of glass fiber reinforced Nylon-6 composites, *Journal of Reinforced Plastics and Composites*, **27**, 985–999 (2008).
18. S. J. Liu, M. J. Lin and Y. C. Wu, An experimental study of the water-assisted injection molding of glass fiber filled polybutylene terephthalate (PBT) composites, *Composites Science and Technology*, **67**, 1415–1424 (2007).
19. W. Michaeli, Water injection techniques. In M. Kamal, A. I. Isayev and S. J. Liu (eds.), *Injection molding: Technology and fundamental*. Munich: Hanser Publishers, 2009, pp. 223–250.
20. S. J. Liu, Water assisted injection molding: a review, *International Polymer Processing*, **14**, 315–325 (2009).
21. M. Defosse, In-mold decoration – harbors huge potentials, *Modern Plastics*, **78**, 76–77 (2001).
22. J. L. Throne, *Thermoforming*. New York: Hanser Publishers, 1986.
23. S. C. Chen, S. T. Huang, M. C. Lin and R. D. Chien, Study on the thermoforming of PC films used for in-mold decoration, *International Communications in Heat and Mass Transfer*, **35**, 967–973 (2008).
24. A. C. Y. Wong and K. Z. Liang, Thermal effects on the behaviours of PET films used in the in-mould-decoration process involved in plastics injection molding, *Journal of Materials Processing Technology*, **63**, 510–513 (1997).
25. Y. W. Leong, M. Kotaki and H. Hamada, Effects of the molecular orientation and crystallization on film-substrate interfacial adhesion in poly(ethylene terephthalate) film-insert moldings, *Journal of Applied Polymer Science*, **104**, 2100–2107 (2007).
26. Y. W. Leong, U. S. Ishiaku, M. Kotaki and H. Hamada, Interfacial characteristics of film insert molded polycarbonate film/polycarbonate-acrylonitrile-butadiene-styrene substrate, Part 1: influence of substrate molecular weight and film thickness, *Polymer Engineering Science*, **46**, 1674–1683 (2006).
27. S. J. Baek, S. Y. Kim, S. H. Lee, J. R. Youn and S.H. Lee, Effect of processing conditions on warpage of film insert molded parts, *Fibers and Polymers*, **6**, 747–754 (2008).
28. C. A. Puentes, O. I. Okoli and Y. B. Park, Determination of effects of production parameters on the viability of polycarbonate films for achieving in-mold decoration in resin infused composite components, *Composites Part A: Applied Science and Manufacturing*, **40**, 368–375 (2009).
29. S. C. Chen, H. M. Li, S. T. Huang and Y. C. Wang, Effect of decoration film on mold surface temperature during in-mold decoration injection molding process, *International Communications in Heat and Mass Transfer*, **37**, 501–505 (2010).
30. V. Kumar and J. E. Weller, *ACS Symposium Series*, **669**, 101 (1997).
31. C. B. Park, Continuous production of high-density and low-density microcellular plastics in extrusion. In S.-T. Lee (ed.). *Foam Extrusion: Principle and Practices,* Philadelphia: Technomic Publishing Co., 2000, Chapter 11, pp. 263–306.
32. A. I. Cooper, Polymer synthesis and processing using supercritical carbon dioxide, *Journal of Materials Chemistry*, **10**(2), 207–234 (2000).
33. W. Michaeli, O. Pfannschmidt and S. Habibi-Naini, Routes to microcellular foam: Principles of material and process control, *Kunststoffe*, **92**(6), 48–51 (2002).
34. P. Rachtanapun, S. E. M. Selke and L. M. Matuana, Microcellular foam of polymer blends of HDPE/PP and their composites with wood fiber, *Journal of Applied Polymer Science*, **88**, 2842–2850 (2003).

35. J. Xu and D. Pierick, Microcellular foam processing in reciprocating-screw injection molding machines, *Journal of Injection Molding Technology*, **5**, 152–159 (2001).
36. M. Shimbo, D. F. Baldwin and N. P. Suh, The viscoelastic behavior of microcellular plastics with varying cell size, *Polymer Engineering Science*, **35**, 1387–1393 (1995).
37. M. Yuan, L. S. Turng, S. Gong, D. Caulfield, C. Hunt and R. Spindler, Study of injection molded microcellular polyamide-6 nanocomposites, *Polymer Engineering Science*, **44**, 673–686 (2004).
38. R. Rothon, *Particulate-filled polymer composites*. Harlow: Longman, 1995.
39. K. Ashbee, *Fiber composites*. Lancaster: Technomic Publishers, 1993.
40. F. R. Jones, *Handbook of polymer-fiber composites*. New York: Langman, 1994, p. 78.
41. A. K. Bledzki and O. Faruk, Effects of the chemical foaming agents, injection parameters, and melt-flow index on the microstructure and mechanical properties of microcellular injection-molded wood-fiber/polypropylene composites, *Journal of Applied Polymer Science*, **97**, 1090–1096 (2005).
42. S. Panthapulakkal, A. Zereshkian and M. Sain, Preparation and characterization of wheat straw fibers for reinforcing application in injection molded thermoplastic composites, *Bioresource Technology*, **97**, 265–272 (2006).
43. L. P. Tan, C. Y. Yue, K. C. Tam, Y. C. Lam and X. Hu, Factors that affect fibrillation of the liquid crystalline polymer (LCP) phase in an injection moulded polycarbonate/LCP blend, *Key Engineering Materials*, **312**, 133–138 (2006).
44. S. Pisharath and S.C. Wong, Processability of LCP-nylon-glass hybrid composites, *Polymer Composites*, **24**, 109–118 (2003).
45. J. L. White, *Twin-screw extrusion*. Munich: Hanser, 1991.
46. 'Modern Plastics', *World Encyclopedia*, pp. 160–217 (2007).
47. W. Michaeli, A. Brunswick and M. Gruber, Step on the gas with water injection, *Kunststoff Plastics Europe*, **89**, 20–21 (1999).
48. H. X. Huang and Z. W. Deng, Effects and optimization of processing parameters in water-assisted injection molding, *Journal of Applied Polymer Science*, **108**, 228–235 (2008).
49. S. Lang and M. J. Parkinson, Two-component water assisted injection moulding, *Plastics, Rubber and Composites*, **34**, 232–235 (2005).
50. J. L. White, *Principles of polymer engineering rheology*. New York: Wiley, 1990.
51. A. Vaxman, M. Narkis, A. Siegmann and S. Kenig, Short fiber reinforced thermoplastics. II. Interrelation between fiber orientation and rheological properties of glass fiber-reinforced Noryl, *Polymer Composites*, **10**, 78–83 (1989).
52. S. R. Tremblay, P. G. Lafleur and A. Ait-Kadi, Effects of injection parameters on fiber attrition and mechanical properties of polystyrene molded parts, *Journal of Injection Molding Technology*, **4**, 1–7 (2000).
53. M. R. Kamal, L. Song and P. Singh, Measurement of fiber and matrix orientations in fiber reinforced composites, *Polymer Composites*, **7**, 323–328 (1986).
54. M. Vincent, Flow induced fiber micro-structure in injection molding of fiber reinforced materials. In M. Kamal, A. I. Isayev and S. J. Liu (eds.), *Injection molding: Technology and fundamentals*. Munich: Hanser Publishers, 2009, pp. 253–272.
55. M. Akay and D. Barkley, Fibre orientation and mechanical behaviour in reinforced thermoplastic injection mouldings, *Journal of Materials Science*, **26**, 2731–2742 (1991).
56. N. Dontula, N. S. Ramesh, G. A. Campbell, J. D. Small and A. L. Fricke, An experimental study of polymer-filler redistribution in injection molded parts, *Journal of Reinforced Plastics and Composites*, **13**, 98–100 (1994).

57. S. Y. Yang and S. J. Liou, Development of moldability diagrams for gas-assisted injection molding, *Advances in Polymer Technology*, **14**, 197 (1995).
58. H.-C. Ludwig, G. Fischer and H. Becker, A quantitative comparison of morphology and fibre orientation in push-pull processed and conventional injection-moulded parts, *Composites Science and Technology*, **53**, 235–239 (1995).
59. B. Mlekusch, E. A. Lehner and W. Geymayer, Fibre orientation in short-fibre-reinforced thermoplastics I. Contrast enhancement for image analysis, *Composites Science and Technology*, **59**, 543–545 (1999).
60. P. Gerard, J. Raine and J. Pabiot, Characterization of fiber and molecular orientations and their interaction in composite injection molding, *Journal of Reinforced Plastics and Composites*, **17**, 922–934 (1998).
61. S. G. Advani and C. L. Tucker III, The use of tensors to describe and predict fiber orientation in short fiber composites, *Journal of Rheology*, **31**(8), 751–784 (1987).
62. P. F. Bright, R. J. Crowson and M. J. Folkes, A study of the effect of injection speed on fiber orientation in simple mouldings of short glass fibre-filled polypropylene, *Journal of Materials Science*, **13**, 2497–2506 (1978).
63. R. D. Leaversuch, Now they want plastics to be heavy?, *Plastics Technology*, September (2001).
64. A. Hassan, R. Yahya, A. H. Yahaya, A. R. M. Tahir and P. R. Hornsby, Tensile, impact and fiber length properties of injection-molded short and long glass fiber-reinforced polyamide 6,6 composites, *Journal of Reinforced Plastics and Composites*, **23**, 969–986 (2004).
65. T. Sakai and K. Kikugawa, Injection molding machines, tools, and processes. In M. Kamal, A. I. Isayev and S. J. Liu (eds.), *Injection molding: Technology and fundamentals*. Munich: Hanser Publishers, 2009, Chapter 2, p. 129.
66. A. Clarke and C. Eberhardt, *Microscopy techniques for materials science*. Cambridge: Woodhead Publishing, 2002.
67. M. Vincent, T. Giroud, A. Clarke and C. Eberhardt, Description and modeling of fiber orientation in injection molding of fiber reinforced thermoplastics, *Polymer*, **46**, 6719–6725 (2005).
68. F. Folgar and C. L. Tucker, Orientation behavior of fibers in concentrated suspensions, *Journal of Reinforced Plastics and Composites*, **3**, 98–119 (1984).
69. G. B. Jeffery, The motion of ellipsoidal particles immersed in a viscous fluid, *Proceedings of the Royal Society London*, **A102**, 161–179 (1922).
70. X. Fan, N. Phan-Thien and R. Zheng, A direct simulation of fibre suspensions, *Journal of Non-Newtonian Fluid Mechanics*, **74**, 113–135 (1998).
71. N. Phan-Thien, X. Fan, R. I. Tanner and R. Zheng, Folgar-Tucker constant for a fibre suspension in a Newtonian fluid, *Journal of Non-Newtonian Fluid Mechanics*, **103**, 251–260 (2002).
72. J. Wang, J. F. O'Gara and C. L. Tucker III, An objective model for slow orientation kinetics in concentrated fiber suspensions: theory and rheological evidence, *Journal of Rheology*, **52**, 1179–1200 (2008).
73. J. H. Phelps and C. L. Tucker III, An anisotropic rotary diffusion model for fiber orientation in short-and long-fiber thermoplastics, *Journal of Non-Newtonian Fluid Mechanics*, **156**, 165–176 (2009).
74. K. Oda, J. L. White and E. S. Clark, Jetting phenomena in injection mold filling mold filling, *Polymer Engineering Science*, **16**, 585–592 (1976).
75. S. J. Liu and J. H. Chang, The occurrence of surface roughness in gas assist injection molded nylon composites, *Polymer Composites*, **21**, 322–331 (2000).

76. Y. Ide and Z. Ophir, Injection molding of thermotropic liquid crystal polymers, *Polymer Engineering Science*, **23**, 792–796 (1983).
77. D. Done and D. G. Baird, The effect of thermal history on the rheology and texture of thermotropic liquid crystalline polymers, *Polymer Engineering Science*, **27**, 816–822 (1987).
78. R. M. Criens and H. G. Mosale, The influence of knit-lines on the tensile properties of injection molded parts, *Polymer Engineering Science*, **23**, 591–596 (1983).
79. S.-G. Kim and N. P. Suh, Performance prediction of weldline structure in amorphous polymers, *Polymer Engineering Science*, **26**, 1200–1207 (1986).
80. A. Meddad and B. Fisa, Weldline strength in glass fiber reinforced polyamide 66, *Polymer Engineering Science*, **35**, 893–901 (1995).
81. S. Fellahi, A. Meddad, B. Fisa and B. C. Favis, Weldlines in injection-molded parts: a review, *Advances in Polymer Technology*, **14**, 169–195 (1995).
82. A. Savadori, A. Pelliconi and D. Romanini, Weld line resistance in polypropylene composites, *Plastics and Rubber Processing and Applications*, **3**, 215–221 (1983).
83. B. Fisa and M. Rahmani, Weldline strength in injection molded glass fiber-reinforced polypropylene, *Polymer Engineering Science*, **31**, 1330–1336 (1991).
84. B. Fisa, J. Dufour and T. Vu-Khanh, Weldline integrity of reinforced plastics: effect of filler shape, *Polymer Composites*, **8**, 408–418 (1987).
85. A. Vaxman, M. Narkis, A. Siegman and S. Kenig, Weld-line characteristics in short fiber reinforced thermoplastics, *Polymer Composites*, **12**, 161–168 (1991).
86. V. M. Nadkarni and S. R. Ayodhya, The influence of knit-lines on the tensile properties of fiberglass reinforced thermoplastics, *Polymer Engineering Science*, **33**, 358–367 (1993).
87. S. J. Liu, J. Y. Wu, J. H. Chang and S. W. Hung, An experimental matrix design to optimize the weldline strength in injection molded parts, *Polymer Engineering Science*, **40**, 1256–1262 (2000).
88. S. J. Liu, *Polymer Engineering Science*, Modeling and simulation of thermally induced stress and warpage in injection molded thermoplastics, **36**, 807–818 (1996).
89. H. Kikuchi and K. Koyama, Material anisotropy and warpage of nylon 66 composites, *Polymer Engineering Science*, **34**, 1411–1418 (1994).
90. K. T. Chiang and F. P. Chang, Analysis of shrinkage and warpage in an injection-molded part with a thin shell feature using the response surface methodology, *International Journal of Advanced Manufacturing Technology*, **12**, 468–479 (2007).

3
Compression molding in polymer matrix composites

C. H. PARK, University of Le Havre, France and W. I. LEE, Seoul National University, Korea

Abstract: Compression molding is a popular manufacturing technique for composite parts. In particular, the development of high-strength sheet molding compounds drove wide adoption of compression molding process in automotive and appliance applications. In this chapter, we present some advantages and disadvantages of compression molding. We also introduce molding materials for compression molding such as sheet molding compound and bulk molding compound. To obtain high quality products, it is important to optimize mold design and processing conditions. Process modeling, such as flow and cure analysis, is especially useful to predict the knit line formation, part curing, fiber orientation and separation in the final product.

Key words: compression molding, sheet molding compound, mold design, flow and cure analysis, fiber orientation and separation.

3.1 Introduction

3.1.1 Process description

Compression molding is one of the most common processing techniques to fabricate plastic and polymer composite products. Traditionally, beginning from the early twentieth century, the compression molding process has been used for molding thermoset powders (e.g., phenolic and alkyd) and rubber compounds. Especially, this process has seen a remarkable growth since the early 1950s in automotive and appliance applications by virtue of the development of sheet molding compounds. For example, the first application of glass fiber reinforced polymer in the automotive industry is the front panel of the GM Corvette that was developed in 1953 (Davis *et al.*, 2003). Today, the sheet molding compound materials used for compression molding are the most popular form of composite material by dint of their low cost (the cheapest composite material in terms of the price per unit mass: about two times more expensive than steel per unit mass) as well as their molding efficiency.

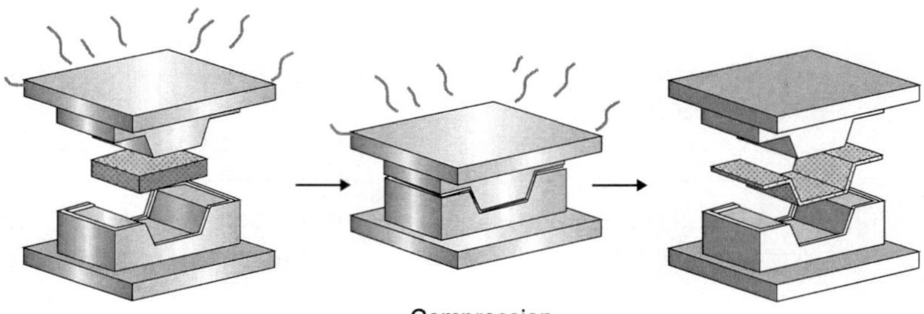

3.1 The compression molding process.

In the compression molding process, the desired quantity of material is placed between the upper and lower molds, and the material then is formed into the final shape by pressing the mold as illustrated in Fig. 3.1.

The compression molding process is composed of four basic phases.

1. *Precharge preparation and placement*: A stack of molding material (e.g., Sheet Molding Compound [SMC] sheets) is placed in the hot mold. This material is called the precharge or, simply, charge. The precharge dimensions are selected to cover about 50% of the mold surface area, and the precharge weight is measured before it is placed in the mold. The position of the precharge in the mold is a key factor that affects the part quality since it influences the fiber orientation, knit line formation and void content.
2. *Mold closure*: Once the precharge is placed in the mold, the upper mold moves down quickly to touch the top surface of the precharge material. Then, the upper mold continues to move down slowly (usually at 5–10 mm/s) to compress the precharge. As the molding pressure builds up with the mold closure, the precharge material flows to fill the mold cavity and causes air to escape through the shear edges or air vents of the mold. The mold closing speed and mold temperature are key factors that affect process performance and product quality.
3. *Curing*: After the mold cavity is completely filled with the precharge material, the mold is kept closed while the molding pressure is maintained for a pre-assigned period of time. During this period, the resin is cured and the part is consolidated. The curing time depends on the resin mixture formulation, part thickness and mold temperature.
4. *Part release*: Once the resin is cured and the part is solidified, the part is removed from the mold with the aid of ejector pins. Then, the part is cooled down outside the mold, while the mold surfaces are cleaned, and an external mold release agent is applied to the mold surfaces for the next molding cycle.

Generally, the process cycle time ranges from 1 to 3 minutes depending on the part thickness (see Fig. 3.2). During the process, the material temperature is in the range of 130°C to 160°C under a relatively low molding pressure of the order of 10 MPa (see Fig. 3.3).

Frequently, compression molding is also called matched-die molding since the equipment is a press that is fitted with both male and female dies. The force developed by the press can reach up to several hundred tons depending on the part size, which is useful to obtain good part uniformity and to collapse the voids that may be generated during the molding process.

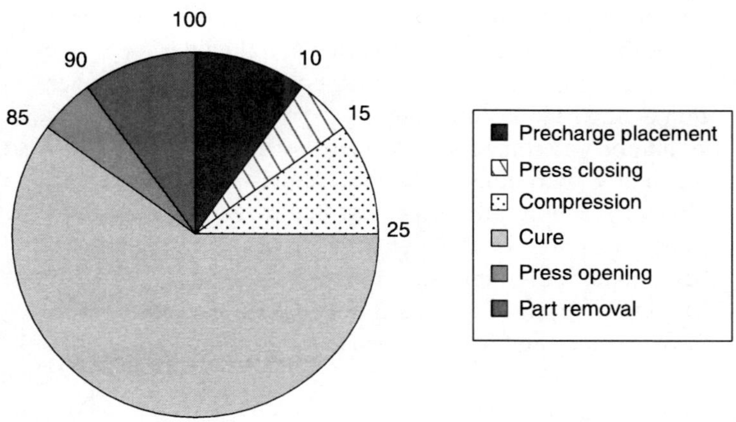

3.2 Typical time in seconds for each operation constituting a cycle of compression molding.

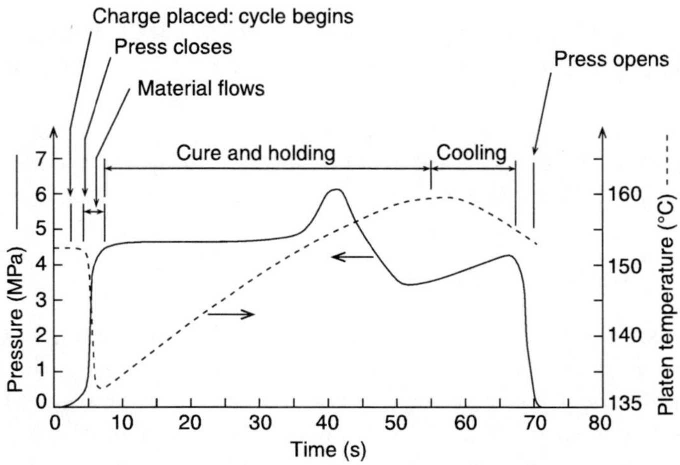

3.3 Temperature and pressure variation during compression molding.

Generally, the molds are made of a hard metal such as tool steel, and can be highly polished and chromium plated to obtain a good surface finish.

The compression molding process has some advantages and disadvantages over the other processes.

Advantages:

- Excellent part reproducibility is obtained.
- The fiber content and type can be easily controlled to yield a wide variety of mechanical and physical properties.
- Both interior and exterior surfaces are finished.
- Excellent electrical performance and flame resistance can be obtained depending on the resin formulation and the filler type.
- Complex shapes (including ribs, bosses, curvatures, holes, inserts and thin details) can be fabricated.
- A number of components can be consolidated into one part without secondary operations or assembly steps.
- High production rates can be obtained.
- Labor costs are low.
- Scrap material is minimal.
- Minimum trimming of parts is needed.
- SMC parts exhibit better dimensional stability than thermoplastic composites. Moreover, the thermal expansion coefficient of SMC can be controlled to match that of steel or aluminum, hence making SMC parts compatible with steel or aluminum parts.

Disadvantages:

- More equipment is needed than in the case of hand lay-up.
- Compression molds and tooling are more expensive than hand lay-up molds.
- Transparent products are not possible.
- There can be surface imperfections such as pitting and waviness.
- Thermoset resins (such as in SMC) have limited shelf-lives.

3.1.2 Industrial applications

The most popular form of material for compression molding is the SMC. The early applications of SMC were in electrical and industrial products such as electrical fixtures, control boxes, light fittings tool boxes and machine guards. In the early 1970s, the automotive industry widely adopted SMCs to produce exterior body components such as grille opening panels and hoods. Since high-strength SMCs were introduced in the mid-1970s, SMCs have been applied to produce semi-structural automotive components such as bumper beams, road wheels, cross-members, and tailgates and exterior body panels. Even though the principal applications of SMC are in the transportation industry, other applications also are found in home appliances (refrigerator doors, washing machine doors), furniture (chairs, table tops) and construction (door panels) (Mallick, 1990).

Today, fiber reinforced thermoplastic resins such as glass fiber reinforced thermoplastics (GMT) and long fiber reinforced thermoplastics (LFT) are attracting interest from the automotive industry. Their excellent toughness and recycling possibility led to a wide application of thermoplastic compression molding for structural components of modern vehicles. A representative example is a front end module (FEM) carrier which is the frame where headlamp, radiator, air bag sensor, hood latch and horn are integrated (Fig. 3.4). Compared with the metal counterparts, this component can be manufactured as a single part and additional assembling steps or secondary operations can be eliminated.

(a)
(b)
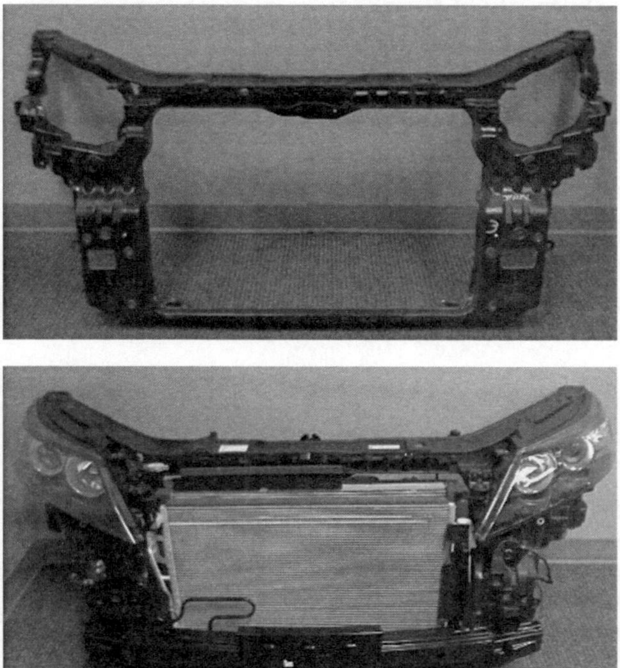

3.4 Automotive FEM (front end module) carrier manufactured by the compression molding of GMT (Courtesy Hanhwa-Azdel, Inc.): (a) FEM carrier before assembling other components and (b) FEM after assembling other components.

Recently, some new applications in the energy industry are molded by compression molding of reinforced polymer. A good example is the bipolar plate of fuel cell. Bipolar plate is one of the most important components of fuel cell. In particular, it accounts for about 60–80% of the total weight of a fuel cell and 30–50% of the total cost of fuel cell. Traditional metal bipolar plate is too heavy and this has been a critical issue for electrical vehicles employing fuel cells. As an attractive alternative to metal counterparts, polymers (thermoset or thermoplastic) containing conductive fillers (graphite powder, carbon fiber, carbon nanotube, carbon black, carbide, etc.) are considered as a bipolar plate manufacturing material by virtue of their lightness. Because of large planar dimensions and small thickness of

3.5 Compression molded polymer-graphite bipolar plates for direct methanol fuel cell (Chen *et al.*, 2010).

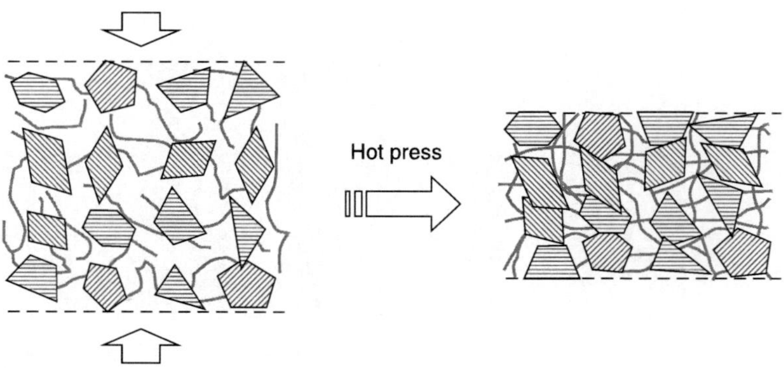

3.6 Illustration of particles forming percolation networks during compression molding of composite bipolar plate (Chen *et al.*, 2010).

bipolar plates, compression molding is widely applied to fabricate polymer bipolar plates. In contrast to metals, polymers offer not only corrosion resistance but also low manufacturing cost from its net shape forming capability. In Fig. 3.5, compression molded bipolar plates (vinyl ester–graphite powder) are shown. During the compression of molding material in the thickness direction, conductive fillers come into contact with one another and form conductive paths (percolation network) in the thickness direction (see Fig. 3.6).

3.2 Molding materials

For the compression molding, 'premix' of reinforcing fibers, fillers and base resin are prepared and used for molding. Premix has been defined as 'a fiber reinforced thermoset molding compound not requiring advancement of curing, drying of volatiles or other processing after manufacture to make it ready for use at the molding press' (White, 1964). Most of the premix can be molded without reaction by-products under sufficiently low pressure to flow and consolidate the material, whereas condensation type resins such as phenolic, melamine or urea compounds have water as a reaction by-product and require high pressure to avoid steam generation and void formation in the molding. In practice, the mold is opened slightly for a few seconds after it is closed in order to allow the gas formed by the heated condensate to escape. This process is called degassing or breathing the mold. There are various forms of premix such as SMC and bulk molding compound (BMC).

As a molding material, thermoplastic resins can also be used. A popular form of fiber-reinforced thermoplastics is GMT (Glass Mat Thermoplastics). Thermoplastic resins offer better toughness compared with thermoset resins. Also, process time can be saved as the material frequently is preheated outside and transferred into the mold for part forming, and thus heating inside the mold is not necessary. In this case, the process is also called thermoforming, hot stamping or sheet forming.

3.2.1 Sheet molding compound (SMC)

SMC is a composite material in a sheet form where thermoset resin is mixed with reinforcing fibers. The manufacturing of SMC is illustrated in Fig 3.7. SMC manufacturing is composed mainly of two phases, namely, compounding and maturation. During compounding, fibers are embedded between two thin layers of a resin paste by chopping fibers (usually glass fibers) or swirling continuous fibers onto a sheet of plastic (usually polyethylene) film on which a resin/initiator/filler mixture has been doctored. Then, another

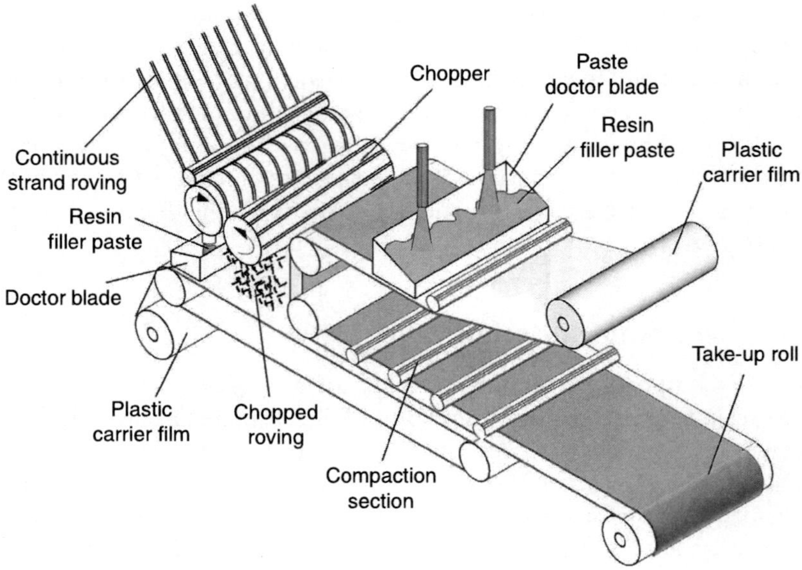

3.7 The SMC manufacturing process. (*Source*: Adapted from McCluskey and Dohert, 1995.)

film that also has the resin mixture doctored onto it is placed on top, and the sandwich of resin mixture and chopped glass is passed between compaction rolls to wet the fibers and mix the constituents. In the maturation phase, the resin is slightly cured to the B-stage, which is an uncured but highly viscous (thickened) state for an easy handling, and rolled-up for shipment.

The width of the SMC sheet is decided by the SMC manufacturing machine whereas the thickness is commonly 3 mm. Machine manufacturers provide a range of width from 0.61 to 1.52 m (2–5 ft), and the most common width is 1.22 m (4 ft).

A typical SMC contains about 30–50% of fiber (25–75 mm long, frequently E-glass fiber 25.4 mm), 25% resin (usually unsaturated polyester), and 25–45% filler (usually, calcium carbonate, alumina or clay) by the weight fraction. In the case of structural applications, 50–70 wt% of fibers can be used in SMCs. In general, fillers are not added if the SMC contains more than 60 wt% of fiber (see Table 3.1, Mallick, 1990).

SMC is refrigerated before use because it contains a curing initiator and thus has a limited shelf-life. SMC should be handled with care because the styrene cross-linker is quite volatile.

The major components of SMC are the resin, fibers and fillers. Even though other ingredients (catalyst, additives, thickener, inhibitor, mold release agent) are used in small amounts, they serve important functions in the compression molding process (see Table 3.1, Mallick, 1990).

Table 3.1 Typical composition of SMC-R in weight fraction

Material	SMC-R30	SMC-R65
Resin paste		
Unsaturated polyester	30.96	31.85
Low profile additive	4.64	0
Catalyst	0.31	0.32
Mold release agent	1.55	0.95
Filler	30.96	0
Thickener (MgO)	1.55	1.91
E-glass fiber (25 mm long)	30.03	64.97

Source: Adapted from Mallick (1990).

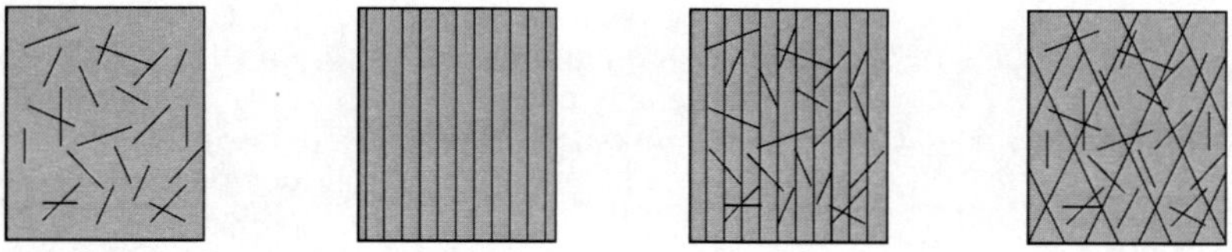

3.8 Different designation of SMC.

SMCs are designated according to the type of reinforcing fibers used (see Fig. 3.8):

- SMC-R: discontinuous fibers with random orientations.
- SMC-CR: continuous fibers with unidirectional orientation on one side of the sheet and randomly oriented discontinuous fibers on the other side of the same sheet.
- Directionally reinforced molding compound (XMC): continuous fibers crossed in an 'X' pattern.

At the end of each letter designation, the fiber content is shown by the weight percentage (wt%). For example, SMC-R40 designates that 40 wt% of randomly oriented discontinuous fibers are included, and SMC-C30R10 designates that 30 wt% of unidirectional continuous fibers and 10 wt% of randomly oriented discontinuous fibers are included. Usually, XMC contains about 70 wt% of continuous fibers with a small quantity of discontinuous fibers.

Fillers are also included in SMCs to reduce the volumetric shrinkage of resin and lower the material cost. The most common filler is calcium carbonate ($CaCO_3$). Kaolin clay, talc, solid or hollow glassbeads and alumina trihydrate ($Al_2O_33H_2O$) are also used depending on the application. Electrically conductive fillers such as carbon blacks, carbon fibers, aluminum or nickel in

Table 3.2 Additives for resin paste formulation

Additive	Function	Example
Catalyst (curing agent)	Initiate the curing reaction at elevated temperature	Benzoyl peroxide (e.g., tertiary butyl perbenzoate)
Inhibitor	Prevent or lessen any curing reaction during mixing, maturation and storage	Hydroquinone, parabenzoquinone (PBQ)
Internal mold release agent	Prevent adhesion between the resin and the mold surface	
Thickener	Increase the resin paste viscosity for the SMC sheet to be handled, cut, stacked and draped in the mold and maintain the resin viscosity sufficiently high to preserve the homogeneity of SMC	Alkaline group IIA metal oxides and hydroxides (e.g., MgO, CaO, $Mg(OH)_2$, and $Ca(OH)_2$)
Low-profile additives	Control the shrinkage of SMC parts	Thermoplastic powders (e.g., polyvinyl acetate, polycaprolactone, polyacrylate copolymers, cellulose acetate butyrate, polystyrene, polyethylene)
Impact modifier	Improve the fracture toughness, impact energy absorption and damage resistance	Acrylonitrile-butadiene copolymers, styrene-butadiene copolymers
Colorant, flame retardant, untraviolet absorber		

fiber, powder and flake forms are sometimes adopted in small quantities to realize the electromagnetic interference shielding function.

Many other ingredients are added to the base resin in order to optimize SMC molding properties. The functions and representative examples of these ingredients are summarized in Table 3.2.

3.2.2 Bulk molding compound (BMC)

Bulk molding compound (BMC) is a bulky mixture of chopped glass fibers, resin paste and fillers. Even though other fibers such as sisal, asbestos, carbon, aramid, chopped nylon rag and wood are used, the most common

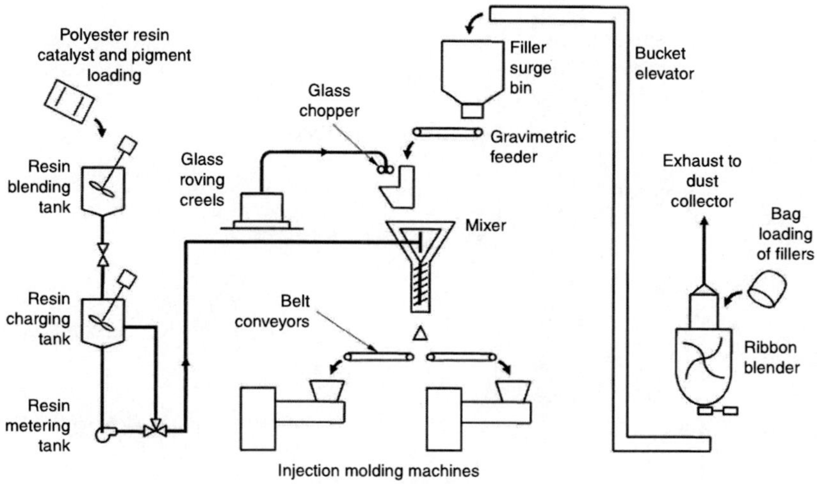

3.9 Schematic of a continuous compounding system for BMC. (*Source*: Courtesy Farrel Machinery Group USM Corp., Adapted from Young, 1982.)

reinforcing fiber in BMC is E-glass fiber. As BMC is usually mixed in a dough-like form rather than as a sheet, it is also called dough molding compound (DMC), and is sold in a log or rope form. The resin formulation of BMC materials is similar to that of SMC materials. Chopped glass fibers are compounded with the resin paste in an intensive mixer and extruded in the form of a continuous log. The extruded log is cut to the desired length by a pneumatic cutter located outside the extruder. Continuous BMC compounding system is shown in Fig. 3.9. The glass fiber content of BMC is generally 5~10% lower than that of SMC, and the fibers are shorter (generally less than 25 mm). Therefore, the mechanical properties of BMC are lower than those of SMC. The impact strength is highly dependent on the fiber length. This improvement in the impact resistance by the increase in fiber length becomes less significant above the fiber length of 12.7 mm (0.5 inch). Moreover, fiber lengths over 12.7 mm result in mixing and molding problems. Therefore, the fiber lengths of BMC seldom exceed 12.7 mm, and the most common fiber length is 6.23 mm (0.25 inch). Sometimes, BMC is mixed at the molding location whereas it also can be purchased from commercial manufacturers that sell the material as a ready-to-mold premix.

3.2.3 Thick molding compound (TMC)

Thick molding compound (TMC) is a kind of SMC for which the thickness may range up to 50 mm. Generally, high fiber content and directional continuous fibers are not possible for TMC. TMC sheets are produced by

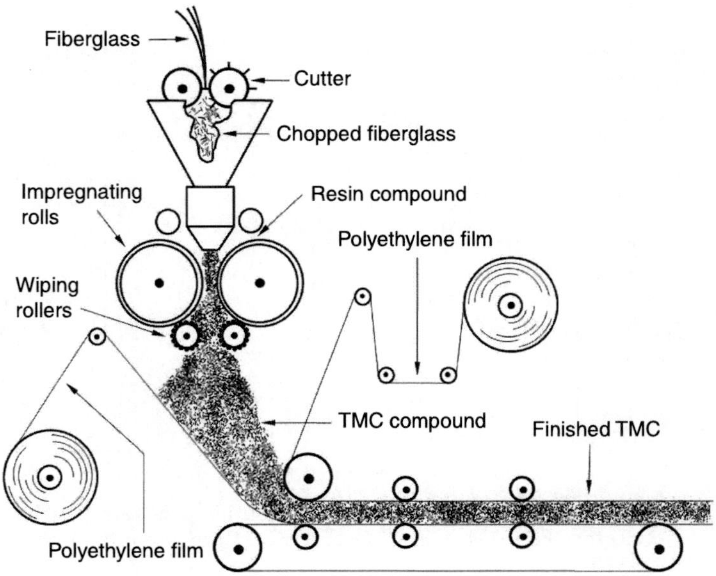

3.10 Schematic of TMC production. (*Source*: Courtesy Polyester Unit, USS Chemicals. Adapted from Young, 1982.)

metering the premixed resin paste and chopped glass fibers directly onto two large impregnation rolls (see Fig. 3.10). Then, the compound is deposited between two moving polyethylene carrier films.

The fibers in BMC are in a three-dimensional random orientation state, whereas the fibers in SMC are in a planar random orientation state. Because fewer layers of TMC are required to achieve the desired thickness of product compared to SMC sheets, the interlaminar crack formation can be reduced.

3.2.4 Directionally reinforced molding compound (XMC)

Directionally reinforced molding compound (XMC) contains continuous reinforcements (up to 70 wt%) arranged in an X pattern with some chopped fibers and the final product has strong directional properties in the continuous fiber direction. XMC can be made on almost any filament winding machine as shown in Fig. 3.11.

3.2.5 Glass mat thermoplastics (GMT)

Recently, reinforced thermoplastic resins are also widely used in the compression molding process. Representative molding materials are GMT and

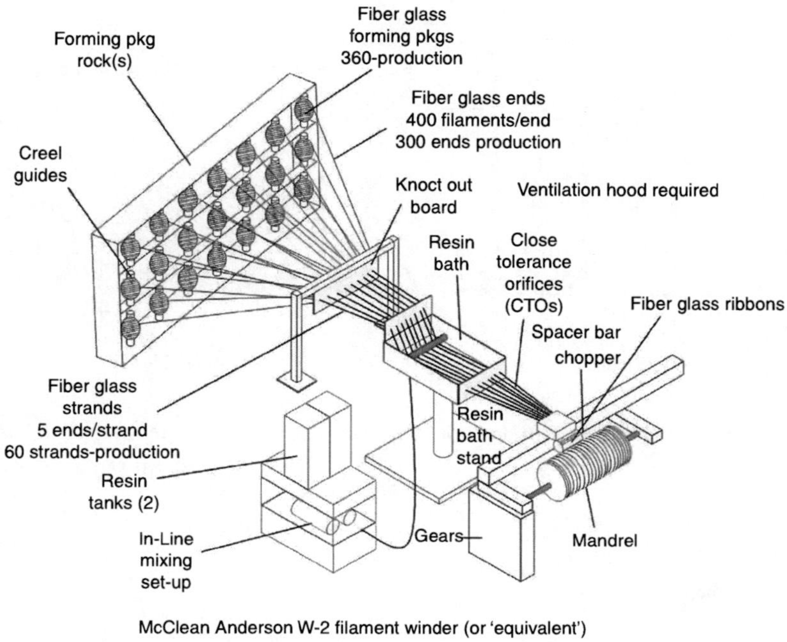

3.11 Schematic of XMC production. (*Source*: Courtesy PPG Industries. Adapted from Young, 1982.)

long fiber reinforced thermoplastics (LFRT or LFT). By virtue of better impact strength, they have become attractive materials for structural components in the automotive industry. Moreover thermoplastic materials have no definite shelf-life limit, unlike the SMC.

Polypropylene is the most popular matrix material for its low cost, even though other thermoplastics such as nylon are also used. The most common reinforcing fiber is the E-glass fiber. Both the discontinuous fibers and continuous fibers can be used. In general, continuous fibers are preferred for the structural components. In Fig. 3.12, a schematic of glass fiber mat production is shown. Chopped glass fibers are stacked on to the unidirectional glass fibers and these two different fiber layers are bound together by needle punching.

This glass fiber mat is impregnated by thermoplastic resin which is compounded with additives, under heat and pressure (Fig. 3.13). A double belt press is usually used to apply heat and pressure for the resin impregnation process. After the consolidation process, glass fiber reinforced sheets which have been conveyed through the double belt press, are cut into a desired size and stocked.

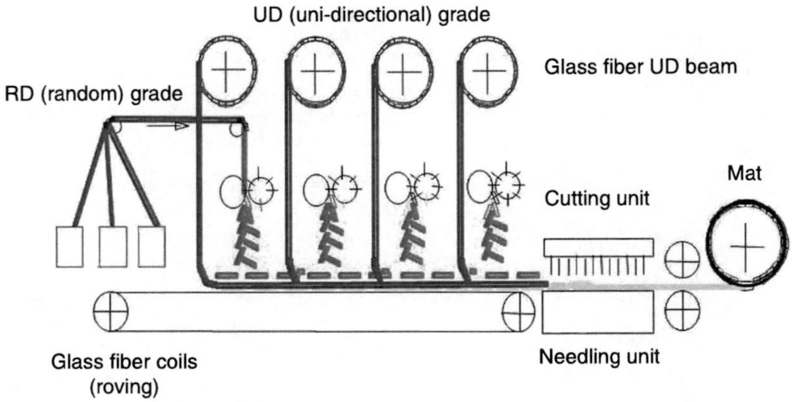

3.12 Glass fiber mat production. (*Source*: Courtesy Hanhwa-Azdel, Inc.)

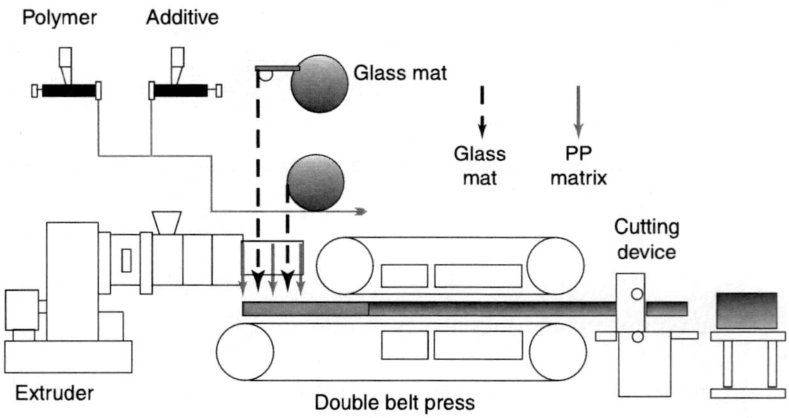

3.13 Glass mat thermoplastic production. (*Source*: Courtesy Hanhwa-Azdel, Inc.)

In general, the preheating of GMT is required before it is placed into the mold. The mold can be heated as in the compression molding of thermoset materials. Frequently, however, the mold is not heated to reduce the part cooling time which accounts for the significant portion of total process cycle time. In this case, this process is called cold compression molding. The schematic of a typical GMT compression molding process is shown in Fig. 3.14. As thermoplastic materials do not need chemical reaction, they can be preheated outside before they are introduced into the mold. Therefore, the cycle time can be considerably reduced compared to thermosetting materials. The reduction in the cycle time is another merit of GMT materials.

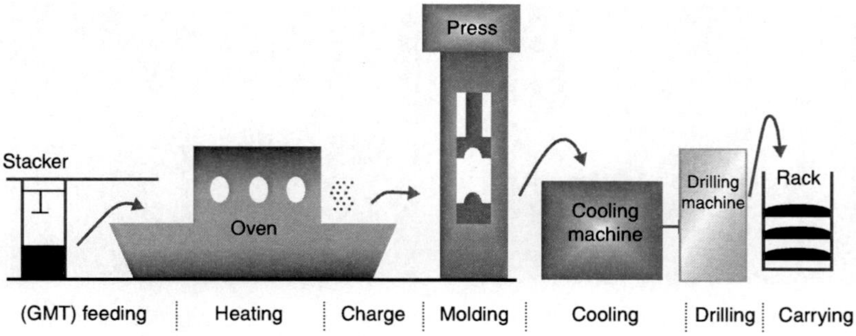

3.14 GMT compression molding process. (*Source*: Courtesy Hanhwa-Azdel, Inc.)

Table 3.3 Defects in compression molded parts

Surface defects	Waviness, sink marks, pinhole, craters, surface roughness, pop-up blisters in painted parts, dark areas
Internal defects	Porosity or voids, blisters, knit lines, delamination, preferential fiber orientation, fiber separation, incomplete curing, resin rich zone, warpage, residual stress, fiber buckling (continuous fiber)

3.3 Process defects and remedies

The defects in compression molded parts can be classified into two categories: surface defects that affect the visual appearance and internal defects that influence the mechanical properties. Surface defects can be masked by post-processing such as painting by spray coating after molding. Being invisible from the part surface, the internal defects are detected by non-destructive testing methods such as radiography or ultrasonic testing methods. Many defects can be avoided or reduced by optimizing the molding material, process conditions and part design. Some defects induced during the compression molding are listed in Table 3.3.

3.3.1 Voids

Voids or porosities are generated if air is entrapped inside the molded part. These voids significantly deteriorate mechanical properties such as interlaminar shear strength and flexural strength. Most of the air bubbles are generated during the mixing and compounding stages in SMC production since most of the entrapped air is driven by the SMC flow and escapes from the mold through the air vents and shear edges in the mold (Himebaugh and Newman, 1985). However, some voids are also generated

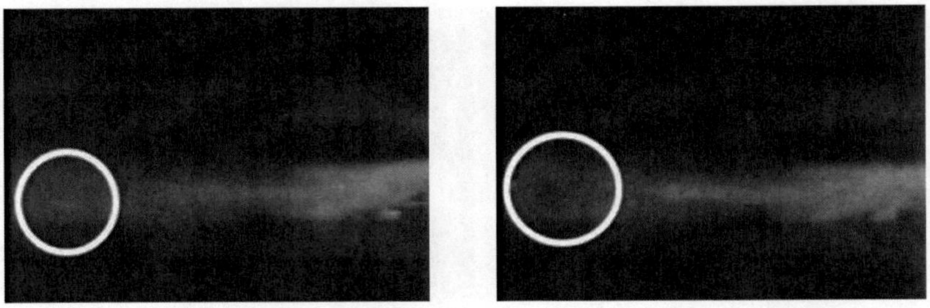

3.15 Snapshots of SMC flow showing void defect generation (a black dot inside the white circle in the right photo) (Odenberger *et al.*, 2004).

during the SMC flow in the mold (see Fig. 3.15). The impregnation during the SMC compounding can be improved by adding active additives to the resin paste to decrease the interfacial surface energy (Slotfeldt-Euingsen *et al.*, 1986).

3.3.2 Blisters

As the internal pressure of volatiles (e.g., styrene monomer) or entrapped air is released at the moment of mold opening, interlayer cracks may be created near the part surface. These cracks are called blisters. They often appear to be dome-shaped bulges. They also can be created at elevated temperatures (e.g., in paint bake ovens) by the expansion of entrapped air.

The formation of blisters can be decreased either by the vacuum application in the mold to minimize the entrapment of air or volatiles or by the use of coupling agents and contaminant reduction in the resin formulation to increase the interlaminar shear strength.

3.3.3 Delamination

Delamination can result from either poor fiber wetting with resin or residual stress that is caused by a non-uniform temperature distribution and/or incomplete curing. Hence, delamination is usually observed near the centerline of thick sections. To lessen this problem, it is most effective to use a relatively slow catalyst in order to obtain uniform curing reaction. This remedy, however, can increase the process cycle time.

3.3.4 Sink marks

Sink marks are small depressions appearing on the part surface that usually are located opposite to the ribs or bosses (see Fig. 3.16). Even though they

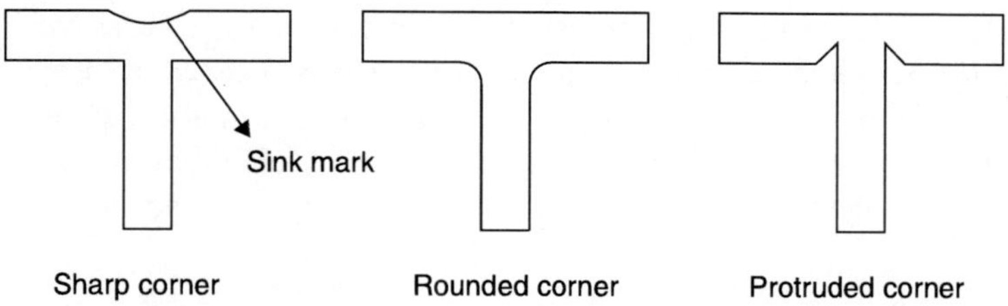

3.16 Illustration of sink mark and various rib corner designs to prevent it (Smith and Suh, 1979).

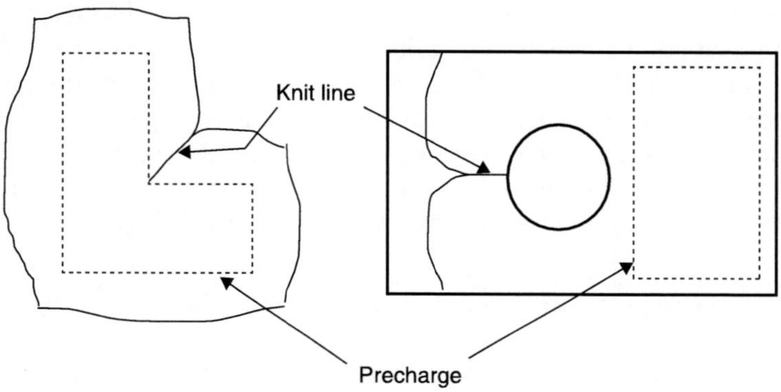

3.17 Formation of knit line. Left: L-shaped perform. Right: Mold with a circular opening.

do not deteriorate the mechanical properties, they affect the visual appearance of the final product leading to varying gloss on the flat surface.

To avoid this problem, a low-profile additive can be added to the resin formulation. However, it is more effective to optimize the part design. It is recommended to avoid thick sections directly near the rib or boss. If possible, it is preferable to use a protruding corner instead of a square or rounded corner at the rib juncture (see Fig. 3.16; Smith and Suh, 1979).

3.3.5 Knit lines

A knit line is generated when two or multiple flow fronts join or merge to form a single flow front (see Fig. 3.17). Along the knit line, the fibers are aligned parallel to the knit line. As a result, the strength in the direction normal to the knit line is significantly lower than in the parallel direction. If the load is applied in the transverse direction relative to the knit line, matrix cracking is easily initiated, and the part can fail. The most common location of knit line formation is behind molded-in holes that are made by core pins in the mold.

Hence, it is desirable to drill holes after the part is molded. Otherwise, molded-in holes should be located in zones of low stress. A key parameter to knit line formation is the precharge location. If possible, multiple precharge or uneven precharge placements should be avoided. Mold filling analysis before manufacturing may be useful to design the precharge specification.

3.3.6 Warpage or residual stress

Warpage or residual stress is created by non-uniform cooling or curing of the molded part. The residual stress can reduce the effective stress that the part can support and can cause distortion after molding. It is often observed at zones where the thickness is suddenly changed. Hence, a uniform thickness is desirable in the part design. Even for a part with a uniform thickness, a preferential fiber orientation also can create anisotropic thermal expansion and corresponding differential resin shrinkage during the cooling stage (Osswald *et al.*, 1994). Hence, it is necessary to minimize the flow path to reduce the preferential fiber alignment along the SMC flow.

3.3.7 Preferential fiber orientation

As the material flows, the discontinuous fibers in SMC change their orientation. Consequently, the final part may exhibit anisotropic mechanical properties due to the directional fiber orientation even though the initial fiber orientation of SMC is random and isotropic (see Fig. 3.18).

In general, the fiber orientation depends on the flow velocity gradient and the flow distance. If the initial coverage of the precharge in the mold is decreased, the anisotropy in orientation becomes more distinctive because the material flow path becomes longer. In Fig. 3.19, we can see that more fibers are aligned into the flow direction (i.e., $\phi = 0$) when the initial coverage of the mold by the precharge is 33% than 67%.

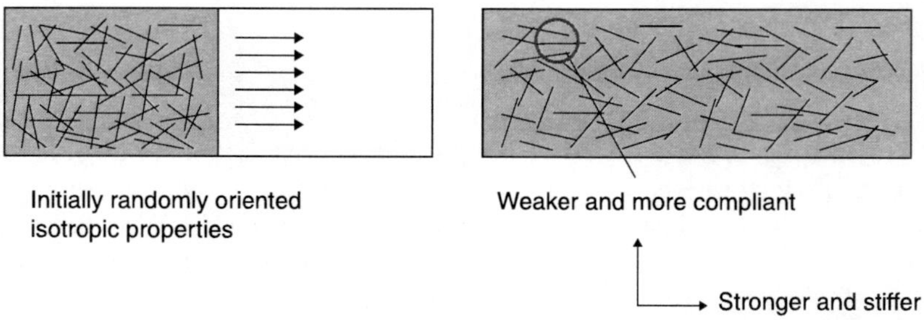

3.18 Change in the fiber orientation during the compression molding process.

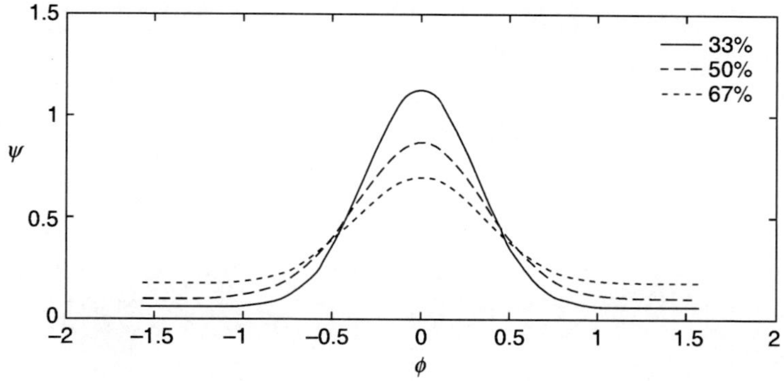

3.19 Distribution of the fiber orientation with respect to the initial coverage of the mold by SMC (Londono-Hurtado *et al.*, 2007).

In real processes, the initial coverage of the mold by precharge material usually is about 50%, and a material flow is not avoidable. Hence, the optimization by process simulation may be useful to avoid problems. For example, precharge placement can be optimized so that the fibers in the final product would be aligned in the loading direction or the anisotropy in fiber orientation would be less significant in the zone of high stress.

3.3.8 Fiber separation

Fiber separation or fiber segregation takes place when the fibers do not migrate with the resin flow in the mold. As a consequence, the fiber content becomes non-uniform leading to a higher than nominal fiber content in the upstream and a lower than nominal fiber content in the downstream (see Fig. 3.20). If SMC material with higher fiber content or longer fibers is used, the interaction between fibers increases. Hence, this interaction causes a network force that disturbs fiber migration along the resin flow. If the flow-induced drag force that is applied to the fibers is smaller than this network force, the fibers fail to move with the resin flow.

In general, the fiber separation is more significant as the fiber content and fiber length are increased. Moreover, the fibers that are oriented parallel to the flow direction are more susceptible to fiber separation than the fibers that are oriented normal to the flow direction. Even in the case of SMC-CR or XMC, this phenomenon often is observed if continuous fibers are aligned in the parallel direction to the principal flow direction (see Fig. 3.21). If continuous fibers are aligned in the perpendicular direction to the principal flow, the fibers tend to buckle or bend due to the flow-induced drag force. Hence, the precharge placement usually covers nearly 90% of the mold surface in the case of SMC-CR and XMC.

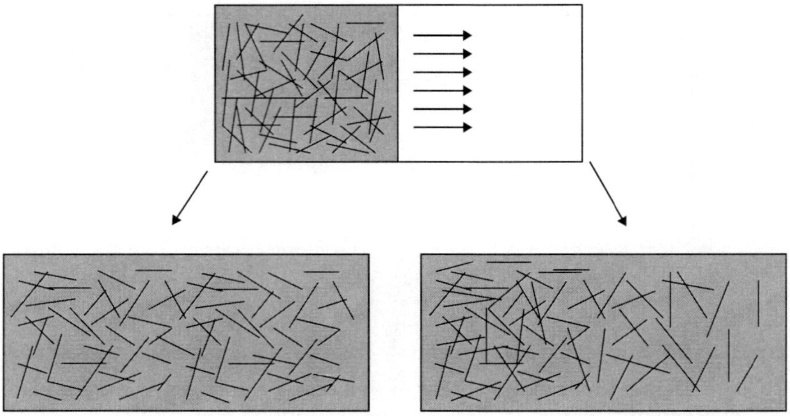

3.20 Fiber separation: homogeneous suspension (left) and heterogeneous suspension (right).

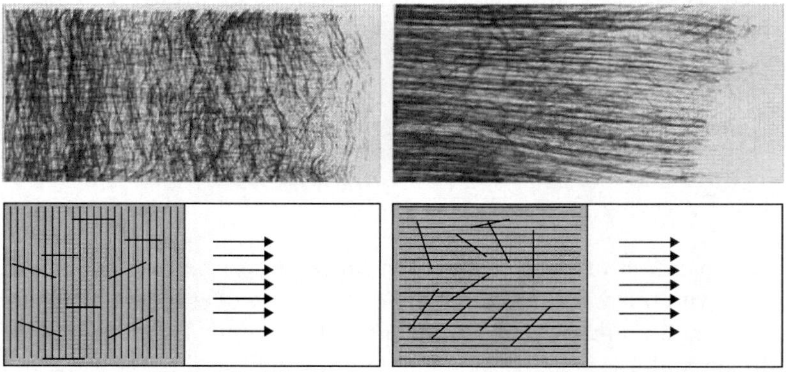

3.21 Dependence of fiber separation on the fiber orientation (Park *et al.*, 2001).

The fiber separation is more likely to occur when the resin viscosity is relatively low. A high viscosity can lessen this problem whereas molding becomes more difficult. Fillers can reduce fiber separation as they decrease the fiber–fiber interaction.

To avoid the fiber separation problem, the initial precharge thickness should be increased and/or the speed of mold closure should be increased. However, if the initial precharge thickness is increased, the flow path should be increased. This change in flow path may generate a preferential fiber orientation state and the corresponding material anisotropy.

The fiber separation is more likely to occur if the fibers are blocked at the entrance of the ribs or bosses where the flow direction is suddenly changed. In general, the fiber length in SMC is much greater than the part thickness.

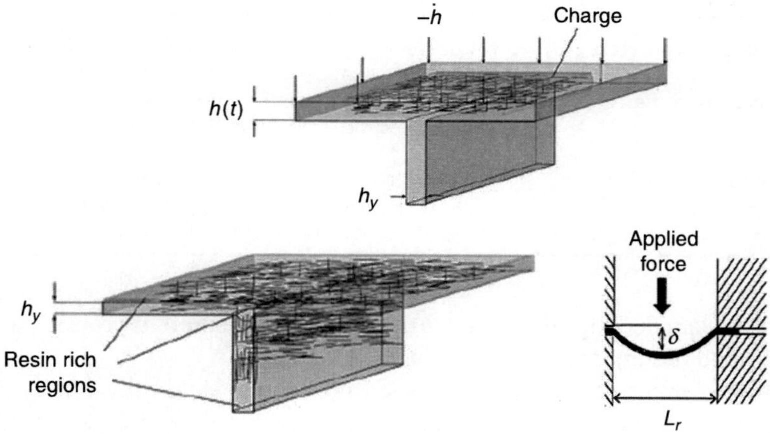

3.22 Fiber separation at the rib: there is a resin rich zone at the end of the rib (Londono-Hurtado *et al.*, 2007).

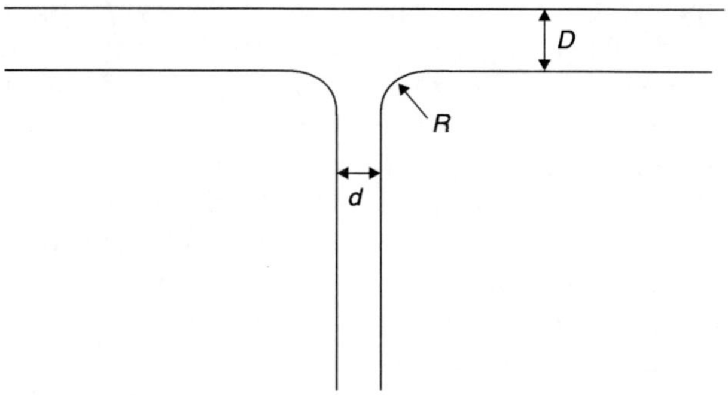

3.23 Dimensions of the rib and flange.

Consequently, resin-rich zones are frequently created at the end of ribs or bosses (see Fig. 3.22). To avoid this shortcoming, the thickness ratio between the flange and the rib should be carefully selected to reduce the flow resistance at the entrance to the rib. In general, a large rib thickness (d) relative to the flange thickness (D) facilitates flow into the ribs (see Fig. 3.23). It is also helpful to increase the entrance radius (R) to the rib to decrease the flow resistance into the rib.

3.4 Recent developments in press design and process optimization

To improve process productivity and part quality, many parameters should be optimized in the compression molding process. The principal parameters

Table 3.4 Principal parameters in the compression molding process

Material parameters	Resin paste formulation, SMC maturation period, SMC sheet thickness, SMC storage temperature
Process parameters	Mold temperature, molding pressure, mold closing speed, precharge specification (size, shape and location)
Tooling parameters	Part design (ribs, bosses, thickness, radius), air vent, shear edge, ejection system, parting line, draft angle, tool material, surface finish

can be classified into three categories: material parameters, process parameters and tooling parameters (Table 3.4).

In real processes, it is important to improve the yield ratio while reducing the process cycle time. Many techniques have been recently developed to improve productivity and part quality. In this section, some efforts for process optimization in terms of press design and process control are presented.

The hydraulic presses that are widely used in the compression molding process often suffer from the problem of eccentric loading. An uneven precharge placement or pressure distribution leads to a slight inclination of the upper platen during the mold closure. This problem of non-parallel platens will cause poor part dimensions (especially in terms of the thickness). Moreover, it also may affect the part quality (non-uniform curing or flow instability). To circumvent this problem, the press is fitted with four independent hydraulic cylinders at each corner of the moving platen. Positional sensors monitor the platen location and transmit the error signal to the hydraulic cylinders to correct the deviational error.

The mold temperature is a key parameter to affect the product quality and process cycle time. The mold is usually heated by circulating hot fluid such as oil or superheated steam through the channels in the mold. To obtain a uniform curing rate in the part, it is important to obtain a uniform mold surface temperature. It is desirable that the deviation of mold temperature is less than 5°C. In general, both the upper and lower molds are heated and maintained at the same temperature. In common practice, the heating liquid channels are uniformly spaced. However, this heating channel design may result in a non-uniform temperature distribution during molding, especially for large-part manufacturing. Specifically, the temperature in the zone initially covered by the cold precharge can become lower than the temperature in the zone that is subsequently filled by the precharge. To maintain the uniformity of the temperature in the mold, it has been proposed to optimize the channel design by employing different channel spacing (Barone and Caulk, 1981). A shorter spacing of heating channels can be employed in the zone that is initially covered by the precharge, whereas a wider spacing can be used in the zone that is filled by the material flow afterwards. Moreover, independent control of each heating circuit is also favorable to obtain a uniform temperature distribution in the mold.

Currently, a number of automated techniques are available for the auxiliary operations as well as for the main mold pressing operation. For instance, automatic cutting, weighing and stacking of SMC sheets are used in the precharge preparation step. An automatic feeding manipulator also is used for the exact positioning of the precharge into the mold and for unloading the molded part from the mold. Computer-controlled automation is widely employed for post-molding operations such as the removal of flash around the part contour, drilling and painting.

Recently, several other techniques have been also used to improve the process productivity and/or the part quality.

1. Precharge preheating: If the precharge is preheated to a temperature just below the resin gel point before it is placed into the mold, the process cycle time can be reduced. Also, this method is generally employed for the thermoplastic thermoforming process.
2. Vacuum-assisted molding: The application of vacuum in the closed mold reduces the air entrapment and the surface porosity to enhance the part strength and surface quality.
3. In-mold coating: This method is used to mask the surface defects of the final products such as pores, sink marks and surface waviness. The most common method is to open the mold by a small distance (0.2–0.5 mm) during the molding cycle and inject a flexible coating material (polyester or polyester–urethane hybrid). Once this coating material covers the entire surface of the part, the mold is reclosed and the curing is progressed under a normal molding pressure.

In Fig. 3.24, the process window is presented in terms of the mold temperature and molding pressure or the speed of mold closure. Practical problems that may occur are also listed when the process conditions are outside the process window. According to the material characteristics and the part size (especially the thickness and planar dimensions), the process parameters should be optimized with respect to the process window. For process optimization, both analytical models and numerical simulations may be utilized.

3.5 Modeling and simulation

In the past, the engineering of part design in compression molding has been done in an empirical manner, that is, 'cut and try'. However, numerical simulations may be useful for process optimization to improve the part quality

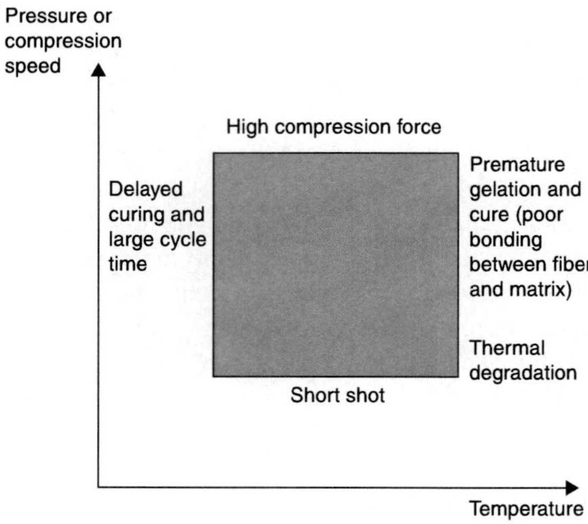

3.24 Process window in the compression molding process.

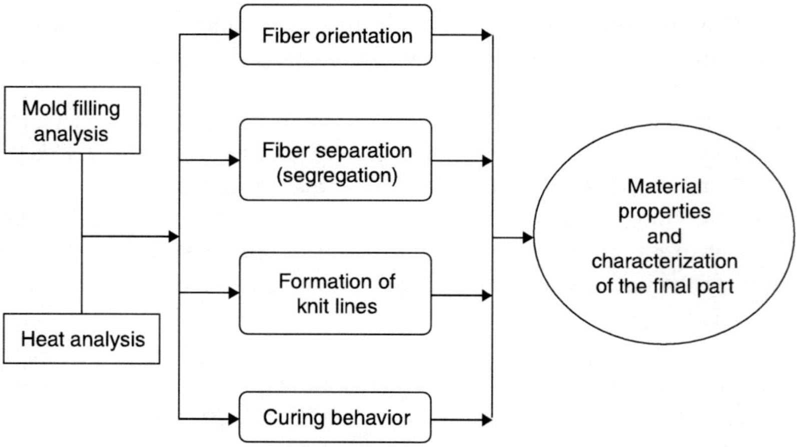

3.25 Main issues in numerical analysis of the compression molding process.

and to save the manufacturing cost. The main topics in process simulation are presented in Fig. 3.25.

The modeling of resin curing is one of the most important issues in the analysis of compression molding because the cure cycle constitutes a major portion of the total process cycle time and an incompletely cured part should be discarded. Material flow in mold filling is also an important topic because knit line formation can be avoided by optimizing the precharge location. Moreover, mold filling analysis is indispensable to predict the fiber orientation and fiber separation. As the mechanical and thermal properties of composites are greatly affected by the fiber orientation and the fiber volume fraction, the prediction of the fiber orientation and separation is an important task to optimize the product quality. The modeling approaches on these issues are introduced in the following sections.

3.5.1 Temperature and cure cycle

When exposed to high temperature, thermoset resin undergoes an exothermic reaction, whereby a number of cross-links of polymer molecules are established in a form of three-dimensional network. This process is called curing. The degree of cure is a measure of the number of cross-links. Hence, it affects the mechanical and physical properties of the final products. The strength, the elastic modulus, the heat resistance, and the chemical resistance of SMC parts increase with the degree of cure, whereas the maximum strain at failure and the impact resistance tend to decrease. In general, a high degree of cure is required to obtain good quality.

The cure cycle is an important part of the process cycle time. Thus, the molding cycle time can be reduced by optimizing the cure cycle. The cure cycle depends on the mold temperature and the resin mixture formulation. Standard SMC and BMC based on unsaturated polyester require 45–60 s at 138–149°C (280–300°F) for curing a part with a thickness of 2.5 mm (0.1 inch).

The cure kinetics of thermoset resin is dependent on heat transfer from the mold to the SMC. In general, heat conduction along the thickness direction is much more important than conduction along the planar directions in the case of the compression molding of thin parts (Lee and Tucker, 1987). Moreover, the viscous dissipation is also negligible because the strain rate and the viscosity are sufficiently low in compression molding. Thus, the heat transfer equation in thickness (z) direction can be obtained as follows.

$$\rho C_P \frac{DT}{Dt} = k_Z \frac{\partial^2 T}{\mathrm{d}z^2} + H_R \frac{\mathrm{d}\alpha_C}{\mathrm{d}t} \qquad [3.1]$$

where ρ is the density, C_P is the specific heat and k_Z is the thermal conductivity in the thickness direction. The last term on the right-hand side in Equation [3.1] represents the rate of internal heat generation due to curing. H_R is the total heat of reaction and α_c is the degree of the cure that can be defined as follows.

$$\alpha_C = \frac{H(t)}{H_R} \qquad [3.2]$$

Table 3.5 Cure model parameters for unsaturated polyester and vinyl ester

Resin	Temperature (°C)	k_1 (min^{-1})	k_2 (min^{-1})	m	n
Unsaturated polyester	50	0.0278	0.793	0.35	1.65
	60	0.0924	1.570	0.40	1.60
Unsaturated polyester with 10 wt% polyvinyl acetate	50	0.0064	0.458	0.29	1.71
	60	0.0500	0.780	0.35	1.65
Vinyl ester	50	0.0151	0.998	0.35	1.65
	60	0.0624	1.590	0.49	1.51

Source: Lem and Han (1984).

In Equation [3.2], $H(t)$ is the amount of heat released up to a certain time.

The evolution of the degree of cure is frequently described by an empirical model:

$$\frac{d\alpha_C}{dt} = \left(k_1 + k_2 \alpha_C{}^m\right)\left(1 - \alpha_C\right)^n \quad [3.3]$$

where α_C is the degree of cure, k_1 and k_2 are the cure rate parameters and m and n are the constants for the order of reaction. For thermoset resins such as unsaturated polyester and vinyl ester, a second-order reaction (i.e., $m + n = 2$) is commonly adopted. The cure rate parameters (k_1 and k_2) are dependent on the temperature, while the exponents for the order of reaction (m and n) remain constant or may change slightly with the temperature (see Table 3.5).

Usually, the dependency of the cure rate parameters (k_1 and k_2) on temperature is modeled through Arrhenius type equations.

$$k_1 = K_1 \exp\left(-\frac{E_1}{RT}\right), \quad k_2 = K_2 \exp\left(-\frac{E_2}{RT}\right) \quad [3.4]$$

In Equation [3.4], K_1 and K_2 are the model constants, E_1 and E_2 are the activation energy constants, R is the universal gas constant and T is the absolute temperature.

Sample thermal and cure kinetic parameters for SMC are presented in Table 3.6 (Barone and Caulk, 1979).

In Fig. 3.26, sample computational results of the temperature profile in the thickness direction are presented at different instants (Lee, 1981). As the precharge material, which is initially at a low temperature, is placed

Table 3.6 Cure model parameters for SMC

K_1	4.9×10^{14}/s
K_2	6.2×10^{5}/s
E_1	140.0 kJ/mol
E_2	51.0 kJ/mol
M	1.3
N	2.7
H_R	84 J/g
ρ	1.9 g/cm^3
C_P	1.0 J/g/K
k_Z	0.53 W/m/K

Source: Barone and Caulk (1979).

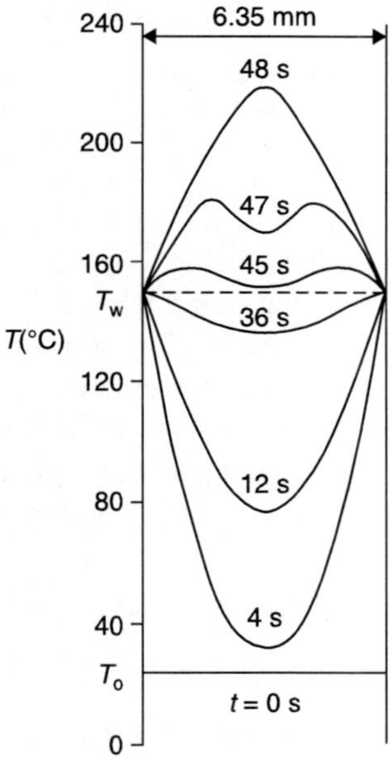

3.26 Change in temperature profile with time in the thickness direction during SMC molding (Lee, 1981).

in the hot mold, the heat is conducted from the mold surface to the precharge. At the early stage of molding, the temperature in the skin zones that are in contact with the mold surfaces is higher than the temperature in the center. However, the temperature in the center increases with time and eventually surpasses the temperature at the outer surfaces. This inversion of the temperature profile can be explained by the exothermic reaction of thermosetting resin. In other results regarding the temperature and the

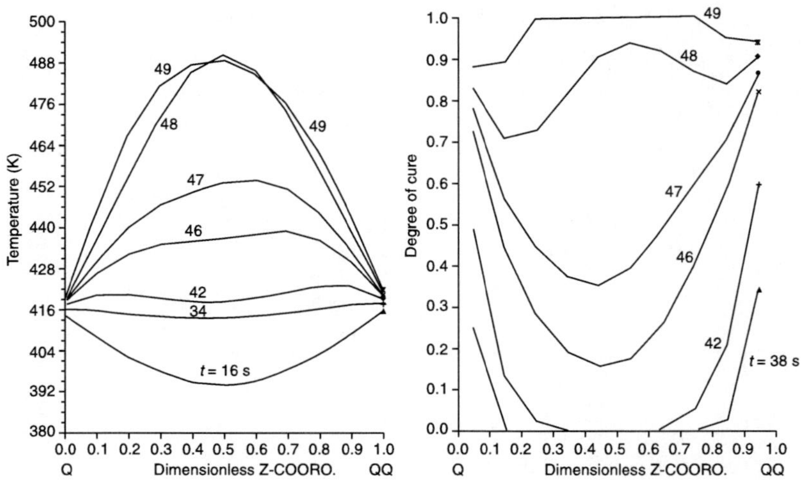

3.27 Change in the temperature and the degree of cure profiles in the thickness direction during SMC molding (Lee, 1989).

degree of cure in the thickness direction, the same inversion of temperature profile as the degree of cure progresses can be observed (see Fig. 3.27; Lee, 1989). Initially, the degree of cure is lowest at the center while it is highest in the outer zones. However, the evolution of curing is more important in the center zone, and the degree of cure eventually becomes relatively uniform along the thickness direction. Therefore, the generation of reaction heat by resin curing is more important in the center zone, and this phenomenon leads to a higher temperature in the center than in the outer zones that are in contact with the hot mold surface.

3.5.2 Material flow and mold filling

The solution to the full set of conservation equations for mass, momentum and energy requires major computational resources and may become very costly. Instead, some assumptions are introduced to obtain simplified flow models. The most popular models are the generalized Hele–Shaw (GHS) model and the Barone–Caulk model.

The material flow in a thin cavity can be modeled by the Hele–Shaw model. Also a lubrication approximation can be used with assumptions such as a negligible inertial force, dominant shear stress, a thin cavity (compared with the planar dimensions), incompressible fluid, a small capillary number and a generalized Newtonian fluid (Hieber and Shen, 1980; Tucker and Folgar, 1983).

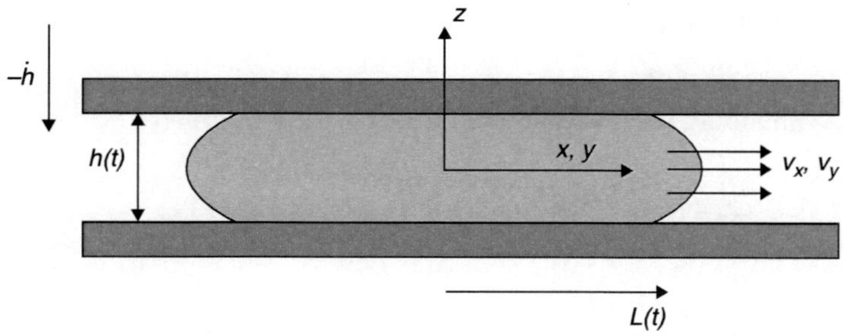

3.28 Illustration of material flow during compression molding.

In the GHS model, the no-slip condition is assumed between the mold surface and the charge surface layer. Hence, the transverse shear stress across the narrow gap between the mold surfaces is the dominant factor that drives material flow, and the in-plane stresses are negligible. The material flow configuration is illustrated in Fig. 3.28.

The gap-wise averaged velocities in the planar directions can be expressed as follows.

$$\bar{v}_x = -\frac{S}{h}\frac{\partial P}{\partial x}, \qquad \bar{v}_y = -\frac{S}{h}\frac{\partial P}{\partial y} \qquad [3.5]$$

where S is the flow conductance which is defined as the following relation:

$$S = \int_{-h/2}^{+h/2} \frac{(z-\lambda)^2}{\mu}\,dz \qquad [3.6]$$

where h is the gap height, μ is the viscosity and λ is the value of z where the shear stress is null.

We can obtain the governing equation for the pressure field (P), if the relations for the gap-wise averaged velocities are integrated into the continuity equation:

$$\frac{\partial}{\partial x}\left(S\frac{\partial P}{\partial x}\right) + \frac{\partial}{\partial y}\left(S\frac{\partial P}{\partial y}\right) = \dot{h} \qquad [3.7]$$

where $\dot{h}$ is the rate of change in the gap height, which is a negative value during the mold-closure step. It is noted that the pressure in the thickness (z) direction is assumed to be uniform as the cavity is thin.

The governing equation for the pressure is solved using the boundary conditions. At the flow front, the essential condition can be applied.

$$P = 0 \ \text{at} \ L(t) \qquad [3.8]$$

At the mold wall, natural conditions are applied:

$$\frac{\partial P}{\partial n} = 0 \ \text{at the mold walls} \qquad [3.9]$$

where n represents the normal direction to the mold wall.

The GHS model is simple and computationally efficient because the formulation consists of only one governing equation in terms of one variable, namely, pressure. Whereas the GHS model applies to thin cavities, the Barone–Caulk model better predicts the flow of thick charge (Erwin and Tucker, 1995; Lee, 1984). The Barone–Caulk model characterizes the flow by uniform extension though the cavity thickness with a slip boundary condition between the charge surface layers and the mold surfaces instead of the no-slip condition that is adopted in the GHS model (Barone and Caulk, 1985, 1986). For further information on the Barone–Caulk model, the readers are recommended to see the references providing comprehensive description and simulation examples (Davis *et al.*, 2003; Tucker, 1987).

Both models can be solved by different numerical schemes. The boundary element method has the advantage that the velocity gradient can be obtained more accurately than the finite-element method, which is important for predicting the fiber orientation (Barone and Caulk, 1986; Osswald and Tucker, 1988). However, it is restricted to single charge, flat parts and cases of uniform thickness. Hence, many simulation efforts have converged to the use of the finite-element/control-volume method (Erwin and Tucker, 1995; Osswald and Tucker, 1990) coupled with the volume of fluid (VOF) method to track the position of flow front (Hirt and Nichols, 1981).

Recently, Dumont *et al.* improved the Barone–Caulk model considering the non-Newtonian behavior of SMC (Dumont *et al.*, 2007). Assuming SMC to be an incompressible and purely viscous material, they proposed a transversely isotropic rheology model. The basic governing equations for the SMC flow are the mass conservation and the momentum balance equations.

$$\nabla \cdot \underline{v} = 0 \qquad [3.10]$$

$$\nabla \underline{\underline{\sigma}} = 0 \qquad [3.11]$$

where $\underline{v}$ is the velocity vector and $\underline{\underline{\sigma}}$ is the stress tensor.

Assuming the plug flow in the mold cavity of plate geometry, the material flow velocity vector can be described as follows.

$$\underline{v} = v_x(x,y)\underline{e}_x + v_y(x,y)\underline{e}_{yx} + v_z(x,y,z)\underline{e}_z \quad [3.12]$$

From the mass conservation equation and the boundary conditions at the mold wall $\left(v_z(z=+h/2)=\dot{h}/2, v_z(z=-h/2)=-\dot{h}/2\right)$, we can obtain the strain rate tensor.

$$\underline{\underline{D}} = \begin{pmatrix} \frac{\partial v_x}{\partial x} & \frac{1}{2}\left(\frac{\partial v_y}{\partial x}+\frac{\partial v_x}{\partial y}\right) & 0 \\ \frac{1}{2}\left(\frac{\partial v_x}{\partial y}+\frac{\partial v_y}{\partial x}\right) & \frac{\partial v_y}{\partial y} & 0 \\ 0 & 0 & \frac{\dot{h}}{h} \end{pmatrix} = \underline{\underline{\tilde{D}}} + \frac{\dot{h}}{h}\underline{e}_z \otimes \underline{e}_z \quad [3.13]$$

where $\underline{\underline{\tilde{D}}}$ is the planar strain rate tensor.

The stress tensor can be decomposed into the hydrodynamic pressure and the viscous stress tensor.

$$\underline{\underline{\sigma}} = -p\underline{\underline{\delta}} + \underline{\underline{\sigma}}^V\left(\underline{\underline{D}}\right) \quad [3.14]$$

where p is the hydrodynamic pressure, $\underline{\underline{\delta}}$ the identity tensor and $\underline{\underline{\sigma}}^V$ the viscous stress tensor.

Considering the SMC as an incompressible power law fluid exhibiting a transverse isotropy, viscous stresses in the planar direction and the transverse direction can be expressed as follow.

$$\underline{\underline{\tilde{\sigma}}}^V = \mu_{eq}\frac{2}{1+2H}\left(\frac{D_{eq}}{D_o}\right)^{n-1}\underline{\underline{\tilde{D}}} \quad [3.15]$$

$$\sigma_{zz}^V = \mu_{eq}\frac{2H}{1+2H}\left(\frac{D_{eq}}{D_o}\right)^{n-1}\frac{\dot{h}}{h} \quad [3.16]$$

$$D_{eq}^2 = \frac{2}{1+2H}\left(\underline{\underline{\tilde{D}}}:\underline{\underline{\tilde{D}}} + H\left(\frac{\dot{h}}{h}\right)^2\right) \quad [3.17]$$

where H is the rheological function accounting for the fiber network anisotropy, n is the power law index of SMC and μ_{eq} is the elongational viscosity at an equivalent strain rate D_{eq} of D_o (arbitrarily fixed to 1 s^{-1}).

For the power law index n, 0.44 for SMC ($0.034 < V_f < 0.234$) and 0.6 for the resin paste without fibers have been suggested (Dumont *et al.*, 2007). For the rheological function H and the elongational viscosity μ_{eq}, the following relations have been suggested (Dumont *et al.* 2003).

$$H = \frac{1+98V_f+980V_f^2}{0.5+67V_f+670V_f^2}-1 \quad [3.18]$$

$$\mu_{eq} = \mu_o\left(1+98V_f+980V_f^2\right)\left(2\frac{1+H}{1+2H}\right)^{-\frac{n+1}{2}} \quad [3.19]$$

where μ_o is the viscosity of resin paste without fibers and can be expressed as an Arrhenius function of temperature.

$$\mu_o = \mu_{oo}\exp\left(b\left(\frac{1}{T}-\frac{1}{T_o}\right)\right) \quad [3.20]$$

For the model constants, $\mu_{oo} = 0.18$ MPa·s, $b = 4500$ K^{-1} and $T_o = 296$ K have been proposed (Dumont *et al.*, 2007).

For the BMC rheology, the same approach can be adopted (Guiraud *et al.*, 2010). In particular, the strain-hardening and nonlinear viscous effect of BMC can be considered as the equivalent stress is defined by the following relation.

$$\sigma_{eq} = \eta_o \exp\left(k\varepsilon_{eq}\right)D_{eq}^n \quad [3.21]$$

where η_o is the consistency and k is the strain-hardening coefficient.

The equivalent strain ε_{eq} can be defined by the equivalent strain rate D_{eq}:

$$\varepsilon_{eq} = \int D_{eq}\,dt \quad \text{or} \quad \frac{d\varepsilon_{eq}}{dt} = D_{eq} \quad [3.22]$$

The equivalent strain rate can be obtained by the following form:

$$D_{eq}^2 = \alpha_o\left(\underline{\underline{D}}:\underline{\underline{D}}+\alpha_1\left(\underline{\underline{M}}:\underline{\underline{D}}\right)^2+\alpha_2\left(\underline{\underline{D}}\cdot\underline{\underline{M}}\right):\underline{\underline{D}}\right) \quad [3.23]$$

$$\begin{cases} \alpha_o = \dfrac{2}{1+2H} \\ \alpha_1 = 1+H-2\dfrac{1+2H}{3L} \\ \alpha_2 = 2\left(\dfrac{1+2H}{3L}-1\right) \end{cases} \quad [3.24]$$

where $\underline{\underline{M}} = \underline{e}_z \otimes \underline{e}_z$ is the microstructure tensor characterizing the transverse isotropy. Two material constants H and L can be determined by simple tests. For example, BMC flow takes place when the equivalent stress reaches a critical value for each of the following cases.

Simple compression along z: $\sigma_{eq} = \sigma_{zz}$ [3.25]

Simple compression along x: $\sigma_{eq} = \sigma_{xx}\sqrt{\dfrac{(1+H)}{2}}$ [3.26]

Shear in the x-z plane: $\sigma_{eq} = \tau_{zx}\sqrt{3L}$ [3.27]

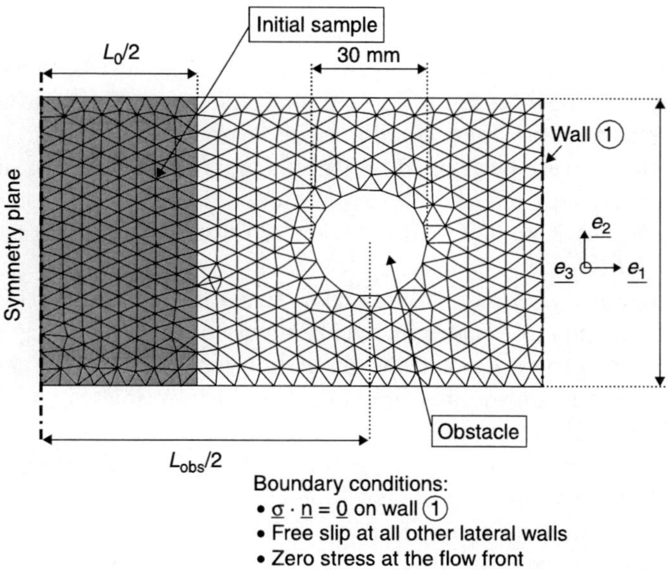

3.29 Mold geometry, mesh, initial charge position and boundary conditions (Guiraud *et al.*, 2010).

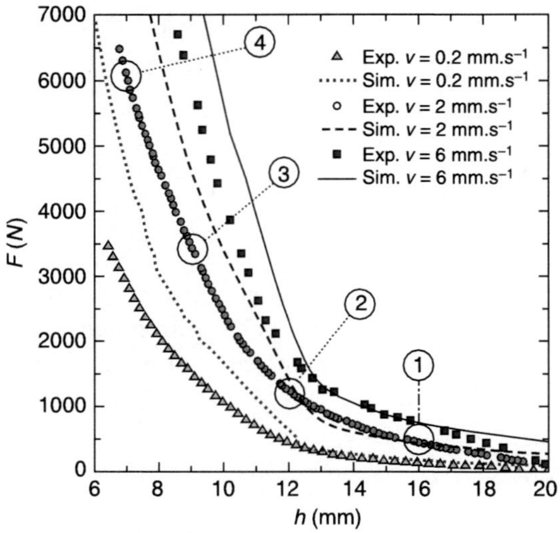

3.30 Comparison between simulations and experiments: mold closing force against mold cavity height for different mold closing speeds of 0.2, 2 and 6 mm/s (Guiraud *et al.*, 2010).

Finally, the viscous stress can be expressed by the following relation.

$$\underline{\underline{\sigma}}^{v} = \alpha_o \eta_o \exp\left(k\varepsilon_{\mathrm{eq}}\right) D_{\mathrm{eq}}^{n-1}\left(\underline{\underline{D}} + \alpha_1\left(\underline{\underline{M}} : \underline{\underline{D}}\right)\underline{\underline{M}} + \frac{\alpha_2}{2}\left(\underline{\underline{D}} \cdot \underline{\underline{M}} + \underline{\underline{M}} \cdot \underline{\underline{D}}\right)\right) \qquad [3.28]$$

Guiraud *et al.* (2010) integrated this rheology model into the finite-element code for SMC/BMC compression molding simulation. The mold geometry, the mesh and the boundary conditions are shown in Fig. 3.29.

Simulations were performed for various mold closing speeds. Simulation results were compared with the experimental data obtained using a CCD camera. In Fig. 3.30, mold closing force (F) was plotted against the mold cavity height (h).

Photographs were taken in the real mold at different instants (1, 2, 3, 4 in Fig. 3.31) for the mold closing speed of 2 mm/s. In Fig. 3.31, the experimental results are shown on top and the simulation results are shown on bottom. Moreover, the results of fiber orientation prediction (that will be explained in the following section) are also presented.

We can observe a fairly good agreement between the numerical prediction and the experimental observation. However, the current model does not consider the planar anisotropy in SMC/BMC rheology which is developed by flow-induced anisotropic fiber orientation in the planar direction. This issue should be addressed in the forthcoming work to improve the model.

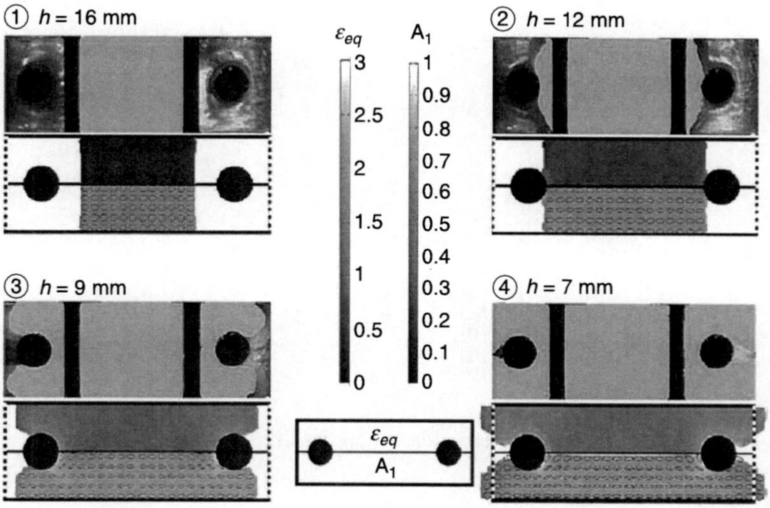

3.31 Comparison of mold filling patterns and fiber orientations between experimental observations (top) and simulation results (bottom) (Guiraud *et al.*, 2010).

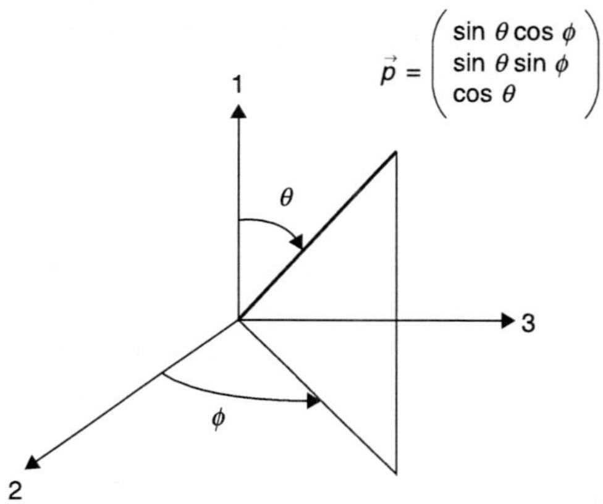

3.32 Representation of a single fiber orientation in spherical coordinates.

3.5.3 Fiber orientation

The orientation of a single fiber can be represented by two angular values ($0 < \theta < \pi$ and $0 < \phi < 2\pi$, see Fig. 3.32) in spherical coordinates. For SMC compression molding processes, the usual part thickness is much smaller than the fiber length. Hence, the orientation state in compression molding processes can be regarded as two-dimensional or planar (i.e., $\theta = 0$).

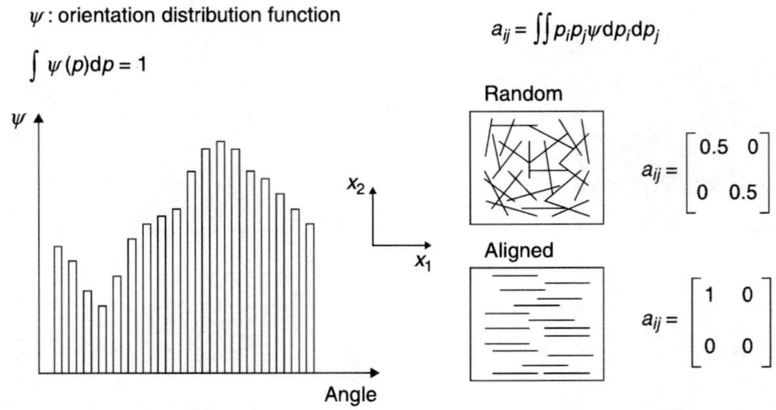

3.33 **Orientation distribution function (left) and orientation tensor (right).**

To represent the orientation states of multiple fibers, however, statistical approaches are used. One of the common approaches is the orientation distribution function, which is a kind of probability density function (see Fig. 3.33). Because one end of a fiber is indistinguishable from the other end, the orientation distribution function (ψ) is periodic.

$$\psi(\phi) = \psi(\phi + \pi) \quad \text{and} \quad \int_{-\pi/2}^{+\pi/2} \psi(x, y, \phi)\,\mathrm{d}\phi = 1 \qquad [3.29]$$

A more efficient alternative method is the orientation tensor approach developed by Advani and Tucker (1987, 1990a). Advani and Tucker proposed orientation tensors that are defined by the following relations (see Fig. 3.33).

$$\begin{aligned} a_{ij} &= \langle p_i p_j \rangle = \int \psi(\vec{p})\, p_i p_j \mathrm{d}\vec{p} \\ a_{ijkl} &= \langle p_i p_j p_k p_l \rangle = \int \psi(\vec{p})\, p_i p_j p_k p_l \mathrm{d}\vec{p} \end{aligned} \qquad [3.30]$$

In Equation [3.30], a_{ij} and a_{ijkl} are the second-order and fourth-order orientation tensors, respectively.

The motion of a single fiber in the flow can be successfully computed by Jeffery's equation (Jeffery, 1922). However, most composite systems can be considered as a concentrated suspension where the interaction between fibers is not negligible. Folgar and Tucker proposed a model introducing the rotary Brownian diffusivity (D_r) to consider the fiber–fiber interaction (Folgar, 1983; Folgar and Tucker, 1984).

Combining Jeffery's equation with the definition of orientation tensors, the equation for the orientation tensor in concentrated suspensions can be obtained considering fiber–fiber interaction.

$$\frac{Da_{ij}}{Dt}+\frac{1}{2}\left(\omega_{ik}a_{kj}-a_{ik}\omega_{kj}\right)=\frac{1}{2}\lambda\left(\dot{\gamma}_{ik}a_{kj}+a_{ik}\dot{\gamma}_{kj}-2\dot{\gamma}_{kl}a_{ijkl}\right)+2C_I\dot{\gamma}\left(\delta_{ij}-\alpha a_{ij}\right) \quad [3.31]$$

where δ_{ij} is the Kronecker delta function, and α is two for two-dimensional and three for three-dimensional cases. λ is the fiber shape factor defined by the fiber aspect ratio.

$$\lambda=\frac{\left(L_f/D_f\right)^2-1}{\left(L_f/D_f\right)^2+1} \quad [3.32]$$

where L_f and D_f are the length and the diameter of fiber, respectively. Furthermore, ω_{ij} and $\dot{\gamma}_{ij}$ are the vorticity tensor and the rate-of-deformation tensor, respectively.

$$\begin{aligned}\omega_{ij}&=\frac{\partial v_j}{\partial x_i}-\frac{\partial v_i}{\partial x_j}\\ \dot{\gamma}_{ij}&=\frac{\partial v_j}{\partial x_i}+\frac{\partial v_i}{\partial x_j}\end{aligned} \quad [3.33]$$

To solve the orientation tensor equation (Equation [3.31]), a closure approximation equation is needed to replace the fourth-order orientation tensor by a function of the second orientation tensor. A widely accepted one is the hybrid closure approximation (Advani, 1990b) whereas many other closure approximations have been recently proposed (Cintra and Tucker, 1995).

For more details, the readers are recommended to see the references (Advani and Tucker 1987, 1990b; Polymer Composites; Osswald and Tseng, 1994).

A sample case of fiber orientation in compression molding is presented in Fig. 3.34. Fiber orientation is expressed in terms of the second orientation tensor, a_{11}. On the other hand, a more general method to represent the planar orientation state is to use ellipsoids. From the second orientation tensors,

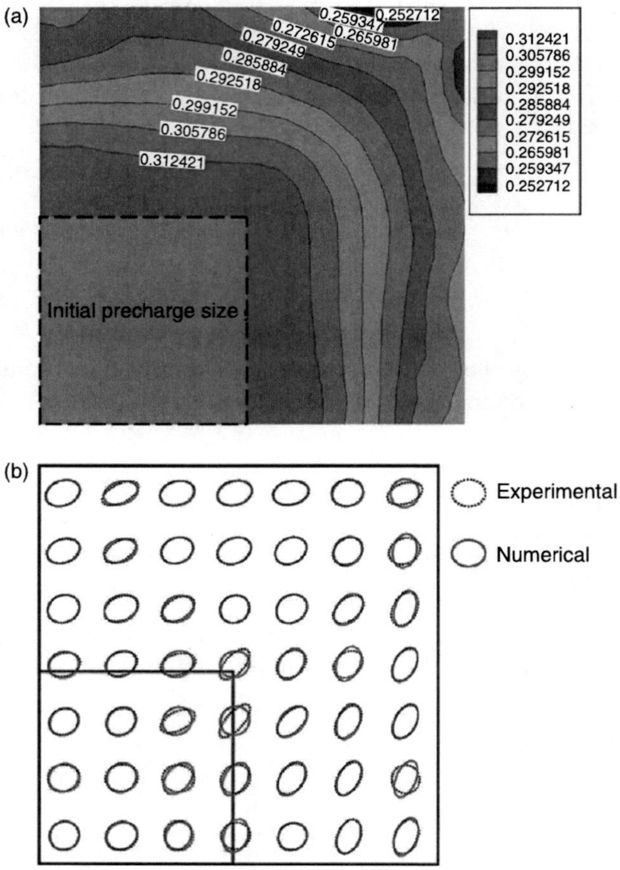

3.34 Experimental and simulation results of fiber orientation in compression molding (Kim *et al.*, 2009; Park *et al.*, 2001): (a) Component of the orientation tensor a_{11} and (b) representation of the planar orientation by ellipsoids.

the principal directions where the off-diagonal terms vanish are found. Also, two principal values of the second-order tensors are computed. Then, an ellipsoid with the minor and major axes of the same lengths as the principal values of the second-order tensors that are tilted in the directions of the principal angles can be traced. In Fig. 3.34b, the planar orientation state is represented by ellipsoids. The random orientation state is expressed by a circle, and the aligned orientation state is represented by an ellipsoid with a long major axe and a short minor axe. We can see that the prediction of fiber orientation state by the orientation tensor model is fairly good in practical molding cases.

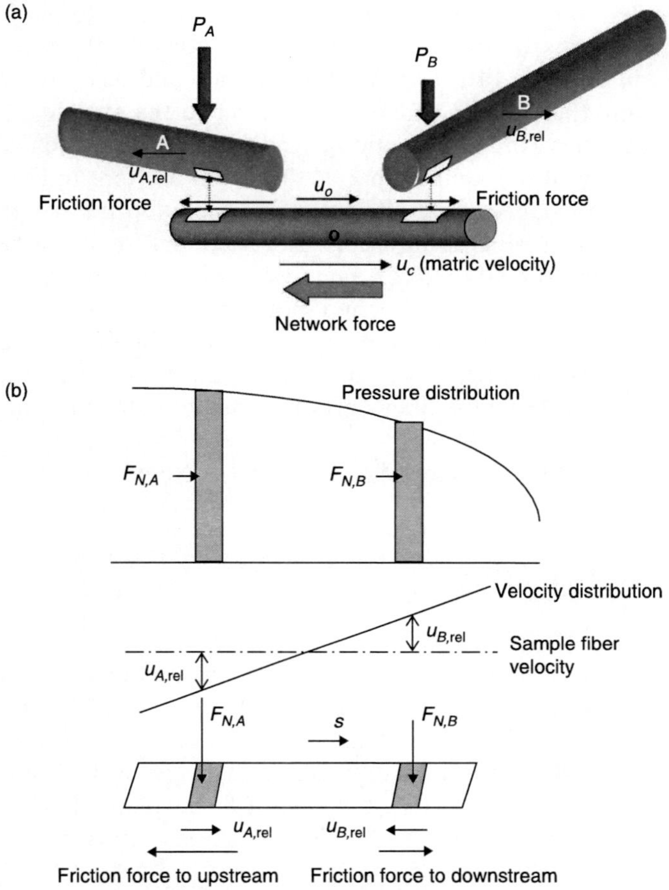

3.35 (a) Illustration of fibers interacting with the sample fiber. (b) Forces acting on the sample fiber.

3.5.4 Fiber separation

The interaction among the fibers is the principal reason for the fiber separation (a.k.a. fiber segregation) which results a non-uniform fiber volume fraction in the final part as stated previously (Hojo *et al.*, 1988). The interactions among the fibers can be considered as a frictional force. Consider a single fiber in the fiber network as a sample fiber (see Fig. 3.35). The fibers that are in contact with the sample fiber can be classified into two categories: the upstream fibers and the downstream fibers. In the squeezing flow, the resin velocity in the downstream is greater than that in the upstream. Hence, the downstream fibers move faster than the sample fiber, while the upstream fibers move slower than the sample fiber. This difference in fiber velocity gives rise to a frictional force on the sample fiber in different directions in the upstream and downstream. In the downstream, the frictional force tends to move the sample fiber along the material flow, while the friction in the upstream fibers tends to delay the translation of the sample fiber. If these two frictions in the opposite directions are balanced, then the net force applied to the sample fiber is zero, and no fiber separation occurs. The frictional force can be considered to be proportional to the normal force which, in turn, is proportional to the resin pressure in compression molding. As the upstream pressure is greater than the downstream pressure, the upstream friction is greater than the downstream friction. Consequently, the net force that is applied to the sample fiber by the fiber network tends to delay the sample fiber migration.

The fiber network force (F_{NW}) can be obtained by the following relation:

$$F_{NW} = \kappa\left(r_{up}\int_{x_O-L_f/2}^{x_O} P(x)D_f\mathrm{d}x - r_{dw}\int_{x_{Of}}^{x_O+L_f/2} P(x)D_f\mathrm{d}x \right) \quad [3.34]$$

In Equation [3.34], x_O is the coordinate of the mid-point of the sample fiber, L_f and D_f are the fiber length and the fiber diameter, respectively, and r_{up} and r_{dw} are the coefficients that represent the degree of contact with other fibers in the upstream and downstream, respectively. The latter two coefficients can be defined as a function of fiber volume fractions in the upstream and downstream. Finally, κ is the friction coefficient between the fibers.

If there is a difference between the fiber and resin velocities, the fiber is submitted to a drag force (see Fig. 3.36). The fiber separation velocity, which is the velocity difference between the fiber and the resin, can be obtained from the equilibrium condition between the fiber network force and the drag force.

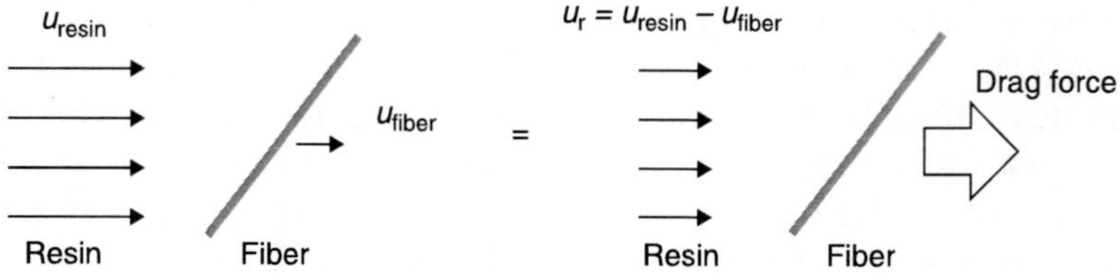

3.36 Drag force applied to the fiber.

The drag force can be computed by the following relation:

$$F_D = \frac{1}{2} C_D \rho_m u_r^2 A_P \quad [3.35]$$

where ρ_m is the matrix density, u_r is the relative velocity ($u_{\text{resin}} - u_{\text{fiber}}$), A_p is the projection area and C_D is the drag coefficient that can be defined as a function of the Reynolds number.

From the force balance between the fiber network force and the drag force, we can obtain the relative velocity.

$$\frac{1}{2} C_D \rho_m u_r^2 A_p = \kappa\left(r_{\text{up}} \int_{x_O - L_f/2}^{x_O} P(x) D_f \mathrm{d}x - r_{\text{dw}} \int_{x_{Of}}^{x_O + L_f/2} P(x) D_f \mathrm{d}x \right) \quad [3.36]$$

Once the relative velocity (u_r) is known, we can obtain the fiber migration velocity (u_{fiber}) because the resin velocity (u_{resin}) is computed from the mold filling analysis. Then, we can obtain the distribution of fiber volume fraction by solving the conservation equation of fiber content.

$$\frac{\partial V_f}{\partial t} + \nabla \vec{u}_{\text{fiber}} \cdot V_f = -\frac{\dot{h}}{h} V_f \quad [3.37]$$

We can verify the dependence of fiber separation on the fiber orientation in Equation [3.36]. If a fiber is aligned in the perpendicular direction to the

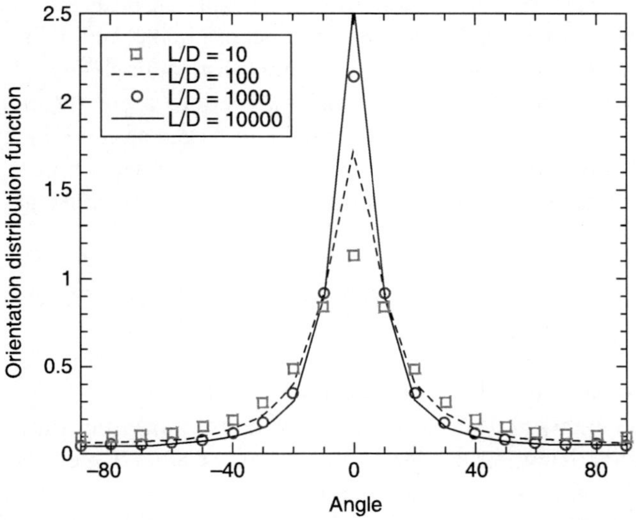

3.37 Orientation distribution function of separated fibers with respect to the angle between the fiber and flow directions for different fiber aspect ratios (Park, 1998).

flow, the fiber network force can be ignored because there is no difference between upstream and downstream. Hence, the relative velocity is zero and there is no fiber separation. If the fiber is aligned parallel to the flow, however, the fiber network force is no longer negligible because the difference in the fiber network between upstream and downstream is significant. Moreover, the projection area with respect to the resin flow is reduced as the fiber is oriented in the flow direction. Subsequently, the relative velocity should be increased because the fiber network force on the right-hand side is significant and the projection area (A_p in the left hand side) is small. As a result, fiber separation becomes significant.

The orientation distribution function of the separated fibers is plotted for different fiber aspect ratios. In Fig. 3.37, it can be seen that the orientation distribution function is maximum at 0° where the fibers are oriented in the flow direction, whereas the orientation distribution function becomes minimum at ± 90° where the fibers are aligned normal to the flow. In general, the fiber separation becomes significant as the fiber aspect ratio increases because the difference in the fiber network that is applied to the sample fiber between the upstream and downstream increases (see Fig. 3.37).

As the fiber volume fraction increases, the fiber separation becomes more significant because the interactions between the fibers increase (see Fig. 3.38a and c). To enhance the mechanical performance of composite structures, it is necessary to increase the fiber volume fraction and the fiber length. However, this condition may increase the fiber separation and leads to a non-homogeneous fiber distribution in the final products. To avoid this problem, the speed of mold closure should be increased. From Fig. 3.38, the influence of the speed of mold closure on fiber separation can be seen: a higher mold-closure speed results in a more homogeneous fiber content distribution (Fig. 3.38a and b). However, excessively fast mold closure may induce a greater mold closing force. Moreover, air can be entrapped and void defects can be generated if the mold is not properly designed. Hence, a proper selection of the process parameters is required to obtain good quality and process efficiency.

3.5.5 Coupling between structural performance and process parameters

As we have seen, the fiber orientation and the fiber volume fraction in the part can be highly non-uniform and these non-uniformities may affect the mechanical and physical properties of the final product. Especially, the position and dimension of precharge in the mold are key parameters to product quality, while the other process parameters such as mold closing speed and mold temperature are also important.

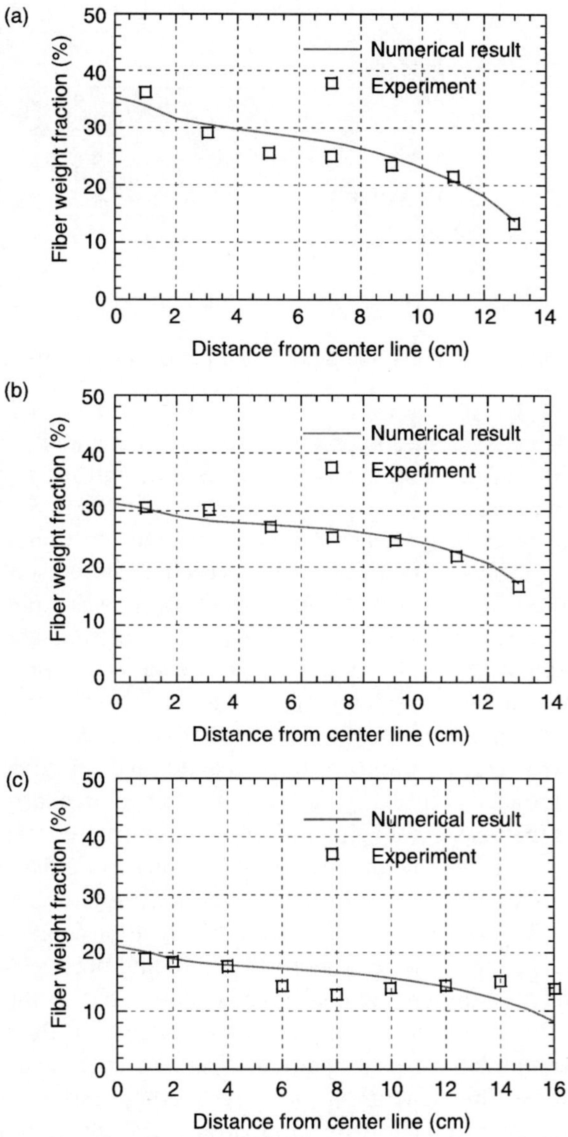

3.38 Fiber content distribution in the compression molded part (fiber length: 40 mm): (a) 25 wt%, 5 mm/min closing speed; (b) 25 wt%, 20 mm/min closing speed and (c) 15 wt%, 5 mm/min closing speed (Yoo, 1997).

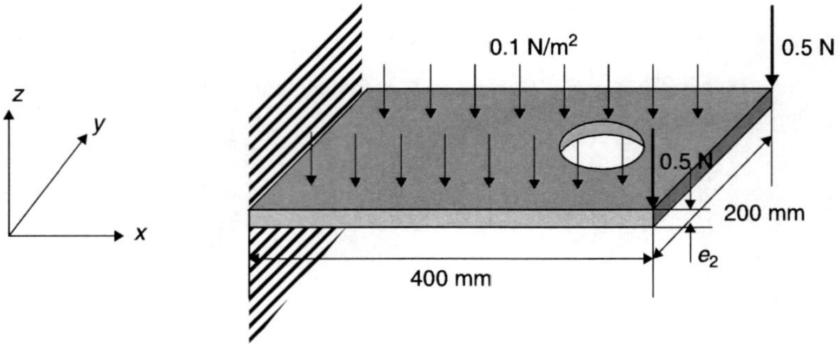

3.39 Dimensions and loading and supporting conditions of the plate.

As a sample case, consider a compression molded rectangular plate with a circular hole (Kim, 2007). The dimensions and loading and supporting conditions are shown in Fig. 3.39.

Consider three cases for precharge dimensions and locations as shown in Fig. 3.40. It is a common practice to place precharge in the center of the mold (Case 3) to reduce the flow path. A shorter flow path can minimize the non-uniformity of fiber orientation and fiber volume fraction and result in smaller mold clamping force. In some cases, however, eccentric precharge location (Case 1) can align the fibers into the main flow direction and the mechanical performance can be enhanced in the fiber direction. If the principal load is applied in the fiber direction, the maximum displacement under the load can be reduced (Fig. 3.41). However, eccentric precharge loading can lead to an inclination of mold platen. Hence, a compromise between the molding condition and the structural performance can be found if the reduction of structural performance and process efficiency is tolerable (Case 2).

Kim *et al.* (2009, 2011) proposed a structural optimization considering processing parameters in the case of compression molding process. While the conventional structural design and analysis has been based on the assumption of isotropic properties in the final part, they integrated the structural analysis with the simulation of fiber orientation and separation into the optimization scheme. This structure–process coupled optimization is useful to further improve the product quality compared to the conventional separate structural design and process optimization procedures.

3.6 Future trends

Recently, thermoplastic composites such as GMT and LFT are becoming more and more popular in the automotive industry whereas thermoset molding compounds have been the principal molding materials. Carbon fibers are increasing their share even in the automotive industry in spite

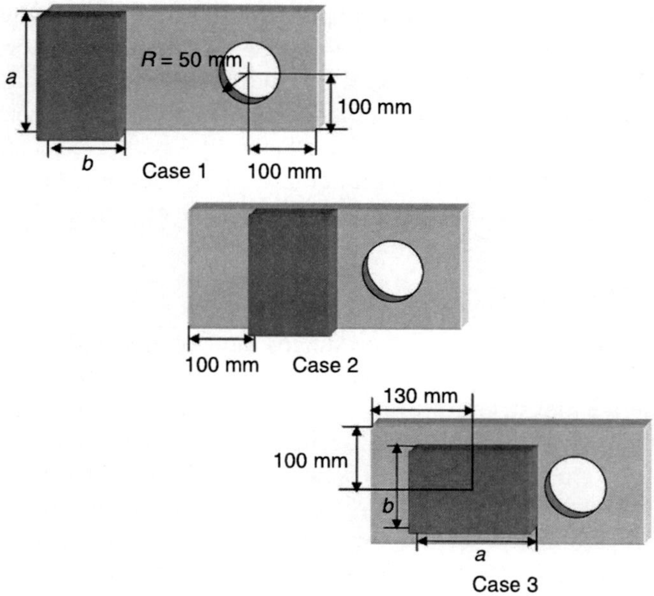

3.40 Precharge location and dimensions (a = 200 mm, b = 120 mm).

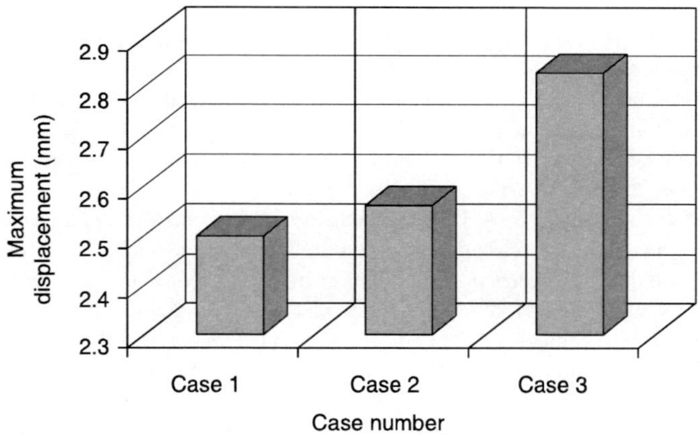

3.41 Maximum displacements under the load for three cases.

of their high cost, because the weight reduction is becoming a paramount objective. Moreover, to meet specific requirements, reinforcing fibers and resin can be compounded *in situ* and directly transferred into the compression press. An extruder or transfer molding equipment is used together with a compression molding setup for this purpose. These methods are frequently employed for LFT and are effective for small batch applications.

The compression molding process can be considered to be an optimal manufacturing option for natural fiber composites. A high molding temperature and pressure of injection molding process is a main obstacle to using natural fibers. Thanks to relatively low molding pressure and temperature, the compression molding process can preserve good mechanical properties of natural fibers. Moreover, SMC compounding by randomizing the fiber orientation suggests potential to circumvent the property scattering issue of natural fibers.

3.7 Conclusions

Compression molding is a very productive and economical way of manufacturing composite parts. It is suited for large volume applications such as automotive industry. However, the process requires large capital investment for the press and the mold. Hence, it is very important to improve yield ratio and to decrease process cycle time for this manufacturing process to be cost effective. With proper selection of material and process design, product quality and process performance can be improved. In particular, process modeling to predict resin flow and curing may be helpful to reduce the product development time.

3.8 References

Advani, S. G. and Tucker III, C. L. (1987), 'The use of tensors to describe and predict fiber orientation in short fiber composites', *Journal of Rheology*, **31**, 751–784.

Advani, S. G. and Tucker III, C. L. (1990a), 'A numerical simulation of short fiber orientation in compression molding', *Polymer Composites*, **11**(3), 164–173.

Advani, S. G. and Tucker III, C. L. (1990b), 'Closure approximations for three-dimensional structure tensors', *Journal of Rheology*, **34**, 367–386.

Barone, M. R. and Caulk, D. A. (1979), 'The effect of deformation and thermoset cure on heat conduction in a chopped-fiber reinforced polyester during compression molding', *International Journal of Heat and Mass Transfer*, **22**(7), 1021–1032.

Barone, M. R. and Caulk, D. A. (1981), 'Optimal thermal design of compression molds for chopped-fiber composites', *Polymer Engineering and Science*, **21**(17), 1139–1148.

Barone, M. R. and Caulk, D. A. (1985), 'Kinematics of flow in sheet molding compounds', *Polymer Composites*, **6**(2), 105–109.

Barone, M. R. and Caulk, D. A. (1986), 'A model for the flow of a chopped fiber reinforced polymer compound in compression molding', *Journal of Applied Mechanics*, **53**(2), 361–371.

Chen, W., Liu, Y. and Xin, Q. (2010), 'Evaluation of a compression molded composite bipolar plate for direct methanol fuel cell', *International Journal of Hydrogen Energy*, **35**, 3783–3788.

Cintra, J. S. and Tucker III, C. L. (1995), 'Orthotropic closure approximations for flow-induced fiber orientation', *Journal of Rheology*, **39**(6), 1095–1122.

Davis, B. A., Gramann, P., Rios, A. C. and Osswald, T. A. (2003), *Compression molding*. Munich: Hanser Gardner.

Dumont, P., Orgéas, L., Le Corre, S. and Favier, D. (2003), 'Anisotropic behavior of sheet molding compounds (SMC) during compression molding', *International Journal of Plasticity*, **19**, 625–646.

Dumont, P., Orgéas, L., Favier, D., Pizette, P. and Venet, C. (2007), 'Compression moulding of SMC: *In situ* experiments, modeling and simulation', *Composites Part A*, **38**, 353–368.

Erwin, W. L. and Tucker III, C. L. (1995), 'A finite element method for flow in compression molding of thin and thick parts', *Polymer Composites*, **16**(1), 70–82.

Folgar, F. P. (1983), 'Fiber orientation distributions in concentrated suspensions: a predictive model', PhD thesis, Department of Mechanical Engineering, University of Illinois, Urbana–Champaign.

Folgar, F. P. and Tucker III, C. L. (1984), 'Orientation behavior of fibers in concentrated suspensions', *Journal of Reinforced Plastics and Composites*, **3**, 98–119.

Guiraud, O., Dumont, P., Orgéas, L., Vassal, J. P., Le, T. H. and Favier, D. (2010), 'Towards the simulation of mould filling with polymer composites reinforced with mineral fillers and short fibres', *International Journal of Material Forming*, **3**(S2), S1313–S1326.

Hieber, C. A. and Shen, S. F. (1980), 'A finite-element/finite-difference simulation of the injection-molding filling process', *Journal of Non-Newtonian Fluid Mechanics*, **7**, 1–32.

Himebaugh, D. C. and Newman, S. (1985), *Proceedings of 38th Annual SPI Technical Conference*, Session 9-E, Houston, Texas, February 1983.

Hirt, C. W. and Nichols, B. D. (1981), 'Volume of fluid (VOF) method for the dynamics of free boundaries', *Journal of Computational Physics*, **39**, 201–225.

Hojo, H., Kim, E. G., Yaguchi, H. and Onodera, T. (1988), 'Simulation of compression moulding with matrix-fibre separation and fiber orientation for long fibre-reinforced thermoplastics', *International Polymer Processing*, **3**, 54–61.

Jeffery, G. B. (1922), 'The motion of ellipsoidal particles immersed in a viscous fluid', *Proceedings of the Royal Society: A. Mathematical, Physical and Engineering Sciences*, **102**, 161–179.

Kim, M. S. (2007), 'Optimization of composite structures considering the manufacturing process', PhD thesis, School of Mechanical and Aerospace Engineering, Seoul National University, Korea.

Kim, M. S., Lee, W. I., Han, W. S., Vautrin, A. and Park, C. H. (2009), 'Thickness optimization of composite plates by Box's complex method considering the process and material parameters in compression molding of SMC', *Composites: Part A*, **40**, 1192–1198.

Kim, M. S., Lee, W. I., Han, W. S. and Vautrin, A. (2011), 'Optimisation of location and dimension of SMC precharge in compression moulding process', *Computers and Structures*, **89**, 1523–1534.

Lee, C. C. (1984), 'Numerical model for compression mold filling', PhD thesis, Department of Mechanical Engineering, University of Illinois, Urbana-Champaign.

Lee, C. C. (1989), 'Reaction and thermal analysis for SMC (Sheet Molding Compound) molding in complicated geometries', *Polymer Engineering and Science*, **29**(15), 1051–1058.

Lee, C. C. and Tucker III, C. L. (1987), 'Flow and heat transfer in compression molding', *Journal of Non-Newtonian Mechanics*, **24**, 245–264.

Lee, L. J. (1981), 'Curing of compression molded sheet molding compound', *Polymer Engineering and Science*, **21**(8), 483–492.

Lem, K. W. and Han, C. D. (1984), 'Thermokinetics of unsaturated polyester and vinyl ester resins', *Polymer Engineering and Science*, **24**, 175–184.

Londono-Hurtado, A., Hernandez-Ortiz, J. P. and Osswald, T. A. (2007), 'Mechanism of fiber–matrix separation in ribbed compression molded parts', *Polymer Composites*, **28**(4), 451–457.

Mallick, P. K. (1990), 'Sheet molding compound'. In Mallick, P. K. and Newman, S. (eds.), *Composite materials technology: Process and properties*. Munich: Hanser Gardner, pp. 25–65.

McCluskey, J. J. and Doherty, F. W. (1995), 'Sheet molding compounds, constituent material forms'. In *Composite Handbook. ASM Handbook Series*, pp. 157–160.

Odenberger, P. T., Andersson, H. M. and Lundström, T. S. (2004), 'Experimental flow-front visualization in compression moulding of SMC', *Composites: Part A*, **35**, 1125–1134.

Osswald, T. A. and Tseng, S. C. (1994), 'Compression molding'. In Advani, S. G. (ed.), *Flow and rheology in polymer composites manufacturing*. Amsterdam: Elsevier, Chapter 10.

Osswald, T. A. and Tucker III, C. L. (1988), 'A boundary element simulation of compression mold filling,' *Polymer Engineering and Science*, **28**(7), 413–420.

Osswald, T. A. and Tucker III, C. L. (1990), 'Compression mold filling simulation for non-planar parts', *International Polymer Processing V*, **2**, 79–87.

Osswald, T. A., Sun, E. M. and Tseng, S. C. (1994), 'Experimental verification on simulating shrinkage and warpage of thin compression moulded SMC parts', *Polymer & Polymer Composites*, **2**(3), 187–198.

Park, C. H. (1998), 'A study on the fiber orientation in the compression molding using fiber reinforced polymer composite material', Master's thesis, Department of Mechanical Engineering, Seoul National University, Seoul, Korea.

Park, C. H., Lee, W. I., Yoo, Y. E. and Kim, E. G. (2001), 'A study on fiber orientation in the compression molding of fiber reinforced polymer composite material', *Journal of Materials Processing Technology*, **111**(1–3), 233–239.

Slotfeldt-Euingsen, D., Magnus, E., Ekern, E., Holtmon, E. and Corneliussen, L. (1986), 'Air entrapment in sheet molding compound (SMC)', *Polymer Composites*, **7**, 431–434.

Smith, K. L. and Suh, N. P. (1979), 'An approach towards the reduction of sink marks in sheet molding compound', *Polymer Engineering and Science*, **19**, 829–834.

Tucker III, C. L. (1987), 'Compression molding of polymers and composites'. In Isayev, A. I. (ed.), *Injection and compression molding fundamentals*. New York: Marcel Dekker, Chapter 7.

Tucker III, C. L. and Folgar, F. P. (1983), 'A model of compression mold filling', *Polymer Engineering and Science*, **23**, 69–73.

White, R. B. (1964), *Premix molding*. New York: Reinhold Book Corp.

Yoo, Y. E. (1997) 'A study on the analysis of compression molding process of composite material structures', PhD thesis, Department of Mechanical Engineering, Seoul National University, Seoul, Korea.

Young, P. R. (1982), 'Thermoset matched die molding'. In Lubin, G. (ed.), *Handbook of composites* (A82–42651 21–24). New York: Van Nostrand Reinhold Co., pp. 391–448.

4

Processing of polymer nanocomposites

J. GOU, J. ZHUGE and F. LIANG, University of Central Florida, USA

Abstract: Polymer nanocomposites processing requires incorporating nanoparticles into polymer matrix in a controllable fashion in order to successfully transfer the outstanding properties of nanoparticles to the final nanocomposite. This chapter first reviews various processing techniques to fabricate nanoparticles reinforced polymer nanocomposites. It then discusses some critical processing-related issues for property improvement, including the selection of nanoparticle and polymer matrix, quality of dispersion, alignment and functionalization of nanoparticles.

Key words: polymer nanocomposites processing, carbon nanoparticles, buckypaper, quality of dispersion, alignment.

4.1 Introduction

Composite material containing at least one phase with constituents of 1–100 nm in size can be termed nanocomposites. Nanoparticles commonly used in the nanocomposite include single-walled carbon nanotube (SWCNT), double-walled carbon nanotube (DWCNT), multi-walled carbon nanotubes (MWCNT), carbon nanofiber (CNF), graphite nanoplatelet (GNP), montmorillonite (MMT), nanoclay and polyhedral oligomeric silsesquioxanes (POSS). Other nanoparticles, such as SiO_2, Al_2O_3, TiO_2 and nanosilica are also used in the nanocomposite. The potential benefits of the nanoparticles for structural and multifunctional nanocomposites are summarized below.

- Improved mechanical properties. The addition of small size and low loading of nanoparticles will enhance the matrix-dominated properties such as stiffness, fracture toughness and interlaminar shear strength of conventional fiber-reinforced composites.
- Good electrical conductivity at a lower loading level. CNT, CNF or GNP have proven to be an excellent additive to impart electrical conductivity in the nanocomposite at a lower loading level compared to carbon black, chopped carbon fiber or stainless steel fiber.
- Better flame retardancy. The addition of CNT, CNF, GNP, nanoclay or POSS will promote the char formation, increase the char strength,

reduce the heat release rate of the polymer matrix, and help conduct heat away from the flame zone.

- Smooth surface finish. Nanoparticles have less effect on the part surface quality due to their small size and low loading level.
- Lower part warpage. Nanoparticles are much smaller than other particulate additives, thus are more insensitive to shear. The random distribution results in a reduced chance of part warpage.

4.2 Process description

During the past few years, nanoparticles have been incorporated into the polymer matrix to develop a polymer nanocomposite. Although processing techniques of the nanocomposite are inherently different, all of them have to address the following issues that will directly affect the properties of the nanocomposite: dispersion, alignment and functionalization. Common processing techniques of polymer nanocomposites include:

- Melt-mixing. The melt-mixing of nanoparticles and thermoplastic polymers has been widely used through conventional processing techniques, such as extrusion, internal mixing, injection molding and blow molding due to the speed, simplicity and availability of these processes. These methods are beneficial because they are free of solvents and contaminants, which are present in solution processing and *in situ* polymerization.
- *In situ* polymerization. This method has been used to improve the dispersion and integration between the phases. The main advantage of this process is that the reinforcing effect can be obtained at a molecular scale.
- Solution processing. The solution-based method provides an advantage of low viscosity, which can facilitate mixing and dispersion. In addition, the pre-processing of nanoparticles is necessary to prepare the material for processing at a macroscopic scale. The pre-processing includes purification to eliminate the impurities, de-agglomeration for dispersing individual nanoparticles, and chemical functionalization to improve the nanoparticle–polymer interaction and eventually the property enhancement.

4.2.1 Solution processing

Figure 4.1 shows the schematic of solution processing. Nanoparticles are first dispersed in a solvent or polymer solution. The nanoparticle/polymer/solvent solution is mixed by energetic agitation such as magnetic stirring,

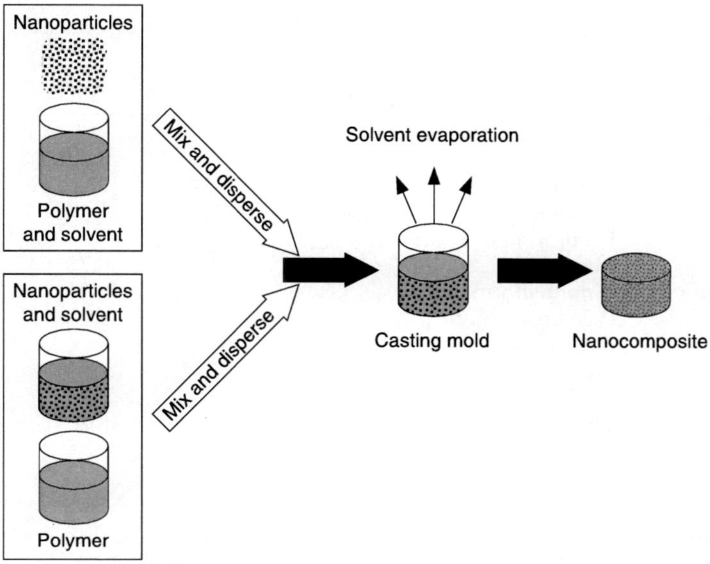

4.1 Schematic of solution processing.

high shear mixing, reflux or sonication. For the sonication, there are two types of equipment: high-power sonication using a tip or horn and mild sonication in a water bath. Nanoparticles usually cannot be dispersed well in most of solvents due to their exceptionally high aspect ratios. The quality of the dispersion can be improved by treating the nanoparticles with some surfactants.[1] After the nanoparticles are homogeneously dispersed, the polymer solution is poured into a mold. The solvent is then evaporated, leaving the composite film or sheet upon the completion of evaporation.[2] Solution processing is commonly used to prepare the nanocomposites with a film or sheet like structure. The selection of solvents is based on the solubility of polymers. Table 4.1 shows the properties of different types of solvents.

Jin *et al.* conducted the experiments to disperse high concentration of MWCNTs into the polymer by solution processing.[3] In their study, MWCNTs produced by an arc discharge were dispersed in chloroform by sonication. The polymer matrix, polyhydroxyaminoether (PHAE) was dissolved into the MWCNT–chloroform solution during the sonication process. The well-dispersed suspension was poured into a Teflon mold and allowed to dry in a fume hood. The solution processing enabled them to prepare the nanocomposite containing 50 wt% of MWCNTs with good quality. Another research group compared the quality of dispersion and mechanical properties of the nanocomposites prepared by different types of processing methods.[4] During the direct mixing, expanded graphite (EG) was dispersed in the hardener of epoxy resin (DGEBA) by a magnetic stirrer. The epoxy

Table 4.1 Properties of typical solvents

Solvent	Boiling point (°C)	Dielectric constant	Density (g/mL)
Non-polar solvents			
Pentane	36	1.84	0.626
Toluene	111	2.38	0.867
Benzene	80	2.3	0.879
Polar aprotic solvents			
Acetone	56	21	0.786
Dimethylformamide	153	38	0.944
Dichloromethane	40	9.1	1.3266
Polar protic solvents			
Water	100	80	1
Methanol	65	33	0.791
Ethanol	79	24.55	0.789

resin was then added into the EG/hardener solution and the mixture was stirred on a hot plate. An accelerator was added to the mixture and stirred with slow agitation followed by degassing. The nanocomposite was made by casting the final solution into a Teflon mold. During the solution processing, EG was first dispersed into acetone by sonication. The DGEBA epoxy resin was introduced into the suspension and sonicated. The solution was stirred on a hot plate until acetone was completely evaporated. The hardener and accelerator were then added, followed by degassing, casting and curing process. The nanocomposite samples prepared with these two different processes were examined by optical microscopy. The experimental results indicated that the quality of the nanocomposite prepared by the solution processing was much better than the nanocomposite prepared by the direct mixing.

Solution processing has been extensively used to prepare clay-based nanocomposites. Most natural clays, known as Bentonite, originates from the alternation and/or deposition of volcanic ash. They can also be obtained by the hydrothermal alteration of volcanic rocks. The main content of natural clay is montmorillonite which is dominated with silica. The montmorillonite essentially has a layered structure, which contains a tetrahedral silicate layer and an octahedral alumina layer. In the silicate layer, a hexagonal network is formed by linking SiO_4 groups and composing a repeating unit of Si_4O_{10}. In the alumina octahedral layer, aluminum atoms are embedded into the center of gallery formed by two layers of closely packed oxygen or hydroxyl. The octahedral layer shares their apex oxygen with the tetrahedral silicate layer and they form a sandwich structure. In its natural state, the montmorillonite clay contains Na^+ cation. The thickness of one sandwich layer is approximately 0.96 nm. The chemical formula of the montmorillonite clay is

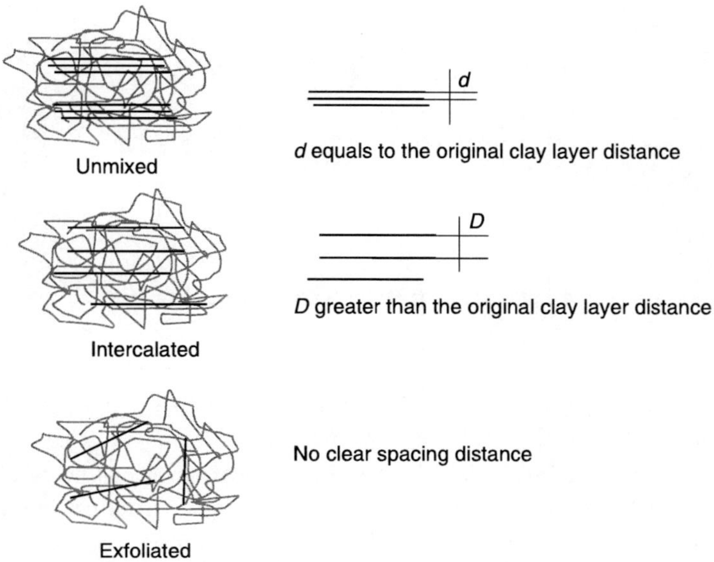

4.2 Structures of clay/polymer composites.

$Na_{1/3}(Al_{5/3}Mg_{1/3})Si_4O_{10}(OH)_2$. Because of the hydrophilic property of natural clay platelets, they are usually organically modified prior to preparing the nanocomposite. Organic treatment is typically accomplished by exchanging inorganic cations (such as Na^+) with the desired organic cations. A number of types of organically modified clay can be developed. For example, Cloisite 20A supplied by Southern Clay Products, Inc. is the natural montmorillonite modified with a quaternary ammonium salt. This product is designed to improve the reinforcing effect, coefficient of linear thermal expansion, heat deflection temperature and barrier properties of plastics.[5]

By using an adequate amount of polymer soluble solvent or a thermosetting reactive pre-polymer for insoluble polymers such as polyimide, nanoclay can be dispersed as individual layers due to the fact that the forces holding these layers together are weak. The polymer matrix is added to the solution, being absorbed into the gaps between the layered silicates. After the solvent is evaporated or the mixture is precipitated, the layered silicates reassemble, sandwiching the polymer to form an ordered, multi-layered structure. Two types of structures can be obtained, namely intercalated and exfoliated, as shown in Fig. 4.2. The intercalated structure is obtained by self-assembled, well-ordered multi-layered structures. Between the gallery spaces (2–3 nm) of those parallel individual silicate layers, extended polymer chains are inserted. When the spaces between individual silicates are large enough so that the interactions between gallery cations of the adjacent layers can no longer be maintained, exfoliated or delaminated nanocomposites

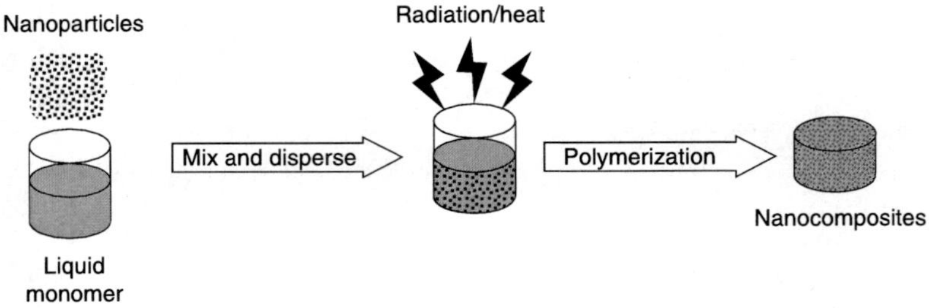

4.3 Schematic of *in situ* polymerization process.

will be formed. The interlayer spacing can be of the order of the gyration radius of the polymer. [6]

The unmixed, intercalated or exfoliated structures can be confirmed by X-ray diffraction (XRD) analysis. The unmixed structure has the same peak as the original nanoclay. However, the peaks for an intercalated structure appear earlier than the original nanoclay. And there is no peak for an exfoliated structure. Rao and Pochan studied the mechanics of clay/polymer nanocomposites prepared by solution processing.[7] 3 wt% of montmorillonite clay (Cloisite Na+) was dispersed in water by vigorous mixing and centrifuging. This process was also expected to remove impurities. The colloidal aqueous remained without settlement for several months. A copolymer latex ($PBSMaSO_3$), which was synthesized by emulsion polymerization, was mixed with the clay aqueous. After the well-dispersed solution was obtained, it was poured into a Teflon cast overlaid by Kapton film and dried at an ambient temperature. After the nanocomposite film was formed, it was peeled off and dried in a vacuum oven. The quality of $PBSMaSO_3$-clay nanocomposite was evaluated by XRD, which suggested that if the loading level of nanoclay was less than 4 wt%, an exfoliated structure was obtained. However, the increase in the loading level of clay (e.g., > 4 wt%) could result in an intercalated structure in the nanocomposite.

4.2.2 *In situ* polymerization process

Figure 4.3 shows the schematic of *in situ* polymerization process. During the *in situ* polymerization process, nanoparticles such as CNTs and layered silicates are first dispersed in a liquid monomer. The polymerization reaction is initiated either by heat or radiation, by the diffusion of a suitable initiator, or by an organic initiator or by catalyst fixed on the surface of nanoparticles.[1] Upon the completion of polymerization, polymer molecules are either wrapped around or covalently bonded to the nanoparticles depending on the surface functionality of the nanoparticles and the polymer being formed.[2]

In situ polymerization can be used to prepare the nanocomposite containing insoluble and thermally unstable polymer, which cannot be processed by solution or melting process. Ring-opening, radical, anionic and chain transfer metathesis polymerizations are used, depending on the required molecular weight and molecular weight distribution of the polymers. The advantages of the *in situ* polymerization process include enabling the grafting of polymer macromolecules onto the surface of nanoparticles, allowing the preparation of the nanocomposite with high loading level of nanoparticles and good miscibility with the polymer.[8] The first commercial application of the nanocomposites (clay/polyamide-6) prepared by this process was developed by Toyota Motor Corp in the early 1990s.[9] Zeng and Lee prepared poly (methyl methacrylate) and polystyrene/clay nanocomposites by *in situ* polymerization.[10] They found that the compatibility of the initiator and monomer with the surface of the clay was a critical factor to affect the dispersion of clay. They claimed that a combination of more polar, less hydrophobic monomer and initiator was preferred in order to obtain high quality of dispersion. Additionally, by introducing polymerizable groups onto the surface of the clay, the dispersion quality was further improved. Exfoliated PMMA and PS/clay nanocomposites were successfully synthesized with a clay concentration of 5 wt%. Kim *et al.* used this process to prepare MWCNT/high density polyethylene (HDPE) nanocomposites.[11] They first attached a metallocene catalyst complex onto the wall of the nanotubes. The polymer branches were then generated on the walls by the surface-initiated polymerization. Through transmission electron microscopy (TEM) and scanning electron microscopy (SEM) analysis, they found that the nanotubes were uniformly dispersed and wrapped by PE molecules. Yang *et al.* studied the dispersion of silica into polyamide 6 by *in situ* polymerization.[12] The silica nanoparticles were modified with aminobutyric acid. They were dispersed in ε-caproamide by stirring. The mixture was polymerized at a high temperature in a nitrogen atmosphere. Their study showed that the modified silica was dispersed more homogeneously than the pristine silica. This process was very useful in preparing an inorganic/organic nanocomposite, which can avoid the agglomeration of inorganic particles and improve the interfacial interactions between the inorganic components and polymer.

4.2.3 Melting process

The melting process is an alternative to deal with the polymer matrix that is insoluble. It is particularly useful in preparing the nanocomposites composed by thermoplastic polymers such as HDPE, polyamide-6,

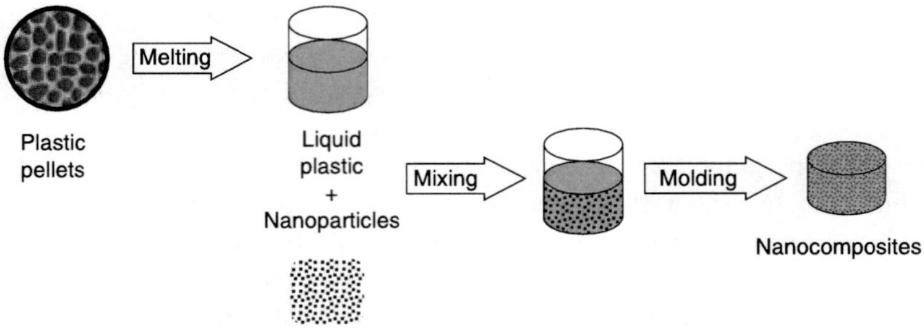

4.4 Schematic of melting processing.

polycarbonate, polypropylene, polystyrene, etc. This process allows a high volume of nanoparticles to be mixed into polymer matrix. It is based on the fact that thermoplastic polymers become soft when being heated up while the properties of polymers remain the same after they are cooled down. In other words, amorphous polymers and semi-crystalline polymers can be processed above their glass transition temperature and melting temperature, respectively. Figure 4.4 shows the schematic of the melting process. During the process, polymer pellets are melted to form a viscous liquid. The nanoparticles are then blended into the liquid polymer by a high shear mixer or in an extruder. The final bulk nanocomposite samples can be obtained by using injection molding, compression molding or extrusion.

The addition of nanoparticles into the melting polymer will affect its viscosity which probably lead to the unexpected polymer degradation under high shear condition. Therefore, processing conditions should be optimized for the whole range of polymer/nanoparticle combinations.[1] Due to its simplicity and speed, this process has been widely applied for large-scale industry production. When the clay-based nanocomposite is prepared with this method, the polymer molecules can enter into the interlayer space between clay particles. The diffusion process tends to peel the clay layers away. Depending on the compatibility between the clay and polymer matrix, and the processing conditions, either an intercalated or an exfoliated structure can be achieved. Cho and Paul prepared Nylon 6/clay nanocomposites using two different types of extruders: single and twin screw extruders.[13] The Killion single screw extruder has a diameter of 25.4 mm. The compounding process was carried out at 240°C using a screw speed of 40 rpm and a residence of 2.35 min. The Haake intermeshing co-rotating twin screw extruder has a diameter of 30 mm and centerline spacing of 26 mm. The barrel temperature during compounding process was 240°C but the screw speed was 180 rpm with a residence time of 5.3 min. The extruded pellets were then injected into an 80°C mold under an injection pressure of 70 bars and a holding pressure of 35 bars. The microstructures of the nanocomposite were examined by XRD, which indicated that the interlayer spacing of the clay was 18 Å. The exfoliated structure nanocomposite was obtained by using a twin screw extruder. The relatively poor dispersion of the nanocomposite using a single screw extruder probably was due to the insufficient shear and residence time. Thostenson and Chou prepared CVD-grown MWCNT/polystyrene nanocomposites.[14] Since the nanotubes tended to stick to the walls of the mixer, it was difficult to disperse them into the polymer through the melting process alone. A technique that combined both the solution method and melting process was developed. They first dispersed the nanotubes into polystyrene/tetrahydrofuran solution. The solution was then cast into a Petri dish and sonicated as the solvent was evaporated. The purpose of the sonication was to ensure that the nanotubes were dispersed at the microscale rather than the nanoscaled so that they should be encapsulated in the polymer after the solvent was evaporated. The dried nanotubes/polystyrene composites were processed by melting compound. By drawing the sample direct from the extruder, the nanotubes could be aligned. The good dispersion of the nanotubes was found from TEM images.

4.2.4 Carbon nanotube paper-based nanocomposites

There are various methods to manufacture carbon nanotube paper. The manufacturing process can be divided into two categories: the one-step method and the two-step method. For the two-step method, carbon nanotubes are synthesized first and the paper is made from the nanotubes. In the one-step method, carbon nanotube paper is grown directly. The two-step method is the main process to make carbon nanotube paper, including vacuum/pressure filtration, spin-coating and domino pushing method.

Vacuum/pressure filtration

The filtration method is a process in which the suspension of carbon nanotubes is filtrated through a filter paper to make a free-standing non-woven mat.[15] The entire process can be divided into four steps: (1) preparation and purification of carbon nanotubes; (2) uniform dispersion of carbon nanotubes into a solvent to form a stable suspension; (3) filtration of the suspension of carbon nanotubes under vacuum or pressure and (4) drying the nanotube paper in a vacuum oven. The advantage of this method is low cost and easy manufacturing. However, the high degree of curviness and easy aggregation of the nanotubes may result in the brittleness of the nanotube paper.

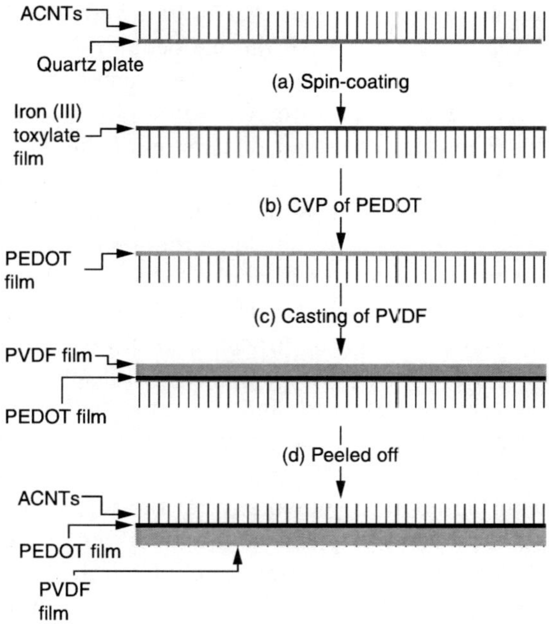

4.5 Schematic of spin-coating process. (From reference 16.)

Spin-coating method

A uniform carbon nanotube paper is produced by spin-coating from a highly concentrated nanotube dispersion, which is stabilized by chemical oxidation.[16] In this process, the solution of purified carbon nanotubes is poured into a high speed spinning glass panel. The thickness of the nanotube paper can be controlled by changing the concentration of the nanotubes and the speed of spinning. High quality nanotube paper can be obtained with uniform surface quality and reduced thickness, which lead to potential applications in the field of solar batteries. Aligned MWCNTs are produced and an aligned CNT-based membrane electrode is used as the anode material in a rechargeable lithium-ion battery. Figure 4.5 shows the procedure of the spin-coating method.

Domino pushing method

A simple but effective macroscopic manipulation of carbon nanotube arrays called 'domino pushing' has been developed to make thicker aligned carbon nanotube papers.[17] The schematic of the domino pushing method is shown in Fig. 4.6. First, the nanotube array is covered with a piece of micro-porous membrane. The nanotubes in the array are then

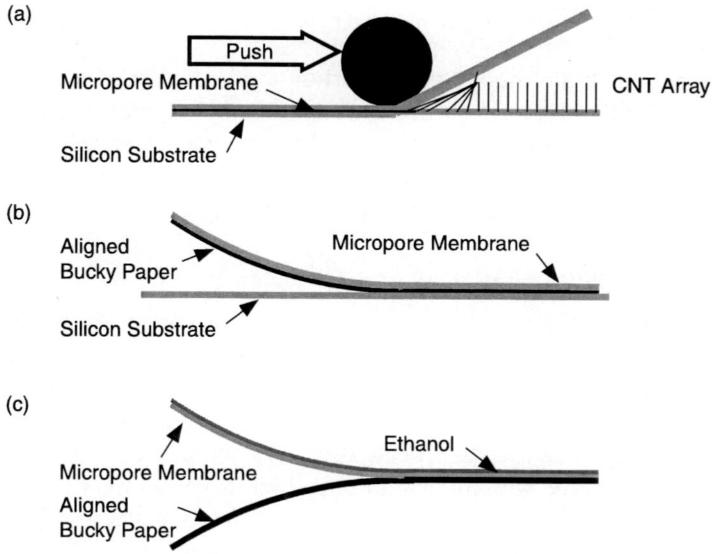

4.6 Schematic of domino pushing method. (From reference 17.)

forced down to one direction by pushing a cylinder, which is placed on the nanotube array with a constant pressure. All the nanotubes in the array are attracted together due to strong van der Waals forces and form an aligned carbon nanotube paper. The aligned carbon nanotube paper is then peeled off the silicon substrate with the membrane. Ethanol is spread on the micro-porous membrane and permeates through the membrane, and then the aligned nanotube paper can be peeled off from the membrane easily. The curviness and aggregation of the nanotubes can be avoided with this method; however, any impurities stay with the nanotubes, and the domino pushing method is more complicated than the filtration and spin-coating process.

Direct synthesis

In this method, carbon nanotube paper is synthesized *in situ* by using trichlorobenzene (TCB) as a precursor. There are some reports on the one-step method in which carbon nanotube paper is grown by the CVD method. Ferromagnetic carbon nanotube paper filled with Fe nanowires (Fe-CNT) was synthesized *in situ* without using any surfactant or acid treatment.[18] The one-step method is much simpler than the two-step method. The organic solvent can be avoided in the one-step process. While, good mechanical properties are obtained. However, high impurity content limits the applications of carbon nanotube paper made by the one-step method.

4.2.5 Graphene paper-based nanocomposites

Graphene is a flat monolayer of carbon atoms tightly packed into a two-dimensional honeycomb lattice. Since graphene has been discovered, it has gained great attention and a number of journal papers have been published on graphene. Graphene conducts electrons extremely well and is known to be exceptionally strong and tougher than diamond.

Vacuum filtration

The first step to fabricate a graphene paper is the exfoliation of graphite. The main methods to exfoliate the graphite are the micromechanical cleavage and chemical exfoliation of the graphene oxide (GO). Andre Geim *et al.*[19] first employed the mechanical exfoliation of graphite to obtain graphene. Cohesive tape was used to repeatedly cleave graphite crystals into less layer graphite. The tape with attached optically transparent flakes was dissolved into acetone. After a few further steps, the flakes, including the monolayers, were sedimented on a silicon wafer; individual atomic planes were then observed in an optical microscope. Ruoff *et al.*[20] developed a solution-based method involving the chemical oxidation of graphite to hydrophilic graphite oxide, which can be readily exfoliated as individual graphene oxide sheets by ultrasonication in water.

The next step is to fabricate the graphene paper from the graphene oxide colloids. The graphene oxide colloids can be directly filtrated with an aid of vacuum. Drzal *et al.* made the graphene paper through a vacuum filtration[21] with an ANODISC membrane. A controlled multi-filtration process (5 mL at a time) was used to minimize the disordered layering structure caused by water flow. The graphene paper was suction dried and then dried in a vacuum oven overnight prior to peeling the graphene paper off the membrane. The free-standing paper was thermally annealed in a furnace, and further cold-pressed at room temperature. The annealing process improves the thermal conductivity by decomposing the PEI molecule that is adsorbed on the graphene particles while still maintaining the porosity of the graphene paper. Mechanically compressing the sample effectively reduces the pore volume of the graphene paper and increases contact area among individual sheets. Due to the strong effect of alignment and larger contact area of the annealed graphene paper, the in-plane thermal conductivity is increased by 80% and the through-plane conductivity is reduced by 10%.

Electrophoretic deposition

An electrophoretic deposition (EPD) process has been developed to fabricate the graphene paper.[22] The film is composed of overlapped and stacked platelets of reduced graphene oxide. Figure 4.7 shows the experiment setup of the

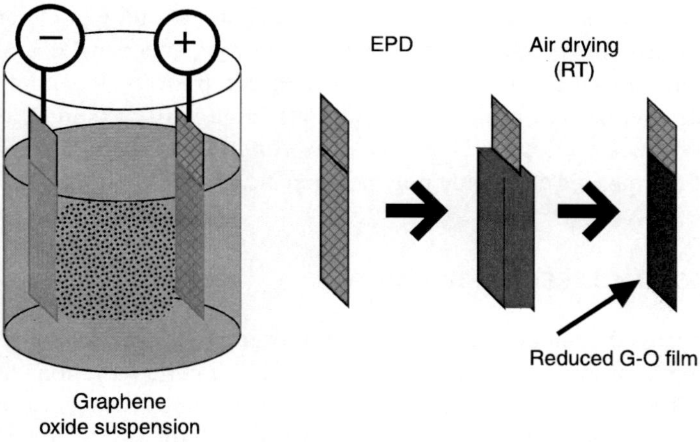

4.7 Schematic of electrophoretic deposition process. (From reference 22.)

EPD-GO film. GO was dispersed in water and sonicated at room temperature. After the preparation of colloidal suspensions of individual GO platelets in purified water, EPD-GO was deposited on stainless steel and various other electrically conductive substrates. Typical concentrations of GO and applied direct current (DC) voltage were 1.5 mg/mL and 10 V, respectively; the deposition time was in the range of 1–10 min. The GO platelets migrated toward the positive electrode when a DC voltage was applied. The deposition rate depended on several factors including the concentration of the GO suspension, the applied DC voltage and the conductivity of the substrate. A smooth film was deposited on the stainless steel positive electrode in a short time period with 10 V of applied potential. After the deposition, the samples were air dried at room temperature. By varying the current and time, films with a thickness in the range between several hundreds of nanometers and tens of micrometers could be deposited.

4.3 Methods to improve process

In the past few years, significant progress has been achieved to develop carbon nanotube-based nanocomposites. Despite these successes, several critical processing issues must be solved before the full potential of carbon nanotubes is realized in the nanocomposite.

4.3.1 Selection of polymer matrix

The polymer matrix should be properly chosen before processing of carbon nanotube/polymer composites. Both thermosetting and thermoplastic polymers have been used as matrix materials in carbon nanotube/polymer composites. Research reports are conflicting regarding mechanical and functional properties of carbon nanotube/polymer composites. Depending on the polymer matrix and processing conditions, large variations in their measured properties are found. Therefore, the properties of carbon nanotube/polymer composites are highly polymer-specific.

4.3.2 Selection of carbon nanotubes

In order to process the nanocomposite with optimized properties, the selection of carbon nanotubes plays a crucial role due to the large variations of the material. The size of the nanotubes such as their diameter, aspect ratio and their statistical distribution has significant impact on the dispersion of carbon nanotubes. It is well known that when the aspect ratio of the nanotubes increases, they become more difficult to disperse. The impurity of the nanotubes can greatly affect the multi-functionalities of the nanocomposite. It is desirable to have the nanotubes with low impurity. However, the cost of such material will be much more expensive. Other aspects of carbon nanotubes such as their chirality, defect type and density, thermal/electrical conductivity and the number of walls (single-walled, double-walled or multi-walled) also play an important role in determining the properties of the nanocomposite. For example, Young's modulus of SWNT, armchair SWNT, zigzag SWNT, chiral SWNT and MWNT are 1~5, 0.94, 0.94, 0.92 and 0.2~0.95 TPa, respectively. Therefore, the reinforcing effect of carbon nanotubes varies from the microstructural characteristics of the nanotubes in the nanocomposite.

4.3.3 Dispersion of carbon nanotubes

Due to the van der Waals interaction, the nanotubes tend to aggregate to form bundles or ropes and further agglomerate when dispersed in the polymer matrix. The high surface area of the nanotubes also results in a high viscosity of the nanotube/polymer mixture particularly when fabricating the nanocomposite with high loading level of the nanotubes, which makes the dispersion of the nanotubes extremely difficult. The dispersion of carbon nanotubes in the polymer matrix can be made by several techniques.

Sonication

Sonication is widely used in the laboratory to disperse nanotubes into the polymer matrix. This process utilizes ultrasound energy to agitate nanoparticles in the polymer matrix. It is usually carried out by an ultrasonic bath or a horn/probe which is also known as the sonicator. During the sonication process, the ultrasound propagates by a series of compression. When it passes through the polymer medium, attenuated waves are induced, promoting the 'peeling off' of the CNTs located at the outer parts of the nanoparticle bundles/agglomerates. As a result, individual nanoparticles are separated and high quality dispersion can be achieved.[23] However, the sonication process is only suitable to disperse nanotubes in solutions that have a very low viscosity, such as water, acetone and ethanol. But most polymers are viscous and it is necessary to dissolve or dilute the polymer prior to the dispersion process. When the duration of the process is longer and the intensity of the input energy is higher, better dispersion quality can be obtained. However, such aggressive treatment could also seriously damage the structure of nanotubes, especially when a probe sonicator is used. For example, the graphene layers of carbon nanotubes could be completely destroyed and the particles become amorphous carbon nanofibers.[24] Consequently, such induced damage would deteriorate the electrical, thermal and mechanical properties of the nanocomposite.

Ball milling

Ball milling is a grinding method that grinds nanotubes into extremely fine powders. During the ball milling process, the collision between the tiny rigid balls in a concealed container will generate localized high pressure. Usually, ceramic, flint pebbles and stainless steel are used.[25] In order to further improve the quality of dispersion and introduce functional groups onto the nanotube surface, selected chemicals can be included in the container during the process. The factors that affect the quality of dispersion include the milling time, rotational speed, size of balls and balls/nanotube amount ratio. Under certain processing conditions, the particles can be ground to as small as 100 nm. This process has been employed to transform carbon nanotubes into smaller nanoparticles, to generate highly curved or closed shell carbon nanostructures from graphite, to enhance the saturation of lithium composition in SWCNTs, to modify the morphologies of cup-stacked carbon nanotubes and to generate different carbon nanoparticles from graphitic carbon for hydrogen storage application.[25] Even though ball milling is easy to operate and suitable for powder polymers or monomers, process-induced damage on the nanotubes can occur.

Shear mixing, extrusion and calendaring

While the sonication and ball milling processes may occasionally introduce damage to the nanotubes' structure, there are alternatives to disperse the nanotubes without damage. The alternative processes include shear mixing, extrusion and calendaring which is also known as the three-roll mills. Shear mixing is commonly used in the laboratory to disperse the nanotubes into the polymer matrix. It is well known that the size and shape of the propeller and its rotational speed determine the dispersion quality. However, it was observed that for some thermosetting polymers, the re-agglomeration of the nanotubes becomes spontaneous under static condition.[26] In this case, a much higher mixing speed is required. The extrusion process which is usually carried out by a twin screw is available in the industry field for large-scale production. The process is only suitable to mix the nanotubes with thermoplastics. Factors that influence the dispersion of the nanotubes include environmental temperature and the configuration and rotational speed of the screw. The calendaring process utilizes the shear force created by rollers to mix, disperse or homogenize the nanotubes in the liquid polymers, oligomers or monomers. Factors such as the rotational speed of rollers and the distance between adjacent rollers significantly affect the quality of dispersion. Under some conditions, the alignment of the nanotubes in the polymer matrix can occur.[25]

4.3.4 Alignment of carbon nanotubes

The properties of carbon nanotubes are not only sensitive to their diameter and chirality but are also highly anisotropic. The stiffness, strength and electrical, magnetic and optical properties of the nanotube-reinforced composites strongly depend on the orientations of the nanotubes.[27] Therefore, it is crucial to obtain the controllable alignment of the nanotubes in the nanocomposite.

Ex situ *alignment of carbon nanotubes*

The term '*ex situ*' means that the alignment of carbon nanotubes has been achieved before fabricating the nanotube/polymer nanocomposites. This method is adopted by the *in situ* polymerization process during which the placements of the nanotubes will not be disturbed. One can obtain aligned nanotubes by a filtration process during which the nanotubes are first dispersed in solutions such as water, ethanol and acetone. Then, they are aligned by high-pressure filtration through a selected medium.[28,29] Another alignment method is to use the so-called plasma-enhanced chemical vapor deposition process. In this process, the nanotubes are synthesized in an aligned direction.[30,31] The nanotubes can be aligned by manipulating the template used in the chemical vapor deposition process. In most cases, porous substrates are required and carbon nanotubes are grown within the pores.[32,33] Yanagi *et al.* reported self-oriented short SWCNTs were deposited on highly oriented pyrolytic graphite that served as a template.[34]

Induced alignment of carbon nanotubes

There are many ways to induce the alignment of carbon nanotubes, one of which is to utilize a force field such as shear force. Ajayan *et al.* developed a simple method to align the nanotubes through the flow-induced anisotropy (the nature of rheology in the composite media). However, their approach is only suitable to fabricate a thin film because the alignment effect becomes less pronounced if the thickness of composite slice is increased.[35] Vigolo *et al.* were able to fabricate meter-long ribbons by injecting low viscous SWCNT solution into a high viscous PVA solution using a syringe.[36] Because of the difference in viscosity, the shear force was induced and the nanotubes were rapidly stacked together as they were injected out from the syringe needle. Other process for shear force-induced alignment can be achieved through stretching or extrusion.[27]

The alignment of carbon nanotubes can be achieved in a magnetic field. The degree of alignment can be customized by controlling the intensity of the magnetic field. Kimura *et al.* utilized a high magnetic field to prepare an anisotropic carbon nanotube/polymer nanocomposite.[37] In their study, a constant magnetic field of 10T was applied to MWCNT/unsaturated polyester solution. The solution was polymerized under a magnetic field so that the aligned nanotubes were frozen in the polyester matrix. Similar to the principal of the magnetic field induced alignment, the nanotubes can also be aligned by electrospinning. In the electrospinning process, there is a high DC voltage generated between a negatively charged carbon nanotube/polymer fluid and a positive metallic fiber collector. The nanotubes are initially randomly oriented. Because of the 'sink-like' flow that has been generated in a wedge-like electrospinning jet, the nanotubes can be gradually aligned along the streamlines.[38] Carbon nanotubes can also be aligned in an electric field. Chen *et al.* tried to align SWCNT with an alternating-current electric field. In their study, the nanotubes were first dispersed in ethanol. They prepared Si/SiO_2 substrates on which gold interdigitated electrodes were deposited in vacuum. The substrates were immersed in the SWCNT suspension and the AC electric field was applied to the electrodes. The suspension was dried in air. They found that the alignment efficiency significantly depended on the frequency and magnitude of the electric field.[39]

In addition to force-induced alignment, carbon nanotubes can be aligned by the geometric constraints (created by a liquid crystalline phase). Due to the unique properties of the liquid crystals, their geometry can be easily modified by the force, magnetic and electric field. For example, Lynch *et al.* aligned

MWCNTs by using a liquid crystal matrix that had been oriented in an electric field.[40]

4.3.5 Functionalization of carbon nanotubes

An efficient load transfer from the polymer matrix to the nanotubes is required to take advantage of the very high Young's modulus and strength of carbon nanotubes in the nanocomposite. Unlike conventional fiber-reinforced polymer composites, large interfacial areas are available for load transfer in nanotube/polymer composites due to high aspect ratio of the nanotubes. However, a lack of strong interfacial bonding between the nanotube and polymer matrix causes the nanotube to be pulled out under stress and leads to inefficiency of load transfer. Functionalization is employed to unbundle nanotubes and increase the interfacial bonding between the nanotube and polymer matrix. The methods of functionalization of nanotubes include covalent and non-covalent. Covalent functionalization is used to link some functional groups onto the nanotubes by covalent bonds. The functional groups can be added on the defect sites which can be on the sidewalls or end caps. Non-covalent functionalization is based on the attractive force between the supramoleculars, such as van der Waals force. The wrapping of carbon nanotubes with surfactants is a good example of non-covalent functionalization.

Covalent functionalization of carbon nanotubes

The covalent functionalization of SWCNTs can occur at the ends, defect sites and sidewalls of the nanotubes. Compared to the sidewalls of the nanotubes, the end caps and defect sites are highly reactive.[41] The typical defects include the open ends, sp^3–hybridized defects and pentagon–heptagon pairs. The first step of the reactions of end and defect functionalization is the oxidization of SWCNTs with strong acids such as HNO_3, H_2SO_4 and $KMnO_4$. The oxygenated functional groups are substituted with different types of molecular moieties, such as metal colloids;[42,43] protein also can be associated with SWCNTs for biological applications. Dekker *et al.* attached DNA onto the end of the nanotubes.[44] With this method, low toxicity nanostructures can be achieved to use as biocompatible transporters.

The covalent sidewall functionalization of SWCNTs has been conducted by a number of researchers. Highly reactive addends are needed for the sidewall functionalization reactions. The outer surface of SWCNTs is more active than the inner surface. There are several methods for covalent sidewall functionalization of SWCNTs, including fluorination, diazonium, radical chemistry, nitrene, azomethine ylides and dichlorocarbene, as shown in Table 4.2. The added functional groups include aryl, fluorine, pyrrolidine, aziridine and cyclopropane. Multiple analytical techniques can be used to

Table 4.2 Common methods of SWCNT sidewall functionalization

Method	Addend	Degree of functionalization	Highest solubility
Diazonium	Aryl	1 addend in every 10 carbons in SDS/water	0.8 mg/mL
Fluorination	Fluorine	1 addend in every 2 carbons	1 mg/mL in 2-propanol
Radical chemistry	Alky	1 addend in every 6 carbons	–
Azomethine ylides	Pyrrolidine	1 addend in every 100 carbons	50 mg/mL in $CHCl_3$
Nitrene	Aziridene	1 addend in every 50 carbons	1.2 mg/mL in DMSO
Dichlorocarbene	Cyclopropane	1 addend in every 25 carbons	–

Source: From Reference 45.

characterize the covalently functionalized SWNTs: Raman spectroscopy is used to verify that the functionalization covalently occurs at the sidewall; thermogravimetric analysis (TGA) and X-ray photoelectron spectroscopy (XPS) can be employed to determine the degree of functionalization; and atomic force microscopy (AFM), SEM and TEM are used to analyze the diameter of the bundles or individual nanotubes.

Fluorination of carbon nanotubes

Due to the high reactivity of fluorine, fluorination was the first option which was used for sidewall functionalizing carbon nanotubes.[43] Hamwi *et al.* carried out the fluorination reaction of nanotubes at 500°C in a fluorine atmosphere for 4 h. Carbon nanotubes were fluorinated and the C-F covalent bond was confirmed by an infrared spectroscopy. Margrave *et al.* reported extensive controlled and non-destructive sidewall fluorinations of SWCNTs using elemental fluorine as the fluorinating agent.[44] 'Fluorocarbon single-walled carbon nanotubes' (F-SWCNT) of approximately C_2F product stoichiometries was obtained at temperatures from 150°C up to 500°C. The fluorinated SWCNTs can be regarded as an exfoliation of nanotube ropes and bundles into individual nanotubes. The major downside of this functionalization is that the high degree of fluorine atoms addition could lead to a great number of defects and damage to the nanotubes. However, the solubility of the fluorinated SWCNTs is significantly increased and can serve as a precursor for a wide variety of sidewall chemical functionalization.

Diazonium of carbon nanotubes

Diazonium coupling is used for the covalent functionalization of carbon nanotubes due to its simplicity and versatility. This method gives the most highly functionalized SWCNTs by adding preformed arene diazonium salts to micelle-coated carbon nanotubes at pH of 10.[45] SWCNTs functionalized with this method can have approximately 1 in 10 carbons on the sidewalls bearing an aryl moiety. The nanotubes have significant decrease in the bundling; therefore, functionalized SWCNTs are easily dispersed as individuals in an organic solvent. Compared to pristine carbon nanotubes, the solubility in DMF of functionalized nanotubes is dramatically increased from 0.07 to 0.8 mg/mL. Such a profound increase in solubility allows the functionalized nanotubes to be solvent blended into the host polymer and achieve good dispersion in the polymer matrix.

Nitrene of carbon nanotubes

Sidewall functionalization of SWCNTs can be achieved by the method of addition of reactive alkyloxycarbonyl nitrenes obtained from alkoxycarbonyl azides. Nitrenes attack the nanotube sidewalls in a [2 + 1] cycloaddition by the way of thermally induced N_2-extrusion, and form an aziridine ring at the nanotube's sidewalls.[46] Nitrene additions led to considerable amount in an organic solvent. The highest solubility of 1.2 mg/mL was achieved for SWCNT adducts with nitrenes containing crown ethers of oligoethylene glycol moieties in DMSO and TCE.[47] The electronic properties of SWCNTs were mostly retained after functionalization, which revealed about 2 wt% functional groups were added onto the carbon nanotube sidewalls.[48]

Carbene addition of carbon nanotubes

During the study of organic functionalization of carbon nanotubes, Haddon *et al.* discovered that dichlorocarbene was covalently bound to soluble SWCNTs.[49] There are two resources to generate the carbine, including chloroform with potassium hydroxide[50] and phenyl (bromodichloromethyl) mercury.[49] The shortcoming is the low degree of functionalization, with the results showing chlorine atom amounts of only 1.6 wt%.[51] Furthermore, due to impure starting material and a large amount of amorphous carbon, the sites of reaction could not be determined.[50]

Non-covalent functionalization of carbon nanotubes

Compared to covalent functionalization, the main advantage of non-covalent functionalization is the preservation of the nanotube's structure by preventing disruption of the intrinsic sp^2 structure and conjugation while significantly improving their solubility. Zhao *et al.* reported extensive research on non-covalent functionalization, including (i) aromatic small molecule based non-covalent functionalization, (ii) biomacromolecule based non-covalent functionalization and (iii) polymer based non-covalent functionalization.[52] Aromatic molecules were employed to interact with the sidewalls of SWCNTs by means of π–π stacking interactions. Biomacromolecules used for the functionalization include simple saccharides, polysaccharides, proteins, enzymes and DNA.[53–57] These molecules have been employed in the non-covalent functionalization of SWCNTs for biomedical applications. Conjugated polymers have been shown to serve as excellent wrapping materials for the non-covalent functionalization of SWNTs as a result of π–π stacking and van der Waals interactions between the conjugated polymer and the surface of SWCNTs.[58–61]

Compared to pristine carbon nanotubes, functionalized carbon nanotubes are more efficient in reinforcing the polymer matrix because the dispersion of the nanotubes is significantly improved and the interfacial bonding between the nanotube and polymer matrix is profoundly increased. Functional groups can provide better interfacial load transfer via bonding or entanglement with the polymer matrix. Frankland *et al.* theoretically predicted that chemical bonding between SWCNT and the polymer matrix with only 0.3% grafting density can increase the shear strength of a polymer–nanotube interface by 1000%.[62] Geng *et al.* obtained a 145% increase in tensile modulus and 300% increase in yielding strength with only 1 wt% fluorinated SWCNT in the polyethylene matrix.[63] In fact, the tensile strength is higher than the value estimated from the rule of mixture. This indicates that the covalent bonding at the nanotube–polymer interface can be very effective in load transfer. In order to well disperse the nanotubes, the agglomeration of the nanotubes must be overcome. The organic functional groups prevent bundling so that the agglomeration is mitigated and the quality of dispersion is improved. In the cross-linked composites, carbon nanotubes with an appropriate addend can be covalently attached to the host polymer matrix to increase the modulus of the nanocomposite while the elongation-to-break is maintained.

4.4 References

1. Ray, S. S. and Okamoto, M., 'Polymer/layered silicate nanocomposites: A review from preparation to processing', *Progress in Polymer Science*, 2003, **28**, 1539–1641.
2. Mallick, P. K., *Fiber-reinforced composites: Materials, manufacturing, and design*, 3rd edn. Boca Raton, FL: CRC Press, 2007, p. 559.
3. Jin, L., Bower, C. and Zhou, O., 'Alignment of carbon nanotubes in a polymer matrix by mechanical stretching', *Applied Physics Letters*, 1998, **73**, 1197–1199.
4. Yasmin, A., Luo, J. and Daniel, I. M., 'Processing of expanded graphite reinforced polymer nanocomposites', *Composites Science and Technology*, 2006, **66**, 1182–1189.
5. Southern Clay product description on Cloisite 20A. Available from: http://www.scprod.com.
6. Gilman, J. W. and Kashiwagi, T., 'Polymer-layered silicate nanocomposites with conventional flame retardant'. In Pinnavaia, T. J. and Beall, G. W. (eds.), *Polymer-clay nanocomposites*. New York: John Wiley, 2000, pp. 193–206.

7. Rao, Y. and Pochan, J. M., 'Mechanics of polymer-clay nanocomposites', *Macromolecules*, 2007, **40**, 290–296.
8. Coleman, J. N., Khan, U., Blau, W. J. and Gun'ko, Y. K., 'Small but strong: A review of the mechanical properties of carbon nanotube-polymer composites', *Carbon*, 2006, **44**, 1624–1652.
9. Yano, K., Usuki, A. and Okada, A., 'Synthesis and properties of polyimide–clay hybrid', *Journal of Polymer Science Part A: Polymer Chemistry*, 1997, 35, 2289–2294.
10. Zeng, C. and Lee, L. J., 'Poly(methyl methacrylate) and polystyrene/clay nanocomposites prepared by *in situ* polymerization', *Macromolecules*, 2001, 34, 4098–4103.
11. Kim, J., Hong, S. M., Kwak, S. and Seo, Y., 'Physical properties of nanocomposites prepared by *in situ* polymerization of high-density polyethylene on multiwalled carbon nanotubes', *Physical Chemistry Chemical Physics*, 2009, **11**, 10851–10859.
12. Yang, F., Ou, Y. and Yu, Z., 'Polyamide 6/slica nanocomposites prepared by *in situ* polymerization', *Journal of Applied Polymer Science*, 1997, **69**, 355–361.
13. Cho, J. W. and Paul, D. R., 'Nylon 6 nanocomposites by melt compounding', *Polymer*, 2001, **42**, 1083–1094.
14. Thostenson, E. T. and Chou, T. W., 'Aligned multi-walled carbon nanotube-reinforced composites: Processing and mechanical characterization', *Journal of Physics D: Applied Physics*, 2002, **35**, L77.
15. Liu, J., Rinzler, A. G., Dai, H. J., *et al.*, 'Fullerene pipes', *Science*, 1998, **280**(5367), 1253–1256.
16. Ago, H., Petritsch, K., Shaffer, M. S. P., *et al.*, 'Composites of carbon nanotubes and conjugated polymers for photovoltaic devices', *Advanced Materials*, 1999, **11**(15), 1281–1285.
17. Li, Y. H., Zhao, Y. M., Roe, M., *et al.*, 'In-plane large single walled carbon nanotube films: *In situ* synthesis and field emission properties', *Small*, 2006, **2**(89), 1026–1030.
18. Lv, R. T., Tsuge, S., Gui, X. C., *et al.*, '*In situ* synthesis and magnetic anisotropy of ferromagnetic buckypaper', *Carbon*, 2009, **47**(4), 1141–1145.
19. Geim, A. K., and MacDonald, A. H., 'Graphene: Exploring carbon flatland', *Physics Today*, 2007, **60**, 35–41.
20. Li, D., Muller, M., Gilje, S., Richard, B. K., and Gordon G. W., 'Processable aqueous dispersions of graphene nanosheets', *Nature Nanotechnology*, 2008, **3**, 101–105.
21. Xiang, J. L. and Drzal, L. T., 'Thermal conductivity of exfoliated graphite nanoplatelet paper', *Carbon*, 2011, **49**, 773–778.
22. Sung, J. A., Zhu, Y. W., Lee, S. H., Stoller, M. D., Emilsson, T., Park, S.J., Velamakanni, A., An, J. H. and Ruoff, R. S., 'Thin film fabrication and simultaneous anodic reduction of deposited graphene oxide platelets by electrophoretic deposition', *Journal of Physical Chemistry Letters*, 2010, **1**, 1259–1263.
23. Sonifer product description on ultrasonics. Available at: http://www.sonifier.com.
24. Mukhopadhyay, K., Dwivedi, C. D. and Mathur, G. N., 'Conversion of carbon nanotubes to carbon nanofibers by sonication', *Carbon*, 2002, **40**, 1373–1376.
25. Ma, P. C., Siddiqui N. A., Marom, G. and Kim, J. K., 'Dispersion and functionalization of carbon nanotubes for polymer-based nanocomposites: A review', *Composites: Part A*, 2010, **41**, 1345–1367.
26. Li, J., Ma, P. C., Chow, W. S., To, C. K., Tang, B. Z. and Kim, J. K., 'Correlations between percolation threshold, dispersion state and aspect ratio of carbon nanotube', *Advanced Functional Materials*, 2007, **17**, 3207–3215.

27. Xie, X. L., Mai, Y. W. and Zhou, X. P., 'Dispersion and alignment of carbon nanotubes in polymer matrix: A review', *Materials Science and Engineering R*, 2005, **49**, 89–112.
28. de Heer, W. A., Bacsa, W. S., Chatelain, A., Gerfin, T., Baker, R. H., Forro, L. and Ugarte, D., 'Aligned nanotube films: Production and optical and electronic properties', *Science*, 1995, **269**, 845–847.
29. Walters, D. A., Casavant, M. J., Qin, X. C., Huffman, C. B., Boul, P. J., Ericson, L. M., Haroz, E. H., O'Connell, M. J., Smith, K., Colbert, D. T. and Smalley, R. E., 'In-plane-aligned membranes of carbon nanotubes', *Chemical Physics Letters*, 2001, **338**, 14–20.
30. Ren, Z. F., Huang, Z. P., Wang, D. Z., Wen, J. G., Xu, J. W., Wang, J. H., Calvet, L. E., Chen, J., Klemic, J. F. and Reed, M. A., 'Growth of a single free standing multiwall carbon nanotube on each nanonickel dot', *Applied Physics Letters*, 1999, **75**, 1086–1088.
31. Cui, H., Zhou, O., Zhu, W. and Stoner, B. R., 'Deposition of aligned bamboo-like carbon nanotubes via microwave plasma enhanced chemical vapor deposition', *Journal of Applied Physics*, 2000, **88**, 6072–6074.
32. Li, W. Z., Xie, S. S., Qian, L. X., Chang, B. H., Zhou, B. S., Zhou, W. Y., Zhao, R. A. and Wang, G., 'Large-scale synthesis of aligned carbon nanotubes', *Science*, 1996, **272**, 1701–1703.
33. Schlittler, R. R., Seo, J. W., Gimzewski, J. K., Durkan, C., Saifullan, M. S. M. and Welland, M. E., 'Single crystals of single-walled carbon nanotubes formed by self-assembly', *Science*, 2001, **292**, 1136–1139.
34. Yanagi, H., Sawada, E., Manivannan, A. and Nagahara, L., 'Self-orientation of short single-walled carbon nanotubes deposited on graphite', *Applied Physics Letters*, 2001, **78**, 1355–1357.
35. Ajayan, P. M., Stephen, O., Colliex, C. and Trauth, D., 'Aligned carbon nanotube arrays formed by cutting a polymer resin–nanotube composite', *Science*, 1994, **265**, 1212–1214.
36. Vigolo, B., Penicaud, A., Coulon, C., Sauder, C., Pailler, R., Journet, C., Bernier, P., and Poulin, P., 'Macroscopic fibers and ribbons of oriented carbon nanotubes', *Science*, 2000, **290**, 1331–1334.
37. Kimura, T., Ago, H., Tobita, M., Ohshima, S., Kyotani, M. and Yumura, M., 'Polymer composites of carbon nanotubes aligned by a magnetic field', *Advanced Materials*, 2002, **62**, 1380–1383.
38. Dror, Y., Salalha, W., Khalfin, R. L., Cohen, Y., Yarin, A. L. and Zussman, E., 'Carbon nanotubes embedded in oriented polymer nanofibers by electrospinning', *Langmuir*, 2003, **19**, 7012–7020.
39. Chen, X. Q., Saito, T., Yamada, H. and Matsushige, K., 'Aligning single-wall carbon nanotubes with an alternating-current electric field', *Applied Physics Letters*, 2001, **78** (23), 3714–3716.
40. Lynch, M. D. and Patrick, D. L., 'Organizing carbon nanotubes with liquid crystals', *Nano Letters*, 2002, **2**, 1197–1201.
41. Niyogi, S., Hamon, M. A., Hu, H., Zhao, B., Bhowmik, P., Sen, R., Itkis, M. E. and Haddon, R. C., 'Chemistry of single-walled carbon nanotubes', *Accounts of Chemical Research*, 2002, **35**, 1105–1113.
42. Banerjee, S., Hamraj-benny, T. and Wong, S. S., 'Covalent surface chemistry of single-walled carbon nanotubes', *Advanced Materials*, 2005, **17**(1), 17–19.
43. Hamwia, A., Alvergnatb, H., Bonnamyb, S. and Béguin, F., 'Fluorination of carbon nanotubes', *Carbon*, 1997, **35**, 723–728.

44. Mickelson, E. T., Huffman, C. B., Rinzler, A. G., Smalley, R. E., Hauge, R. H. and Margrave, J. L., 'Fluorination of single-wall carbon nanotubes', *Chemical Physics Letters*, 1998, **296**(1–2), 188–194.
45. Christopher, A. D., and James, M. T., 'Covalent functionalization of single-walled carbon nanotubes for materials applications', *Journal of Physical Chemistry A*, 2004, **108**(51), 11151–11159.
46. Holzinger, M., Vostrowsky, O., Hirsch, A., Hennrich, F., Kappes, M., Weiss, R. and Jellen, F., 'Sidewall functionalization of carbon nanotubes', *Angewandte Chemie International Edition*, 2001, **40**, 4002–4005.
47. Vencelova, A., Graupner, R., Ley, L., Abraham, J., Holzinger, M., Whelan, P., Hirsch, A. and Hennrich, F., 'Doping of single-walled carbon nanotube bundles by Bronsted acids', *AIP Conference Proceedings*, 2003, **685**, 112.
48. Holzinger, M., Abraham, J., Whelan, P., Graupner, R., Ley, L., Hennrich, F., Kappes, M. and Hirsch, A., 'Functionalization of single-walled carbon nanotubes with (R-) oxycarbonyl nitrenes', *Journal of the American Chemical Society*, 2003, **125**, 8566–8580.
49. Chen, J., Hamon, M. A., Hu, H., Chen, Y., Rao, A. M., Eklund, P. C. and Haddon, R. C., 'Solution properties of single-walled carbon nanotubes', *Science*, 1998, **282**, 95–98.
50. Chen, Y., Haddon, R. C., Fang, S., Rao, A. M., Eklund, P. C., Lee, W. H., Dickey, E. C., Grulke, E. C., Pendergrass, J. C., Chavan, A., Haley, B. E. and Smalley, R. E., 'Chemical attachment of organic functional groups to single-walled carbon nanotube material', *Journal of Materials Research*, 1998, **13**, 2423–2431.
51. Lee, W. H., Kim, S. J., Lee, J. G., Haddon, R. C. and Reucroft, P. J., 'X-ray photoelectron spectroscopic studies of surface modified single-walled carbon nanotube material', *Applied Surface Science*, 2001, **181**, 121–127.
52. Zhao, Y. and Stoddart, J. F., 'Noncovalent functionalization of single-walled carbon nanotubes', *Accounts of Chemical Research*, 2009, **42**, 1161–1171.
53. Chen, R..J., Zhang, Y., Wang, D. and Dai, H., 'Single-walled carbon nanotubes for protein immobilization', *Journal of the American Chemical Society*, 2001, **123**, 3838–3839.
54. Barone, P. W. and Strano, M. S., 'Reversible control of carbon nanotube aggregation for a glucose affinity sensor', *Angewandte Chemie International Edition*, 2006, **45**, 8138–8141.
55. Star, A., Steuerman, D. W., Heath, J. R. and Stoddart, J. F., 'Starched carbon nanotubes', *Angewandte Chemie International Edition*, 2002, **41**, 2508–2512.
56. Ishibashi, A. and Nakashima, N., 'Individual dissolution of single-walled carbon nanotubes in aqueous solution of steroid or sugar compounds and their Raman and Near-IR spectral properties', *Chemistry – A European Journal*, 2006, **12**, 7595–7602.
57. Goodwin, A. P., Tabakman, S. M., Welsher, K., Sherlock, S. P., Prencipe, G. and Dai, H., 'Phospholipid-dextran with a single coupling point: A useful amphiphile for functionalization of nanomaterials', *Journal of the American Chemical Society*, 2009, **131**, 289–296.
58. Star, A., Stoddart, J. F., Steuerman, D., Diehl, M., Boukai, A., Wong, E. W., Yang, X., Chung, S.W., Choi, H. and Heath, J..R., 'Preparation and properties of polymer-wrapped single-walled carbon nanotubes', *Angewandte Chemie International Edition*, 2001, **40**, 1721–1725.

59. Steuerman, D. W., Star, A., Narizaano, R., Choi, H., Ries, R. S., Nicolini, C., Stoddart, J. F. and Heath, J. R., 'Interactions between conjugated polymers and singlewalled carbon nanotubes', *Journal of the Physical Chemistry B*, 2002, **106**, 3124–3130.
60. Star, A., and Stoddart, J. F., 'Dispersion and solubilization of single-walled carbon nanotubes with a hyperbranched polymer', *Macromolecules*, 2002, **35**, 7516–7520.
61. Star, A., Liu, Y., Grant, K., Ridvan, L., Stoddart, J. F., Steuerman, D. W., Diehl, M. R., Boukai, A. and Heath, J. R., 'Noncovalent side-wall functionalization of singlewalled carbon nanotubes', *Macromolecules*, 2003, **36**, 553–560.
62. Frankland, S. J. V., Caglar, A., Brenner, D. W. and Griebel, M., 'Molecular simulation of the influence of chemical cross-links on the shear strength of carbon nanotube–polymer interfaces', *Journal of Physical Chemistry B*, 2002, **106**: 3046–3048.
63. Geng, H., Rosen, R., Zheng, B., Shimoda, H., Fleming, L., Liu, J. and Zhou, O., 'Fabrication and properties of composites of poly(ethylene oxide) and functionalized carbon nanotubes', *Advanced Materials*, 2002, **14**, 1387–1390.

Part II

Manufacturing of polymer matrix composites (PMCs): thermoplastic based processing

5

Sheet forming in polymer matrix composites

T. CREASY, Texas A&M University, USA

Abstract: Key objectives in sheet forming are reduced cost per component, increased production and continuous stepwise processing. Thermoplastic sheet forming turns a flat composite laminate into a structural shape when dies or diaphragms press the laminate while the thermoplastic is liquid. The matrix flows and the fibres move as the laminate adopts its new shape. Designers must accept larger variances in fibre placement than present in other processes. Stretch-broken carbon fibres have renewed interest in sheet forming. Advanced thermoplastics made via *in situ* ring-opening polymerization might reduce processing time because low-viscosity precursors form high molecular weight chains during thermal processing.

Key words: composite sheet thermoforming, fibre-reinforced thermoplastics, die forming, matrix flow, diaphragm forming.

5.1 Introduction: key objectives

The key objectives in composite sheet forming are to reduce structure cost with automation, to increase production volume – when compared to thermoset processing – by eliminating a time-intensive thermal cure process, and to provide composite structures through continuous stepwise processing (Okine, 1989).

Sheet forming produces metal products such as automobile bodies by stamping sheet metal to set a new permanent shape through plastic deformation. This process is successful – it allows manufacturers to automate production while creating complex curvature structures once formed piecemeal by skilled workers. Figure 5.1 shows a metal sheet forming process that forms cups from circular blanks cut from larger metal sheets. The centre image in Fig. 5.1 shows a cut section view with the disk held between a clamping ring and the die. The punch forces the blank into the die as the clamping ring applies pressure to prohibit the blank from wrinkling at the perimeter. Plastic deformation sets the blank to a persistent shape. There is some waste material to cut from the part and recycle. For metals, this process is rapid. Researchers and manufacturers are interested in using sheet forming processes to bring similar economies of scale and labour reductions to composite structures.

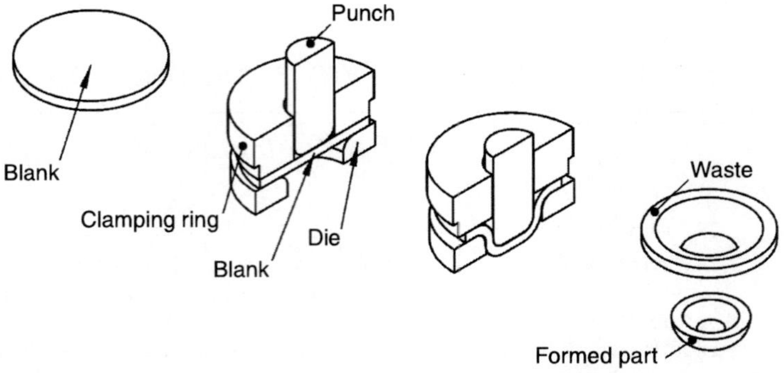

5.1 **Sheet forming turns the flat metal disk at the left into a three-dimensional structure by plastic deformation.**

Although sheet forming works with thermoset materials (Gutowski *et al.*, 1995) there is more interest in forming thermoplastic matrix composites in sheets. For one thing, sheet forming a thermoset matrix requires heating the material so that the matrix can flow. Because the matrix cures into a fixed cross-linked solid, the curing reaction limits the time available to process the composite and reprocessing to add additional features or recycling into new components is not possible. Additionally, processes like resin transfer moulding (RTM) and resin film infusion (RI) produce quality structures from fibre preforms without moving the fibres and the matrix through a tool.

Thermoplastic matrix composites are more difficult to process than thermosets because a high performance thermoplastic in the molten state has a large molecular weight, which typically produces high low-shear rate viscosity. Any flow process must overcome the viscous resistance and this either makes the process slow or generates large shear stress in the fluid. Some novel chemistry (van Rijswijk and Bersee, 2007) might allow initial processing at low viscosity followed by polymerization to a high molecular weight thermoplastic; however, reprocessing or recycling must deal with the high molecular weight product.

Knowing the key objectives and interest in thermoplastic sheet forming, we next address the process and its fundamental considerations.

5.2 Process description

Thermoplastic sheet forming turns a flat composite laminate, which contains reinforcing fibres aligned in one or more directions, into a shaped structure with the characteristics a design needs. Sheet forming can be the final process for the component; however, thermoplastics allow for sheet forming steps and additional processing.

5.2.1 Physical processes during forming

The physical processes that occur during sheet forming are:

- Heating.
- Flow under pressure.
- Internal fibre motion.
- Consolidation.
- Cooling.

First, the process equipment heats the prelaminate, which is typically a flat panel, until the matrix is molten and flowable. Heat arrives by surface convection, surface irradiation or combined convection and irradiation. Second, pressure forces the molten prelaminate to move towards the tool. Pressure sources include tool surfaces and fluid pressure from compressible gas or incompressible liquid. During this motion, the pressure can drive the matrix to flow between and through the fibres as the prelaminate changes to the component shape.

As the prelaminate moves and the matrix flows, the fibres can move under the influence of the matrix flow and the applied pressure. Fibres angles can shift and the process can draw continuous fibres into the new shape as the prelaminate conforms to the tool. Discontinuous fibres can slide relative to one another and accommodate stretching without drawing fibres into the tool from the prelaminate's edges. At high forming rates, the viscous drag induced by shear lag can put fibres into tension and break the fibres.

Once the prelaminate reaches the tool surface and takes on its new shape, the pressure acts to consolidate the plies to make a fully dense component. Finally, the pressure holds the composite against the tool until the component has enough strength to survive as it is removed from the tool.

5.2.2 Design process

For most composite materials and processes, design must incorporate some material and processing effects knowledge. Sheet forming adds complexity because the process moves fibres and matrix from one conformation to another dynamically. Figure 5.2 shows one possible design process that requires iterative numerical design and process development.

The process starts with component design. Current CAD/FEA software helps a designer obtain the shape and performance needed based on known materials properties. For sheet forming, the designer must expect that forming and tooling issues will require several design iterations. First, a materials and process engineer should advise the designer about sheet forming issues before and during the analysis. It might be impossible to use the optimum

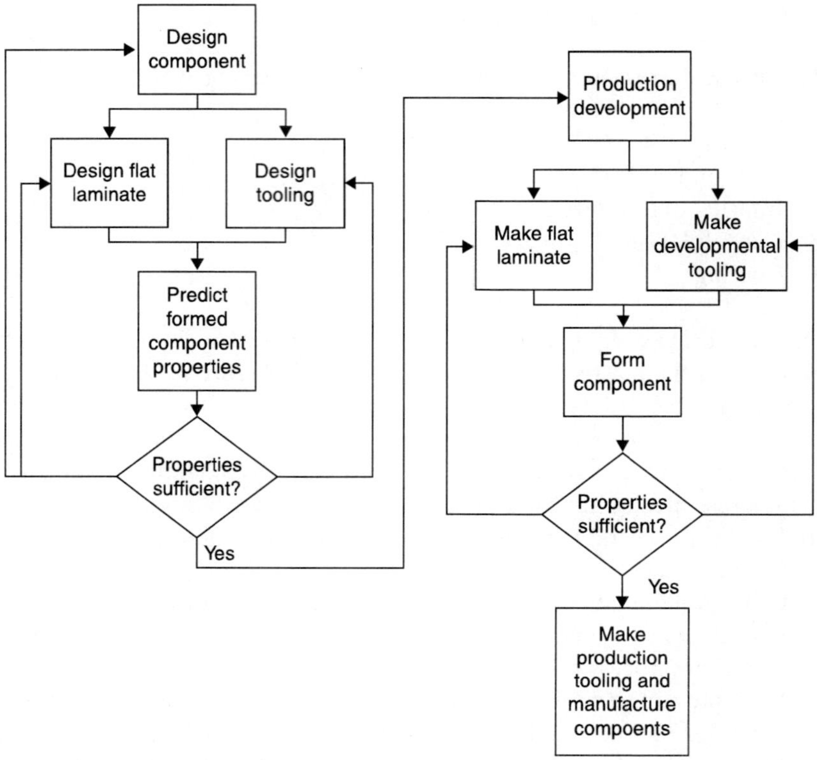

5.2 Successful sheet forming requires interactive design and development.

component design because sheet forming processes might not place the reinforcement into the best position. Although a suboptimal design will increase a component's weight and material cost, labour savings and a reduced scrap must make the component competitive with other materials and processes. Once the designer has a suitable design, the designer should confer with the materials-and-process and tooling engineers to design the initial flat laminate and the tooling that will form the flat laminate and structure.

Flat laminates can be produced using the processes described in other chapters of this handbook. For example, a double belt press or compression moulding can produce the precursor laminate. Figure 5.3a shows thermoplastic composites with unidirectional fibres in each layer consolidated into a laminate with a picture frame mould. A picture frame mould goes into a hot press or autoclave and turns composite lamina into a fused precursor. To some extent, sheet forming adds cost to a final structure because a manufacturer creates two components: the flat precursor and the formed structure.

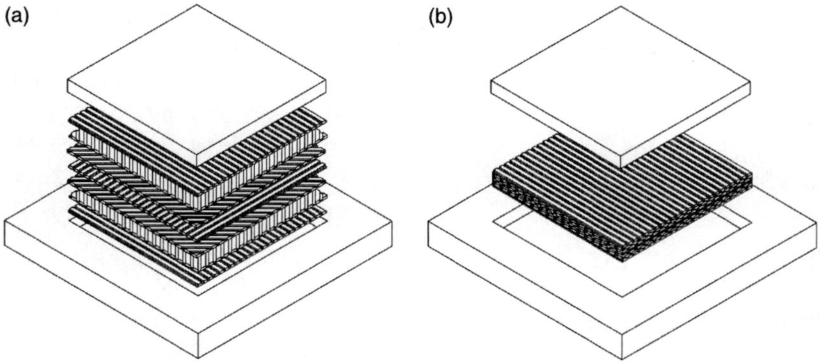

5.3 The picture frame mould at left, which sits on a hot press's lower platen, consolidates multiple thermoplastic matrix composite plies into the laminate at right.

Designing tooling for sheet forming is difficult. Durable tools are expensive and a failed design would introduce added costs to the project. Using software to analyse the forming process and temporary tooling to prove the process will reduce cost in labour and capital should high volume, permanent tooling fail to perform once it is built.

The next step in Fig. 5.2, predicting formed composite properties, employs software to determine that the flat precursor composite and the tooling interact to form a component that meets the design need. At this point all three designs – final component, flat precursor and tooling – are virtual and open to changes at the cost of the labour needed to revise the models. The decision step 'properties sufficient' refers to requirements for shape and mechanical needs. For example, at this point the prediction could indicate that the shape meets the need; however, the reinforcement is not within the expected position range. The designer must analyse the component for mechanical performance and determine what design or designs – component, flat laminate, forming tools – must change to get to the requirements.

Once the virtual component behaves as necessary, the production development stage, which is the second part of Fig. 5.2, begins. First, a tape placement machine can fabricate a trial precursor laminates and the production staff can fabricate a developmental tool built from lower cost processes such as rapid prototyping. With the precursor and tool moved from virtual designs to hardware, staff can produce components and inspect them. Should the components fail to meet specifications, the initial correction loop shown in Fig. 5.2 directs staff to check on the tool and precursor design and fabrication. Corrections can drive the process back up to the tooling, flat laminate and component design stage if the components cannot meet requirements with the designs under trial. Finally, once successful components come from the precursor laminate and developmental tooling, the manufacturer can build durable production tools.

Because design and manufacturing are strongly coupled in sheet forming, engineers working on new components must understand the way that sheet forming functions. The next sections discuss typical sheet forming processes and fundamental issues.

5.2.3 Sheet forming fundamentals

In this section some fundamental issues concerning thermoplastic sheet forming appear. The content here builds on Chapter 1; but the focus is specific to sheet forming.

Short and long fibre composites

Composites with fibre reinforcement may use continuous or discontinuous fibres (Chang and Pratte, 1991; Ford, 2001). Figure 5.4 shows short and long fibres in a polymer matrix. Short fibre composites typically undergo injection or compression moulding rather than sheet forming. Short fibres with random alignments have low fibre volume fractions that limit performance; however, there are benefits to using these composites. Sheet forming with short fibres can lead to uneven drawing and most interest in sheet forming is for making high performance composites, which typically need high fibre volume fractions.

Figure 5.4b shows a long-discontinuous fibre composite. In this material continuous fibre became discontinuous in a stretch-breaking process similar to that used to make staple fibre from synthetic continuous fibres for clothing. Figure 5.4b shows the breaks in fibres; these breaks appear randomly through the tow. By keeping the fibres aligned, the volume fraction remains high; the discontinuous fibre may slip past one another and reduce the wrinkling problems discussed later in this chapter.

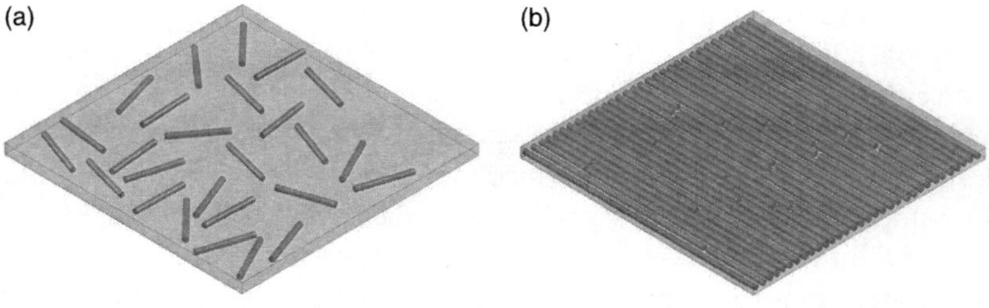

5.4 Short fibre composites (a) tend to have low fibre concentration; sheet forming is suited to continuous fibre composites or (b) long discontinuous fibre composites.

Continuous fibre composites

As noted earlier in this handbook, continuous fibre composites with fibres accounting for 50% or more of the material volume present the highest properties possible. Unfortunately, continuous fibres make forming – and sheet forming in particular – difficult to control. Before specifying continuous fibre components made by sheet forming, a designer should determine whether a discontinuous fibre product could meet the need.

Freidrich *et al.* (1997) formed composites in a system similar to the metal stamping system appearing in Fig. 5.1. That is, a circle laminate was clamped between rings as a die forces it into a hemispherical mould. The clamped flange area is in compression, but it slips between the clamps to allow material to flow into the hemisphere. The die's downward force and friction between the clamping surfaces and the composite generate tension in the sheet. In this process, the fibres were not broken; the fibres had to move into the hemisphere by drawing. The relative fibre motions generate shear stress in the matrix and this stress can contribute to out-of-plane deformations that wrinkle the component.

Continuous fibre reinforcements are available as a fabric. The interlocking fibres force the composite to draw fibres from all directions (Freidrich *et al.*, 1997). Freidrich formed a glass fabric reinforced composite at two clamping stresses. When the composite's flange area received 0.1 MPa pressure, the composite generated enough buckling force to lift the clamping ring and wrinkle the part. These wrinkles reach the edge and ruin the component. With 0.2 MPa clamping pressure, the wrinkles are suppressed. The outcome from this study is that operators must apply the right clamping pressure; too much pressure would increase the friction between the clamping ring and composite and in-plane stress could break fibres and weaken the material.

Thermal issues

Thermoplastic sheet forming is a thermal process and thermal conduction limits the production rate. Typical matrix polymers are poor conductors. Injection moulding and extrusion make polymer components affordable because they do not depend on conduction to provide all the heat needed to make a thermoplastic flow. Sheet forming, however, must bring a composite sheet to a uniform temperature and complete the forming process before any matrix – typically, the matrix at the surface because matrix there touches the cooler tool – becomes too rigid for forming.

Heating to the forming temperature is the rate-limiting step for sheet forming. Slow thermal conduction limits sheet forming suited to thin laminates. The reinforcement can help or hinder this process. For example, high thermal conductivity carbon fibre will aid conduction within a lamina and make the temperature within that lamina uniform. This in-plane effect does

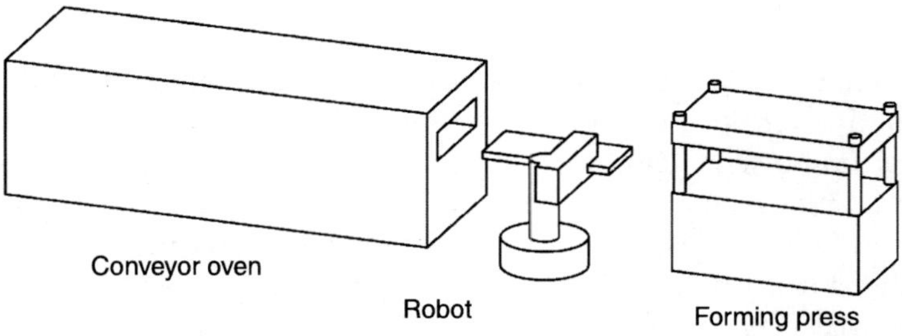

5.5 A conveying oven heats flat laminated material and moves it to the press in the time needed to melt the matrix. A simple robot can move the hot laminates to the press and then remove the formed component.

not help with through-the-thickness conduction because carbon fibres do not lead from one lamina to another. The matrix-rich regions between lamina keep conduction low. Glass – or other – fibres with lower thermal conductivity than the matrix make matters worse. The insulating fibre slows thermal conduction in all directions.

Sheet forming times at the tool are much shorter than the time needed to heat the precursor laminate. Therefore, manufacturers use a separate heating process to warm the precursor before moving it to a press by manual or automatic means. Figure 5.5 shows a simple method: precursor laminates move through a continuous linear oven that brings them to processing temperature by the time they reach the sheet forming press. Some fabricators save on floor space by having multiple heating ovens that discharge the laminate into the press just before receiving a cold laminate. These systems use two or more ovens to support the production rate.

Now that the fundamentals are covered, the next sections present typical sheet forming methods. The methods presented next are common; however, new forming methods are still in development (Zampaloni *et al.*, 2004).

5.2.4 Matched die method

The matched die method, which appears in Fig. 5.6, can produce the highest quality parts because precision machined dies act on both surfaces of a thermoplastic composite sheet. The reader should note that there is no requirement for both – or either – of the tool surfaces to be a precision shape. The designer has two surfaces available to control forming and can use them as needed to fabricate the component.

Matched dies are typically used in a mechanical press rather than in an autoclave. The press and tool surfaces can generate all the stress needed to form the hot laminate. The left image in Fig. 5.6 shows the upper and lower tools separated to allow a machine or operator to insert a hot, flat composite

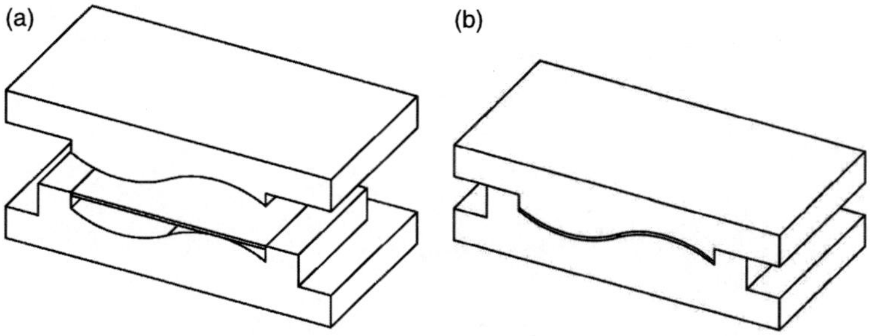

5.6 (a,b) The matched die method uses two tool surfaces to drive the composite sheet into the final shape.

sheet. The tool surfaces match the upper and lower surfaces from the component design and when the tool is closed, these surfaces are in the correct three-dimensional positions to hold the composite in its final shape. Figure 5.6b shows the closed tool with the component sandwiched between them.

All sheet forming tools must remove heat from the formed component. With intimate contact with one or both sides, the tool provides the maximum contact area possible; however, all heat removed while the component is within the matched tool must occur by conduction through the tool. Most tools are machined metal with good to high thermal conductivity. For example, tools can be tool steel or aluminium. During large volume production, each precursor composite laminate will carry heat into the press and leave some heat in the tool. Therefore, the tools have active cooling. A typical cooling system incorporates internal channels that allow a heat transfer fluid to collect heat and remove it to an external heat exchanger the conditions the fluid before the fluid returns to the tool.

Tooling thermal design can be a significant task. A tool's temperature must be lower than the thermoplastic processing temperature so the tool can solidify and stabilize the composite. Yet, too high a temperature will slow the heat transfer and increase the time the tool must hold the composite in the right shape. These contradictory conditions suggest that each tool will have an optimum steady temperature and production rate.

Matched tooling can provide high production rates because two surfaces remove heat; however, the limited thermal conductivity discussed earlier continues to work against high production rates. Heat stored deep in the composite will take some time to remove completely and expensive matched dies spend as much time as possible forming parts rather than stabilizing their shape once formed. A typical procedure is to remove enough heat to allow demoulding and to let the composite cool under natural or forced convection. A secondary, lower cost tool might stabilize the structure until the polymer is rigid again.

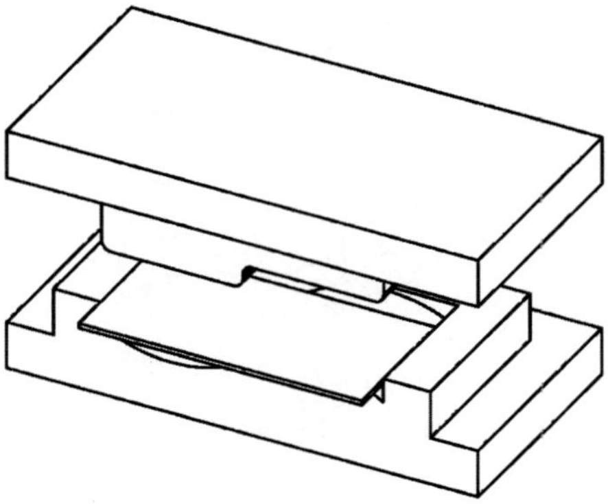

5.7 The rubber pad method has one tool surface that provides a precise shape and a compliant rubber pad that deforms as an incompressible fluid to deliver pressure to the composite.

5.2.5 Rubber pad method

One way to reduce tool cost while applying pressure to both sides of the laminate is to use rubber pad forming. Figure 5.7 shows a tool similar to the one presented in Fig. 5.6, but with the upper, machined tool replaced with shaped rubber pad bonded to a metal plate. The metal plate is thick enough to transfer load from the press platen to the laminate. With a single surface to machine, this tool will cost less than a matched tool.

The rubber pad must be an elastomer that is compatible with the composite processing temperature and the pad can introduce a new design element. With the elastomer acting as an incompressible fluid, the pad can apply all the necessary pressure needed to form a component. However, the designer will need to analyse the pad deformation and ensure that the pressure distribution across the laminate is the distribution needed to shape the part. This design challenge can introduce a method for obtaining a desired distributed pressure. This problem must be handled iteratively with finite element analysis as the inverse problem – that is, finding the initial rubber pad shape from the final shape and pressure distribution – may be intractable. Because the rubber pad method cannot produce a precise shape, this method is attractive for components that have one critical surface and an unimportant second surface.

5.2.6 Diaphragm method

The diaphragm method uses gas pressure to shape components from flat laminates. This method is often run in an autoclave because high-pressure gas is available for forming and consolidation.

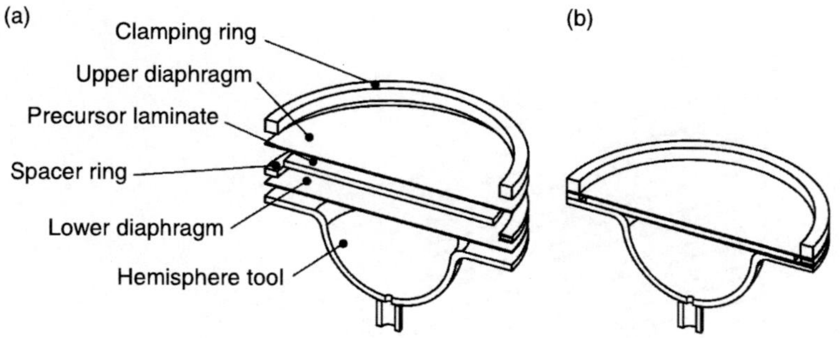

5.8 Diaphragm methods sandwich the precursor between extendable sheets as shown in (a). The right image shows the system assembled. (New drawing based on McGuinness and ÓBrádaigh, 1995.)

Diaphragm forming shapes three sheets into the final shape. Figure 5.8 shows a half-section drawing from a tool that forms a round laminate into a hemisphere (McGuinness and ÓBrádaigh, 1995). The top element is the upper clamping ring that – when clamped to the flange around the hemisphere tool – holds the diaphragm-spacer/laminate-diaphragm sandwich together so that a vacuum may be applied between the diaphragms. The object immediately under the upper clamping ring is the top diaphragm. Beneath the top diaphragm lie the spacer ring and the composite laminate. The spacer ring is close to the laminate's thickness; this helps the diaphragms seal to the clamping ring and flange. The lower diaphragm is the next component in the figure and provides a seal between the tool flange and the spacing ring. The bottom element is the tool. The tool has a bottom vent so that any air trapped under the lower diaphragm can leave as the sandwich forms a hemisphere.

Figure 5.8b shows the components assembled and ready for forming. The spacer ring has vents in one or more places so that a vacuum pump can remove air from the volume between the diaphragms. This pressure differential applies a consolidation pressure to the laminate that helps it remain a contiguous laminate as its shape changes. Once the laminate has melted, the autoclave applies pressure to the assembly and the operator vents the tool to external pressure at one atmosphere. Driven by the applied pressure, the diaphragms and laminate move into the tool.

Tool shapes can be complex. For example, O'Bradaigh *et al.* (1991) formed flat, eight-ply, quasi-isotropic sheets into a rectangular cargo trailer bed with cylindrical wheel wells. The diaphragm can be any material that supports the laminate and that deforms to the component shape under the applied pressure. Composite aircraft structures, which are expensive by any process, are formed between superplastic aluminium diaphragms (Raman and Barnes, 2010). The procedure fabricates three parts: the composite laminate

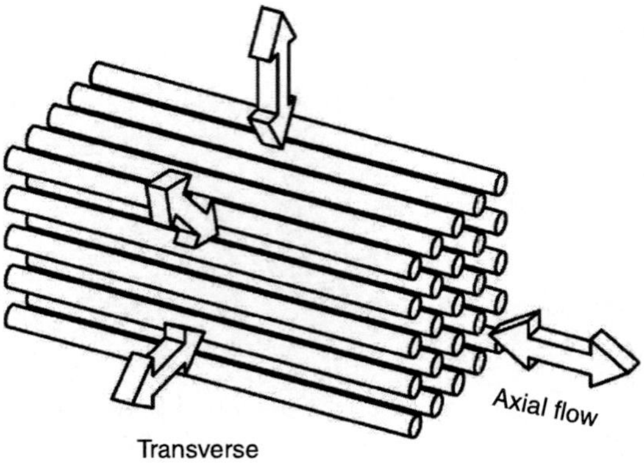

5.9 While the composite sheet deforms, the melted thermoplastic matrix can flow among the fibres in axial and transverse directions.

and two diaphragms. The aluminium is recycled after workers strip it from the assembly. Researchers have used polymer film diaphragms – for example, polyimide film – that remains solid at the composite's processing temperature. These films are usually permanently deformed by the process and discarded once the part is finished. It might be possible to create reusable diaphragms from silicone elastomers that have high elongation to failure.

5.3 Matrix flow and fibre deformation

Thermoforming unfilled polymer sheet is relatively simple if the sheet starts with isotropic properties and remains substantially isotropic throughout the process. Fibre-reinforced thermoplastics with high fibre volume fractions are anisotropic materials and have been called hyper anisotropic during forming because the stiffness and viscosity vary tremendously with direction. These factors complicate the forming processing practically and analytically as matrix and fibres can move as they interact during the forming process.

5.3.1 Matrix flow

Figure 5.9 shows a high volume-fraction fibre array with the matrix removed from the diagram. This image shows that the matrix can flow along the fibre direction – that is, axially – and perpendicular to the fibres, or transversely, during sheet forming. Ideally, one wants the matrix and fibres to remain at good volume fractions and alignments during the process. This might not occur as the fibre/matrix system is susceptible to independent motions especially as the forming rate increases – for example the die closing speed.

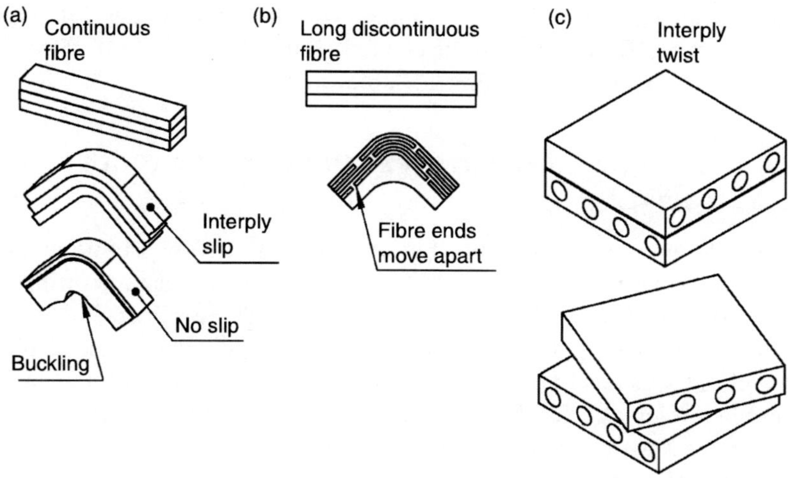

5.10. Fibres can undergo changes in orientation during forming. (Redrawn, based on Schuster and Friedrich, 1994.)

5.3.2 Fibre deformation and placement issues

Another component of matrix flow is fibre deformation and placement. If matrix leaves a region, the fibres may move closer together. Although the higher fibre volume fraction would increase the local properties, the resulting matrix-rich region will be weak. Figure 5.10a shows matrix and fibre migration that contributes to non-uniform properties. A continuous fibre-reinforced thermoplastic when bent needs to have interply slip in order to keep fibres running parallel through the thickness. Figure 5.10a shows an ideal condition: the fibres and matrix retain their ideal distributions when the composite bends 90°. Because continuous fibres cannot change their length, the laminates' edges will show steps as layers must slip relative to one another. As sheet formed components must be cut from the unused flange material, the stepped edge will be discarded. If the plies do not slip, the inner layers will go into compression and the fibres with buckle, which results in poor properties in the bends.

Figure 5.10b shows ideal forming in a long-discontinuous fibre laminate (Lee *et al.*, 2008). As the flat laminate bends, the fibre ends move apart. This extension ability reduces wrinkling. New long-discontinuous fibre forms are available and since about 2001, these are called stretch-broken fibre or SBXF materials with the X replaced with a letter that represents a specific fibre. Carbon fibre material is called SBCF, for example.

Figure 5.10c shows fibre orientation changing during forming. Fibres in each lamina will try to orient to align with the local tensile stress.

The matrix-rich region between lamina will allow this motion, effectively lubricating it. These orientation changes must be incorporated into the design or accommodated as an allowable variance.

5.4 Changes to the process to improve product quality

Research continues to produce better composite sheet forming methods. When faced with a composite sheet forming process, manufacturing staff must work throughout the process to obtain a quality product. The first step is to avoid designing the component as if the reinforcement can be placed as well as that available with prepreg tape. Design allowables must include a greater tolerance for fibre volume fraction and orientation variances. Experiments with existing and new tooling can provide a measure of fibre position variation.

Second, modelling software is under continuous development. New software can be evaluated with experiments on new components or on previous outcomes. That is, known fibre positions obtained by experiment should become part of a database that will support rapid evaluation of new software.

Sheet forming tools can benefit from real-time monitoring for clamping pressure, vacuum level between diaphragms, forming rate and consolidation. Trial runs can combine grid strain analysis (Krebs *et al.*, 1997) with finite element methods to validate performance.

5.5 Future trends

Research continues in sheet forming with rising interest since SBCF arrived made with new processes developed by Hexcel in the United States, Schappe in France and others (Black, 2008; Ross, 2006). These new aligned–discontinuous fibres provide extensibility that reduces or eliminates wrinkling in the laminate. Lee *et al.* (2008) modelled the flow behaviour of SBCF and found that there are three flow contributions: pure shear flow between a fibre and its neighbouring fibres, axial flow in the fibre direction – that is driven by the induced pressure gradient in the stretching directions – and squeeze flow – which fills the gap between fibre ends with the matrix. Although the shear flow created as discontinuous fibres slip past each other contributes, Lee *et al.* find that the axial and squeeze flow are the dominant mechanisms for matrix flow during forming.

Advanced thermoplastic materials might improve their thermal conductivity by a factor of three if they contain carbon nanotubes (Naffakh *et al.*, 2011). Greater thermal conductivity makes sheet forming a faster, more cost effective process; however, at this time adding carbon nanotubes is neither a trivial nor cost effective procedure.

Finally, some components suited to sheet forming might be made by a novel injection moulding process that adds long chopped fibre to a polymer melt and compounds the mixture immediately (Henning *et al.*, 2005). Invented for moulding large automotive panels, the process shows great promise for future sheet structures (Hawley and Jones, 2005).

5.6 Sources of further information and advice

Further information about thermoplastic polymers that make successful composites is available in Sylvie Béland's *High performance thermoplastic resins and their composites*, published by William Andrew Publishing in 1990.

Detailed information about the thermal transport and consolidation in thermoplastic composites appears in Raju Davé and Alfred Loos' *Processing of composites* published by Hanser in 2000.

James Throne's *Technology of thermoforming* (1996) covers general thermoforming practice with some information related to composite thermoforming.

Finally, *Composite sheet forming* (1997), edited by D. Bhattacharyya and published by Elsevier, presents a complete resource showing the outcome of research as high temperature/high performance thermoplastics first became available.

5.7 References

Black, S. (2008), 'Aligned discontinuous fibres come of age', USA: Compositesworld. Available at: http://www.compositesworld.com/articles/aligned-discontinuous-fibers-come-of-age. Accessed 31 May 2010.

Chang, I. and Pratte, J. (1991), 'LDF thermoplastic composites technology', *Journal of Thermoplastic Composite Materials*, **4**, 227–252. DOI: 10.1177/089270579100400302.

Ford, R. (2001), 'Thermo-ductile composites: New materials for 21st century manufacturing – Micro-perforated thermoplastic composite', *Materials & Design*, **22**, 177–183.

Friedrich, K., Hou, M. and Krebs, J. (1997), 'Thermoforming of continuous fibre/thermoplastic composite sheets'. In Bhattacharya, D. (ed.), *Composite sheet forming*. Amsterdam: Elsevier, pp. 91–162.

Gutowski, T., Dillon, G., Chey, S. and Li, H. (1995), 'Laminate wrinkling scaling laws for ideal composites', *Composites Manufacturing (UK)*, **6**, 123–134.

Hawley, R. and Jones, R. (2005), 'In-line compounding of long-fibre thermoplastics for injection molding', *Journal of Thermoplastic Composite Materials*, **18**, 459–464. DOI: 10.1177/0892705705054413.

Henning, F., Ernst, H. and Brussel, R. (2005), 'LFTs for automotive applications', *Reinforced Plastics*, **49**, 24–33.

Krebs, J., Bhattacharyya, D. and Friedrich, K. (1997), 'Production and evaluation of secondary composite aircraft components: A comprehensive case study', *Composites Part A: Applied Science and Manufacturing*, **28**, 481–489.

Lee, K., Lee, S. and Ng, S. (2008), 'Micromechanical modeling of stretch broken carbon fibre materials', *Journal of Composite Materials*, **42**, 1063–1073. DOI: 10.1177/0021998308090449.

McGuinness, G. and ÓBrádaigh, C. (1995), 'Effect of preform shape on buckling of quasi-isotropic thermoplastic composite laminates during sheet forming', *Composites Manufacturing*, **6**, 269–280.

Naffakh, M., Diez-Pascual, A. and Gomez-Fatou, M. (2011), 'New hybrid nanocomposites containing carbon nanotubes, inorganic fullerene-like WS2 nanoparticles and poly(ether ether ketone) (peek)', *Journal of Materials Chemistry*, **21**, 7425–7433.

O'Bradaigh, C., Pipes, R. and Mallon, P. (1991), 'Issues in diaphragm forming of continuous fibre reinforced thermoplastic composites', *Polymer Composites*, **12**, 246–256.

Okine, R. (1989), 'Analysis of forming parts from advanced thermoplastic composite sheet materials', *Journal of Thermoplastic Composite Materials*, **2**, 50–76.

Raman, H. and Barnes, A. (2010), 'Superplastic diaphragm forming: A practical method for forming unique components'. In Sanders, D. (ed.), *Superplasticity in advanced materials*. Zurich: TransTech Publications, pp. 85–91.

Ross, A. (2006), 'Will stretch-broken carbon fibre become the new material of choice?', USA: Compositesworld. Available at: http://www.compositesworld.com/articles/will-stretch-broken-carbon-fiber-become-the-new-material-of-choice (Accessed 31 May 2011).

Schuster, J. and Freidrich, K. (1994), 'Fatigue testing of thermoformed bidirectional LDF composites', *Applied Composite Materials*, **1**, 55–68.

van Rijswijk, K. and Bersee, H. (2007), 'Reactive processing of textile fibre-reinforced thermoplastic composites: An overview', *Composites Part A – Applied Science and Manufacturing*, **38**, 666–681.

Zampaloni, M., Pourboghrat, F. and Yu, W. (2004), 'Stamp thermo-hydroforming: A new method for processing fibre-reinforced thermoplastic composite sheets', *Journal of Thermoplastic Composite Materials*, **17**, 31–50. DOI: 10.1177/0892705704038219.

6
Fabric thermostamping in polymer matrix composites

J. A. SHERWOOD, K. A. FETFATSIDIS and J. L. GORCZYCA, University of Massachusetts Lowell, USA and L. BERGER, General Motors Research and Development Center, USA

Abstract: Thermostamping is a low-cycle, high-volume manufacturing process for continuous fabric-reinforced composites. To ensure high-quality parts are manufactured during the thermostamping process, the mechanical behaviors of the fabric reinforcements and polymer matrix and the critical manufacturing process parameters must be thoroughly understood. Various mechanical tests are conducted to characterize the fabric-reinforcement and polymer-matrix mechanical behaviors, and analytical models (i.e., finite element models) are developed to predict part quality. This chapter describes the thermostamping process and provides examples of typical experimental and analytical methods used to create high-quality composite parts using the thermostamping process.

Key words: thermostamping, thermoforming, woven fabrics, composites, finite element.

6.1 Introduction

The benefits of fiber-reinforced polymer–matrix composite materials relative to metals are well known, for example higher specific strength, higher specific stiffness and tailored energy absorption. Chopped-fiber composites are relatively inexpensive to produce. However, because of the randomness of the fiber directions and the need to use relatively short fibers, the weight savings that can be realized by using this class of composites is not as significant as what can be achieved when using continuous aligned-fiber fabric-reinforced composites. Aligned-fiber composites are made using unidirectional, stitched and/or woven fabrics which are molded to the desired shape. However, the cost in time and labor is an obstacle to the widespread use of these aligned-fiber fabric-reinforced composites. In the absence of a low-cost high-volume manufacturing process, only low-volume production items such as boutique cars and aerospace can justify the relatively high cost that is associated with the benefits that these composites bring to a design.

Common continuous-fiber composite manufacturing processes can be divided into thermoplastic and thermoset molding. A thermoplastic is a polymer that turns to a liquid when heated and solidifies when sufficiently cooled. Thermoset matrices are typically liquid or malleable prior to curing, and once cured, the process is irreversible. Thermoset molding requires a heat cycle to cross-link the polymer, and includes hand lay-up with prepreg, pultrusion, filament winding, tape laying, preform/resin infusion methods such as resin transfer molding (RTM), compression molding and thermostamping. Thermoplastic molding involves preheating the material above its melt temperature, then cooling it in a molding tool, and includes hand lay-up with unconsolidated fabric material and thermostamping, also known as thermoforming, hot pressing or even compression molding. Hand lay-up of fabric layers is the most versatile in that the fabrics can be placed in the mold and smoothed by hand to remove any out-of-plane wrinkles and in-plane waviness in each of the plies. Slits can be made in the fabric sheets to account for mold details that cannot accommodate a continuous length of fabric. The fabric can be either a thermoset or thermoplastic prepreg or a dry fabric to which resin is applied by an infusion process. The disadvantage of the hand lay-up is that it is very time consuming to place the plies and can require a separate resin infusion step. Thus, hand lay-up is not suitable for a low-cost high-volume production situation. Pultrusion and filament winding are relatively fast processes suitable for low-cost high-volume production situations but are only applicable to relatively simple shapes. Pultrusion is limited to constant cross-sections, and filament winding is limited to convex surfaces, that is no recesses. The use of prepreg tape laying machines can expedite the lay-up process, but such machines can be prohibitively expensive and are limited to relatively smooth surfaces such as an aircraft wing. Following the tape laying process, the part must be placed in an autoclave for curing.

Thermostamping has the versatility of hand lay-up to accommodate relatively complex shapes with the added benefit of forming and curing a composite part in essentially a single step. In contrast to a resin infusion process which adds the resin after the fabric has been placed in the mold, the thermostamping process uses a fabric that already includes the matrix material as either a thermoset or a thermoplastic. Because the process can be automated, thermostamping is an attractive choice for high-volume low-cost production of composites. However, the automation of the process does not necessarily allow for the careful inspection for fabric wrinkling and waviness that can be visually observed during the hand lay-up. These wrinkles and waves are defects that compromise the structural integrity due to unwanted changes in the load path and can lead to the formation of voids in the composite (Fig. 6.1). As a result, a major challenge in thermostamping is the need to control the process such that these defects are not developed

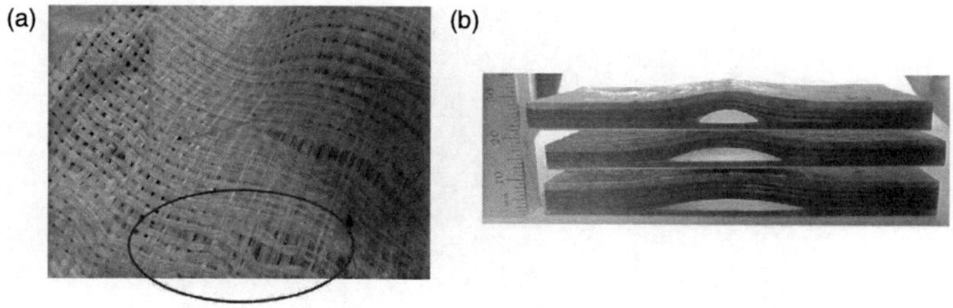

6.1 Composite parts with (a) in-plane yarn waviness and (b) out-of-plane wave defects with resin-rich pockets.

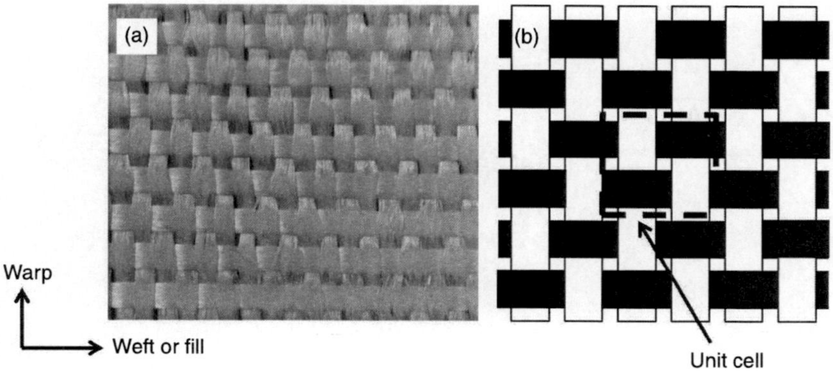

6.2 (a) Plain-weave fabric with (b) representative unit cell (UC).

while the fabric conforms to part geometry details such as compound curvatures, ply drops and deep-drawn corners.

Unlike a metal that can stretch in-plane and change thickness as the metal blank is being stamped to form a part, the yarns comprising the woven-fabric blanks are essentially inextensible. The yarns are typically fiberglass and/or carbon based with a maximum strain of ~1–3% before breaking. Thus, the primary mode of deformation for the fabrics to assume the shape of the die is fabric shearing. Thus, a simulation of the forming process must account for the change in load path associated with yarn rotation.

As an example of fabric shearing, consider the plain-weave fabric shown in Fig. 6.2. Woven fabrics generally consist of two sets of interlaced yarn components, known as warp and weft (or fill) yarns according to the yarn orientation. Each yarn is a bundle of filaments (or fibers), and the yarn size is measured by the number of filaments in the yarn and the diameter of those filaments. Woven fabrics offer many advantages in terms of deformation capabilities, including dimensional stability, good conformability and deep-draw formability. Compared to unwoven-fabric composites,

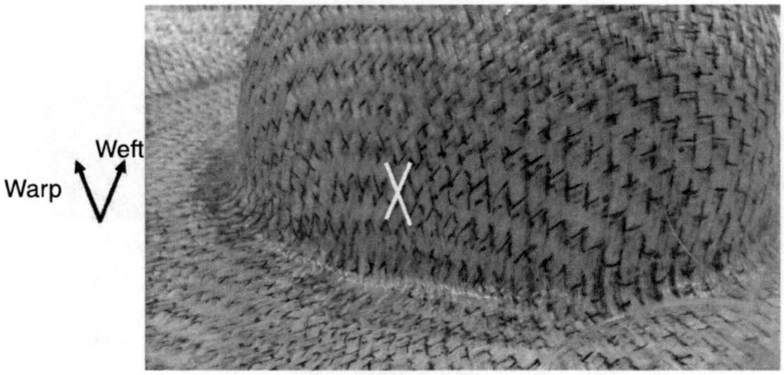

6.3 Shearing of a deformed woven-fabric.

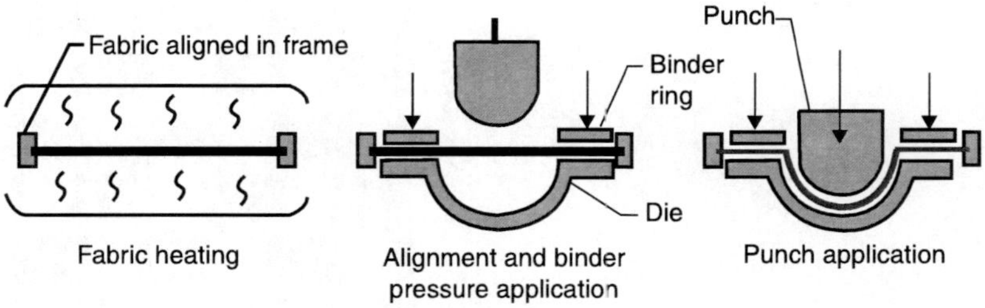

6.4 Schematic of a thermostamping process.

the woven-fabric composites provide more balanced properties, higher impact resistance, easier handling and lower fabrication cost, particularly for parts with complex shapes. Figure 6.2a clearly shows the plain-weave of the fabric where the warp and weft yarns are mutually perpendicular, and Fig. 6.2b denotes a unit cell (UC) of this weave pattern. When using this fabric to form a hemisphere, the yarns rotate relative to one another, that is shear, to allow the fabric to conform to the double curvature of the hemisphere (Fig. 6.3). Note the mutually perpendicular lines in Fig. 6.2a become sheared in Fig. 6.3. The associated shear stiffness and the evolution of this shear stiffness as a function of the deformed state of the fabric are critical to the thermostamping process.

6.2 Process description

The equipment utilized for the thermostamping process is very similar to that used for the sheet-metal stamping process. Figure 6.4 shows a schematic of a thermostamping process. Multiple layers of prepreg fabrics are aligned in a frame and arranged in a prescribed stacking sequence. The prepreg can be a polymer that is combined with the fabric in any one of a number of ways, for example coating on the yarns used for making the fabric, a resin such as that used in a sheet molding compound (SMC) applied to fabric, or thermoplastic polymer-based fibers that are commingled with fiberglass and/or carbon fibers to make the yarns that are subsequently used to make the fabric. In thermoplastic molding scenarios, the stack may then be placed in an oven to preheat the fabrics so as to allow the polymer matrix to flow and to wet the fibers fully. If no preheating is necessary, as is the case with an SMC fabric, then the oven is bypassed, otherwise the stack is then shuttled to the press and aligned with a 'punch and die', or more appropriately termed 'core and cavity' in the composites-forming community, of the desired shape. Prior to the application of the punch force, one or more binder rings, sometimes simply referred to as binders, may surround the punch area to provide in-plane forces that reduce the potential for wrinkles and waves to be formed in the fabric stack as it is punched into the die cavity. The tools, consisting of the core and cavity, and binder(s), are typically heated to control the curing rate. Once punched, the part assumes the shape of the die and results in a solid part after the polymer has cured. The length of time that the formed part sits in the tool is a function of the matrix choice, thermoplastic or thermoset, and the temperature of the tool. Typically the time to cure can vary from a few seconds for a thermoplastic to several minutes for a thermoset matrix.

Figure 6.5 shows the geometry of the tooling used in an international benchmarking program where the objective was to compare the prediction capabilities of various thermostamping simulations for the resulting fabric deformations to physical parts (Sargent *et al.*, 2010). This geometry was chosen because it includes compound curvatures as depicted by the

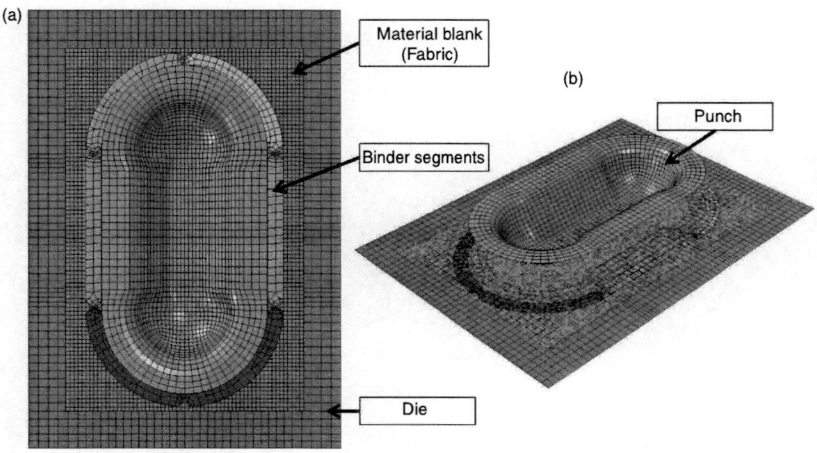

6.5 Finite element model of the double dome tooling (a) top view and (b) isometric view.

hemispheres on either end which transition to a trapezoidal trough in the center. The flat sides of the trough allow for the ability to extract test coupons for subsequent material characterization of the cured composite. This tooling, referred to as the Double Dome, consists of a stationary die above which the fabric stack is placed. A set of six binder segments is applied around the perimeter of the die using a prescribed normal pressure that in turn induces an in-plane tensile membrane force in the fabric. This in-plane stretching of the fabric reduces the potential for the development of waves and wrinkles as the punch pushes the fabric into the die and the fabric layers assume the desired shape. Figure 6.6 shows examples of parts formed from two different fabrics using this double dome tool.

The parts shown in Fig. 6.6 were made using (a) Twintex® balanced plain-weave fabric and (b)Twintex® unbalanced twill-weave fabric. Twintex® is a woven fabric comprised of commingled fiber glass/polypropylene yarns (Fig. 6.7) and is designed for the manufacture of composite parts using a

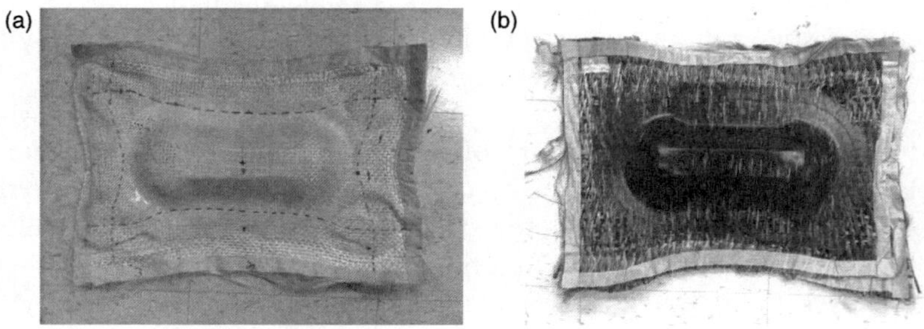

6.6 Formed parts resulting from the thermostamping process using Twintex® (a) balanced plain-weave fabric and (b) unbalanced twill-weave fabric.

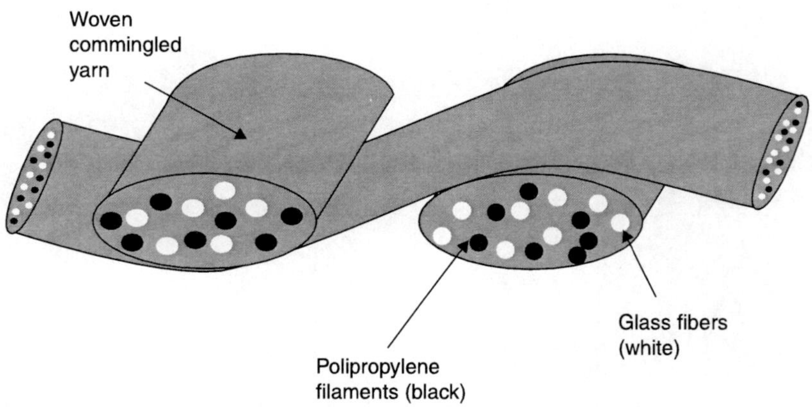

6.7 Twintex® commingled glass/polypropylene yarns.

thermostamping process. Upon heating in the oven, the polypropylene melts, and the liquid polypropylene wets the fibers. Upon cooling, the polypropylene will solidify to become the composite matrix. Often the layers of this type of Twintex® fabric are pre-consolidated either together or separately prior to stamping to accommodate ease of handling. The use of a frame facilitates handling of the fabric.

6.2.1 Fabric architecture

The thermostamping process is dominated by the fabric architecture. Currently, most of the pure and hybrid woven fabrics used in textile composites are simple 2D fundamental weaves, that is, plain, twill and satin weaves, which are identified by the repeating patterns of the interlaced regions in the warp and weft directions. The plain weave is one of the most commonly used basic reinforcements for woven-fabric composites. In a plain-weave structure, one weft yarn goes over and under warp yarns as shown in Fig. 6.2. In a twill-weave structure as shown in Fig. 6.8a, each warp yarn is woven over two consecutive weft yarns and under the following two weft yarns. The UC of a twill-weave fabric is shown in Fig. 6.8b. The satin-weave fabric has good drapability, with a smooth surface and minimum thickness. One warp yarn is woven over N ($N > 2$) successive weft yarns and then under one weft yarn. This weave structure is called an ($N + 1$)-harness satin weave. The satin-weave fabric as shown in Fig. 6.9a is a 4-harness satin-weave fabric, and the associated UC is shown in Fig. 6.9b.

Unidirectional fabrics are also used for the thermostamping process. These fabrics do not exhibit the undulations (crimp) that are present in woven fabrics, and thus, composites made from the non-crimp unidirectional fabrics will have essentially straight-line yarn paths. These unidirectional fabrics may be either stitched or unstitched. The primary role of these stitches is to

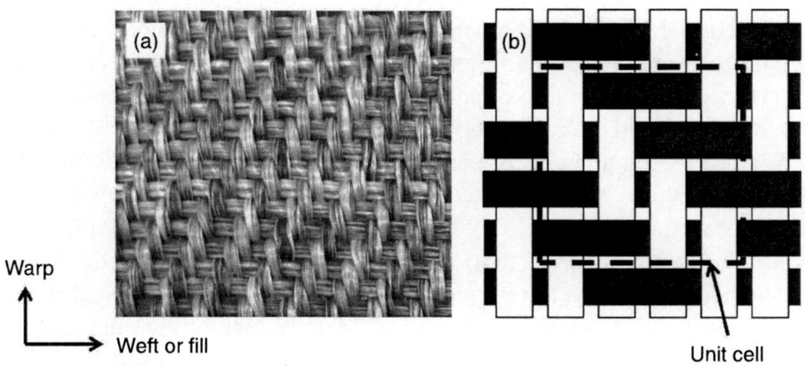

6.8 (a) Twill-weave fabric with (b) representative unit cell (UC).

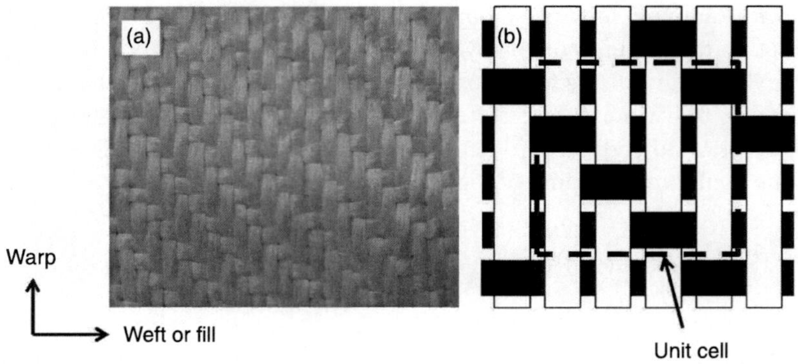

6.9 (a) 4-harness satin-weave fabric with (b) representative unit cell (UC).

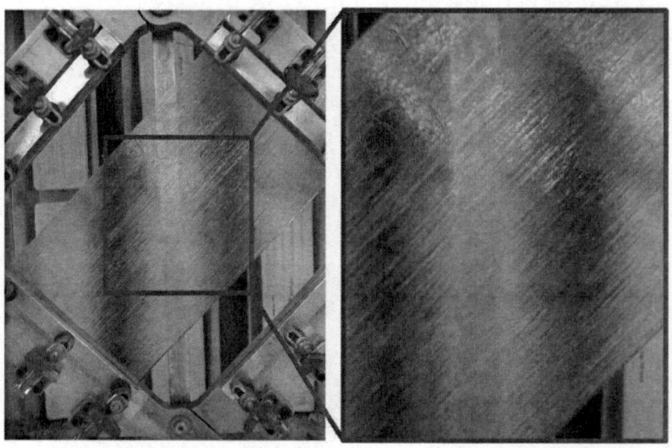

6.10 Unidirectional fabric.

facilitate fabric handling. Figure 6.10 shows a single layer of an unstitched unidirectional fabric that has an uncured polymer film. Figure 6.11 shows two mutually perpendicular layers of unidirectional yarns with cross stitching. Hence, this fabric is similar to a plain-weave fabric but does not have the undulations of a woven fabric.

The processing temperature is critical when using a thermoplastic matrix material. If there is incomplete melting of the thermoplastic matrix, then matrix cracking, fiber buckling, fiber bridging and out-of-plane wrinkling can occur. Compounding the situation is the potential sensitivity of the thermoplastic matrix material to temperature history, influencing induced residual stress, crystallinity, and thus, mechanical performance of the final part. One of the most important processing parameters that significantly influences matrix material crystallinity is the cooling rate after forming (Parlevliet

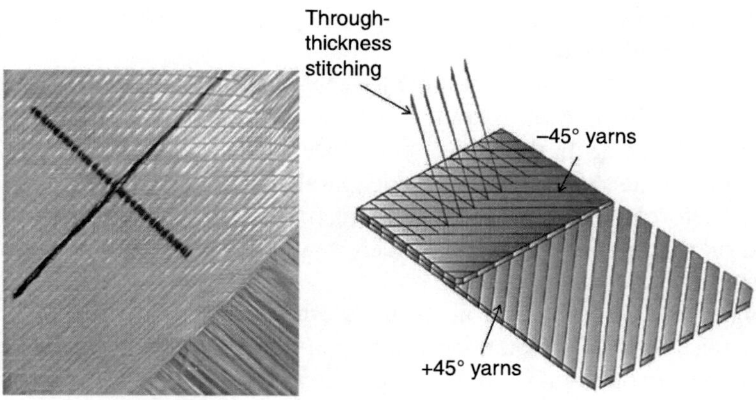

6.11 Stitched double-bias fabric.

et al., 2006) which is directly related to the tool temperature. For semi-crystalline thermoplastics, a high cooling rate may lead to a reduced peak crystallization temperature and reduced crystallinity levels, and less matrix shrinkage. A relatively high level of crystallinity will result in increased levels of static strength but reduced matrix fracture toughness properties. In addition, the tool temperature will impact the cycle time of the process as it is related to the amount of time required for the part to set. Thus, when using a stack of thermoplastic fabric plies, care should be taken to ensure that the temperature gradient within the stack as a result of preheating the material in the oven does not result in an overheating situation on the outer plies and insufficient heating on the inner plies.

6.3 Material characterization

The basic fabric material characterization consists of quantifying the evolution of the shear stiffness as a function of yarn rotation and the tensile behavior. Such characterization should be done over the range of temperatures that can occur during fabric deformation when using prepreg fabrics. For dry fabrics, that is, no prepreg, the characterization can be done at room temperature.

6.3.1 Shear-frame testing

As previously mentioned, shear is the primary mode of deformation experienced by a fabric when it is formed into a part with multiple curvatures. Thus, accurate material characterization is required for modeling of the shear deformation of woven fabrics to obtain a fundamental understanding of the mechanical behavior of the fabric to predict the fabric response during forming processes.

The standard test for measuring the shear behavior of fabrics is the shear-frame test, also known as trellis-frame test, or the picture-frame test, as shown in Fig. 6.12. In this test, a fabric specimen is clamped with the yarns typically directed perpendicular and parallel to the four clamping bars. Shear deformation is developed by fixing one corner and applying a tensile load on the opposing corner. The deformation of the fabric in the shear-frame test is shown in Fig. 6.13.

Figure 6.14 shows a typical shear deformation curve overlaid with visual depictions of yarn rotation for a woven-fabric composite. The yarns are orthogonal to one another in the initial state. Upon initiation of intraply shear deformation, the yarns begin to rotate and possibly to slip. Yarn slip occurs when two interlaced yarns slide over one another. During yarn rotation, it is assumed that the load is initially due to frictional interaction between adjacent yarns at their crossover points and possibly a viscous contribution if a liquid resin is present. Further deformation continues until an angle is reached where yarn compaction initiates. During compaction, the adjacent yarns begin to compress each other, increasing the shear stiffness. The locking angle is reached (Fig. 6.14) when the yarns are no longer able to compact

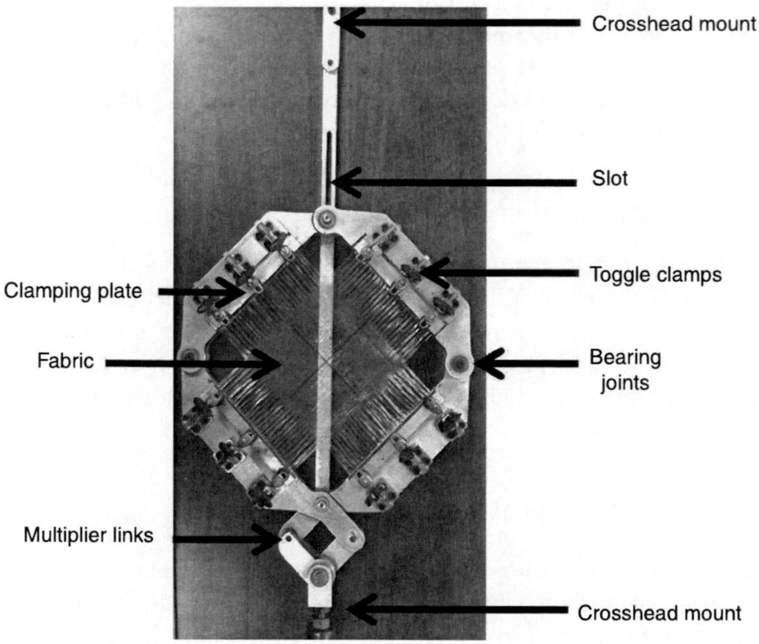

6.12 Shear-frame test (Lussier, 2002).

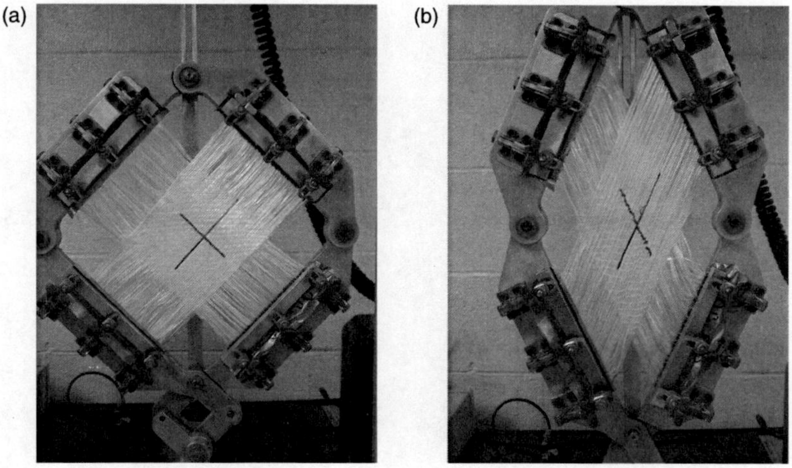

6.13 Deformation of shear-frame test (a) before deformation and (b) after deformation.

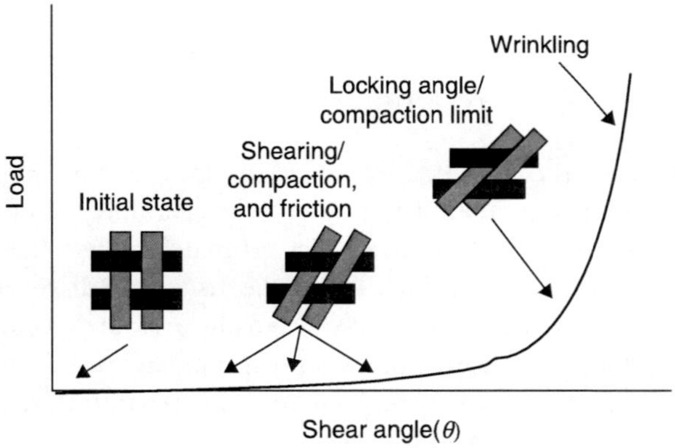

6.14 Typical woven-fabric shear behavior curve (Lussier, 2002).

easily. Deformation beyond the locking angle causes the fabric to buckle out of plane followed by a significant increase in stiffness. At this point, the deformation mode becomes a combination of shear, yarn compaction and tension.

The shear-frame test assumes there is only in-plane pure-shear deformation of the fabric before any out-of-plane buckling occurs. This assumption has been verified using digital image correlation (DIC). Figure 6.15 shows a shear-frame test of a plain-weave fabric where DIC has been used to map the state of shear over the surface if the fabric. There is some variation in the shear angle across the surface, but the variation is in the order of $\pm 1°$, which can be assumed to be essentially uniform.

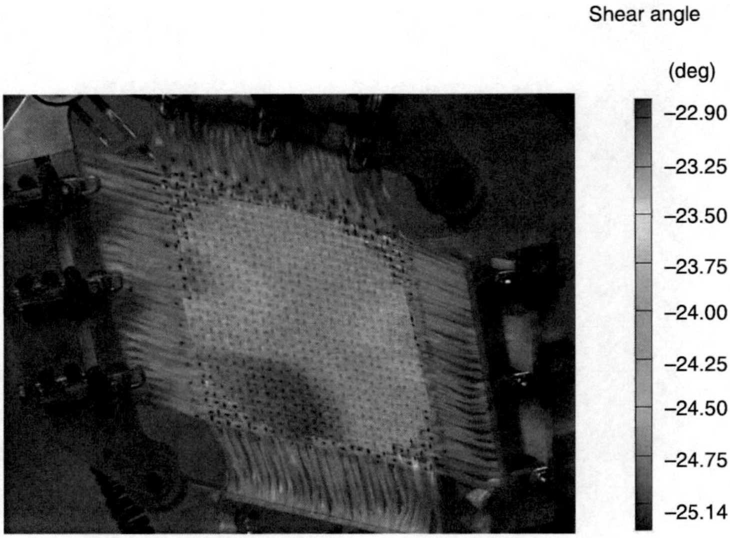

6.15 Digital image correlation contours showing pure shear.

6.3.2 Tensile testing of yarns

As fabric is pressed into a mold, certain yarns may exhibit high tensile stresses as they bridge undulated regions of the geometry, that is, peaks and valleys. These high tensile stresses can ultimately cause the yarns to break, leading to a poor-quality part. Therefore, to characterize the tensile mechanical behavior of the yarns, uniaxial tensile tests are performed on individual yarns. Pneumatic cord and yarn horn capstan grips are used, and the gage length is set to a relatively high value (~ 1 m) to minimize the effect of the deformation of the yarn within the grips.

The tensile tests quantify the fracture load and modulus of the yarn (Fig. 6.16). The modulus of the yarn is obtained from the slope of the associated stress/true-strain curve. The effective cross-section of the yarn A_{eff} can be determined based on the linear density of the yarn ρ_{linear} and the fiber material density ρ_{mat}:

$$A_{\text{eff}} = \frac{\rho_{\text{linear}}}{\rho_{\text{mat}}} \qquad [6.1]$$

In addition, the crimp ratio (i.e., the difference between the length of the yarn and the length of the fabric) can be determined for woven fabrics that feature yarn undulations. The combination of the yarn crimp ratio and the yarn modulus is needed to describe the tensile mechanical behavior of a composite fabric (Fig. 6.17).

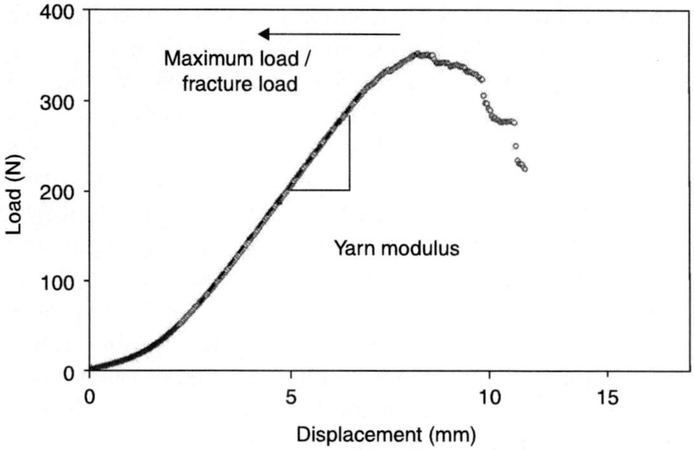

6.16 Typical load-displacement curve from yarn tensile testing.

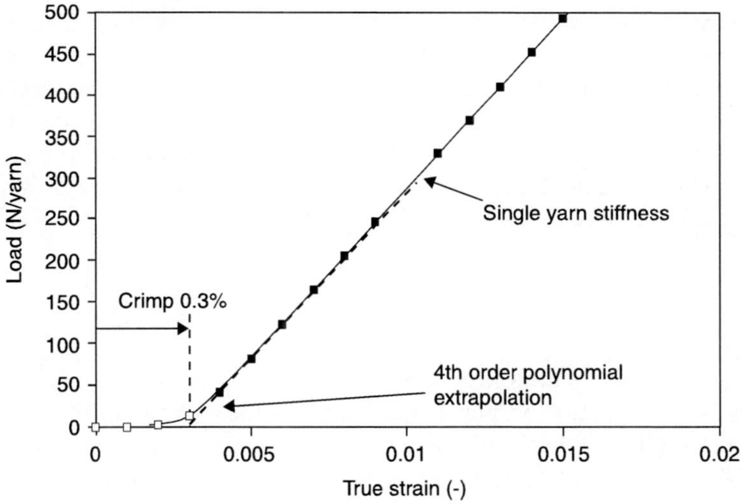

6.17 Determination of the fabric tensile behavior from the yarn modulus and the crimp ratio for a typical yarn.

6.3.3 Friction

Similar to sheet-metal forming, the thermostamping of a composite part also exhibits part-quality challenges like wrinkling and tearing. One of the significant processing parameters that can highly influence the quality of a composite part is the amount of in-plane tension that is induced in the fabric due to the normal forces acting on the fabric from the interaction with the tooling, for example the binders and the punch, and the adjacent fabric layers. If the in-plane forces are too low, then the fabric may develop defects in the form of in-plane waviness and out-of-plane wrinkles during the forming process. Conversely, if the in-plane forces are too high, then the yarns may separate and/or break and/or the fabric may tear (Wilks, 1999). Because the fabric slides under the binder(s) and over the surfaces of the punch and die, understanding the friction between the tool (in particular the binder) and fabric surfaces is critical to the forming of quality composite parts (Akkerman *et al.*, 2007; Chow, 2002; Gorczyca *et al.*, 2003; Vanclooster *et al.*, 2008). Additionally, several layers of fabric are often pressed into the die together which may cause some of the layers to slide relative to one another. Thus, the friction between adjacent layers of fabric must also be quantified to allow for a credible model for simulating the forming process and predicting composite part quality with confidence.

The coefficient of friction between the fabric and the metal tooling and between adjacent layers of fabric during the thermostamping process has been shown to be primarily a function of normal force, fabric velocity and temperature (Akkerman *et al.*, 2007; Chow, 2002; Fetfatsidis, 2009; Gamache, 2007; Gorczyca *et al.*, 2003; Vanclooster *et al.*, 2008). In some cases, fabric orientation and fabric shearing can influence the resulting coefficient of friction. The friction coefficient has been related to the Hersey number, H, which is sometimes referred to as the Stribeck number (Hutchings, 1992; Stachowiak and Batchelor, 2005), and is a function of resin viscosity, η, speed, U and normal load, N (Equation [6.2]).

$$H = \frac{\eta \cdot U}{N} \quad [6.2]$$

Figure 6.18 shows a generic coefficient of friction vs. Hersey number curve. During the thermostamping process, the polymer matrix material can act as a lubricating fluid film as the fabric is drawn into the die. The outer fabric layers slide against the metal tooling and the inner layers may slide relative to one another. Typically, for a high-volume process such as thermostamping, it is advantageous to have a low-viscosity matrix material to allow rapid flow throughout the part, and a low melting/curing temperature such that minimal heating is required. Polypropylene is often used as a thermoplastic matrix material due to its relatively low viscosity and low melting temperature. During the thermostamping process, as the temperature of the fabric changes, the viscosity and shear rate of the polypropylene will vary (Fig. 6.19). Changes in these properties will affect the thickness of the fluid film that separates the contacting surfaces, which in turn will affect the resulting part quality. Depending on the thickness of the fluid film, the sliding friction has been shown to be based on hydrodynamics (Gorczyca *et al.*, 2003)

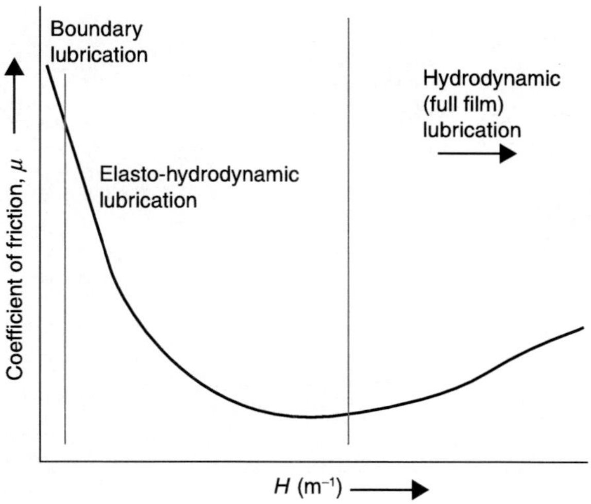

6.18 Theoretical Stribeck curve.

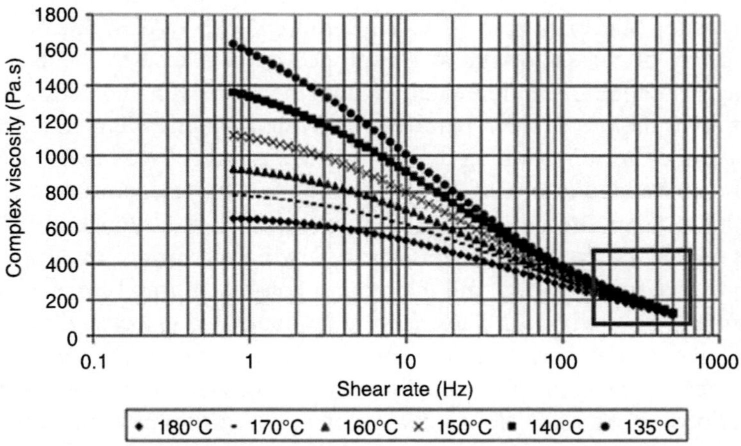

6.19 Flow curves for polypropylene at different temperatures (Vanclooster *et al.*, 2008).

or on elasto-hydrodynamics (Chow, 2002). Hydrodynamic lubrication assumes the contacting surfaces are fully separated by the fluid film and elasto-hydrodynamic lubrication accounts for deformation of the surfaces. To capture the frictional behavior of woven-fabric composites, a friction-testing apparatus capable of varying and measuring ranges of temperature, velocity and normal force (pressure) relevant to the thermostamping process is used.

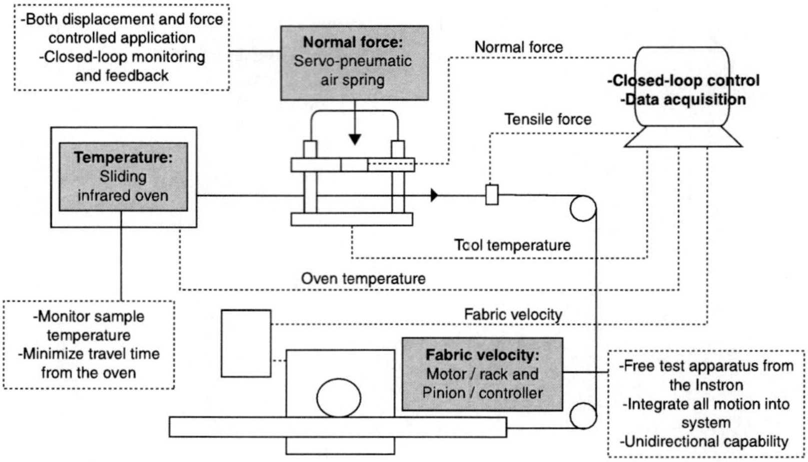

6.20 Schematic of a fabric friction-test apparatus.

A schematic of the fabric friction-test setup used in the Advanced Composite Materials and Textile Research Lab (ACMTRL) at the University of Massachusetts at Lowell (UML) is shown in Fig. 6.20. The fabric is clamped in a holder, heated in an infrared oven and subsequently shuttled to the press, which features two heated platens – analogous to the steps of the thermostamping process as shown in Fig. 6.4. A thermocouple can be embedded into each sample to monitor the temperature throughout the test. An alternative to using the oven is to use the platens to heat the samples. The normal force is applied by an air spring. To control the air pressure in the spring and maintain a constant normal force, a servo-pneumatic system is used. The system is implemented in a closed-loop control configuration which continuously monitors the normal-force load cells and updates the command signal to obtain a desired force level which is prescribed by the LabVIEW software. In addition to prescribing the command signal for the air spring, the LabVIEW program also continuously acquires the signals from the normal-force load cells, a tension-force load cell that is in line with the pullout cable, and a linear transducer, which is used to track fabric displacement. A DC motor drives a rack and pinion to pull a cable which in turn pulls the sample through the press. To measure the fabric/fabric friction, two pieces of fabric are clamped together on one end, and these two pieces of fabric sandwich the fabric sample mounted in the fabric holder (Fig. 6.21). As the platens close shut and the fabric inside the holder begins to pull, the 'sandwich' clamp is blocked by the closed platens, thus allowing only the middle layer of fabric to be pulled through the test device.

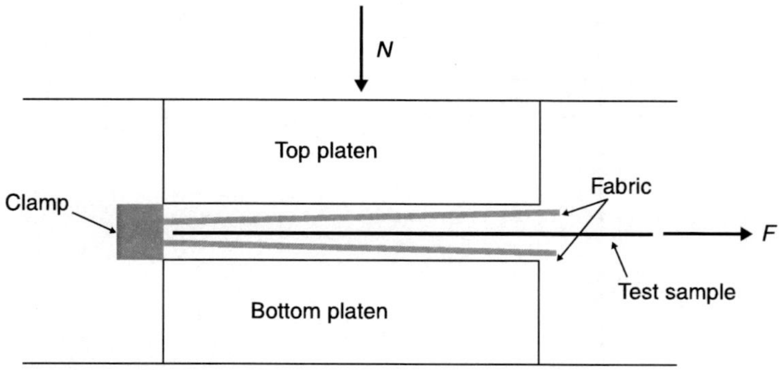

6.21 Schematic of fabric/fabric setup.

Typically, when plotting the pull out force, *F*, as a function of displacement, an initial peak force needed to initiate slipping is observed. This initial peak corresponds to the static coefficient of friction. Following the initial peak is an essentially steady-state value of the pull out force corresponding to the dynamic coefficient of friction. The effective coefficient of friction is calculated using:

$$\mu_{\text{eff}} = \frac{F}{2N} \qquad [6.3]$$

where the normal force, *N*, is multiplied by a factor of two to account for the two contacting surfaces on each side of the fabric sample. Figure 6.22 depicts a typical coefficient of friction response for three tests at the same conditions. These data reflect the excellent repeatability of the test.

6.3.4 Additional deformation modes

Depending on the thermostamping process parameters and fabric/matrix type, unwanted defects may appear in the part. Particular attention is focused on preventing fabric wrinkling in the form of in-plane and out-of-plane waves. The in-plane shearing behavior is often used as an indicator via the 'locking angle' as to when wrinkles and/or waves may develop. As yarns rotate relative to one another, they eventually reach a point where they can no longer rotate in-plane and must buckle out-of-plane. However, it has been shown that simply relating wrinkling to the shear angle is not sufficient. The formation of wrinkles and/or waves depends on the combination of in-plane shear, tension and bending behaviors of the fabric (Boisse *et al.*, 2010).

When characterizing the dry fabric bending stiffness, a cantilever 'beam' method is often used (de Bilbao *et al.*, 2009), where the fabric is allowed to

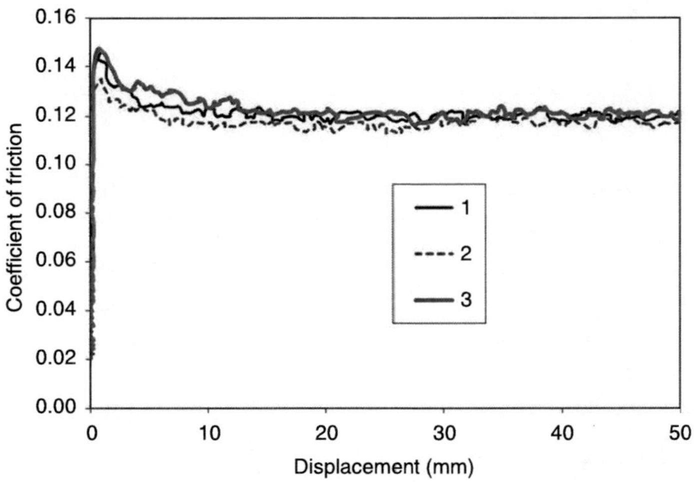

6.22 Typical coefficient of friction vs. displacement curves resulting from friction experiments.

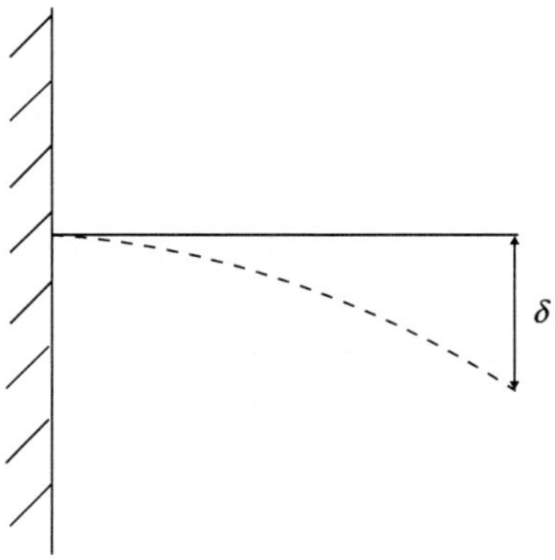

6.23 Cantilever experimental setup.

bend due to its own weight (Fig. 6.23). However, depending on the length of the sample, the effective direction of the distributed load on the sample changes due to the large deformations, and thus, a single value for the fabric bending stiffness cannot be concluded for all lengths despite efforts to compensate for the effect of gravity.

An alternative method for characterizing the bending stiffness is to align the length of the beam with gravity and thereby reduce the nonlinear loading effects (Soteropoulos *et al.*, 2011). Fabric samples are clamped at one end and hung vertically. A horizontal load is applied to displace the tip of

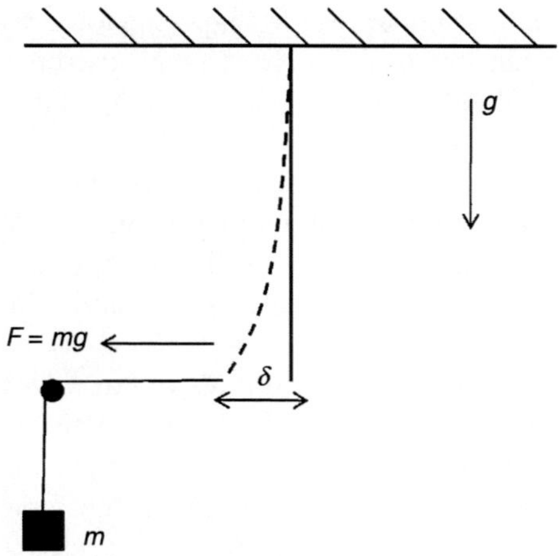

6.24 Schematic of bending stiffness test setup.

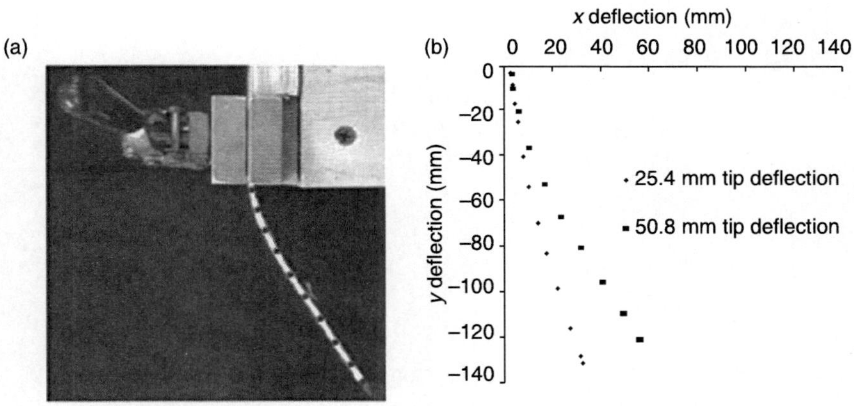

6.25 Post-processing of experimental data including (a) generation of *x* and *y* data points and (b) plotting of data points.

the fabric a known amount. This load is applied by attaching masses to a string tied to the tip of the fabric sample (Fig. 6.24).

With the effective fabric length positioned in the clamp, various tip displacements can be applied using an appropriate load, and a digital image of the fabric is captured. Individual data points can be generated along the fabric length (Fig. 6.25a) and plotted (Fig. 6.25b).

Finite element models similar to the experimental fabric samples are used to determine the fabric bending stiffness. Equivalent tip displacements are prescribed in the models and a range of bending-stiffness values are used to replicate the experimental profile shape of the sample. While several values of bending stiffness may lead to a similar profile shape, the force required to displace the tip a known amount can be compared to the experimental force to ensure that the correct bending stiffness is being used in the model. Note that for the effective bending stiffness as used here the unit for moment, M, is N-mm and for curvature, κ, is mm^{-1}. Thus, the bending stiffness M/κ to be defined in the finite element model should be given as N-mm^2. As a means of verifying that the bending stiffness is valid, the bending stiffness can be applied to models with different tip displacements and different fabric sample lengths (Fig. 6.26).

As mentioned previously, the formation of wrinkles and/or waves depends on the combination of in-plane shear, tension and bending behaviors of the fabric. It is certainly possible to predict defects based on the in-plane shear and tension behaviors alone. However, the bending stiffness more accurately defines the wrinkle/wave shape. For example, compressing a fabric sample model from two sides will yield a different number of out-of-plane waves with different amplitudes depending on the magnitude of the bending stiffness. A much stiffer fabric will result in fewer waves with greater

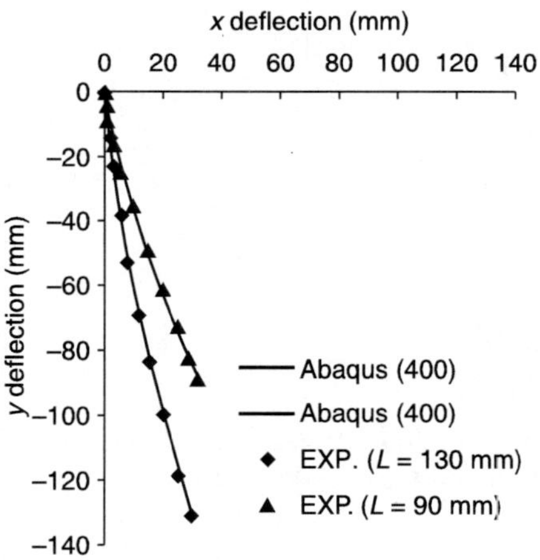

6.26 Comparison of experimental data to FE model results for varying tip displacements and fabric lengths.

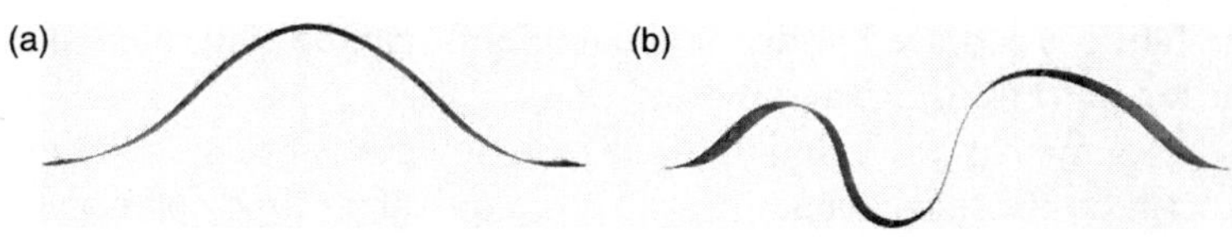

6.27 FE model of a fabric sample compressed from each side with (a) high bending stiffness and with (b) low bending stiffness.

amplitude (Fig. 6.27a) than a fabric with a low bending stiffness which will show multiple waves with smaller amplitudes (Fig. 6.27b).

As the in-plane and out-of-plane deformation modes are better understood, research is moving in the direction of characterizing compaction of the fabric yarns (Grujicic *et al.*, 2004). As a fabric blank is pressed into a mold during the thermostamping process, the shearing, sliding and bending of different yarns may lead to different levels of compaction, although the overall thickness of the part is mostly controlled by the molds. This compaction can lead to a variable fiber volume density in certain locations, as well as influence the permeability of the fabric. Changes in permeability can affect the flow path of the matrix fluid which may lead to voids and dry spots. These voids and dry spots can have detrimental effects on the mechanical properties of the formed part, as they introduce discontinuities into the part. Voids and dry spots can be avoided by heating the matrix to a temperature that would increase flow and reduce viscosity or by increasing pressure exerted on the fabric blank by the binders and the punch as the fabric blank is pressed into the mold. However, if the matrix flows too easily or the pressure is too high, the matrix can be squeezed out of the fabric during the forming process thereby altering the fiber volume. Additionally, large pressures can also cause the woven yarns to separate and/or break during forming. Thus, formation of a quality part free of dry spots and voids involves a careful balance among many processing parameters including tool temperature, fabric temperature, binder ring pressure and mold pressure. Research is currently underway to provide answers to the optimal combination of processing parameters to avoid voids and dry spots for a variety of formed shapes.

6.4 Modeling

To reduce manufacturing cost and time and to achieve an acceptable composite design without expensive prototype testing, it is desirable to simulate the thermostamping process. There is currently no widely accepted modeling approach that can accurately capture all the important aspects of fabric deformation and effectively predict both the macroscopic mechanical response of the fabric as well as the response of the yarns at the mesostructural level. Specialized models employing various approaches have been proposed.

Several families of modeling methods have been developed. Boisse *et al.* (2008) have reviewed many of these methods and have classified them as continuous, discrete and semi-discrete approaches. In the continuous approach, the fabric is homogenized and considered as a continuum. Conventional shell or membrane elements are used, but special considerations of solid mechanics are used to track the evolution of the principal load paths over the fabric within each element. A large volume of work has also been done on discrete modeling, that is, models where individual components of the fabric such as the fibers or yarns are considered. To maintain reasonable computing time, the models are generally limited to the scale of the yarns. A trellis of truss, beam or spring elements is used to describe the woven fabric. The incorporation of the in-plane shearing resistance over this trellis can be done by the addition of diagonal truss elements (Sharma and Sutcliffe, 2004; Skordos *et al.*, 2007) or shell or membrane elements (Li *et al.*, 2004; Sidhu *et al.*, 2001). The semi-discrete approach has been introduced recently and shows promise. Specific 4- and 3-node finite elements that consider the mesoscale components of the woven fabric (the yarns) have been developed to model the specific mechanical behavior of the fabric (Boisse *et al.*, 2001; Hamila and Boisse, 2007). The continuous and semi-discrete models can be very challenging to implement into commercially available finite element packages. In contrast, discrete methods are generally easier to set up (Sharma and Sutcliffe, 2004; Skordos *et al.*, 2007). Apart from a few exceptions, for example a model by Yu *et al.* (2003, 2005) and a recent continuous model developed by Willems (2008), material parameters cannot generally be directly deduced from testing results, and an identification step is needed (Willems, 2008). This identification step involves the simulation of tensile and shear tests to tune the material parameters to match the experimental data.

In this chapter, a discrete approach based on an explicit finite element formulation using a hypoelastic description is used to present examples of modeling the thermostamping process to form parts. The explicit formulation is chosen because it is currently the best suited formulation for completing forming simulations due to its computational time efficiency and relatively robust contact algorithms. In developing this method, the objective was to keep a relatively simple description and to have a scheme that can be implemented within popular commercially available explicit finite element packages that allow for the linking of user-defined material subroutines with the explicit solver. The model allows a direct identification of the material parameters from simple tensile and shear tests of the fabric.

A discrete description of the fabric is built using a mesh of 1D and 2D elements (Fig. 6.28). The 1D elements account for the tensile contribution of the yarns to the fabric material behavior and automatically capture the evolution of the orientation of the principal load paths as the yarns rotate to conform to the compound curvatures of the surface of the tooling. The 2D elements account only for the shearing resistance of the fabric and have no tensile stiffness. Appropriate nonlinear constitutive equations are associated with the 1D and 2D elements via user-defined material subroutines to capture the mechanical behavior of the fabric. The mathematical details of the model are summarized in Jauffres *et al.* (2009).

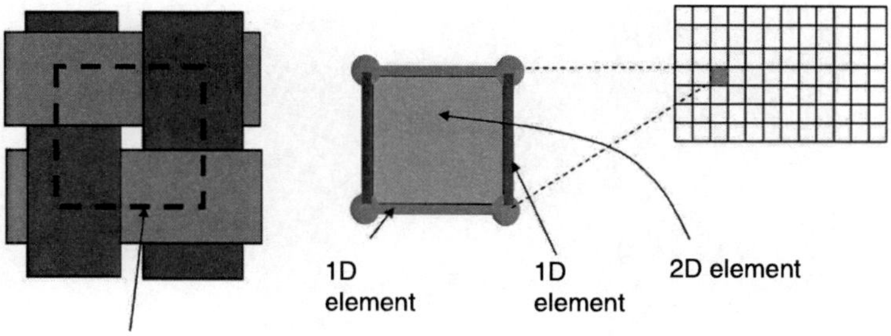

6.28 Principle of the discrete mesoscopic model using a combination of 1D and 2D elements to model a plain-weave fabric.

To capture the weave-pattern effect, the mesh is defined to match the fabric UC dimensions. Consequently, in the case of a large part, the fabric mesh could be extremely fine, which can lead to very long computation times. However, it is possible to use a coarser mesh by reducing the number of 1D elements by a given ratio in each direction and multiplying the 1D element cross-section by the same ratio to have an overall similar stiffness of the fabric. The directions of the 1D elements follow the respective yarn directions.

6.4.1 Bias extension

The ability of the finite element method to simulate a combined state of deformation, that is, tension and shear, is demonstrated via a bias-extension test simulation. A bias-extension test consists of a tensile test of the fabric pulled in its bias direction, that is, the yarns oriented at 45° to the tensile direction (Fig. 6.29). A typical sample geometry is a rectangle having its length equal to twice its width. This geometry leads to a specific strain field involving three different zones with 0°, $\gamma/2$ and γ shearing angles. It is sometimes used as an easy way to evaluate the shearing properties of fabrics (Cao *et al.*, 2008; Lebrun *et al.*, 2003).

The bias-extension test is of particular interest for validation purposes as it involves combined shearing and tension, as well as rigid rotation of some shell/membrane elements. Consider the LS-DYNA plot in Fig. 6.30. The triangular regions at the two ends are in pure tension, the middle diamond region is in pure shear and the remaining regions are a combination of tension and shear. Recall that the pure tension and pure shear test data are used to conclude the material constants for the fabric constitutive models for tensile and shear, respectively. The bias-extension test then allows for a validation that the combination of the two deformations can be captured by the finite element model, thereby giving credibility of the finite element

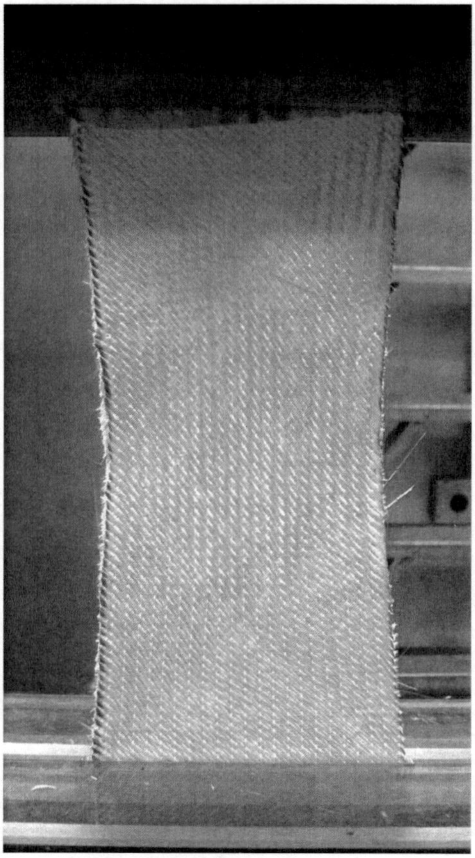

6.29 Bias extension test setup.

model to capture the mechanical behavior of the fabric in a forming simulation. The overall tensile force vs. displacement is compared between the experiment and the finite element model, and the two curves are a measure of the relative credibility of the model.

In a manner similar to that shown for the shear-frame test, DIC can be used to capture the shear-angle contours developed in the fabric during a bias-extension test, which can be compared to the shear angles predicted by the finite element model (Fig. 6.30). It is noted that the boundaries for the three different theoretical zones as viewed in the experiment (DIC image) do not exhibit as sharp a transition as is shown for the boundaries in the simulations. The sharp transitions observed in the model are due to the pin-jointed connections between the elements whereas in reality, the yarns are continuous and are allowed to bend along a smooth curve (Fig. 6.31).

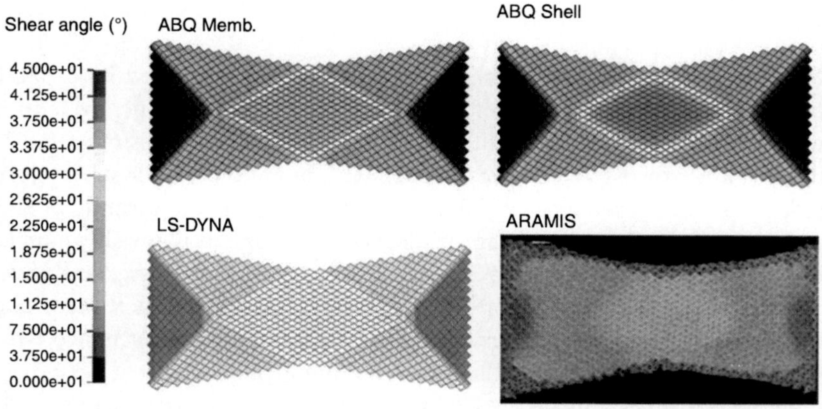

6.30 Shear angle contours in degrees during bias extension from finite element simulations and DIC measurements.

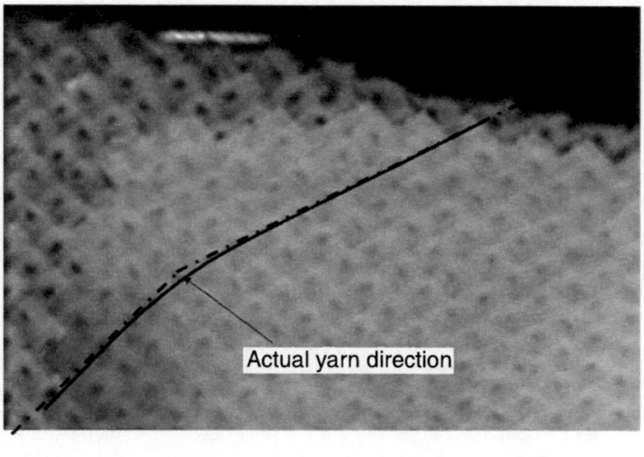

6.31 Detail of the actual direction of the yarns from the DIC.

The solid line in Fig. 6.31 denotes the actual yarn path, and the dot-dash line denotes the yarn path as predicted by the finite element model.

At large crosshead displacements in the bias-extension tests, experimental data typically indicate lower shear angles relative to the finite element model that is being presented here. Again, due to the pin-jointed nature of the model, the yarns are not allowed to slide as is commonly observed during bias-extension testing (Creech and Pickett, 2006; Harrison *et al.*, 2004). During a bias-extension test, the yarns are clamped only on one side, and consequently some of them can slide. Only the friction prevents them from sliding, and when the load to overcome this friction is developed within the fabric, sliding occurs.

6.4.2 Forming examples

Now that the finite element method has been discussed and its capabilities to capture the complex deformation mechanisms as applied to a plain-weave fabric have been presented via the bias-extension test, simulations of the forming process will be presented. Such simulations can be used to provide feedback regarding the fabric deformation during the forming process and give insight as to if and where defects such as high in-plane shear angles, in-plane and out-of-plane waves and high yarn tensile stresses are developed in the part. With this information, the designer can then reconsider the ply lay-up so as to achieve the desired structural performance while minimizing the potential for the formation of defects.

In the actual forming process, fabric blanks are initially large enough such that the desired part can be formed. However, after the fabric has been drawn into the mold there will usually be excess fabric around the part that must be removed in an additional step. To ensure that the forming process will be a low-cost, high-volume process, the simulation can also be used to determine the initial blank shape such that no excess fabric will remain after being pressed into the mold. A possible approach used to determine the blank size consists of first running a simulation with an oversized blank. From the deformed shape of this oversized blank, the 'ideal' blank geometry is extracted. Then, a second simulation is done to validate the 'ideal' blank size. Figure 6.32 shows the initial and final blank shapes with contours of shear angles (in degrees) for a 90/0 orientation using the double-dome geometry that was previously presented in Fig. 6.5.

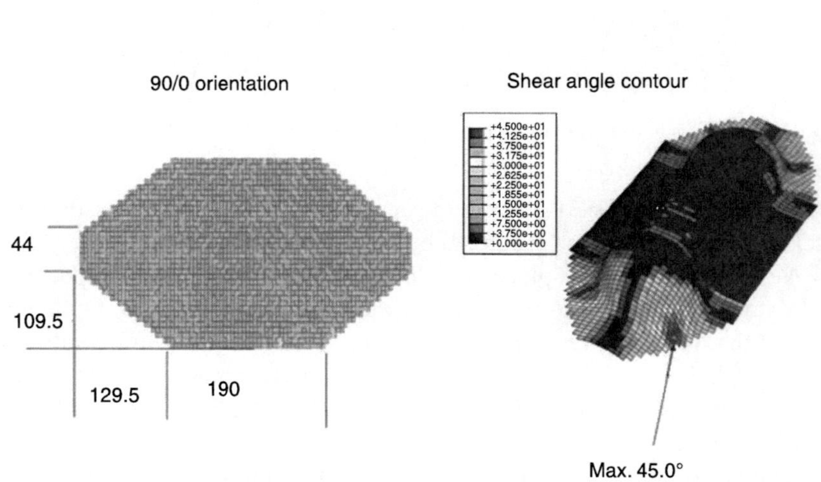

6.32 The 90/0 fabric orientation: fabric blank and shear angle contour from the simulation.

Visual comparisons of yarn orientations can qualitatively evaluate the validity of the modeling. In Fig. 6.33, the double-dome stamping of an actual 90/0 sheet is shown alongside its corresponding finite element model result. The contours shown on the deformed model are the in-plane shear strains in degrees. The overall shapes of the model and the molded part are in good agreement. A zoomed-in view on one quarter of one hemisphere of the double-dome (Fig. 6.34) confirms the ability of the model to capture the resulting yarn orientation in the molded part. One visible difference between the model and the trial molding is that more draw-in is predicted by the model than is observed in the formed part, especially in the width. A close examination of the molded parts shows a very large spacing between the yarns on the lateral low sheared zone of the double-dome rim (Fig. 6.35). This observation implies that the yarns in the fabric have slipped within the fabric and the current formulation of the model is unable to consider this phenomenon, resulting in less fabric draw-in in the actual forming of the part in comparison to that observed in the model.

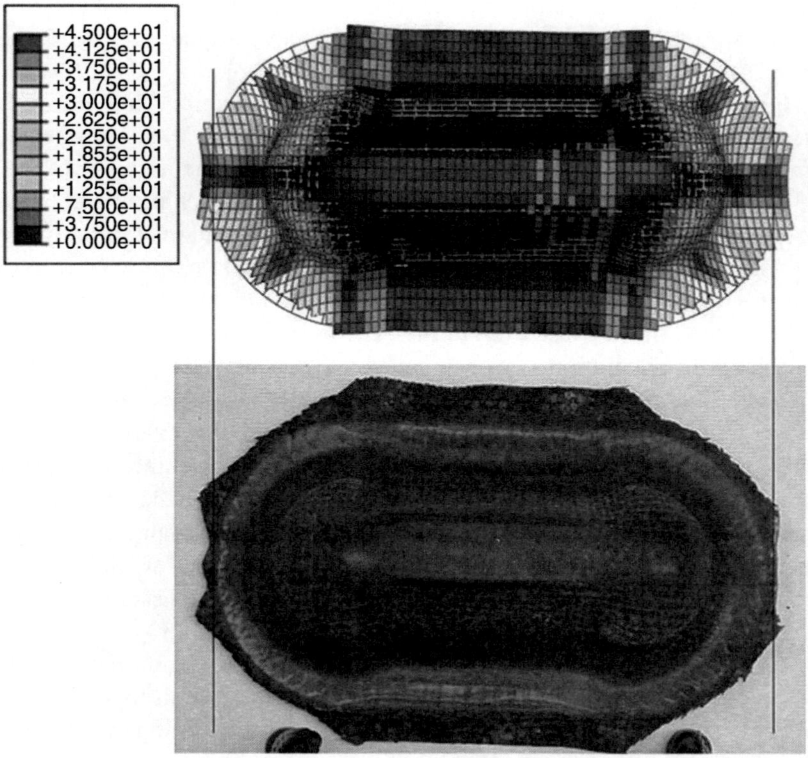

6.33 Experimental-simulation comparison of the final shapes for a 90/0 fabric orientation.

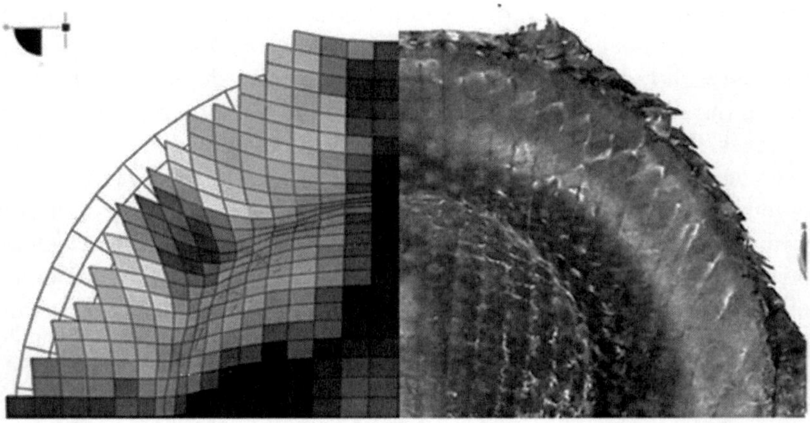

6.34 Detail of experimental-simulation comparison of the final shapes for 90/0 orientation for a quarter of a hemisphere on the double dome.

Figures 6.36 and 6.37 show a rear tub and a floorpan, respectively. Forming simulations show that either the fabric orientations must be changed or the geometry of the tool must be changed to achieve a quality part for these shapes.

First consider the geometry. The ± 45° orientation exhibits better drapability to the tub geometry than the 90/0 layout as shown by the lower shear angles in Fig. 6.38 (–45/45) than in Fig. 6.39 (90/0). The high shear angles (> 60°) in Fig. 6.39 suggest the fabric must exceed its 'locking angle' to form the shape, at which point the loads may increase beyond the fabric strength, or wrinkling may occur. Figure 6.40 shows two alternate fabric orientations for the tub. However, due to the presence of the deep-drawn corners, it is difficult to form the part without high shear angles in the fabric regardless of the fabric orientation.

Because of the relatively more complex geometry of the floorpan with its many small details, the floorpan fabric orientations also exhibit zones with shear angles greater than 60°. However the 45/–45 orientation (Fig. 6.41) draped relatively better than the 90/0 orientation (Fig. 6.42) and the high shearing zones are located in places where a cut in the blank fabric can be used to avoid this high shearing. Such cuts were made and smaller pieces of fabric were used to form a demonstration part. The demonstration part was made using SMC woven fabric and can be seen in Fig. 6.43.

As previously mentioned, it is also important for the simulation to be able to predict high tensile stresses in the yarns that may result in yarns breaking and subsequent fabric tearing. The contour of the tension in the yarns is depicted in Fig. 6.44 for the 22.5/112.5 orientation using the tub geometry. High yarn tension at the bottom of the tub can be seen in this figure. The yarn

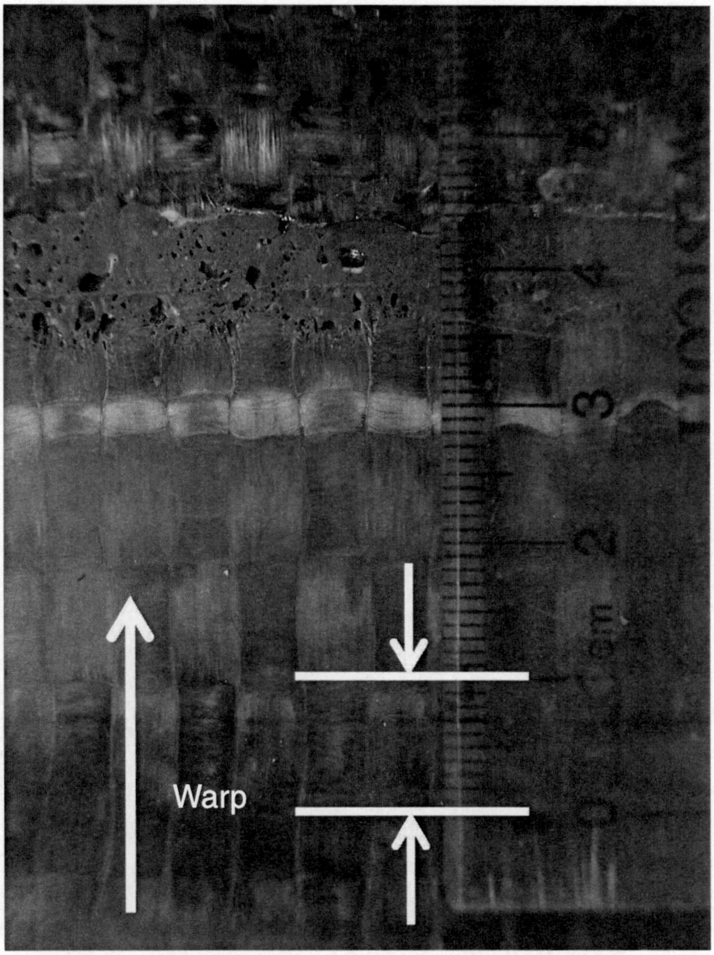

6.35 Detail of the 90/0 part showing large spacing of the weft yarn. The initial interval of the yarns is 1.4 yarns/cm, while this image shows intervals closer to 1 yarn/cm. Note that the area above the gray line is off the tooled surface of the part, and would be cut off.

fracture stress could be reached in an actual stamping, leading to fabric tearing in this zone. The high tensile stresses may be a result of the geometry design that features sharp ribs at the bottom. Either the tool design can be changed or smaller patches of fabric can be formed over these areas rather than trying to use one large piece of fabric over the entire part.

Figure 6.45 shows a contour of the yarn tensile forces in the floorpan for a 90/0 fabric orientation. The floorpan does not exhibit marked ribs such as the ones present at the bottom of the tub, which explains the lower yarn

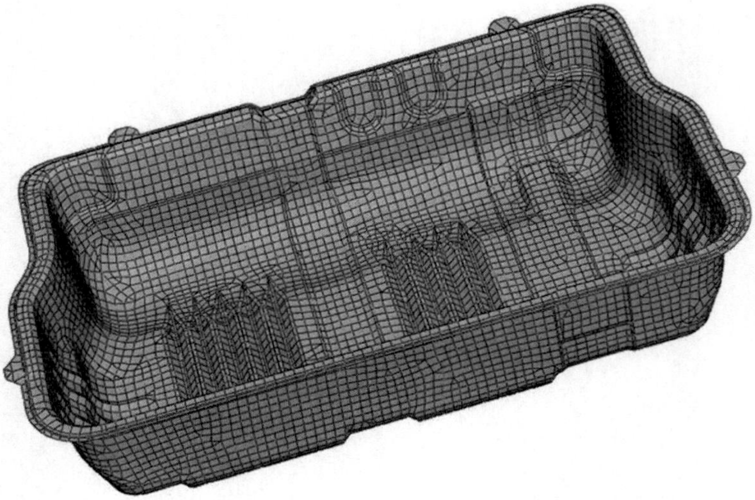

6.36 Finite element mesh of the tub punch.

6.37 Finite element mesh of the floorpan punch.

tension observed for the floorpan. It is also observed that there is compression build up in some yarns, in particular for the 45/–45 orientation (Fig. 6.46). This compression could be due to the complexity of the floorpan geometry, compared to the tub.

Molding of the tub and the floorpan confirmed these model predictions, at least qualitatively. For the tub, the tooling was built before this modeling was done. Molding attempts showed that the 45/–45 plies were able to be

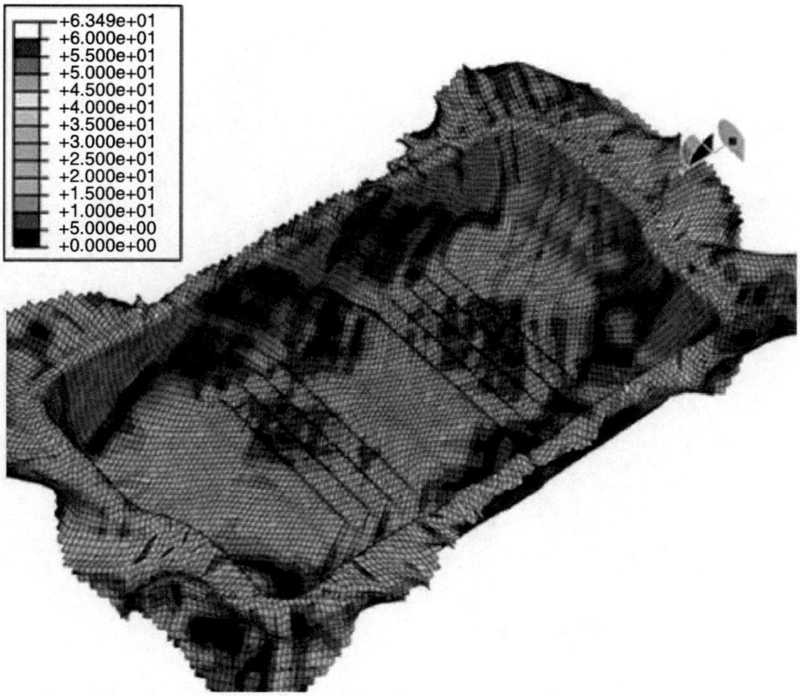

6.38 Shear angle contour in degrees for the 45/–45 fabric orientation.

formed successfully, but the 0/90 plies were not moldable without significant darting. Note that for structural property reasons, it is not feasible to only mold 45/–45 plies, as part loads will be quasi-isotropic. The floorpan modeling was done before the tooling was designed. Thus, knowing the areas of concern allowed the design to be modified, primarily by softening radii in the areas shown to have high shear angles in Figs. 6.41 and 6.42. The final molding of the floorpan required blanks of different sizes in different plies because the floorpan has several areas of different thicknesses to maximize mass savings. Thus, none of the plies was required to cover the entire floorpan as one blank. This use of various sized plies relaxed the need for shear deformations for the fabrics to conform to the shape of the mold, thereby reducing the shear angles. The molded floorpan is shown in Fig. 6.43.

6.5 Methods of improving the process to improve product quality

To realize the weight-saving benefits that composites can contribute to a design relative to metal parts, aligned-fiber composites must be used, as opposed to chopped-fiber composites. The thermostamping process can

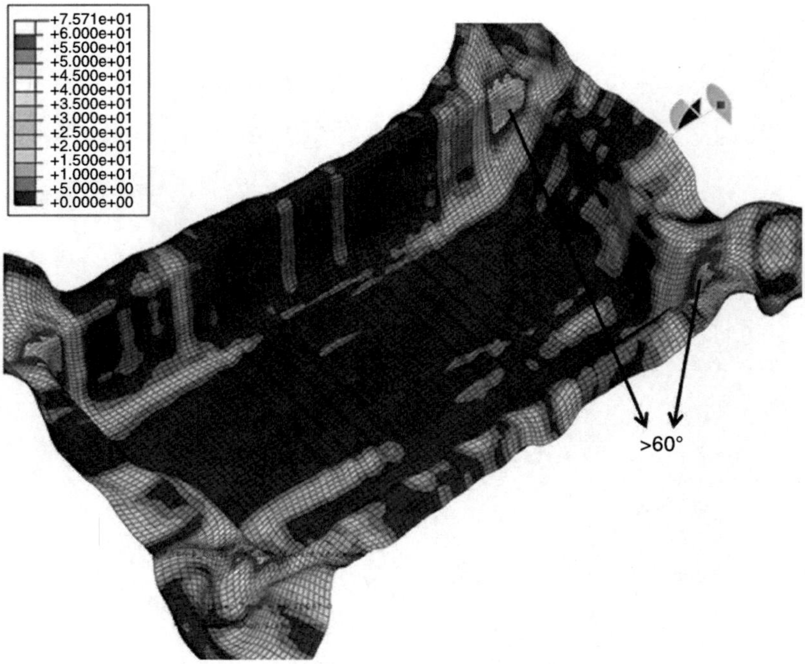

6.39 Shear angle contour in degrees for the 90/0 fabric orientation.

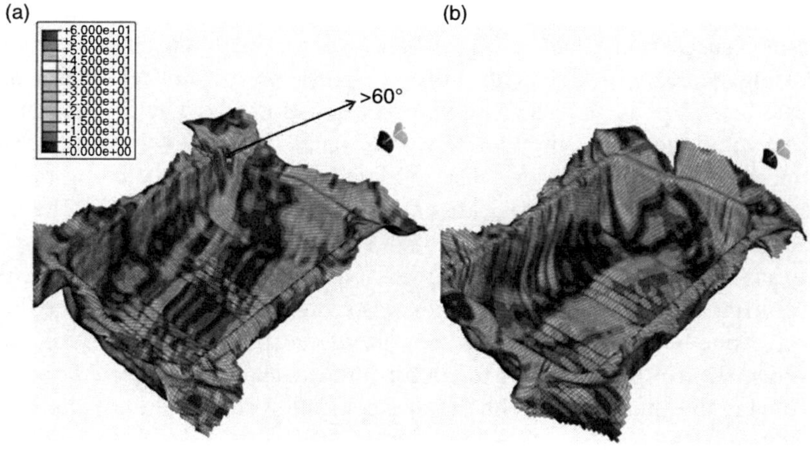

6.40 Additional orientations. Shear angle contour. (a) 15/105; (b) 60/–30.

make aligned-fiber composites, but the final fiber directions must be known with high confidence. Thus, design tools that link the final yarn positions (and directions) between the manufacturing process and the final part geometry must be used by the composite-part design engineers. As a result, credible

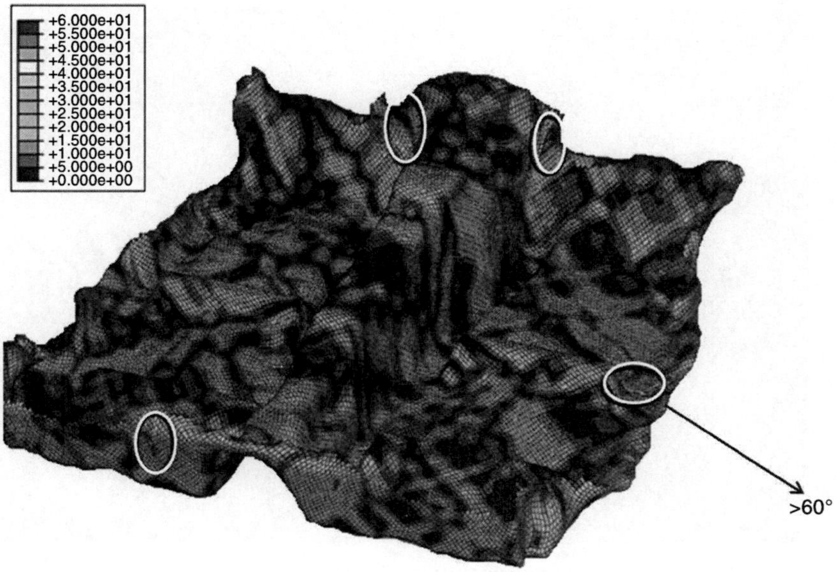

6.41 Shear angle contour in degrees for the 45/–45 fabric orientation.

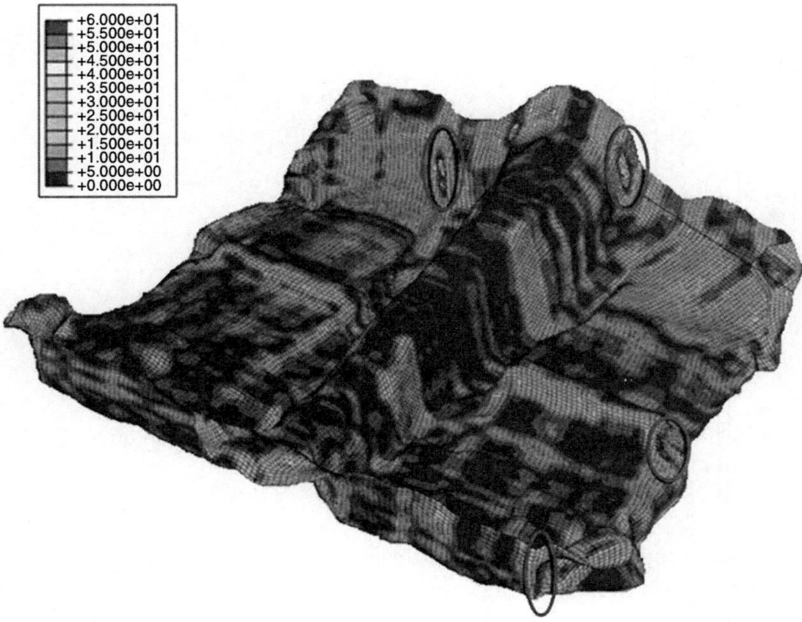

6.42 Shear angle contour in degree for the 90/0 orientation.

6.43 Plan view of floorpan demonstration part.

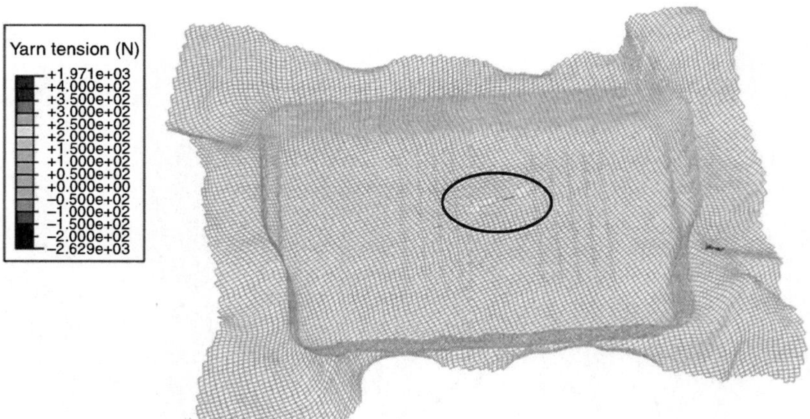

6.44 Yarn tension contour for the 22.5/112.5 fabric orientation. Oval denotes area of high tensile stress in yarns.

design tools must be available, and design engineers must have a high-level of comfort with the use of these design tools and the skill, that is education and experience, to use the tools correctly so as to achieve parts with the optimal structural integrity.

Any future improvements in the use of the thermostamping process will require developing further the theory of the mechanical behavior of fabric

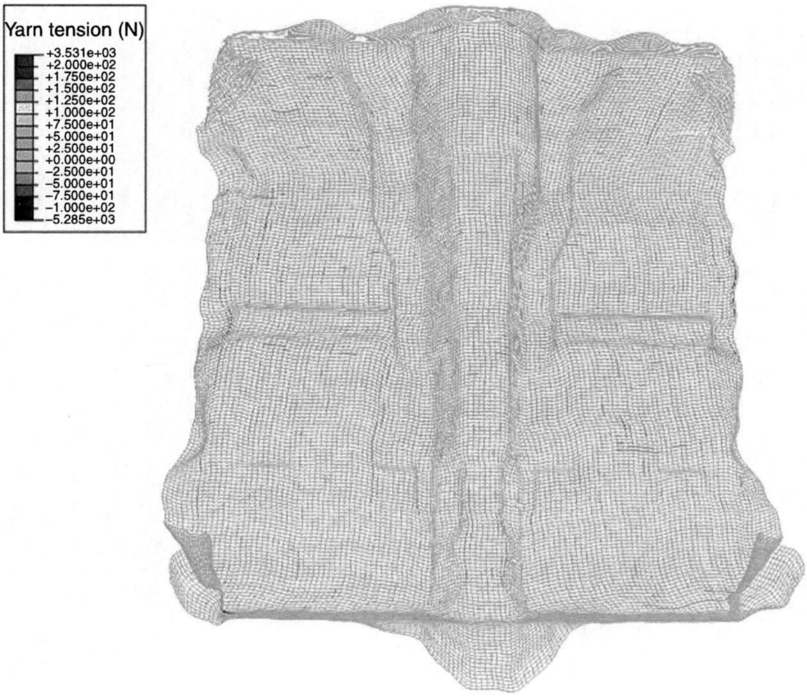

6.45 Yarn tension contour for the 90/0 fabric orientation.

architectures than is currently available and continuing development of the simulation tools to implement the theory, for example the roles of material parameters such as yarn size, yarn spacing and weave pattern must be well understood. Ideally, the design tool should be able to follow the evolution of the composite part from the initial stacking of the fabrics to the final shape, overall structural stiffness and thickness or fiber content variations across the part. The thickness or fiber content variations will be a consequence of the nesting of the adjacent layers of fabric and yarn compactions, as well as the molding tool design and process.

Significant research efforts have been devoted to the development of forming simulations as part of the overall design tool. The geometric models give a general idea of the final yarn positions, but the results are fairly crude. The finite element method looks to be most promising in that it can account for the kinematics of the forming process and for the associated mechanical behavior of the fabrics used in the forming process. Currently, the finite element method best suited to these forming simulations is the explicit formulation, which then requires the engineer to account for inertial effects that may or may not be of any consequence to the finite element solution.

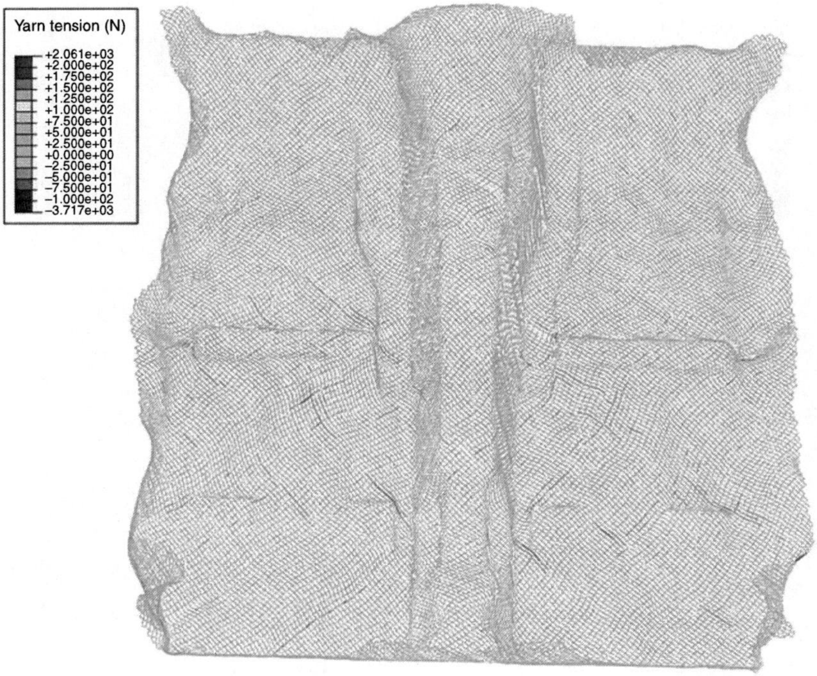

6.46 Yarn tension contour for the 45/−45 fabric orientation.

A more efficient simulation tool would be to use an implicit solver, but that solver would require improved implicit contact algorithms. A scheme for extending the capabilities of the simulation from using a pin-jointed analogy to consider yarn crossovers to allowing for yarn sliding would be another positive step in the simulation moving closer to the observed mechanical behavior of fabrics.

Methods for controlling the material and processing costs must be well known. Ideally, any simulation tool should be able to predict the initial blank shape so as to minimize material waste. Areas that are not drapable must be identified and allowances must be made for formability.

In the case of using the thermostamping process to make a preform that will subsequently be used to make a composite part using an infusion process, the resulting yarn orientations and effective porosity distribution should be well known. Such knowledge can be used to determine the location and number of gates that are needed to successfully infuse the part.

Ideally, any design tool should be seamless from start to finish, that is, from the conception of the desired final form of the composite design to the associated manufacturing process to the selection of materials to be used. In turn, the methodology should assist in the design of the molding tools.

6.6 Future trends

In this chapter, the thermostamping process was presented as stacking several flat layers of fabrics and forming the final geometry by a matched pair of tools, that is, cavity and core. Going forward, fabric-based composites will be increasingly important to industry, particularly transportation industries as the price of fuel and environmental concern about energy usage rise. To maximize mass savings in vehicles, the design of fabric-based composites will need to be optimized. This optimization will require further understanding of fiber direction and the utilization of variation in part thickness. Thus, a flat stack of fabric blanks each with 100% coverage will need to be modified. For some part geometries, a stack of blanks with varying shapes may work. For other geometries, a preform composed of pieces built up on a preforming tool will be needed. To increase the efficiency of this process, it will be necessary to use automation to place and form the pieces, and to move the preform of several layers to the molding tool.

As composite modeling improves, there will be greater understanding of the load path within the composite, allowing more precise design of fabric direction. This improvement, in turn, will take advantage of better precision in fabric draping simulations, which will allow the use of fabrics in more complex parts, and more structural parts. The further development of processes for molding the fabric, including both infusion methods and prepreg methods will be necessary.

6.7 Sources of further information and advice

Much effort has been put into developing models for capturing the mechanical behavior of fabrics. One of the simplest approaches used to model the fabrics is to homogenize the behavior of the underlying mesostructure to approximate the fabric as an anisotropic continuum. Continuum models typically allow great computational efficiency and are easily integrated into relatively large or multi-component systems. However, the challenge of continuum models is to identify the appropriate homogenized material parameters.

Various researchers have approached this challenge in different manners. Xue *et al.* (2003) described continuum models for woven-fabric composites and established the constitutive equation based on the stress and strain analysis in the global orthogonal coordinates and the non-orthogonal material coordinates, together with the rigid-body rotation matrices. The relationship between the stress and the strain is updated at the end of each increment. The change of shell thickness is considered. The equivalent material properties of the shell element in the material coordinates are determined by fitting the force-displacement curves obtained from the experimental data under biaxial-tension and pure-shear loadings. Yu *et al.* (2002) developed a non-orthogonal constitutive equation for woven-fabric reinforced thermoplastic composites based on a homogenization method by considering the microstructures of composites including both the mechanical and the structural orientation on anisotropy. Ivanov and Tabiei (2001) presented a computational material model for a plain-weave fabric which utilizes the micromechanical approach and homogenization technique. The UC of the micromechanical model consists of four subcells, two anti-symmetric subcells containing the undulated fill yarn and two other anti-symmetric subcells containing the warp yarn. Carvelli and Poggi (2001) proposed a procedure for the numerical evaluation of the mechanical properties of woven-fabric laminates. Three-dimensional finite element models were used in two steps to predict both the stiffness and the strength of woven-fabric laminates. Liu *et al.* (2005) developed a picture-frame model to account for the large shear deformation and to calculate the stress and stiffness of the fabrics. An analytical solid mechanics model was also proposed to predict the shear properties of woven fabrics and to reduce or ultimately eliminate the need for fabric-level experimental characterization.

Unfortunately, most traditional continuum models proposed in the literature do not account for the effects of the interactions between the yarn families, such as locking, resistance to relative yarn rotation and change in the load path direction with the yarn rotation. Locking and yarn rotation are the dominant mechanisms for the response of the fabric to in-plane shear, which is the major deformation mechanism during the thermostamping process. Thus, the continuum models may lose the capability to simulate potentially important behaviors in some fabric applications where both the macroscopic behavior at the continuum level and the yarn interactions at the mesostructural level may be important.

To study the interactions of yarns, a large number of analytical models have been developed. Most of these analytical models use mathematical relations to predict the mechanical response of yarns and fabrics in specific modes of deformation, such as the load-displacement relationship of the fabric under uniaxial and biaxial extension along the warp or weft directions.

Kawabata *et al.* (1973) proposed analytical models for the biaxial, uniaxial and shear deformation behaviors of fabrics based on a simple pin-jointed truss geometry. Other researchers have subsequently employed this Kawabata model to develop improved analytical models, including Realff *et al.* (1997) who modified Kawabata's uniaxial model to include more complex behaviors. Original fabric geometry, bending behavior, consolidation response and flattening response are considered in capturing the tensile properties along the yarn direction.

Kato *et al.* (1999) proposed a trapezoidal fabric lattice model to express the characteristics of the fabric structure of woven-membrane sheets where warps and wefts are interspun with glass fibers and coated with PTFE coatings.

Analytical models of woven fabrics can be incorporated into anisotropic continuum formulations to yield models that track the fabric mesostructure as the continuum deforms. Boisse *et al.* (2001) and Buet-Gautier and Boisse (2001) developed a material model to describe the behavior of fabrics and simulated the response of the plain-weave fabric during forming processes using the finite element method. The yarn directions evolve as the elements deform, and the yarn-direction behaviors are based on Kawabata's analytical model, thereby allowing a prediction of the biaxial fabric behavior. King *et al.* (2005) have proposed an approach for developing a continuum model for mechanical behavior of woven fabrics in planar deformation. They selected a geometric model for the fabrics which is similar to Kawabata's analytical model, coupled with constitutive models for the yarn behaviors. The fabric structural configuration is related to macroscopic deformation through an energy minimization method and is used to calculate the internal force carried by the yarns. The macroscopic stresses are determined from the internal forces using equilibrium arguments. The King *et al.* model can accurately predict the outcome of in-plane uniaxial and bias extension, but lacks the capability to predict of out-of-plane deformations.

Other researchers have used the finite element method to capture the mechanical behavior of woven fabrics by directly modeling every yarn in the fabrics. This method used by Boisse *et al.* (2001) has the advantage of capturing all yarn interactions and providing a detailed description of all mechanisms of fabric deformation. However, its very large computational requirements limit it to relatively small systems. This approach is not suitable to the analysis of large systems, and is generally used to characterize the interactions of the yarns or estimate the homogenized properties for continuum models.

Wang and Sun (2001) developed another numerical method to simulate textile processes and to determine the micro-geometry of textile fabrics. They called it a digital-element model. It models yarns by pin-connected digital-rod-element chains. As the element length approaches zero, the chain becomes fully flexible, imitating the physical behavior of the yarns. The interactions of adjacent yarns are modeled by contact elements. If the distance between two nodes on different yarns approaches the yarn diameter, contact occurs between them. The yarn microstructure inside the fabric is determined by process mechanics, such as yarn tension and interyarn friction and compression. The textile process is modeled as a nonlinear solid mechanics problem with boundary displacement (or motion) conditions. This numerical approach was identified as digital-element simulation rather than as finite element simulation because of a special yarn discretization process. With the conventional finite element method, the element preserves the physical properties of the discretized body. In contrast, with this model, the element does not preserve physical properties. Physical properties are imitated by the element link.

To achieve great computational efficiency, a mix of simple, efficient finite elements (such as beams or bars) can be used to directly model the entire fabric. Cherouat and Billoët (2001) used a combination of truss elements and membrane elements to describe the deformation of woven composites during the shaping process. In the model, each warp and weft yarn was described by truss elements with elastic properties, and the connecting points of truss elements were hinged. The resin was modeled by membrane elements with viscous behavior. The latter was kinematically coupled to the fabric at those connecting points. This model can account for the specific deformation of the composite fabric and large angular variations of the yarns as well as the viscosity of the resin, but the model cannot consider the bending behavior.

6.8 Acknowledgments

The authors would like to acknowledge the input of our industrial colleague Dr Patrick Blanchard of Ford Research Lab. The National Science Foundation supported the research and development of the finite element methods for completing the forming simulations through Award #DMII-0522923. The application of the modeling simulations of the double dome, tub and the floorpan and physical molding of the double dome were completed in cooperation with the ACC (Advanced Composites Consortium) through DOE award DE-FC26-02OR22910.

6.9 References

Akkerman, R., Ubbink, M. P., de Rooij, M. B. and ten Thije, R. H. W. (2007), 'Tool-Ply Friction in Composite Forming', *10th ESAFORM Conference on Material Forming*, Zaragoza, Spain, 18–20 April 2007, pp. 1080–1085.

Boisse, P., Gasser, A. and Hivet, G. (2001), 'Analyses of fabric tensile behaviour: Determination of the biaxial tension-strain surfaces and their use in forming simulations', *Composites: Part A: Applied Science and Manufacturing*, **32**, 1395–1414.

Boisse, P., Hamila, N., Helenon, F., Hagege, B. and Cao, J. (2008), 'Different approaches for woven composite reinforcement forming simulation', *International Journal of Material Forming*, **1**, 21–29.

Boisse, P., Hamila, N., Vidal-Sallé, E. and Dumont, F. (2010), 'Simulation of wrinkling during textile composite reinforcement forming: Influence of tensile, in-plane shear and bending stiffnesses', *Composites Science and Technology*, **71**, 683–692.

Buet-Gautier, K. and Boisse, P. (2001), 'Experimental analysis and modeling of biaxial mechanical behavior of woven composite reinforcements', *Experimental Mechanics*, **41**(3), 260–269.

Cao, J., Akkerman, R., Boisse, P., Chen, J., Cheng, H. S., DeGraaf, E. F., Gorczyca, J., Harrison, P., Hivet, G., Launay, J., Lee, W., Liu, L., Lomov, S., Long, A., Deluycker, E., Morestin, F., Padvoiskis, J., Peng, X.Q., Sherwood, J., Stoilova, T., Tao, X. M., Verpoest, I., Willems, A., Wiggers, J., Yu, T. X. and Zhu, B. (2008), 'Characterization of mechanical behavior of woven fabrics: Experimental methods and benchmark results', *Composites: Part A: Applied Science and Manufacturing*, **39**, 1037–1053.

Cao, J. and Blanchard, P. (2005), personal communication.

Carvelli, V. and Poggi, C. (2001). 'A homogenization procedure for the numerical analysis of woven-fabric composites', *Composites*, **32**, 1425–1432.

Cherouat, A. and Billoët, J. L. (2001), 'Mechanical and numerical modeling of composite manufacturing processes deep-drawing and laying-up of thin pre-impregnated woven-fabrics', *Journal of Materials Processing Technology*, **118**, 460–471.

Chow, S. (2002), 'Frictional interaction between blank holder and fabric in stamping of woven thermoplastic composites', MS Thesis, Department of Mechanical Engineering, University of Massachusetts Lowell.

Creech, G. and Pickett, A. K. (2006), 'Meso-modelling of non-crimp fabric composites for coupled drape and failure analysis', *Journal of Materials Science*, **41**, 6725–6736.

de Bilbao, E., Soulat, D., Hivet, G., and Gasser, A. (2009), 'Experimental study of bending behaviour of reinforcements', *Experimental Mechanics*, **50**, 333–351.

Fetfatsidis, K. (2009), 'Characterization of the tool/fabric and fabric/fabric friction for woven fabrics: Static and dynamic', MS Thesis, Department of Mechanical Engineering, University of Massachusetts Lowell.

Gamache, L. (2007), 'The design and implementation of a friction test apparatus based on the thermostamping process of woven-fabric composites', MS Thesis, Department of Mechanical Engineering, University of Massachusetts Lowell.

Gorczyca, J., Sherwood, J. and Chen, J. (2003), 'Friction between the tool and the fabric during the thermostamping of woven co-mingled glass-polypropylene composite fabrics', *18th Annual American Society for Composites Conference*, Orlando, Florida, 19–22 October 2003, pp. 196–205.

Grujicic, M., Chittajallu, K. M. and Walsh, S. (2004), 'Effect of shear, compaction and nesting on permeability of the orthogonal plain-weave fabric performs', *Materials Chemistry and Physics*, **86**, 358–369.

Hamila, N. and Boisse, P. (2007), 'A meso-macro three node finite element for draping of textile composite preforms', *Applied Composite Materials*, **14**, 235–250.

Harrison, P., Clifford, M. J. and Long, A. (2004), 'Shear characterization of viscous woven textile composites: A comparison between picture frame and bias extension experiments', *Composites Science and Technology*, **64**, 1453–1465.

Hutchings, I. M. (1992), *Tribology: Friction and wear of engineering materials*. Ann Arbor: CRC Press, pp. 62–67.

Ivanov, I. and Tabiei, A. (2001), 'Three-dimensional computational micro-mechanical model for woven-fabric composites', *Composites Structures*, **54**, 489–496.

Jauffres, D., Sherwood, J. A., Morris, C. D. and Chen, J. (2009), 'Discrete mesoscopic modeling for the simulation of woven-fabric reinforcement forming', *International Journal of Forming*, **3**(Suppl 2), 1205–1216.

Kato, S., Yoshiro, T. and Minami, H. (1999), 'Formulation of constitutive equations for fabric membranes based on the concept of fabric lattice model', *Engineering Structures*, **21**, 691–708.

Kawabata, S., Masako, N. and Kawai, H. (1973), 'The finite deformation theory of plain-weave fabrics. Parts I–III', *Journal of the Textile Institute*, **64**, 21–83.

King, M. J., Jearanaisilawong, P. and Socrate, S. (2005), 'A continuum constitutive model for the mechanical behavior of woven fabrics', *International Journal of Solids and Structures*, **42**, 3867–3896.

Lebrun, G., Bureau, M. N. and Denault, J. (2003), 'Evaluation of bias-extension and picture-frame test methods for the measurements of intraply shear properties of PP/Glass commingled fabrics', *Composite Structures*, **61**, 341–352.

Li, X., Sherwood, J., Liu, L. and Chen, J. (2004), 'A material model for woven commingled glass-propylene composite using a hybrid finite element approach', *International Journal of Materials and Product Technology*, **21**, 59–70.

Liu, L., Chen, J., Li, X. and Sherwood, J. (2005), 'Two-dimensional macro-mechanics shear models of woven fabrics', *Composites Part A: Applied Science and Manufacturing*, **36**, 105–114.

Lussier, D. (2002), 'Shear characterization of textile composite formability', MS Thesis, Department of Mechanical Engineering, University of Massachusetts Lowell.

Parlevliet, P. P., Bersee, H. E. N. and Beukers, A. (2006), 'Residual stresses in thermoplastic composites – A study of the literature. Part I: Formation of residual stresses', *Composites Part A: Applied Science and Manufacturing*, **37**, 1847–1857.

Peng, X. Q. and Cao, J. (2005), 'A continuum mechanics-based non-orthogonal constitutive model for woven composite fabrics', *Composites Part A: Applied Science and Manufacturing*, **36**, 859–874.

Realff, M. L., Boyce, M. C. and Backer, S. (1997), 'A micromechanical model of the tensile behavior of woven-fabric', *Textile Research Journal*, **67**(6), 445–459.

Sargent, J., Chen, J., Sherwood, J., Cao, J., Boisse, P., Willem, A., Vanclooster, K., Lomov, S. V., Khan, M., Mabrouki, T., Fetfatsidis, K. and Jauffres, D. (2010), 'Benchmark study of finite element models for simulating the thermostamping of woven-fabric reinforced composites', *International Journal of Material Forming*, **3** (Suppl 1), 683–686.

Sharma, S. B. and Sutcliffe, M. P. F. (2004), 'A simplified finite element model for draping of woven material', *Composites: Part A: Applied Science and Manufacturing*, **35**, 637–643.

Sidhu, R. M. J. S., Averill, R. C., Riaz, M. and Pourboghrat, F. (2001), 'Finite element analysis of textile composite preform stamping', *Composite Structures*, **52**, 483–497.

Skordos, A. A., Monroy Aceves, C. and Sutcliffe, M. P. F. (2007), 'A simplified rate dependent model of forming and wrinkling of pre-impregnated woven composites', *Composites: Part A: Applied Science and Manufacturing*, **38**, 1318–1330.

Soteropoulos, D., Fetfatsidis, K., Sherwood, J. and Langworthy, J. (2011), 'Digital method of analyzing the bending stiffness of non-crimp fabrics', *Proceedings of the 14th ESAFORM Conference*, Belfast, UK, pp. 913–917.

Stachowiak, G. W. and Batchelor, A. W. (2005), *Engineering tribology*, 2nd edition. Boston: Butterworth Heinemann, pp. 182–198.

Vanclooster, K., Lomov, S. V. and Verpoest, I. (2008), 'Investigation of interply shear in composite forming', *Proceedings for the 11th ESAFORM Conference on Material Forming*. Lyon, France, pp. 957–960.

Wang, Y. and Sun, X. (2001), 'Digital-element simulation of textile processes', *Composites Science and Technology*, **61**, 311–319.

Wilks, C. E. (1999), 'Characterization of the Tool/Ply interface during forming', PhD Dissertation, School of Mechanical, Materials, Manufacturing Engineering and Management, University of Nottingham, UK.

Willems, A. (2008), *Forming simulation of textile reinforced composite shell structures*. Faculteit Ingenieurswetenshapen Arenbergkasteel. Leuven (Belgium): Katholieke Universiteit Leuven.

Xue, P., Peng, X. and Cao, J. (2003), 'A non-orthogonal constitutive model for characterizing woven composites', *Composites Part A: Applied Science and Manufacturing*, **34**, 183–193.

Yu, W. R., Harrison, P. and Long, A. (2005), 'Finite element forming simulation for non-crimp fabrics using non-orthogonal constitutive equation', *Composites: Part A: Applied Science and Manufacturing*, **36**, 1079–1093.

Yu, W. R., Pourboghrat, F., Chung, K., Zampaloni, M. and Kang, T. J. (2002), 'Non-orthogonal constitutive equation for woven-fabric reinforced thermoplastic composites', *Composites*, **33**, 1095–1105.

Yu, W. R., Zampaloni, M., Pourboghrat, F., Chung, K. and Kang, T. J. (2003), 'Sheet hydroforming of woven FRT composites: non-orthogonal constitutive equation considering shear stiffness and undulation of woven structure', *Composite Structures*, **61**, 353–362.

7

Filament winding process in thermoplastics

J. MACK, Institut für Verbundwerkstoffe GmbH, Germany and
R. SCHLEDJEWSKI, Montanuniversitaet Leoben, Austria

Abstract: Thermoplastic filament winding offers the possibility to build up continuous fiber reinforced lay-ups with well-defined fiber orientation and accurate positioning of the fibers. Typical part geometries are axially symmetric ones, for example tubes or vessels. Due to the specific characteristics of the thermoplastic material, the process can run with *in situ* consolidation and no curing step after running the winding process is necessary. Various different process setups and types of raw materials are used. The aim of this chapter is to give an overview about the thermoplastic filament winding process and to explain the most important specific process aspects.

Key words: *in situ* consolidation, impregnation, main heater, thermoplastic filament winding.

7.1 Introduction

In most cases composite materials are used due to light weight reasons. Polymeric based composite materials offer favorable light weight potential. The highest specific mechanical properties can be reached if high load-bearing fibers are embedded well aligned in the polymeric matrix. Consequently, processes allowing the composite structure to build up in a defined manner are needed to reach highest mechanical performance of the component. A processing technique offering the possibility to build up the structure having a well-defined position, orientation and alignment of the fibers is the winding technique. In the 1940s the first attempts were made to reinforce components by use of the filament winding technique. The method was further developed since 1947 by Young Development Labs, Princeton, Canada. The first commercially available filament winding machines were produced there. In the 1960s series production of winding equipment started, for example at McClean Anderson, USA and Bolenz & Schäfer, Germany. Over the years the machines were equipped with individual motors allowing winding of adjustable angles. Nowadays most filament winding machinery is based on CNC, which can produce complex structures combined with a high productivity.[1] Filament winding machines are offered by manufacturers as individual machines or as continuous production lines. Common dimensions for

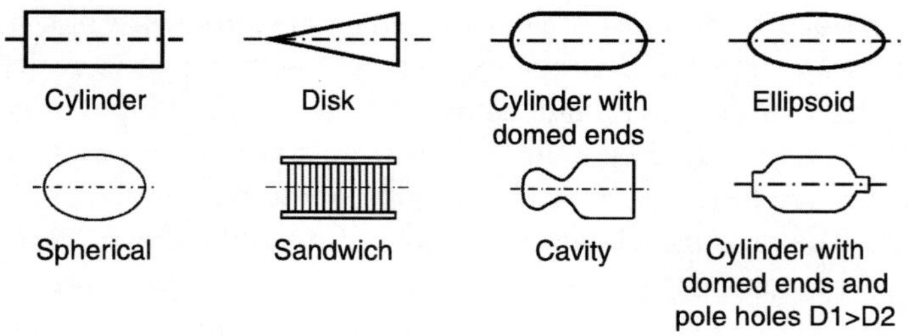

7.1 Possible winding forms.

components have a diameter of up to 1 m and a length of up to 5 m, but even larger component dimensions are possible. Typically axially symmetrical components like pressure vessels, tanks, piping, reactor kettles, picker arms and so on are produced with filament winding technology. In Fig. 7.1 the basic winding geometries are shown.

In the beginning the filament winding technique was used for manufacturing pipes and rocket parts. Reinforcing fibers are impregnated with resin to produce axially symmetrical components.[2] If a thermoset resin is used, the impregnation is done with the uncured polymer. After finalizing the lay-up of the component, a curing procedure is necessary. Depending on the resin type, the curing takes place at room temperature or at elevated temperature. For the latter an oven is necessary which represents a further processing step and often limits the maximum component size. Different to thermoset systems thermoplastic materials do not have to be cured. Instead, thermoplastic materials have to be consolidated, that is, build-up of intimate contact, polymer healing, material compaction and finally solidification. Intimate contact is required to have a contact of the polymeric surfaces. In the contact region the macromolecules can start to penetrate and polymer healing will result. The compaction is necessary to eliminate entrapped gas volumes and voids. The compacted material has to be cooled and solidification will conserve the consolidation level reached. In case of thermoplastic resin systems a one-step process involving a so-called *in situ* consolidation, that is, complete consolidation during the winding, is possible and no further consolidation cycle, for example in an oven process, is necessary. In such a case the winding machine itself is the only component size limiting issue. As a result of consolidation during the winding process, non-geodesic reinforcing paths can be wound with thermoplastic materials. Consequently, increased design freedom is given. Further advantages of thermoplastics compared to thermoset materials include high fracture toughness, unlimited shelf life, reparability and weldability. The interest in thermoplastic filament winding has increased over the last decades.[3,4]

In contrast to thermoset materials, which are processed in a non-cured, low viscosity state, the impregnation process with a high viscosity thermoplastic material is more challenging. The thermoplastic polymer must be melted before the impregnation process can start and the molten high viscosity thermoplastic polymer has to impregnate properly each reinforcing filament. The process velocity, the temperature and the consolidation pressure (as a result of winding tension and compaction pressure, e.g. using a roller, applied) are the parameters which most influence the impregnation process.

7.2 Winding basics

In Fig. 7.2 a thermoplastic filament winding process consisting of a material rack, a tension controller, a preheating zone, a main heater, a consolidation roller and a mandrel is shown schematically. Beside the material used, the most adjustable process parameters are: winding velocity (which is defined by mandrel rotation and crosshead movement), roving tension, temperature, winding angle and mandrel/liner (material and temperature control).

The winding angle (indicated as α in Fig. 7.3) is defined as the angle between the winding path and the rotation axis of the mandrel. Three different winding patterns (helical, polar and hoop), each representing geodesic paths, can be distinguished for both thermoplastic polymer and thermoset resins winding. Helical and polar winding are characterized by crossing laydown paths during the winding process. This is often also called cross winding. Due to the crossing points the reinforcing fibers are undulated, which will, depending on the specific winding pattern, result in slightly changing mechanical performance.[5,6] The differences between helical and polar winding are also shown in Fig. 7.3 (the winding path is straight at polar winding and can be curved at helical winding). Cross winding results in a balanced

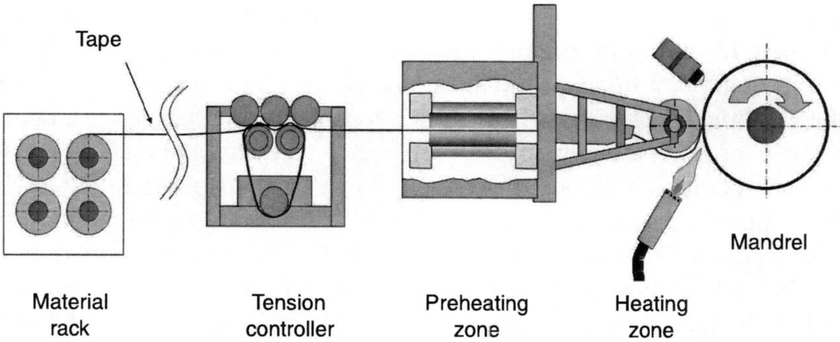

7.2 Schematic of filament winding process chain (includes preheating zone and a consolidation roll).

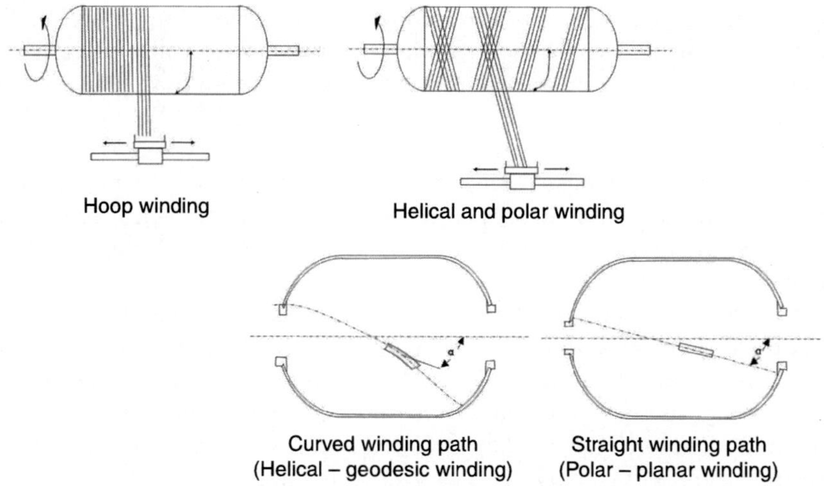

7.3 Comparison of hoop, helical and polar winding (α: winding angle).

Table 7.1 Main differences between filament winding and a lay-up process

	Winding process	Lay-up process
Process	Continuous	Non-continuous
Components	Axially symmetrical; closed convex surfaces	Convex or concave surface; local reinforcement
Fiber orientation	15–88° without additional tools (0° are a lay-up process)	Variable

Source: Schlottermüller *et al.*[1]

lay-up, that is, a double layer consisting of identical amount of fibers having a positive as well as a negative winding angle. In the case of hoop winding an individual layer is wound for each winding angle. These unbalanced layers do not have any undulations.

Whilst winding processes require a rotating axis to wind on, the main process principles of winding are quite similar compared to placement processes. Table 7.1 presents the differences between filament winding and a placement process. The main distinction is a continuous or non-continuous fiber process. The filament winding process works with continuous fibers whereas the placement process can work non-continuously.

In a composite material the reinforcing component is embedded in a matrix. To reach this, the matrix material must impregnate each individual filament. The time required to complete impregnation depends on several factors, for example polymer viscosity, surface tension, permeability and driving forces. Bringing together a polymeric melt and a dry filament bundle (roving), the melt will first macroscopically impregnate the roving, that is, cover the roving. Then the matrix will penetrate transversally into the roving and microscopical impregnation will take place until all filaments are fully embedded by the matrix and saturation is reached. Due to the high viscosity of thermoplastic materials impregnation is very time sensitive. To reduce flowing distances during impregnation, there are several possibilities to distribute the matrix between the reinforcing fibers already before the impregnation takes place, for example in form of powder or polymeric filaments.

Consolidation is the most important aspect regarding winding of thermoplastic fiber reinforced materials. At a given velocity a pressure and a certain temperature profile are applied to the incoming material as well as to the already wound material at the nip point, that is, the contact point of both materials. The aim is to build up an intimate contact[7–9] between the material surfaces and to force polymer healing, that is, interdiffusion of the polymer chains.[10–12] In order to reach this, material deformation[13,14] and viscous flow[15] have to take place, entrapped air has to be squeezed out or at least has to be compressed. On the other hand, care must be taken not to heat the material too high or to expose it for too long a period. Otherwise, oxidative reactions might lead to degrading the polymeric matrix.[16,17]

During the winding process significant consolidation takes place only in a very limited region in which the parameters consolidation pressure and temperature are on a level allowing forcing polymer healing. Since the process is running continuously, the time available depends on the process velocity. Consequently, increasing process velocity will result in reduced degree of consolidation.

7.3 Winding process

As shown in Fig. 7.2 the main steps of the filament winding process chain are the material supply, tension build-up, heating and consolidation. Depending on the individual winding equipment configuration and winding head technology several different setups are possible.

7.3.1 Process steps

Material rack

The first step in a filament winding process chain is the material rack. The fiber material, whether already impregnated, that is tape material, or not, that is, roving material, delivered on spools is mounted here. The number of spools depends on the winding head technology used and the components dimension. In addition to the material storage, a spool creel can also be equipped with a break system and apply tensile force on the roving/tape. If the tension is not appliquéd in the material rack, it can also be done in a separated tension controller.

Tension controller

Through uncoiling of the spools on the material rack a tensile force (intentionally or unintentionally) is acting on the filaments. With a tension controller the tension will be measured, analyzed and controlled. In most cases a constant tension is used to achieve constant winding conditions. Inhomogeneous tension might result in inhomogeneous residual stress within the product, but can also be used to create a specific residual stress configuration which can support the load-bearing behavior in service.[18,19] The tension controlling and tension adjustments (by the tension controller or by an adjustable device) homogenize and regulate the tension.

Heating zone

To produce a product with a thermoplastic matrix the polymer needs to be heated up over the melting point to enable the impregnation and consolidation. The heat source is one of the main modules in the process.[20] Both consolidation quality and processing costs are strongly affected. A detailed description about heat source selection for *in situ* consolidation processes is presented by Schledjewski and Miaris.[21] Highest efficiency is given for laser, flame and hot gas torch. The process efficiency depends on the design of the heat source. But, for heat source selection the total process efficiency has to be analyzed individually. Furthermore, different heating methods affect different heating zone formation and expansion, which are shown in Fig. 7.4. Both maximum temperature reached and exposure time have to be analyzed very carefully to prevent damaging effects, for example oxidation/degradation.[16]

Different types of winding processes require different heating systems. Laser,[22,23] hot gas torch,[24] flame and infrared radiation[25] are the main heating methods used for commingled yarns, bicomponent fibers, powder impregnated and fully consolidated tapes. In some cases, for example high process velocity, a high melting temperature of the polymer, a low performance of the main heating source or the thickness of the input materials (thermal conductivity), a preheating zone is necessary. The temperature distribution within the whole heating zone is shown in Fig. 7.5 (including the preheating zone).

At the online-impregnation process (see section 'Direct impregnation technology') the polymer is heated and melted with an extruder (external) and transported to an impregnation tool near the mandrel or directly on the mandrel. To reach a good bonding between the incoming molten material

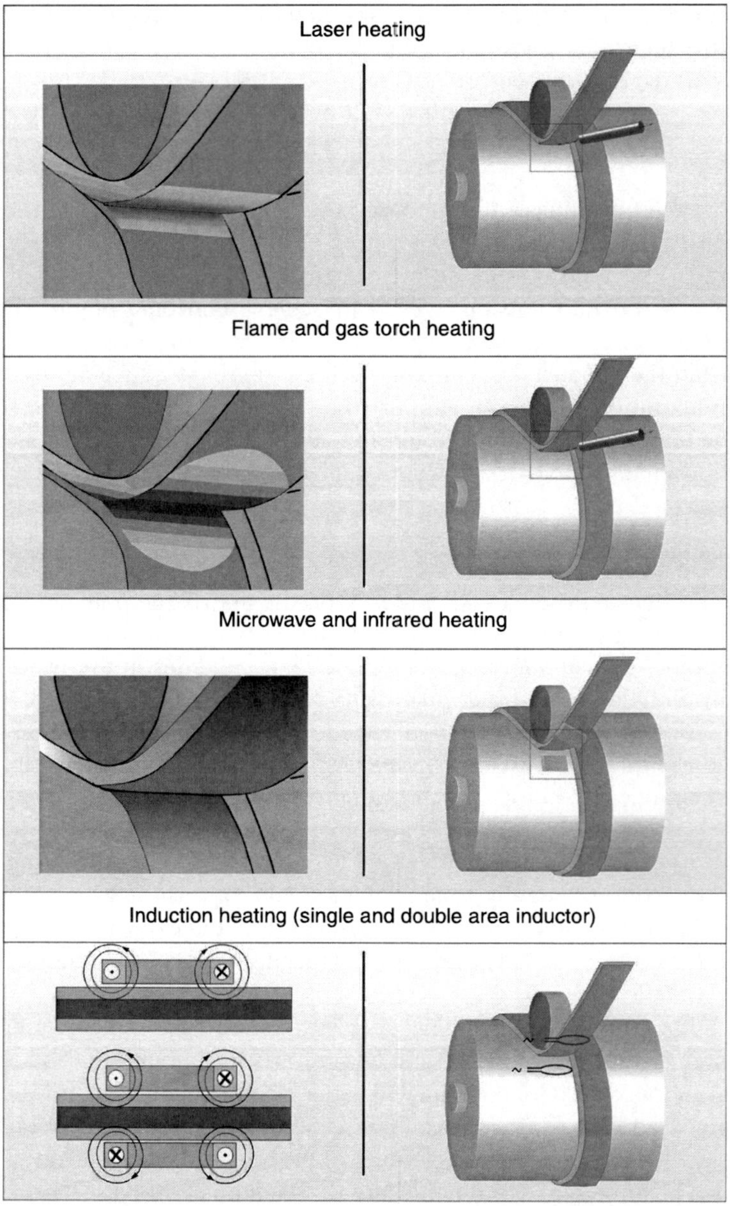

7.4 Heating methods and their thermal expansion in the tape.

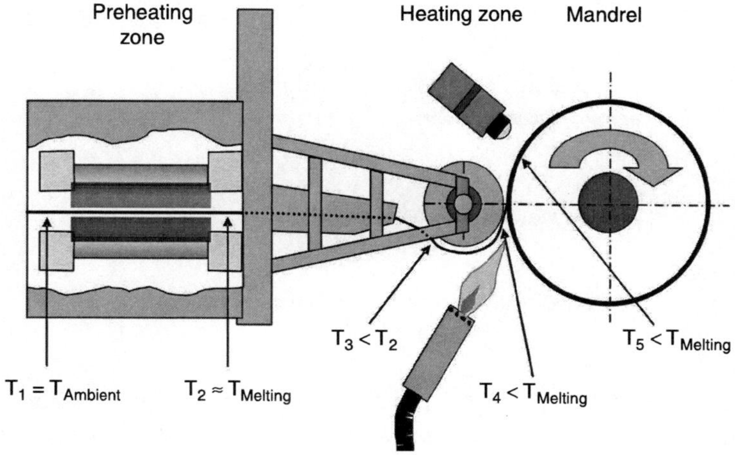

7.5 Heating zones.

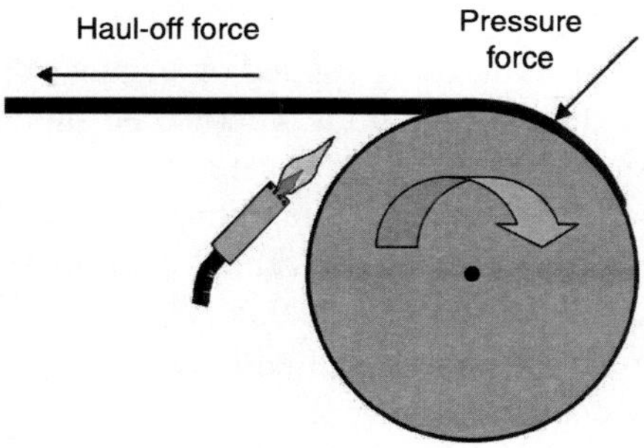

7.6 Consolidation force only by filament break.

and the already wound material on the mandrel, a preheating system for the fiber material is used additionally.[26]

Consolidation force

For an optimal consolidation it is necessary to apply a consolidation pressure in the winding process. After the polymer is molten the rovings/tapes are wound (with consolidation pressure) on the mandrel. Generally, there are two different ways to apply the required consolidation pressure. In the first system the consolidation force is applied only with the tape tension as a result of the break system used (Fig. 7.6). The acting consolidation pressure depends on several factors, besides the acting tension, for example material temperature and viscosity, and the consolidation zone is not well defined. Furthermore, the applied tape tension can induce significant residual stresses in the component.[27] The consolidation roller of the second system applies

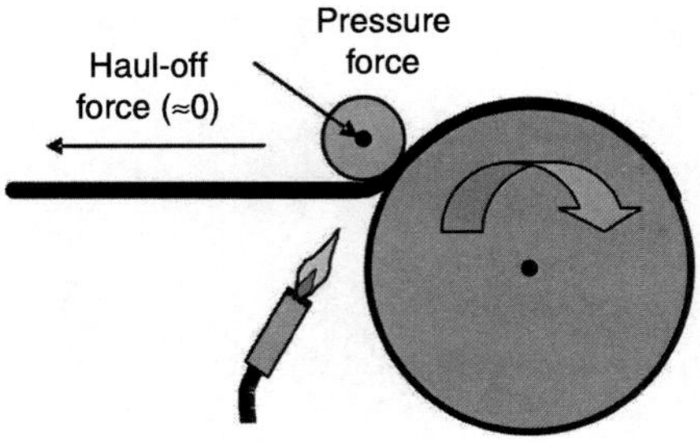

7.7 Consolidation force by using a compaction roller.

the consolidation pressures directly at the nip point (Fig. 7.7). For a homogeneous lay-up temperature the consolidation roller should have a controlled temperature, typically in the range 60–90°C, allowing to run it stationary.[28]

Mandrel lay-on

In a filament winding process the filament is applied on a mandrel. Mandrel types can be distinguished in heated mandrels, non-heated mandrels, dissolvable cores and liners. Using a heated or non-heated mandrel the quality of the mandrel surface is an important factor due to the fact that the finally wound component must be stripped (pulled or pushed) over the surface. By using a liner or a dissolvable core the surface quality is not so relevant, due to the fact that the liner remains in the product and the dissolvable core is removed after the winding process.

Heated mandrels

As described before, the process temperature is an import factor for the component quality. If the mandrel is temperature controlled and can be heated up (to a temperature lower than the melting point of the polymer) the difference of the temperature between the winding material and the mandrel is lower than by using a non-heated mandrel and the polymer healing is much easier. Using a well selected mandrel temperature the mechanical performance can be increased significantly.[29] Additionally, Kugler and Moon[30] demonstrated that, with an adjustable cooling rate (lower cooling rate), the quality of a winding product can be raised.

Non-heated mandrels

The temperature distribution in a winding process with a non-heated mandrel is not homogeneous over the whole winding duration. At the beginning of the process the mandrel has an ambient temperature which increases during the winding due to the heating by the main heater and the heated material wound (a high number of layers results in more pronounced heating of the mandrel). Inhomogeneous product quality, consolidation level and residual stresses might be resulting effects.

Dissolvable cores

Some applications are closed shapes or have a local reduction, like some pressure vessel types or rocket motor cases. If the products have no liner the core must be dissolvable. For dissolvable materials the following are usable: sand (soluble, water soluble), plaster (soluble, breakout), salt (meltable, eutectic) and alloy (with a low melting temperature). All dissolvable cores are precast in an extra process step before the winding process starts. After the winding process the core is nearly complete covered with filaments. Depending on the dissolvable material used, different techniques to dissolve the core are required such as rinsing with water or heating up to a certain temperature.[2] All dissolvable cores are non-heated.

Liners

In some cases, especially for pressure vessels (connection quality), or the usage of slightly volatile mediums, a liner as mandrel is used. The liner can be a vital part of the product which is also considered by the requirements. The filaments are wound on the liner, which remains in the application. Materials for the liner manufacturing depend on the end-use of the application. Liners can be made, for example, of polymer, aluminum alloy or steel. Also like the dissolvable core the liner cannot be temperature controlled.

Mandrels consisting of different components

Peters *et al.*[2] described how, for huge winding applications, the usage of one complete (fully) mandrel is not useful, based on the weight and costs. The mandrel for winding applications with large diameters consists of different components. The inner part is a hollow framework construction and only the surface of the mandrel is closed with wood, aluminum alloys, steel or a similar material (Fig. 7.8).

7.3.2 Winding configurations

Three main winding configurations – flat bed winding, gantry winding and robotized winding systems – can be differentiated and are explained in the following.

Flat bed and gantry winding systems

The most frequently used winding systems are the flat bed or gantry winding systems (horizontal and vertical). All system axes are shown and explained

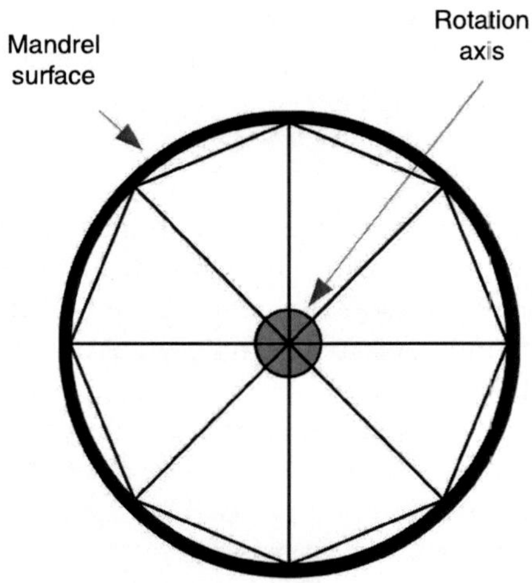

7.8 Mandrel that can be disassembled.

in Table 7.2 and Fig. 7.9. The difference between a flat bed and a gantry winding system is the number of axes. In a flat bed winding system axis number six (*w*-axis) is missing; in some cases also axis five (*v*-axis) is not available compared to a gantry winding system.

The number of axes used has grown over the past years. At the beginning two axes (mandrel rotation and winding head) were sufficient for a winding process. Nowadays six and even more axes are used to enable a highly accurate winding and allow the winding of more complex shapes, for example a pipe T-joint.

Robotized winding

Due to technological change and the widespread application of computer systems, filament winding is not only possible on special winding systems. A standard industrial robotic, with a thermoplastic filament winding head, is also usable like a winding machine. The most important advantage is the available technique (industrial robotic) enhanced by a special winding head. With a rotatable external axis a full winding system is complete. In comparison with flat bed or gantry winding systems the investment costs are significantly lower, but winding accuracy is lower, as shown by Carrino *et al.*[31]

7.3.3 Winding head technologies

For thermoplastic filament winding two main technologies can be differentiated. First is the direct impregnation technology in which fibers and

Table 7.2 Explanation of the different axes (see Fig. 7.9)

First axis	X	Mandrel rotation
Second axis	Y	Motion (parallel to the mandrel) of the payout eye
Third axis	Z	Distance between the payout eye and the mandrel rotation
Fourth axis	U	Payout eye rotation around horizontal axis
Fifth axis	V	Payout eye rotation around vertical axis
Sixth axis	W	Changes the level of the payout eye

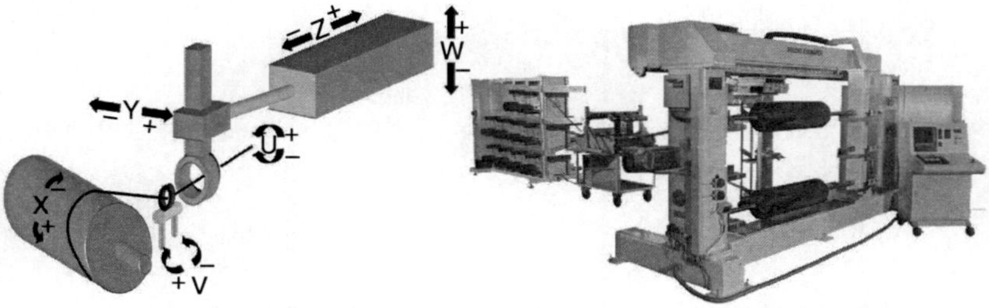

7.9 Winding axis and a gantry winding system. (Courtesy of EHA Composite Machinery GmbH, Germany; previously: Bolenz & Schäfer FW-Technology.)

polymer are fed separately and combined at the winding process. Second are the techniques requiring semi-finished products. Here, fiber and matrix are already combined in the right portion.

Direct impregnation technology

Direct impregnation technique is used with two different impregnation methods. In the first case, for example shown by Henninger and Friedrich,[32] the impregnation of fibers by the matrix is done directly at the winding head. Provided by an external single extruder, the melted polymer is pressed through a heated impregnation unit, for example a porous ring consisting of sintered metal, and the fibers are impregnated with matrix. Depending on the porosity and the wrap angle the impregnation length is changeable. With a variable impregnation length the winding speed can be adjusted to the required winding velocity. The whole direct impregnation process chain is shown schematically in Fig. 7.10.

The second case, for example shown by Christen and coworkers,[33,34] involves a direct impregnation process in which the polymer coating is directly extruded and placed at the mandrel. The principle of the technique

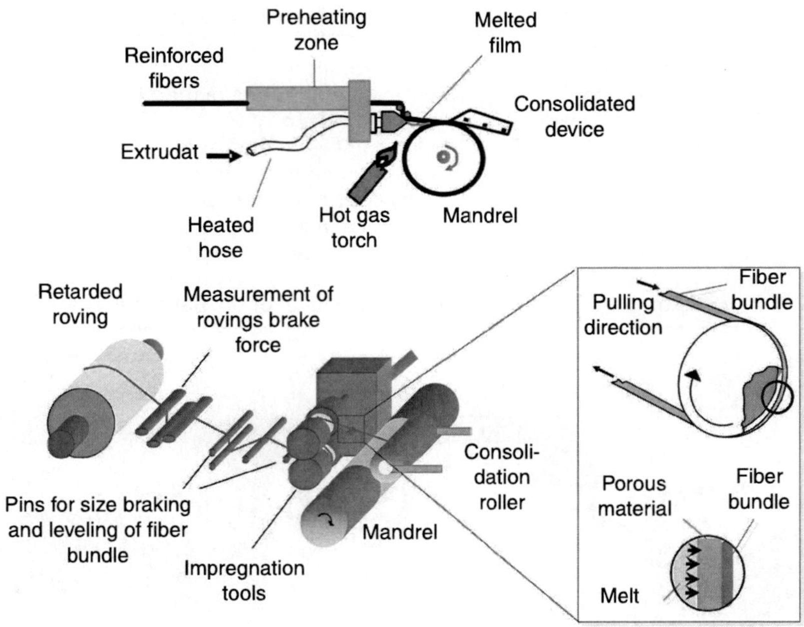

7.10 Direct impregnation process chains (top: polymer extruded and placed on the mandrel; bottom: melted polymer is pressed through a heated impregnation unit).

is to impregnate the fibers with the thermoplastic matrix on the mandrel during the winding and to consolidate them to the final structure. The temperature distribution of the melted polymer and the winding velocity is crucial for the impregnation. If the velocity is too high the impregnation process will not be finished behind the consolidation roller and further impregnation is impossible due to the fact that the temperature (viscosity raises) is too low.

Direct impregnation thermoplastic winding processes have some advantages compared to winding technologies which use semi-finished products. A lower material price is the most important point, because the semi-finished process step (conversion of the material) is not needed. The raw material costs (fibers and polymer) are equal for all processes but further processing steps are not necessary. Another advantage is the possibility to change the fiber volume content online at the winding process.[32] As limiting aspects the need of an extrusion unit and the limited winding velocity have to be mentioned.

To overcome the latter, the use of a multi-feed configuration can be used. Figure 7.11 illustrates the general principals of the process. Using such a setup, a continuous tube production with high material throughput rate is possible and the winding process can be used for series production, for example pressure vessels for the truck industry (Fig. 7.12).

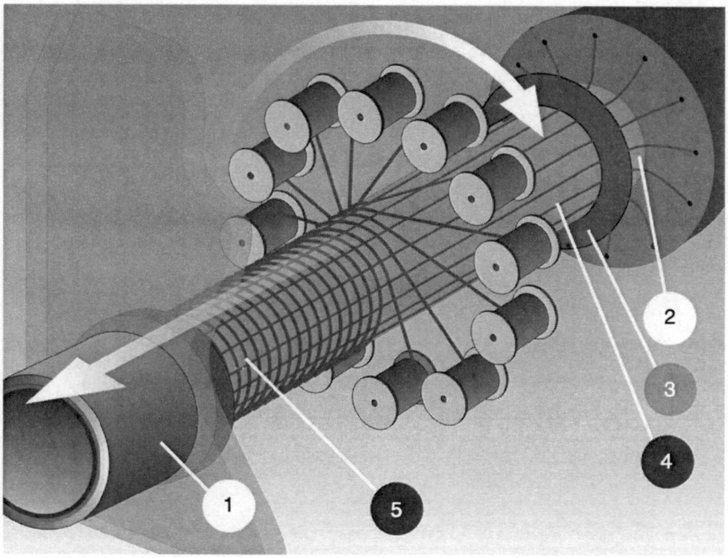

7.11 Direct impregnation with multi feed configuration: 1: Outer liner, 2: Inner liner, 3: Guiding tool, 4: Fibers axial, 5: Fibers circumferential. (Courtesy of COMAT Composite Materials GmbH, Germany.)

Processing of semi-finished material

Semi-finished materials for the thermoplastic winding process can be differentiated in already impregnated fibers in the form of fully consolidated tapes and several different materials delivering fiber and matrix without impregnation of the fibers by the matrix. The latter group is represented by commingled yarns, bicomponent fibers and powder impregnated fibers. All semi-finished materials have the right desired fiber volume content. Consequently, no variation of the fiber volume content in the winding process can be made without changing a new semi-finished material.

The high viscosity of a thermoplastic polymer needs to be considered for commingled yarns, bicomponent and powder impregnated fibers. The impregnation process, as investigated for example by Bernet *et al.*,[35] must be accomplished in the step after the heating zone, when the melting temperature is reached, and after the 'nip point', when the temperature is lower than the melting temperature.[36]

Material and process parameters influences the impregnation time. For commingled yarns (identical to all semi-finished products) the process parameters are: temperature, maximum pressure, pressure rate and time.[37]

7.12 Air pressure vessels for truck application manufactured using direct impregnation technique. (Courtesy of COMAT Composite Materials GmbH, Germany.)

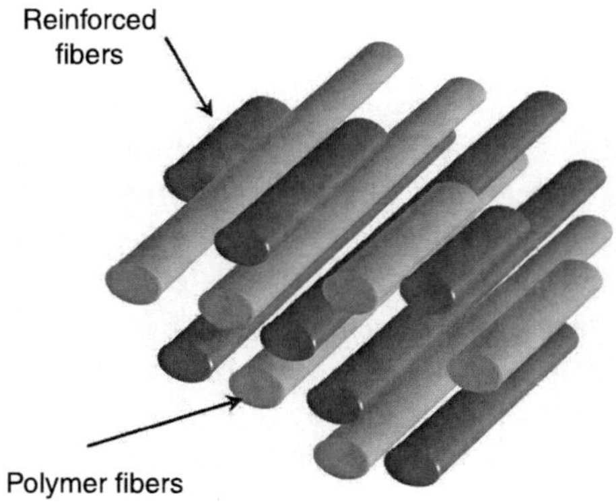

7.13 Commingled yarns.

Commingled yarns

Thermoplastic fibers aligned with reinforcing fibers are combined to form so-called commingled yarns, which are schematically shown in Fig. 7.13. A hot gas torch is the most common heating method for filament winding of commingled yarns. Experiments with laser based heating methods and commingled yarns have also been made. The winding process uses a compaction roller to press the commingled yarns together on the mandrel, as Fig. 7.7

shows.[36] Typical filament wound components based on commingled yarns are lightweight LPG cylinders.[38]

Bicomponent fibers

The thermoplastic filament winding process with bicomponent fibers is not as easy as the other winding methods. Huang and coworkers[39] demonstrated that the fibers, which consist of a polymer (lower melting point) and a liquid crystalline polymer (higher melting point), are very sensitive to the processing temperature. The thermoplastic winding process is easier if the melting point difference between the polymers is high. With the main heater the melting temperature of the low melting polymer can be reached safely without melting the second polymer. The melted polymer filaments are pressed with a compaction roller to the mandrel and the consolidation is reached. One type of bicomponent fiber is shown in Fig. 7.14.

Powder impregnated fibers

Reinforced fibers with thermoplastic polymers in form of powder belong to two categories. In the first method a polymer hose surrounds the fibers and the polymer powder (Fig. 7.15). Both hose and the not fixed powder material consist of the same polymer.[40] The second method is to strew powder over the fibers. Thereby the polymer is temporarily heated in order to adhere to the fibers.

The winding process per se is equal to the commingled yarns and fully consolidated winding process. Heating technologies where reflection is possible by the filaments cannot be used with pre-melted powder filaments. Due to different heat absorptions (polymer and fibers) a homogeneous temperature cannot be reached.

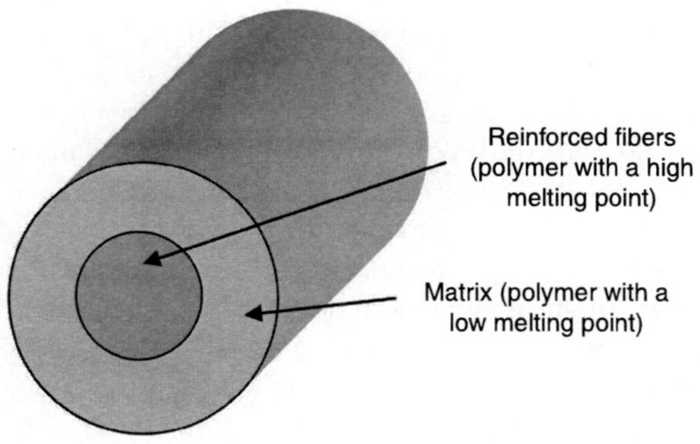

7.14 Bicomponent fibers.

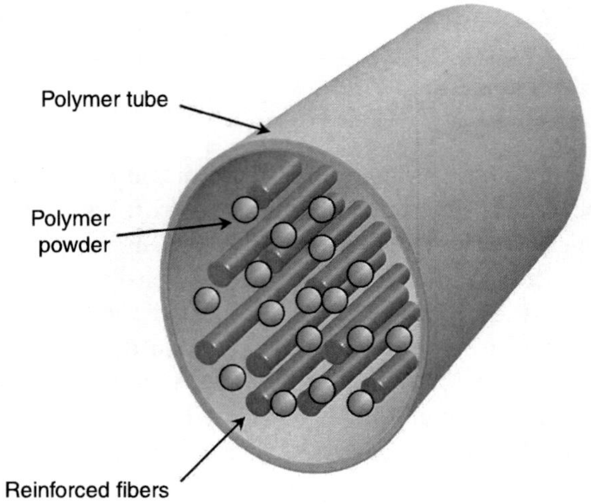

7.15 Powder impregnated fibers.

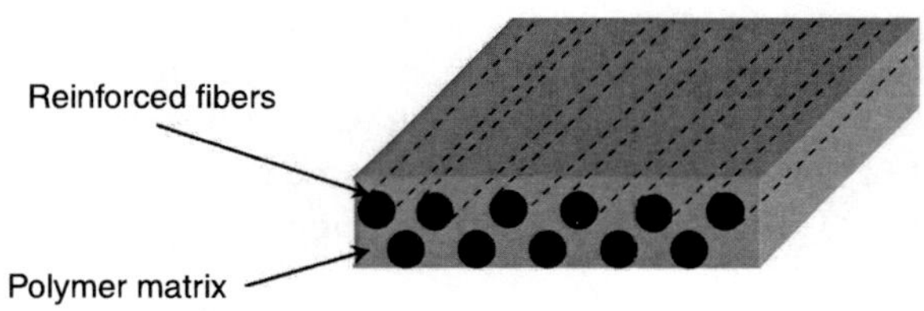

7.16 Fully consolidated fibers (tapes).

Fully consolidated fibers (tapes)

Compared with the three other semi-finished materials, the tapes are fully consolidated (Fig. 7.16), that is, the micro and macro impregnation are completely done. Fully consolidated tapes are available on the market with fiber volume contents of up to 60% and even higher.[41] Due to the not applicable impregnation process the winding speed can be higher in comparison to the other winding patterns. The heating zone melts the polymer and the tension (Fig. 7.6) or compaction roller (Fig. 7.7) presses the filaments on the mandrel.

By the reason of the full consolidation, the quality parameters for tapes must be exact. That is why the process needs a homogeneous tape width, homogeneous fiber volume content, a smooth surface, a homogeneous tape thickness and rectangular edges. The tape width and the rectangular edges are important for a continuous winding side by side without laps and gaps.

Fully consolidated fibers are used for winding of high performance components, for example bearing systems (Fig. 7.17) based on CF/PEEK.[42]

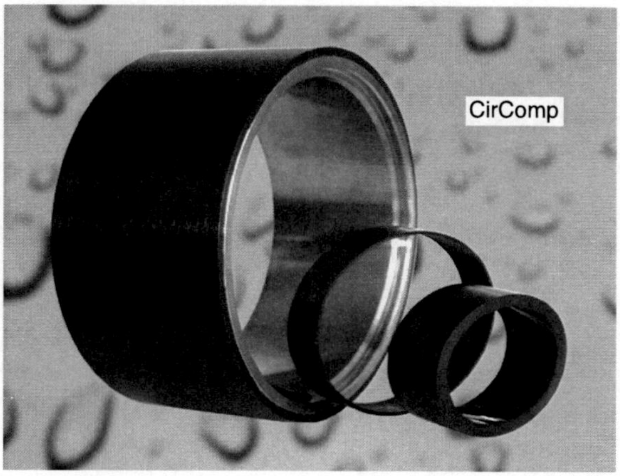

7.17 Thermoplastic tape wound bearing exhibiting a unique performance. (Courtesy of CirComp GmbH, Germany.)

7.3.4 Comparison of the different techniques

A general vote as to which kind of semi-finished material is preferred is not possible. Depending on application and based on the process equipment available and required component quality an individual selection has to be made. Table 7.3 summarizes key benefits, advantages/disadvantages and applications. Material costs and availability for fiber matrix combinations required are often the key factors in the selection decision.

7.4 Simulation tools

7.4.1 Path generation

For all described winding configurations (flat bed, portal and robotized winding systems) the generation of a winding path is required. Three rotations and translations are essential for a path generation: the rotated mandrel, the rotational symmetry axis of the mandrel (not equal to the rotated mandrel) and the winding angle.[43] The path is the winding way of the filaments. In most cases the path is built up so that every place on the mandrel is coated with filaments. With process simulation tools like CADWIND,[44] COMPOSICAD[45] or CADPATH[46] the winding process can be simulated based on the original mandrel information. For every winding machine the resulting code is applicable. With state of the art simulation tools 'axes symmetrical parts' can be simulated for a thermoplastic winding process.

Table 7.3 Comparison of different supply forms of semi-finished materials for thermoplastic filament winding

	Key benefits	Advantages / disadvantages	Applications
Direct impregnation	• Low material costs (no essential pre-process)	• Different fiber volume contents are possible • Higher equipment complexity	• Rings and tubes (nearly every winding application is possible)
Commingled yarns	• Lower material costs (polymer mingling around the fibers)	• Homogeneous fiber volume content • Mostly used for lower requirements with glass fiber	• Pressure vessels • Cost-efficient components with less receivables
Bicomponent fibers	• Recyclable (wholly thermoplastic product)	• Lowest application requirements • High equipment requirements (heating)	• Rings and tubes (currently no volume production)
Powder impregnated	• Lower price compared to fully consolidated tapes	• Good application quality	• Components with a high quality, where the velocity plays a minor role
Fully consolidated	• Good homogeneous quality	• Very good application quality	• Pressure vessels • Shaft drives

The path generation needs to consider the filament widening in the winding process. If there is no adjustment, laps or gaps between the filaments can accrue which have negative quality effects.[47]

7.4.2 Process simulation

Winding process simulations for thermoplastic filaments are important for the component quality. Commercial simulation tools are not available yet. Research institutes are developing different process simulation programs to understand the influences of the material and the process on the component quality. Simulation programs from the CCM (University of Delaware, Center for Composite Materials, USA) and from the IVW (Institut für Verbundwerkstoffe GmbH, Germany) are presented hereafter.

Tierney and Gillespie[48] presented the 'Composite Design and Simulation Software' (CDS), developed at the CCM. Ten main tools are combined in one simulation suite which is structured in five process steps:

1. Materials and database management and generation.
2. Thermal modeling.
3. Process/environment.
4. Microstructure.
5. Structure Mechanics.

Each of these steps depends on the process before. The software suite starts with the material selection, in which also an individual material can be defined. Fiber orientations and the number of plies can be selected in a following step. All material parameters must be chosen before the analysis can start. Six structural analysis modules can be differentiated with the CDS software suite. These solid mechanic modules are thick-walled cylinder, thin plate, thin plate impact-fastener modeling, thick plate, discontinuous tile modeling and compliant beam interlayer analysis. The CDS software suite allows changing the parameters of the manufacturing process or the laminate structure in real time. Four result sections for those parameter changes are provided by the software: the effective properties, thermal-processing response, stress–strain results and the failure response. The CDS software suite is a complete analysis tool kit which is easy to use for the client. The software also allows export into an external simulation tool.

Schlottermüller *et al.*[18] and Lü *et al.*[19] presented a simulation tool analyzing the internal stresses (residual stress) which are influenced by the following winding parameters: numbers of layers, winding velocity, heating and cooling conditions, mandrel temperature, melt viscosity and the crystallization. Four submodels are needed to solve the highly nonlinear characteristics of the analyzed parameters. The first – the 'fiber motion model' – describes the tension, position and displacement of the fibers, the layer thickness and the fiber volume fraction of the whole component. In the 'thermal submodel' the temperature distribution of the wounded composite, with the properties of the raw material as input, is analyzed. Kinetic and rheological parameters are described in the third 'kinetic and rheological submodel'. Viscosity in relation to the temperature is the main parameter of this item. The last submodel – 'Stress–Strain' – describes the stress and strain at the layer interphase. Submodels 2–4 depend on information from each other and from the first submodel. With the results of a winding process simulation the quality of a winding component can be analyzed before the production (winding process) starts.

For processing fully consolidated tape materials the process simulation tool ProSimFRT has been developed at IVW.[28,29] ProSimFRT enables simulation of the component quality depending on the void content and the surface roughness. The simulation can be used for tape winding as well as tape placement processes using a compaction roller.

7.5 Component quality

High input material qualities (fibers, polymer, comingled yarns, tapes and so on) are the first step for a high component quality. Failure, non-constant material parameters,[49,50] and machinery tolerances have negative influences on the application, but these are not considered in this chapter.

Relevant quality factors for components made out of fiber reinforced plastics are: consolidation degree, void content, residual stress, degradation, fiber breaks and inhomogeneous fiber matrix distribution. These factors are influenced by the winding velocity, heating, tension, mandrel material, cooling rate and the number of layers, that is, application thickness.[30]

7.5.1 Consolidation degree

Depending on the process parameters a specific consolidation degree is reached. In the best case it will be on the same level as received, for example by autoclave consolidation. But, the consolidation region is very limited during the process and due to the limited time available for build-up of a pronounced interdiffusion of the polymer chains the degree of consolidation is in most cases below the maximum reachable value.[20,29]

7.5.2 Void content

Beside consolidation degree one of the most important quality factors for fiber reinforced thermoplastic materials is the void content inside the application.

Due to the manufacturing process fully consolidated tapes mostly have a low incoming void content. When used for thermoplastic winding, the heating can lead to a blow-up effect and increased void content of the component. Further reasons for a higher application void content are: intra ply voids (dissolved or entrapped air), water or through releases of fiber stresses by the softening of the matrix.[51] Void formation reasons are almost the same for all types of semi-finished materials used for thermoplastic winding. But, for all non-consolidated semi-finished materials the void content in the component produced is often much higher.

7.5.3 Residual stress

In all winding processes residual stress is present, due to the rising thickness on a rotating mandrel. Residual stress can cause premature failure or a dimensional deviation. Results of research (simulations and practical tests) have shown that the mandrel temperature, an annealing after the winding process, the winding angle, and the number of layers have an effect on the residual stress.[18,19] Thermoplastic winding with a hot mandrel in combination with an additional annealing or only an annealing leads to positive stress at the surface and negative stress in the interface. Slower cooling rates during an additional annealing process reduce the residual stress. The numbers of layers have a large effect on the temperature storage. Thick applications keep the temperatures longer, which results in higher residual stress as in thin applications.

7.5.4 Oxidation/degradation

Very high temperatures and long exposure times might lead to oxidation/degradation. Several different effects, for example polymer chain shortening or cross-linking,[52,53] might result and in many different ways, for example, hindered interdiffusion, reduced mechanical performance or increased viscosity.

7.5.5 Inhomogeneous fiber matrix distribution

Whether coming already from the consolidated tape material or as a result of the impregnation process during winding, inhomogeneous distribution has to be avoided. Otherwise the resulting component will behave inhomogeneously under mechanical load.

7.6 Future trends

Three main topics will determine the future developments in the field of thermoplastic filament winding.

7.6.1 Winding machinery and materials

Robotized winding systems will become increasingly important in the future since they offer reduced plant costs and they are well known in industry. Generally, the need for further automation and integrated manufacturing plants will push developments in the future.

The increasing environmental awareness requires that the used material have a high or full recyclability. Self-reinforced polymers are one of these fully recyclable materials.[54] The manufacturing of such bicomponent materials is not so easy in comparison to other reinforced materials and the winding head technology has to be adapted to reach the required high accurate temperature management during the winding process.

Regarding materials, the main focus will be on commingled rovings and fully consolidated tapes.

7.6.2 Component quality

The incoming material quality is an important factor for the component quality. With new sensor techniques the process monitoring can increase the quality. In combination with new nondestructive component tests the number of destructive tests can be reduced to a minimum.

7.6.3 Simulation

To improve process stability and resulting component performance further development of simulation tools is needed. One challenging topic will be to develop inline working process simulation tools which allow the process to be controlled based on inline acquired material data. In addition to the development of inline quality assurance tools and new analysis methods an increased computing power is required. The latter might also allow more physical aspects to be included in the simulation. In the future the complexity of such simulation tools will increase as well as the results.

7.7 Sources of further information and advice

A lot of filament winding machine suppliers are active on the market. In many cases the machines supplied are produced on special customer request. A list of winding machine suppliers is give here:

- McClean Anderson, 300 Ross Avenue, Schofiled, Wisconsin 54476, USA, http://www.mccleananderson.com/
- EHA Composite Machinery GmbH (formerly: Bolenz & Schäfer (BSD) – Filament winding technology), Niedereisenhausen Bauhofstrasse 2, 35239 Steffenberg, Germany. http://www.eha-maschinenbau.de
- Entec Composite Machines, 2975 S 300 W, Salt Lake City, Utah 84115, USA. http://www.entec.com
- Mikrosam – Control – Automation – Robotics. Mikrosam, Krusevski pat b.b., 7500 Prilep, R. Macedonia. http://www.mikrosam.com/

- AFPT – Advanced fiber placement technology. AFPT GmbH, Trinkbornstraße 15–17, 56281 Dörth, Germany. http://www.afpt.biz
- Addax Addax, 6040 Fletcher Avenue, Lincoln, Nebraska 68507, USA. http://www.addax.com
- Pultrex Pultrex Ltd, 18–20 Riverside Avenue West, Lawford, Manningtress, Essex CO11 1UN, England. http://www.pultrex.com/

7.8 References

1. Schlottermüller, M., Neitzel, M., Schledjewski, R. and Beresheim, G., 'Wickel- und Legetechnik'. In M. Neitzel and P. Mitschang (eds.), *Handbuch Verbundwerkstoffe: Werkstoffe, Verarbeitung, Anwendung*. Munich: Hanser Verlag, 2004, pp. 245–270.
2. Peters, S. T., Humphrey, W. D. and Foral, R. F., *Filament winding: Composite structure fabrication*. Covina, CA: Society for the Advancement of Material and Process Engineering, Medium: X, 1991.
3. Lauke, B. and Friedrich, K., 'Evaluation of processing parameters of thermoplastic composites fabricated by filament winding', *Composites Manufacturing*, 1993, **4**(2), 93.
4. Romagna, J., Ziegmann, G. and Flemming, M., 'Thermoplastic filament winding: An experimental investigation of the on-line consolidation of poly(ether imide) fit preforms', *Composites Manufacturing*, 1995, **6**(3–4), 205.
5. Rousseau, J., Perreux, D. and Verdière, N., 'The influence of winding patterns on the damage behaviour of filament-wound pipes', *Composites Science and Technology*, 1999, **59**, 1439.
6. Päßler, M. and Schledjewski, R., 'Effect of ring winding technology on NOL ring testing and accompanying characteristics', *Proceedings of European Conference on Composite Materials (ECCM 14)*, 7–10 June 2010, Budapest, Hungary.
7. Dara, P. H. and Loos, A. C., 'Thermoplastic matrix composite processing model', *Virginia Polytechnic Institute Report*, 1985, CCMS-85-10.
8. Lee, W. I. and Springer, G. S., 'A model of the manufacturing process of thermoplastic matrix composite', *Journal of Composite Materials*, 1987, **11**(21), 1017.
9. Mantell, S. C. and Springer, G. S., 'Manufacturing process models for thermoplastic composites', *Journal of Composite Materials*, 1992, **26**(16), 2348.
10. De Gennes, P. G., 'Reptation of a polymer chain in the presence of fixed obstacles', *Journal of Chemical Physics*, 1971, **55**(2), 572.
11. Wool, R. P., 'Molecular aspects of tack', *Rubber Chemistry and Technology*, 1983, **57**, 307.
12. Wool, R. P. and O'Connor, K. M., 'Theory of crack healing in polymers', *Journal of Applied Physics*, 1981, **52**(10), 5953.
13. Muzzy, J., Norpoth, L. and Butt, A., 'Quantitative analysis of APC2 consolidation', *Proceedings of 33rd International SAMPE Symposium*, 1988, 1331.
14. Pitchumani, R., Ranganathan, S., Don, R., Gillespie, J. W. and Lamontia, M. A., 'Analysis of transport phenomena governing interfacial bonding and void dynamics during thermoplastic tow-placement', *International Journal of Heat and Mass Transfer*, 1996, **39**(9), 1883.
15. Cogswell, F., *Thermoplastic aromatic polymer composites*. Oxford: Butterworth/Heinemann, 1992.

16. Nicodeau, C., *Modelisation du soudage en continu des composites a matrice thermoplastique*. Paris: Ecole Nationale Supérieure d'Arts et Metiers de Paris, 2005.
17. Nicodeau, C., Cinquin, J., Regnier, G. and Verdu, J., 'In situ consolidation process optimization for thermoplastic matrix composites', *Proceedings of International SAMPE Symposium*, Long Beach, CA, USA, April–May, SAMPE, Paper 212, 2006.
18. Schlottermüller, M., Lu, H., Roth, Y., Himmel, N., Schledjewski, R. and Mitschang, P., 'Thermal residual stress simulation in thermoplastic filament winding process', *Journal of Thermoplastic Composite Materials*, 2003, **16**(6), 497.
19. Lü, H., Schlottermüller, M., Himmel, N. and Schledjewski, R., 'Effects of tape tension on residual stress in thermoplastic composite filament winding', *Journal of Thermoplastic Composite Materials*, 2005, **18**, 469.
20. Schledjewski, R., 'Thermoplastic tape placement process – in situ consolidation is reachable', *Plastics, Rubber and Composites*, 2009, **38**(38), 379.
21. Schledjewski, R. and Miaris, A., 'Thermoplastic tape placement by means of diode laser heating', *Proceedings of SAMPE International Symposium*, 18–21 May 2009, Baltimore, USA.
22. Rosselli, F., Santare, M. H. and Güçeri, S. I., 'Effects of processing on laser assisted thermoplastic tape consolidation', *Composites Part A: Applied Science and Manufacturing*, 1997, **28A**, 1023.
23. Beyler, E. P. and Güçeri, S. I., 'Experimental investigation of laser assisted thermoplastic tape consolidation', *Journal of Thermoplastic Composite Materials*, 1988, **1**, 107.
24. Kim, H. J., Kim, S. K. and Lee, W. I., 'A study on heat transfer during thermoplastic composite tape lay-up process', *Experimental Thermal and Fluid Science*, 1996, **13**, 408.
25. Yousefpour, A. and Ghasemi Nejhad, M. N., 'Experimental and computational study of APC-2/AS4 thermoplastic composite C-rings', *Journal of Thermoplastic Composite Materials*, 2001, **14**, 129.
26. Christen, O., 'Entwicklung eines neuartigen Thermoplast-Direktimprägnierverfahrens für die wickeltechnische Herstellung von Druckbehältern', Düsseldorf: VDI. Simult.: PhD Thesis, Kaiserslautern, 1999.
27. Schlottermüller, M., 'Zur Eigenspannungsausbildung bei der wickeltechnischen Verarbeitung thermoplastischer Bandhalbzeuge', Kaiserslautern: Institut für Verbundwerkstoffe, IVW Schriftenreihe Band 62, 2005.
28. Schledjewski, R. and Latrille, M., 'Processing of unidirectional fiber reinforced tapes: Fundamentals on the way to a process simulation tool (ProSim FRT)', *Composites Science and Technology*, 2003, **63**, 2111.
29. Khan, M. A., Mitschang, P. and Schledjewski, R., 'Identification of some optimal process parameters to achieve higher laminate quality through tape placement', *Advances in Polymer Technology*, 2010, **29**(2), 98.
30. Kugler, D. and Moon, T. J., 'The effects of Mandrel material and tow tension on defects and compressive strength of hoop-wound, on-line consolidated, composite rings', *Composites Part A: Applied Science and Manufacturing*, 2002, **33**(6), 861.
31. Carrino, L., Polini, W. and Sorrentino, L., 'Modular structure of a new feed-deposition head for a robotized filament winding cell', *Composites Science and Technology*, 2003, **63**(15), 2255.
32. Henninger, F. and Friedrich, K., 'Thermoplastic filament winding with online-impregnation. Part A: Process technology and operating efficiency', *Composites Part A: Applied Science and Manufacturing*, 2002, **33**(11), 1479.

33. Christen, O., Neitzel, M. and Rasche, C., 'Filament winding thermoplastic composite cylinders with on-line impregnation', *Proceedings 42nd International SAMPE Symposium & Exhibition*, 4–8 May 1998, Anaheim, California, USA.
34. Christen, O., Beresheim, G., Neitzel, M. and Rasche, C., 'Thermoplastic winding with direct impregnation: Cost-effective production of pressure cylinders', *Kunststoffe Plast Europe*, 1999, **89**(4), 18.
35. Bernet, N., Michaud, V., Bourban, P. E. and Manson, J. A. E., 'An impregnation model for the consolidation of thermoplastic composites made from commingled yarns', *Journal of Composite Materials*, 1999, **33**(8), 751.
36. Kim, S. K., Kim, G. M., Kim, H. J. and Lee, W. O., 'An experimental study on the thermoplastic filament winding process using commingled yarns', *Advanced Composites Letters*, 2002, **11**(2), 67.
37. Van West, B. P., Pipes, R. B. and Advani, S. G., 'The consolidation of commingled thermoplastic fabrics', *Polymer Composites*, 1991, **12**(6), 417.
38. N.N., 'Spirit of conquest', *JEC-Composites*, 2005, **16**, 82.
39. Huang, J. H., Baird, D. G., Loos, A. C., Rangarajan, P. and Powell, A., 'Filament winding of bicomponent fibers consisting of polypropylene and a liquid crystalline polymer', *Composites Part A: Applied Science and Manufacturing*, 2001, **32**(8), 1013.
40. Sala, G. and Cutolo, D., 'Heated chamber winding of thermoplastic powder-impregnated composites. Part 1: Technology and basic thermochemical aspects', *Composites Part A: Applied Science and Manufacturing*, 1996, **27A**, 387.
41. Suprem SA, Product Information Suprem T. Available at: http://www.suprem.ch/en/10080/Suprem-T-narrow.html. Accessed 29 July 2011.
42. Victrex plc, 'Bearings made of Carbon Fibre Reinforced VICTREX® PEEK' Polymer Improve Wear and Friction Performance', Press release, 29 March 2011. Available at: http://www.victrex.com/en/victrex-library/press-releases/detail/detail.php?id=663. Accessed 29 July 2011.
43. Koussios, S., Bergsma, O. K. and Beukers, A., 'Filament winding. Part 1: Determination of the wound body related parameters', *Composites Part A: Applied Science and Manufacturing*, 2004, **35**(2), 181.
44. Weaver, A., 'Design for winding', *Reinforced Plastics*, 1996, **40**(10), 28.
45. Kandola, A., 'Processing software for easy filament winding pattern generation', *Advanced Composites Bulletin*, 2010, August, 6.
46. Johansen, B. S., Lystrup, A. and Jensen, M. T., 'CADPATH: A complete program for the CAD-, CAE- and CAM-winding of advanced fibre composites', *Journal of Materials Processing Technology*, 1998, **77**(1–3), 194.
47. Wang, E. L. and Gutowski, T. G., 'Laps and gaps in thermoplastic composites processing', *Composites Manufacturing*, 1991, **2**(2), 69.
48. Tierney, J. and Gillespie, J. W. J., *Composite Design and Simulation (CDS) Software: A comprehensive toolkit for real-time design of composite processes and structures*. Baltimore: SAMPE, 2009.
49. Beresheim, G., 'Thermoplast-Tapelegen – ganzheitliche Prozessanalyse und –entwicklung', Institut für Verbundwerkstoffe, IVW Schriftenreihe Bd. 32, Dissertation, 2002.
50. Schledjewski, R. and Schlarb, A. K., '*In situ* consolidation of thermoplastic tape material – Effects of tape quality on resulting part properties'. Baltimore, *Proceedings of Sampe 2007*, USA, 2007.
51. Khan, M. A., Mitschang, P. and Schledjewski, R., 'Tracing the void content development and identification of its effecting parameters during *in situ* consolidation of thermoplastic tape material', *Polymers and Polymer Composites*, 2010, **18**(1), 1.

52. Phillips, R., Glauser, T. and Manson, J. A. E., 'Thermal stability of PEEK/carbon fiber in air and its influence on consolidation', *Polymer Composites*, 1997, **18**, 500.
53. Cole, K. C. and Casella, I. G., 'Fourier transform infrared spectroscopic study of thermal degradation in films of poly (etheretherketone)', *Thermochimica Acta*, 1992, **211**, 209.
54. Zhang, J. M. and Peijs, T., 'Self-reinforced poly(ethylene terephthalate) composites by hot consolidation of Bi-component PET yarns', *Composites Part A: Applied Science Manufacturing*, 2010, **41**(8), 964.

8

Continuous fiber reinforced profiles in polymer matrix composites

P. MITSCHANG and M. CHRISTMANN, Institut für Verbundwerkstoffe GmbH, Germany

Abstract: Profiles made of continuous reinforcement fibers with defined orientation provide a much better weight/length ratio than other light weight profiles, for example those made of aluminium. But only the pultrusion process and the continuous compression molding process are capable of large-scale production of such profiles. Both technologies are described in this chapter in detail. The chapter will give an overview of the equipment, the operating principle of both technologies, processable matrix and the reinforcement material. Furthermore, different process combinations of the pultrusion process are introduced. A process model for the continuous compression molding technology is explained. The chapter will conclude with a short comparison of the preferred areas of application of both technologies.

Key words: pultrusion, pull-winding, PAZ, continuous compression molding, powder-prepreg, B-factor.

8.1 Introduction

Continuous profiles rank among the most important semi-finished goods used in various fields of industry. They provide the basis for constructions offering high strength and stiffness as well as little cost. Especially profiles made of fiber reinforced thermoplastic polymers (FRTPs) have some additional benefits, for example they can be easily reshaped into various geometries. Furthermore, when they are made of continuous reinforcement fibers with defined orientation, they provide a much better weight/length ratio than profiles made of aluminium.[1]

Although there are convincing aspects in favor of FRTP profiles, there are only very few technologies that are capable of producing continuous FRTP profiles with defined fiber orientation. Moreover, there are only two technologies which enable large-scale production with sufficient process stability. These are the pultrusion process and the continuous compression molding process. Both technologies will be described in detail in further chapters.

8.2 Pultrusion

Pultrusion is a fully automated process for production of continuous fiber reinforced profiles with constant cross-section. It was developed by W. Brandt Goldsworthy and filed for patent in 1951. The technology enables large-scale production with low labor cost, low rejections and without any consumables like, for example, release films.[2,3] Furthermore, pultrusion enables the production of highly filled composites with fiber weight contents up to 85%, which offer superior strength and stiffness in longitudinal direction.[4] The two major differences between thermoset and thermoplastic pultrusion are the higher viscosity of the matrix material and the high temperatures, which are necessary to melt the polymer. Processing temperatures of thermoplastics used in aerospace (e.g., PEEK) exceed 400°C.[5]

In order to produce a profile, the material is pulled through the manufacturing equipment by a pull-off unit, which is located at the end of the manufacturing process. The reinforcing fibers are provided as separate rovings, which are stored on creels. Guides are used for forming and positioning of the material into a preliminary shape before they enter the die.[5] About two thirds of the die are heated, whereas the rest is cooled. Basically, the reinforcement fibers are impregnated with matrix material while they are moving through the heated section of the die. Thereby the matrix material can be added to the process in different forms, for example as commingled yarns. Additionally to the impregnation, the forming and consolidation takes place. Before leaving the die, the material is solidified in the cooled section. In order to assemble the semi-finished parts to a desired length, a pultrusion unit is normally equipped with a cut-off unit, which is located behind the pull-off unit. In contrast to today's horizontal pultrusion units, at the beginning some machines were designed in a vertical orientation in order to remove all effects of gravity from the fiber arrangement. Another advantage of this previous design is that it is easy to angle internal mandrels properly. But the used equipment of both versions is very similar.[6] Figure 8.1 shows a pultrusion unit.

The first application area was in the 1960s, when fiber reinforced profiles were applied to concrete.[2] Over the past years the production speed has been enhanced due to faster impregnation and modern methods tend to increase the strength and stiffness in off-axis direction. Today, pultrusion profiles are applied in many fields of industry such as window and door frames, marine constructions, ladders, railway carriages and stairways. Furthermore, many companies are associated in the European Pultrusion Technology Association.[8]

8.2.1 Material in use

There are only a few reinforcement materials that are relevant for usage in practical application. These are glass fibers and different types of carbon

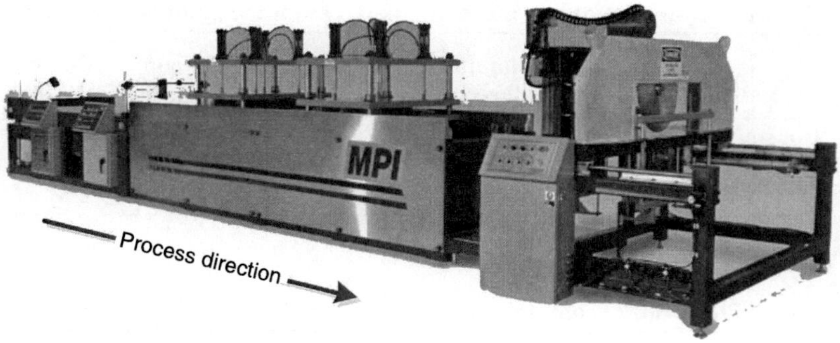

8.1 Horizontal pultrusion unit manufactured by MPI-Ft. Worth Texas. (From reference 7.)

fibers. Commonly, they are applied to the process as endless rovings, which is the cheapest form of reinforcement material available. In this process, the design and fabrication of forming guides attached before the pultrusion die is simplified and the friction inside the die is minimized.[5] Unidirectional reinforcements are available in different, single and multiple strand configurations. For example, carbon fibers are typically available with different numbers of filaments in the range of 3 up to 320K.[6] In order to reduce the space necessary for the storage of the fibers, the use of rovings with a high number of filaments is advantageous. In addition, the time needed to set up the pultrusion line is reduced.[5] Recent research activities have successfully demonstrated the processability of natural fibers, too.[9] Natural fibers are not available as endless filaments, so they are used as yarns. Generally, every material that is available as fibers and is able to sustain the existing pulling forces can be used for the pultrusion process.

A major disadvantage of pultrusion parts consisting only of unidirectional fibers is the minor strength and stiffness in cross-direction. Thus, woven fabrics can be applied to the process. In order to prevent warping of the textile, the woven fabrics have to be placed between some pull-force resistant fibers (e.g., some unidirectional fibers orientated in process direction).

This effect also can be seen when applying reinforcement fibers which are not orientated in process direction. Thus, wound or braided fibers in any fiber orientation between 0° and 90° should be covered by unidirectional fibers in process direction. Otherwise, the fibers will be warped during the process, which reduces the product's quality and furthermore can lead to process interruption. The processability of profiles with sharp edges is also limited due to the minimum bending radius of the circumferential reinforcement fibers. In order to meet special performance requirements, it is possible to apply hybrid mixtures of the reinforcing materials.[5] Table 8.1 gives a short overview on properties of reinforcement materials.

Table 8.1 Properties of reinforcement materials for the pultrusion process

Properties	Natural fibers (flax)	E-glass	Carbon (HT)	Carbon (HM)
Density (kg/m³)	1.48	2.54	1.78	1.85
Tensile strength (MPa)	750	3500	3750	2450
Tensile modulus (GPa)	30	80	240	400

Source: Mitschang and Neitzel (2004).

Table 8.2 Properties of common polymer materials for the pultrusion process

Polymer material	Density (g/cm³)	Modulus (GPa)	Long-term service temperature (°C)	Processing temperatures range (°C)
PP	0.9	1.1	100	190–240
PA6	1.13	1.4	100	240–300
PA12	1.02	1.6	80	230–290
PBT	1.28	2.0	100	230–280
PMMA	1.19	3.2	90	150–180
PPS	1.35	4.0	230	300–350
PEEK	1.32	3.7	300	360–400

Source: Mitschang and Neitzel (2004).

The selection of a suitable thermoplastic matrix material mostly depends on the desired mechanical properties and the desired long-term service temperature. Depending on the application area, there are other decision criteria known from thermoset matrix materials, which can be of major interest, for example the chemical resistance or the water absorption properties. In contrast, thermoplastic composites are normally featured with an improved toughness compared to their thermoset competitors.[11,12] Table 8.2 gives an overview on properties of different common polymers which are used for the pultrusion process.

It is possible to use non-reactive and reactive thermoplastic matrix materials for the pultrusion process. Both groups of material offer specific advantages and disadvantages for the process. The biggest disadvantage of non-reactive thermoplastic polymers is their high melt viscosity, which affects the process speed negatively due to long impregnation times. In contrast, it is quite easy to add the non-reactive matrix material to the process. Reactive thermoplastic matrix materials have been widely neglected in research, though there are only some exercisable reactive matrix systems available (e.g., PA6, PA12, PMMA or TPU).[13–16] The biggest advantage of reactive thermoplastic matrix material is the low viscosity, which is similar to thermoset resins. Thus, the impregnation is completed after a short period of time with the result that high rates of production are possible. However, once the reaction is started, it cannot be stopped. Hence, special system engineering is necessary to handle

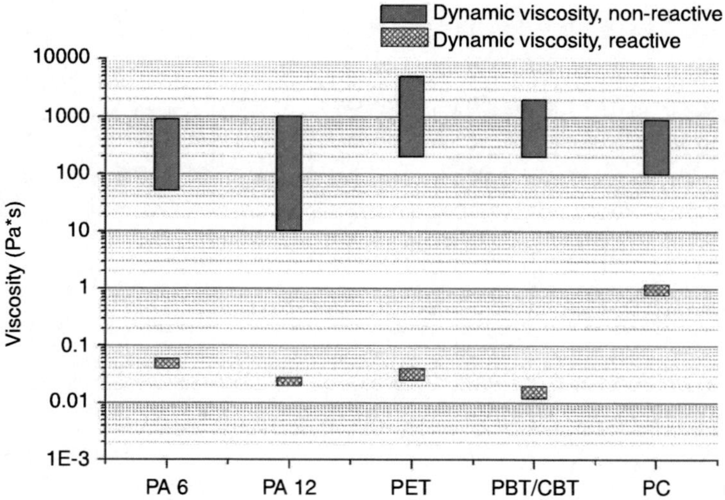

8.2 Viscosity of reactive and non-reactive thermoplastic matrix materials. (From reference 17.)

reactive thermoplastic materials. Figure 8.2 compares the different viscosity levels of reactive and non-reactive matrix materials. One disadvantage of thermoplastic polymers is their huge shrinkage, which is mostly bigger than the shrinkage of thermoset materials. This effect becomes very important for the dimensioning of the pultrusion die in order to avoid bending and spring back or forward of the profiles.

Thermoplastic matrix materials are commonly provided in form of prepreg materials.[6] There are various types of prepregs that are used for the pultrusion process and that offer the possibility of an easy combining of fibers and matrix materials. For example, tape materials, which are the simplest to use in the pultrusion process, are very thin, fully consolidated, and contain an exact ratio of fibers and polymer. Thus, there is no problem due to bad impregnation. During the process, the tapes are molten and forced together. On the other hand, tapes are one of the most expensive semi-finished products to use in the pultrusion process and they have limited usability for complex geometries. In contrast, when using commingled thermoplastic yarns it is easier to realize complex geometries, but the impregnation and consolidation of the fibers has to occur during the process, too.[5] Generally, prepregs have positive effects on different aspects of the process. They simplify the handling, the complexity of the pultrusion facility is reduced, and the desired fiber volume content is preset. Beyond, the process speed can be increased due to a faster impregnation. Figure 8.3 shows different prepreg types that are commonly used for the pultrusion process.

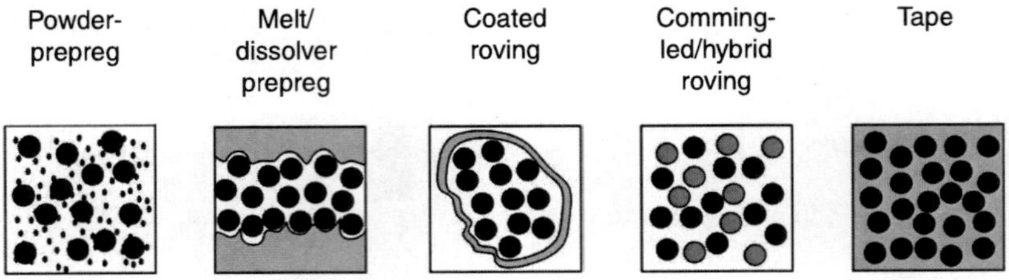

8.3 Prepreg types in use.

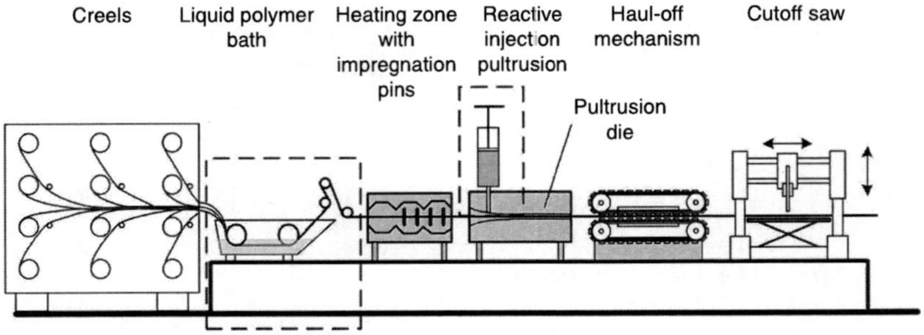

8.4 Chart of the pultrusion process.

8.2.2 The pultrusion process

The following chapter will describe the pultrusion process in detail. Figure 8.4 shows a chart of the pultrusion process.

Material supply

At the beginning of the pultrusion process the material is stored on coils (see Fig. 8.5), which are mounted on a creel and applied to the pultrusion process in the correct sequence. Ceramic holes fixed at the creel assure a desired feeding position of the fibers to the process. Furthermore it is important to control the alignment of the single rovings in order to prevent knotting and twisting of the fiber reinforced profile. To reduce the influence of fiber twist due to the unwinding of the coils, some fibers are applied without twist by a center pull. The reinforcement materials should be provided in a maximum length configuration on each coil in order to enable long process runs without disruption.[6]

Another major aspect of the creel is to ensure that the reinforcement material is provided with the correct tension. Thus it can be equipped with different tension straps, which are either individually controlled or adjusted all together.[6] The correct tape tension is required to minimize the

8.5 Creel assembled with glass fiber (TEXMER GmbH&Co. KG). (From reference 18.)

8.6 Tension control of the reinforcement fibers. (From reference 18.)

deformation of the produced part due to non-uniform shortening. The tension can be provided by, for example, pneumatic attached tension straps (see Fig. 8.6).

Pultrusion die

The die of a pultrusion unit (see Fig. 8.7) is normally made of hardened tool steel or steel alloy with a hardness of 28–30 HRC. Due to the high abrasive properties of the contacting reinforcement fibers, the surface of

8.7 Pultrusion die (manufactured by MPI-Ft. Worth Texas). (From reference 19.)

the tool is coated with a ceramic film or hard chrome plate. The thickness of this protection layer is only few microns, but it has a hardness of up to 70 HRC. Especially the lifetime of the inlet area of the tool can be increased significantly due to such coatings. Normally, chrome plated dies have a lifetime of about 150 000 m. Once the die is worn, it can be reprocessed again to the original dimension. Therefore it is important to do the polishing of the internal surface of the die in longitudinal direction because of the wall-friction.[3,19] In order to minimize bending and distortion due to inhomogeneous temperature distribution, the thickness of the wall should be about ten times the cross-section of the pultrusion part.[6]

Impregnation, consolidation and solidification

In the basic process the impregnation takes place in the pultrusion die due to temperature and due to pressure applied by the diminution of the die. The temperature profile is controlled by the use of distributed heating elements along the length of the die and the pressure profile is generated by the shape of the die. Thus, the only way to make the process faster is the extension of the pultrusion die. The longer the die the longer the time available for the impregnation of the fibers with polymer. In the past years many research studies have been conducted to accelerate the impregnation of the fibers, because the impregnation is the bottleneck of the production rate, and the extension of the die length comes along with a significant raise in tooling cost. All efforts have in common that they try to equalize the process by shifting the impregnation previous to the molding process. One possibility is the preheating of the used prepreg before entering the die in order to

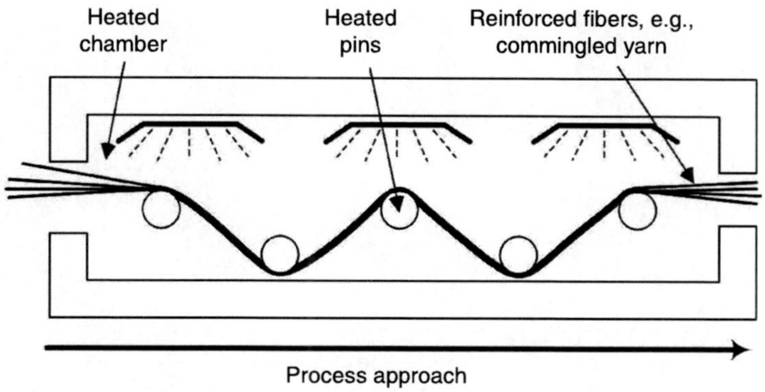

8.8 Pin impregnation of the reinforcement material.

reduce the necessary heating time to melt the polymer fully in the die.[20–23] Other methods raise the preheating temperature above the melting point of the matrix and strip the roving through several pins, whereby the fibers are fully impregnated (see Fig. 8.8). Afterwards, only the consolidation of the fibers, saturated with molten matrix, takes place in the die.[24]

There are different methods for heating the prepreg material before entering the die. The heating can be effected due to infrared,[24,25] convective air,[26,27] hot gas,[28] self-heated pins[26,29] or a combination of these different methods.[13,29] The most effective methods are based on contact heating, because they enable the fastest heat transfer. The necessary heating time can be reduced up to ten times compared to other methods.[29]

There are only a few methods for applying the matrix material to the process when using reactive thermoplastic matrix material. The reinforcement materials can be pulled through a monomer bath, which is quite similar to thermoset material processing.[30] Alternatively the monomers can be injected directly into the die. This method is commonly called Reactive Injection Pultrusion (RIP).[31] In addition to the impregnation, the consolidation and forming of the profile takes place in the die before the matrix is transformed into the solid state. The feasible profile geometries are multifaceted and almost only limited to a constant shape in process direction. Feasible profile geometries can be divided into standard and special profiles. For example U-, L-, O-profiles or different shaped hollowed profiles like round or square pipes rank among the most common profiles (see Fig. 8.9). These profiles are characterized by a simple design and a constant wall thickness, and they can be manufactured by every pultrusion company around the world. They are used for different applications, which do not require an optimized construction.

In contrast, profiles with a special geometry are designed, specified and produced for particular applications. They have, for example, a complex

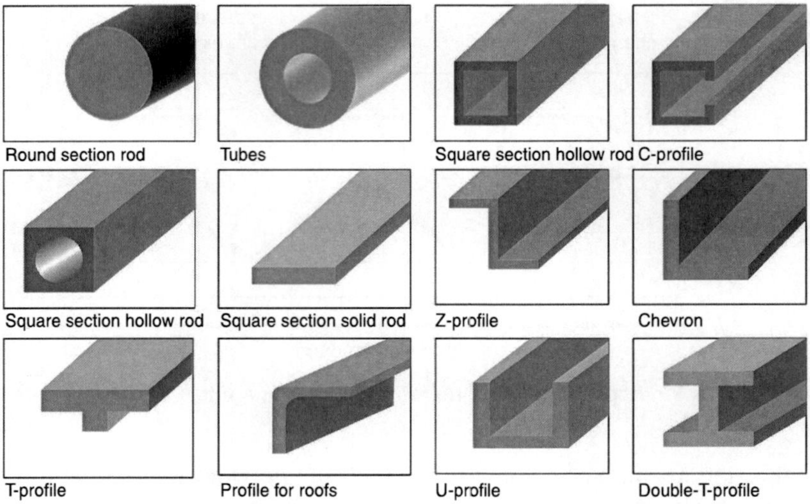

8.9 Standard profile geometries. (From reference 8.)

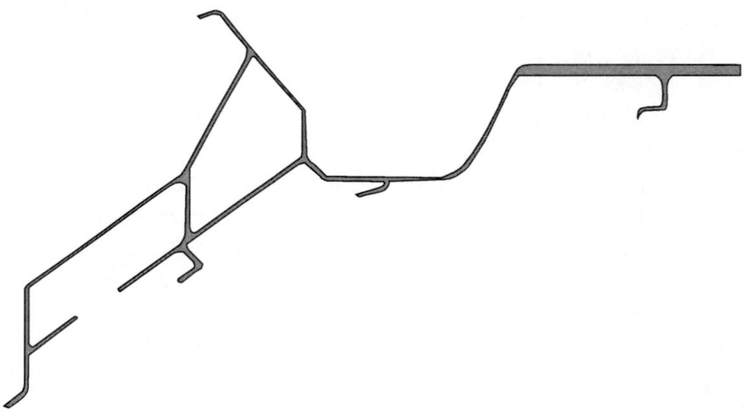

8.10 Special profile geometries. (From reference 8.)

multi-walled structure with different wall thickness (see Fig. 8.10). Although there are nearly no limits for the geometry of the profiles' cross-section, the shape of a profile always has to be constant over the whole length. Thus, it is not possible to produce profiles that have local constrictions or bulges without using an additional thermoforming process or a process combination.

Pulling force

The investigation of pulling forces during the thermoplastic pultrusion process has been widely neglected in research, although it is an important factor for the productivity, the understanding and the control of the process. The main influencing factors on the pulling force are the temperature and the pressure distribution in the die. Especially the pressure can be treated back on many other effects, which interact with each other. These are, for example, the fiber volume fraction, the pulling speed, the fiber and matrix continuity or the collimation force due to upstream resistance.[5,32] Scientific investigations of the pulling forces have shown that there are three main factors contributing the resistance per unit area of a composite in contact with the die. These are the viscous resistance, the compaction resistance and the friction resistance.[32] Figure 8.11 gives a qualitative distribution of these three factors.

In general, there are some accepted statements, which are listed below:

- The pulling force is proportional to the pulling speed,[5] that is, the higher the pulling force, the lower the maximum rate of production.[27]
- The pulling force is proportional to the circumference of the composites contacting the die, that is, the bigger and the more complex the profile geometry, the lower the maximum rate of production due to increasing pulling forces.[32]
- Higher temperatures lead to reduced pulling forces. This indicates that viscous shear forces are more significant than thermal expansion effects.[33]
- Increasing fiber volume content leads to increasing pulling forces.[34]

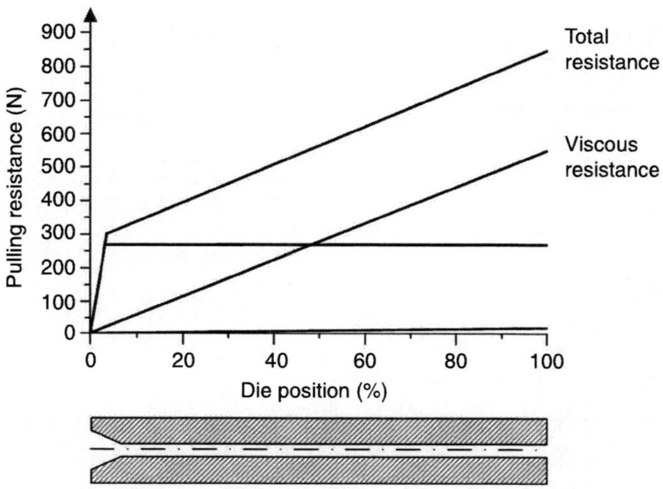

8.11 Elements of resistance in the die. (From reference 32.)

Pull-off and cut-off unit

The pull-off unit, which pulls the composite material through the whole pultrusion unit, is a mechanism located behind the pultrusion die. Because the constructive implementation is of least importance for a successful pultrusion process, it can be any mechanism that pulls the composite through the process.[5] However, there are only two versions of pull-off units in practice – a continuous and a reciprocating unit.[2]

The earliest pultrusion machines were equipped with hydraulic clamps, which gripped the composite and pulled it forward for an appointed distance. Thereby the clamp exactly fits to the pultrusion part. Subsequently, the part is released, the pull-off unit moves back to the beginning and the process starts again (see Fig. 8.12). Because this mechanism cannot provide a constant pull-off speed, today's pultrusion units are equipped with more, separately controlled, clamps, which work in series. One clamp moves forward and pulls the product forward and another clamp moves back to its beginning and starts moving at the same time the first clamp reaches the end and releases the product (see Fig. 8.12).[6] Other pull-off units are equipped with two continuous revolving belts, which clamp the composite in between. Because the clamping area is usually bigger in comparison to reciprocating pull-off units, the applied pressure is lower (see Fig. 8.12).

The cut-off unit is positioned directly behind the pull-off unit. It enables automatic cutting of the product to a desired length. There are many different methods as to how the design can look. All methods have in common that they have to move with the same velocity as the pultrusion line during the cut. For example, it can be made as a radial arm tape equipped with

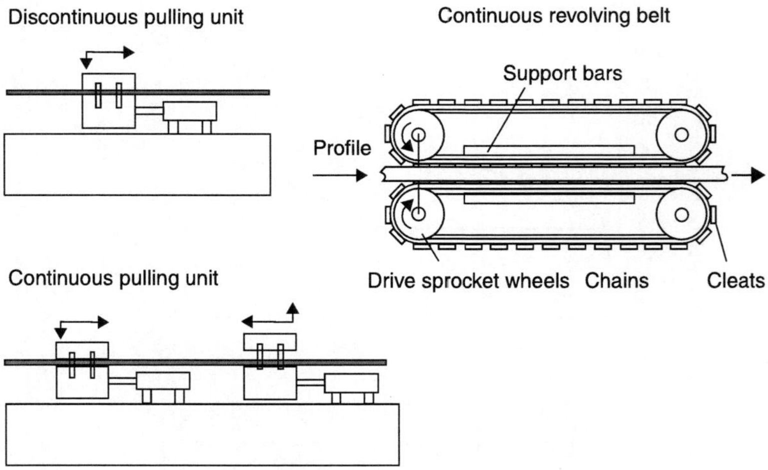

8.12 Different pull-off units.

diamond tipped saw blades. Chop saws, orbital saws or band saws are very common, too. A very elegant solution is the use of a high pressure water jet. It enables clean edges without any fiber pull-out and it is possible to cut every fiber reinforced material.[6]

Advantages and disadvantages of the pultrusion process

The following list gives an overview of the advantages and disadvantages of the pultrusion technology, which have been discussed in the previous explanations of the basis pultrusion process.

Advantages:

- Can be used for thermoset and thermoplastic matrix materials.
- Nearly every type of reinforcement material and prepreg is processable. The raw material is added in form of rovings, which is the cheapest form of reinforcement materials.
- Low rejections.
- High fiber volume content.
- Good alignment of the fibers orientated in process direction.
- Complex shapes with undercuts can be realized.
- Continuous process enables the production of continuous semi-finished goods.
- Automated process with low labor costs.
- Industrialized process with good theoretical background.

Disadvantages:

- Restrictions regarding the fiber orientation.
- There are always fibers necessary, which are aligned in process direction, in order to carry the pulling force.
- Constant profile geometry.
- Cost-intensive tooling due to abrasive fibers.
- Production speed is linked to the geometry of the product, that is, the more complex a geometry is, the slower is the production speed.

8.2.3 Process combinations and variations

The variety of variations, extensions and process combinations for the pultrusion process is very great. Thus, the following paragraphs will describe only some of them.

A combination of the pultrusion with the winding process is commonly known as the pull-winding process (Fig. 8.13). The fibers are laid to the

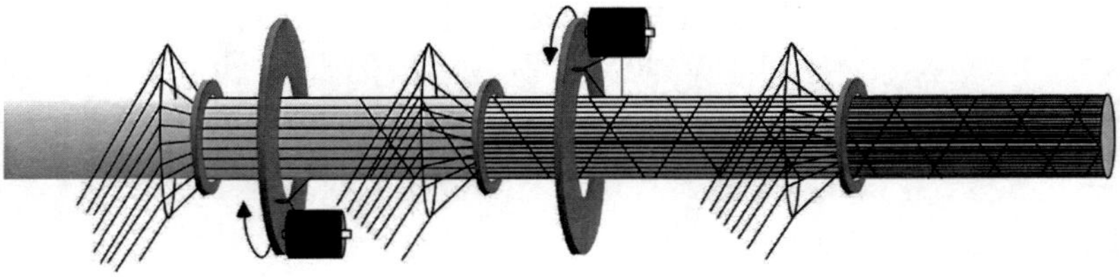

8.13 Pull-winding technology. (From reference 6.)

axial fiber of the pultrusion process by rotating spools. Thus, the axial fibers carry the tensile and bending forces, whereas the rotational fibers carry the torsion forces. This process combination provides a high flexibility in fiber orientation. The interplay of rotating and pulling speed enables the realization of fiber orientations from nearly 0° up to 85°. Because of the high pulling forces due to different resistances inside the die, it is necessary for the practical application to cover the rotational fibers by axial fibers. In general, the pull-winding process combines the advantages of both technologies: it is a fully automated process that enables the production of round or elliptical parts with the desired fiber orientation. Furthermore, the surface quality and the dimensional accuracy are very good.[5,6]

The in-line braiding process is very similar to the pull-winding process. In contrast, no winding equipment is positioned in front of the die, but a vertical braider. Thus, it enables the production of profiles that have a woven radial reinforcement. Again, the fiber angles are a result of the interaction of the braider and the pultrusion line speed. A big advantage of the in-line braiding process is the possibility to integrate unidirectional fibers into the braiding process. Thus, the part exhibits highly stable triaxial reinforcements so that no manufacturing problems occur during the pultrusion process.[5]

Another option for the pultrusion process is the so-called PAZ.[1] This is a combination of the extrusion and pultrusion process (see Fig. 8.14). The process starts with the extrusion of a thermoplastic pipe, which is reinforced with endless fibers in longitudinal direction. This pipe is likewise pushed and pulled out of the extruder. Subsequently, it serves as a winding core. Any desired fiber orientation from 0° to 90° can be realized. The special aspect of this process is that the impregnation and consolidation take place in the die due to the radial expansion of the extruded pipe. This is realized by an expanding mandrel located in the heating section of the die. The main advantage of this method is that the fibers are tautened during the impregnation leading to very low undulations of the fibers in contrast to externally applied pressure. Pull-off speeds of up to 30 m/min can be realized.

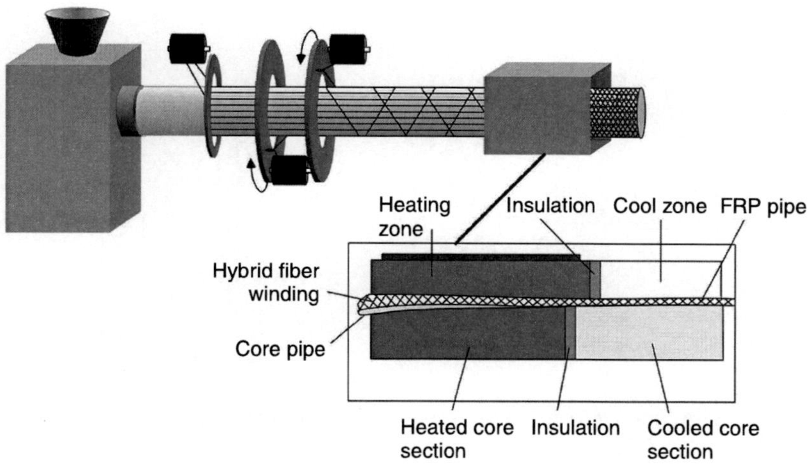

8.14 Schematic drawing of the PAZ process. (From reference 1.)

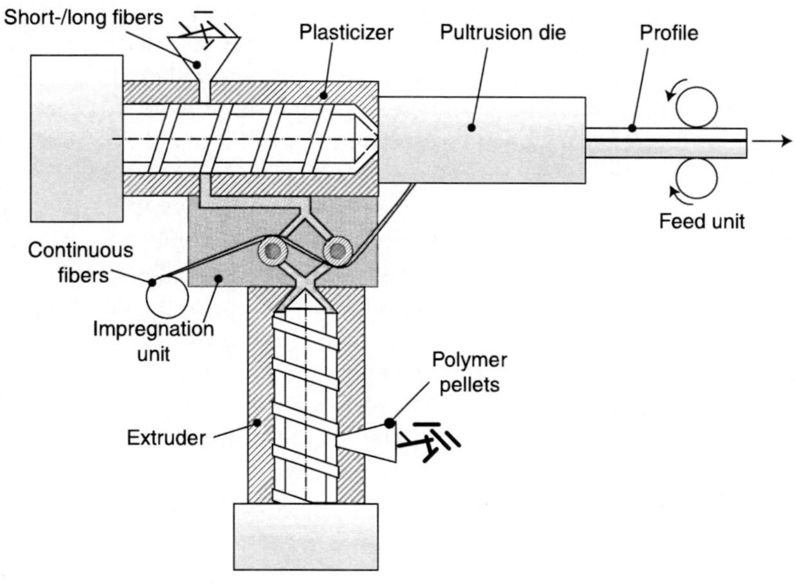

8.15 The pull-extrusion process. (From reference 35.)

Similar to the described combination of pultrusion and extrusion is the so-called pull-extrusion process, which is illustrated in Fig. 8.15.[35] It enables the production of products which are reinforced with both continuous and discontinuous fibers in one profile. A thermoplastic polymer is molten in an extrusion unit. In a first step, the extrudate is combined with a continuous fiber strand. In the second step the remaining polymer is added with

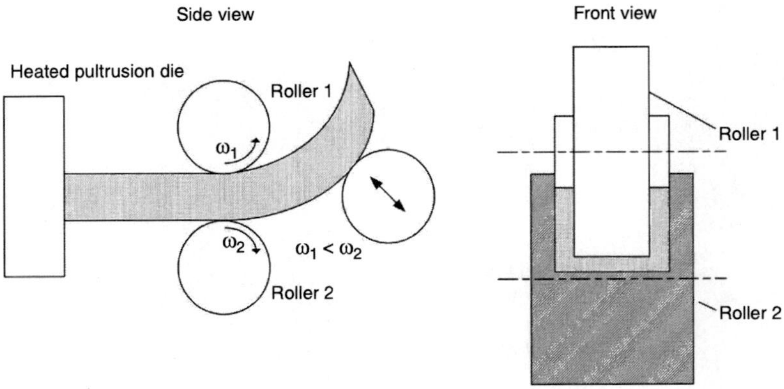

8.16 On-line forming of profiles. (From reference 26.)

some short fibers to a plasticizer. Afterwards, the short-fiber extrudate (fully impregnated) and the fiber strand are combined in a pultrusion die. While moving through the die, the fiber strand is also impregnated fully. This technology enables the production of profiles which have a much better stiffness compared to profiles that are only reinforced with short fibers. Furthermore, the complexity of profiles can be increased significantly. Due to the inhomogeneous thermal expansion of both materials, significant deformation of the products can occur.

For practical application it often can be advantageous to have not just straight profiles but also curved profiles. The so-called online forming process uses driven rollers instead of the cooling zone of the die. They are formed as male and female roller pairs corresponding to the desired profile geometry. Because of that, this process is limited to open profiles without any undercut. These rollers have several main functions. First, they provide the necessary pulling force for the process. Second, they rotate at different speed; this leads to an interlaminar sliding of the fibers (fiber layers) and the profile is bent to the direction of the roller, which rotates at slower speed. Furthermore, the profile is cooled and solidified by the rollers (see Fig. 8.16).[30]

8.2.4 Commercial products and applications

Pultrusion parts can be found in nearly every field of industry, for example in the consumer and construction markets. They are used because of their specific properties and advantages over their metal competitors such as extremely light weight or corrosion resistance. Most products are composed of standard profiles. Figure 8.17 shows some application examples.

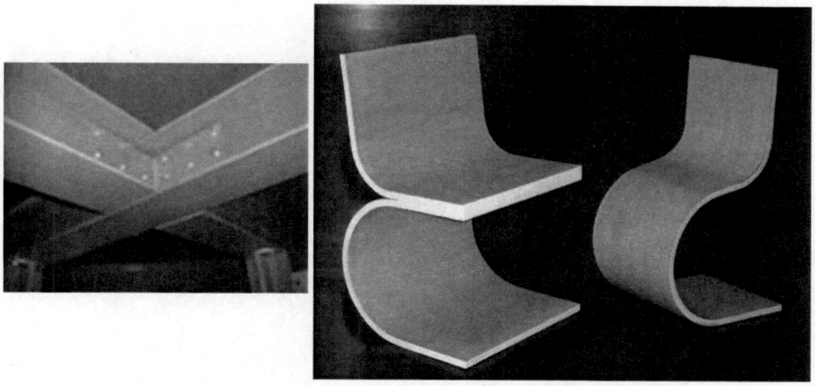

8.17 Industrial and consumer applications (Top Glass S.p.A.). (From reference 36.)

8.3 Continuous compression molding

The continuous compression molding (CCM) machine (see Fig. 8.18) works on the basis of a semi-continuous process. It was first industrialized for the production of FRTP material by Advanced Composites and Machines GmbH, Friedrichshafen, Germany (today[38]). The production of profiles using the CCM technology was filed for patent in 1991.[39]

Basically, the process consists of several pressing tools, which are connected in series. These pressing tools can either be mounted on several single pressing units, where the material is transported from one pressing tool to the next. Alternatively, a continuous tool that is mounted on a press can be used. In this case the material is semi-continuously fed forward for an appointed distance. The idea of this process is the decoupling of the heating and the cooling zone of the tool in order to enable fast heating and cooling rates. Thus, the impregnation and consolidation take place in the heated section at the beginning of the tool. In the cooling section at the end of the tool, the consolidation and solidification take place. In general, the continuous compression molding process comprises the following four steps: closing the mold, applying the pressure and temperature, lifting the mold and transporting the laminate. During the compression stage a constant pressure is applied to the laminate. During the subsequent stage the laminate is released and a feeding unit pulls the material one step forward to start the next compression step. Once these four steps are completed, a new cycle begins.

8.3.1 Material in use

There is a big variety in reinforcement material that can be used in the CCM process, because there are no special requirements as to the material.

8.18 Continuous compression molding (CCM) machine (IVW GmbH). (From reference 37.)

Nevertheless, the most common materials for industrial application are glass and carbon fibers. They can be provided in different forms of woven or non-woven fabrics or as random orientated fibers. Furthermore, it is possible to process different kinds of natural fibers (e.g., flax, kenaf, hemp) in any desired fiber orientation. Each fiber orientation has specific advantages and disadvantages. Woven textiles, for example, offer good drapability but the strength and stiffness are reduced in comparison to non-woven fabrics because of the fiber undulations. Non-woven fabrics enable the production of composites with higher fiber volume content. Depending on whether the interloop is compact or bulk, the drapability or the integrity is advantaged. For use in a large-scale process, the total, continuous length of the reinforcement fabric is important in order to avoid costs for setting up the process again.

Like the pultrusion process, the selection of a suitable thermoplastic matrix material mainly depends on the desired mechanical properties and the desired long-term service temperature. The range of usable matrix materials starts with standard polymers such as polyethylene (PE) or polypropylene (PP) and ends with high performance polymers such as polyetherimide (PEI) or poyletheretherketone (PEEK). Recent developments have shown that the processing of reactive thermoplastic materials is possible as well (CBT). For some physical properties of common matrix materials see Table 8.2.[39,40]

For the *in situ* production of panels or profiles designed as sandwich structures, there are only some core materials that can be used because of the high temperatures during the process. PMMA foam (product name:

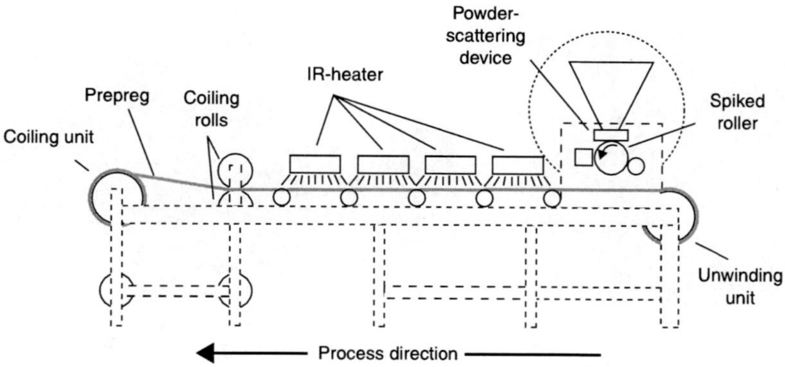

8.19 Schematic of a powder-prepreg unit (IVW GmbH). (From reference 37.)

Rohacell®) is one of the widely used core materials for an *in situ* process. Although its heat resistance is about 200°C, it enables only the production of sandwich structures using polymer matrix materials with a melting temperature in the same range (e.g., PE, PP, PA12).

In contrast to the pultrusion process, it is always necessary to provide the reinforcement material already combined with the matrix material, for example by using the film-stacking method or in the form of plane prepregs. An injection process is not developed yet. Basically, there are two production methods for the prepregs: weaving of different types of prepreg rovings (e.g., commingled rovings) lead to plane prepregs; alternatively the prepregs can be produced after the weaving or interlooping process using one of the most economical methods, the powder-prepreg technology. Figure 8.19 shows a schematic drawing of a powder-prepreg unit.[41]

At the beginning of the process, the reinforcement fibers (woven, non-woven or random orientated fibers) are pulled though a spreading unit with constant speed and powder is homogeneously applied. Subsequently, the polymer powder is heated above the melting point for a short time, for example by infrared heaters. Thereby the matrix wets the fibers superficially in order to enable good handling as well as easy storing. Before coiling the prepreg material again, the polymer is cooled down below the melting point by cooled rolls. Figure 8.20 shows the surface of a powder-prepreg made of glass fiber and polybutylene terephthalate (PBT) and the used powder-prepreg unit of IVW GmbH.[37]

8.3.2 The continuous compression molding process

The following section will give a detailed description of a continuous compression molding process using a pressing unit which is equipped with one continuous tool.

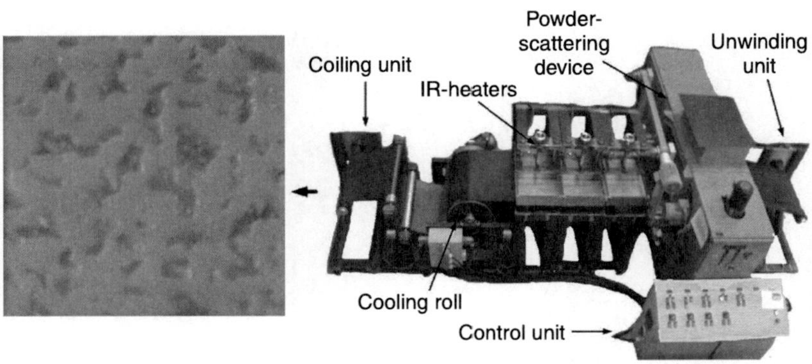

8.20 Surface of a powder-prepreg (GF/PBT) and drawing of the powder-prepreg unit (IVW GmbH). (From reference 37.)

8.21 Creel of the CCM press (left picture) and strap brake (right picture) (IVW GmbH). (From reference 37.)

Material supply/input

For continuous production it is important to provide the raw materials to the process in a sufficient amount. Thus, the reinforcement materials are stored on creels in front of the continuous compression molding machine (see Fig. 8.21). Therefore it is useful to have one unwinding unit for each reinforcement layer. Depending on the used prepreg type the necessary number of unwinding units is increased. For example when film-stacking prepregs are used, each polymer film is stored on a separate unwinding unit. In contrast, when powder-prepregs are used, there are no additional

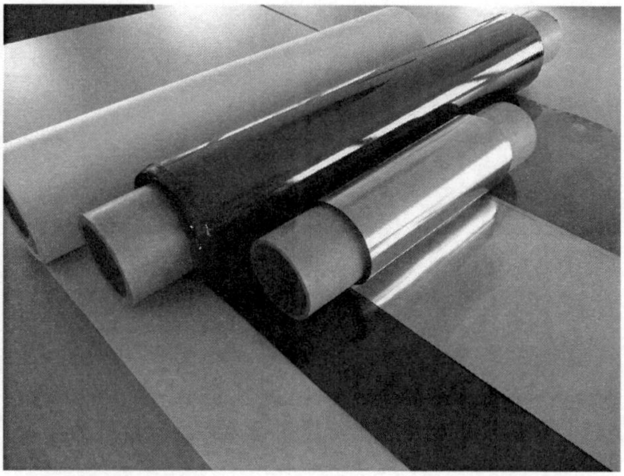

8.22 Different release films made of paper, polyetherimide and steel (IVW GmbH). (From reference 37.)

unwinding stations required. When pulling off the raw material from the creel and carrying it through the CCM machine, it is helpful to keep the material under tension in order to prevent lateral movement. Thus each unwinding unit should be equipped with an individual strap brake.

In addition to the raw material, a release film has to be stored on the creel too. For lower temperatures (up to 220°C) special release paper can be used. For higher temperatures release films made of polyetherimide or steel are commonly used (see Fig. 8.22). The biggest advantage of the paper film is the low price compared to other materials. In contrast, it is only possible to carry tensile forces when polyetherimide or steel is used. This becomes important when reinforcement layers are to be processed, which do not contain any fibers orientated in process direction (0°). Thus, it may be necessary to use the more expansive films although the temperatures are quite low. Another advantage of steel and polyetherimide is that they can be used more than once without providing additional release agent. Paper has to be disposed of after it has been used once. After removing the release film from the laminate behind the feeding unit, it is coiled again.

Pressing unit

The pressing unit (see Fig. 8.23) of a CCM press is conventionally equipped with a mold, which is divided into a lower and an upper part. Furthermore, each half of the mold is divided into a heating and a cooling section (see Fig. 8.24), and the mold can be designed continuously or separately. Both designs offer specific advantages and disadvantages, for example the accurate heating of a separated mold is easier, energy consumption is lower and

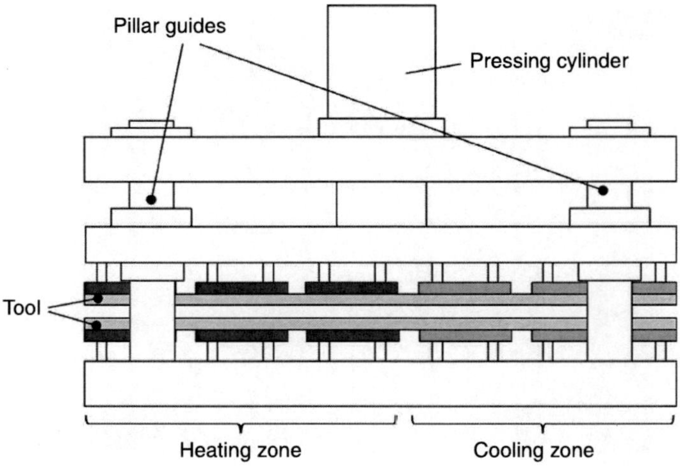

8.23 Sketch of a pressing unit of a CCM machine.

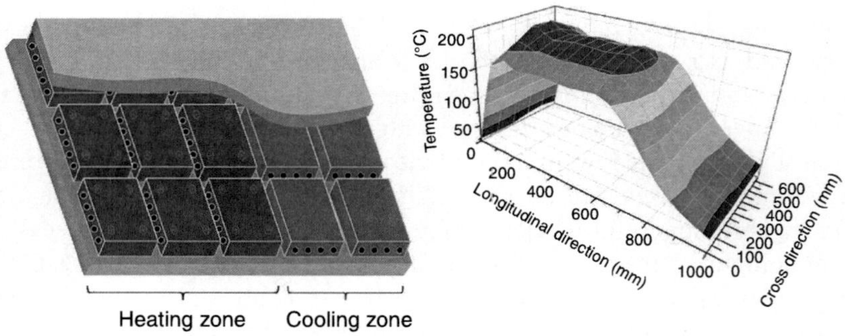

8.24 Heating and cooling section of a CCM machine with corresponding tool temperature (IVW GmbH). (From reference 37.)

there is less thermal stress inside the mold. But the temperature transition of a continuous mold is smoother and there is no gap that can become apparent on the laminate surface. Depending on the working temperature, the tools can be made of steel or aluminium. In order to ensure high variability and low costs for the used tools, the heating and cooling elements are installed on additional steel plates located on the outer face of the tools. The heating is usually realized by electric cartridge heaters and the cooling is done by water. Thus it is possible to generate nearly any desired temperature profile on the surface of the tool.

The application of pressure is done by one or more hydraulic cylinders similar to static pillar presses. Depending on the application area, CCM machines can be designed according to an isochoric or an isobaric operating principle. For the production of flat panels and profiles without core material inside, mostly isobaric presses are used. In contrast, for the *in situ* production of sandwich panels or profiles with compressible core material, it is necessary to use a press working to an isochoric operating principle. In order to enable parallel movement and closing of the mold, the presses are equipped with pillar guides. Additionally, there are some counteractive pressure cylinders which force the moving part of the mold parallel to the stationary part during the closing and pressing phase.

The impregnation mechanism during the CCM process is quite similar to the impregnation in a static pillar press. Most of the matrix flow takes place in through thickness direction. But there is a different impregnation stage of the laminate at each position of the mold in process direction for defined process conditions. Impregnation starts at the beginning of the tool after the polymer is heated above the melting point and is completed just before the laminate reaches the cooling section (using settings for maximum output). In the cold section the polymer is solidified again. Heating and cooling rates up to 200 K/min are reached. Usually, the upper and lower tool is closed by some shearing edges at the side. Thus, the entrapped air mainly has to be displaced out of the laminate against the process direction. Furthermore, the matrix flow sideways in the tool is minimized so that the production of laminates with uniform edges is ensured. Especially when producing unidirectional reinforcement materials only, this effect disables the movement of the fibers in the lateral direction too. In contrast, when using tools without shearing edges the entrapped air is also displaced to the side. At the end of the tool there is a fully impregnated and consolidated composite.[42,43]

The pull-off unit is mostly designed as hydraulic clamps, which grip the composite and pull it forward for an appointed distance semi-continuously (see Fig. 8.25). Thus, the resulting process speed is defined by the appointed pulling distance and by the periods of time for pulling the composite, for pressing the laminate, and for opening and closing the tool. The period of time for pressing is determined by the required time to impregnate the reinforcement material fully.

Optionally, a CCM press can be equipped with a cut-off unit in order to automatically tailor the profiles. Although the composite is stationary during the pressing phases, it is necessary to have a moving cut-off unit in order to cut any desired length of the composite. Otherwise it only would be possible to cut profiles in a multiple length of the pull-off distance.

8.3.3 Profile production

The manufacturing of continuous fiber reinforced profiles (see Fig. 8.26) with the CCM process is principally comparable to the manufacturing of

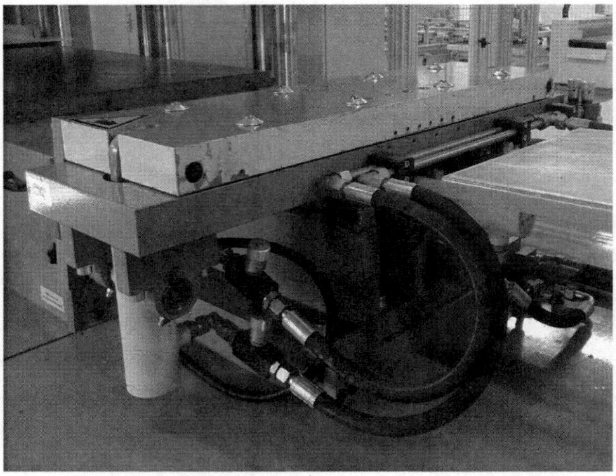

8.25 Pull-off unit (IVW GmbH). (From reference 37.)

8.26 CCM profiles (xperion Aerospace GmbH). (From reference 38.)

organic sheets (flat panels made of continuous FRTP materials). Main differences are the complexity of the tool and the mechanism of feeding the material to the process.

For the manufacturing of profiles a special preforming unit is required, which is adapted to the specific profile geometry (see Fig. 8.27). In the simplest case the preforming unit is designed as two steel sheets which border the raw material below and above. The plane textile semi-finished product is stepwise formed to the geometry of the profile straight in front of the tool in this unit. Alternatively, the preforming unit can be equipped, for example, with some rolls arranged in a line comparable to the roll-forming process. The release films, which are also necessary for the profile production, are supplied to the process in front of the preforming unit. Thus, the release film is formed to the profile geometry by the preforming unit, too.

8.27 Pre-forming unit located in front of the CCM machine (xperion Aerospace GmbH). (From reference 38.)

The CCM process enables the production of profiles with various geometries. It is possible to produce open profiles as well as closed profiles with a hollowed core in a single step process. Figure 8.28 gives an overview of possible profile geometries. The freedom in profile design is mainly limited by the following aspects:

- Reinforcement material is provided to the process in form of one or more stacked, continuous plane textile structures. Thus, the profile's geometry must be representable with this material by shaping. Alternatively, special preforms can be used so that the geometry is limited by the performance of the preforming process.
- The application of pressure to the laminate must be guaranteed at each position of the profile. For that, the number of directions in which the pressure can be applied by the pressing unit (or by several pressing units) is another limiting aspect.
- Supply of release film must be possible.

For the realization of complex geometries it may be necessary to use a CCM press which is equipped with pressing units in vertical and in horizontal direction. Alternatively, the pressure application can be realized with a complex tool design. However, there will always be an inhomogeneous pressure distribution inside the tool. This is affected by the profile angles which are mostly different to the direction of the applied pressure.

The production of hollowed cored profiles is carried out using a core which is mounted buoyant in the tool (see Fig. 8.29). The core usually is equipped with a separate heating and cooling system in order to enable full impregnation of the interior reinforcement fibers. In some cases, for example when polypropylene is used as matrix material, it is not necessary to use a release film around the core. In this case, the temperature distribution of

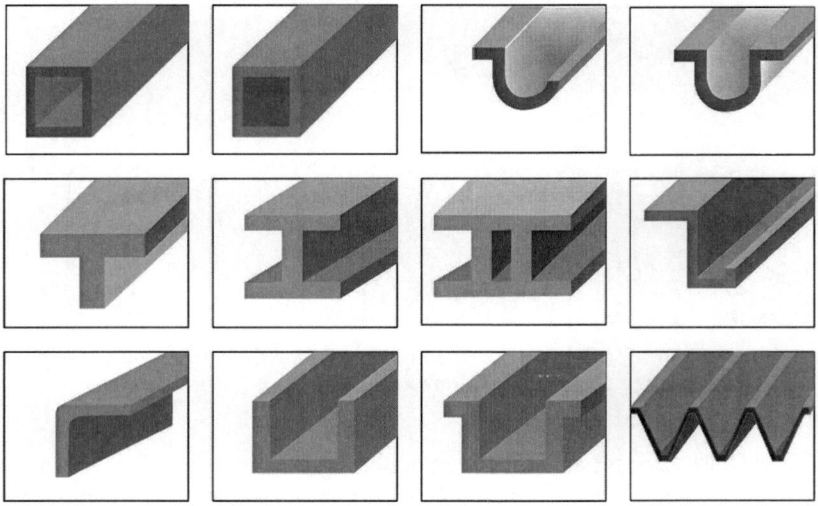

8.28 Capable profile geometries.

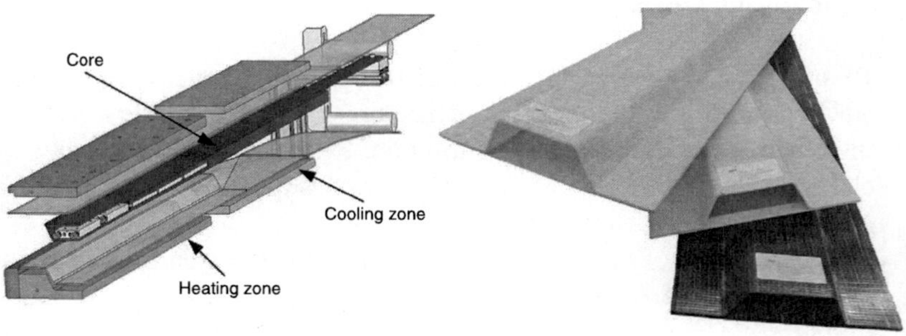

8.29 Closed cap profiles with different reinforcements and matrix materials (IVW GmbH). (From reference 37.)

the core has also a strong influence on the adhesion of the matrix material to the core. In order to minimize pulling forces and fiber drafting, the adhesion of the polymer to the core should be the lowest possible.

For good impregnation quality it is important that the cavity of the tool perfectly matches the thickness of the composite. Otherwise, there is an inhomogeneous compaction of the reinforcement fibers leading to inhomogeneous fiber volume content across the profile. Having a closer look at the fibers across the wall thickness, this irregular fiber distribution becomes apparent. During the phase of forming and pressing of the laminate, the fibers are pulled above the radii of the profile. Thus, the fibers are drawn to the inside of the curves and there is a resin-rich zone in the outside radius. Figure 8.30 illustrates this fact.

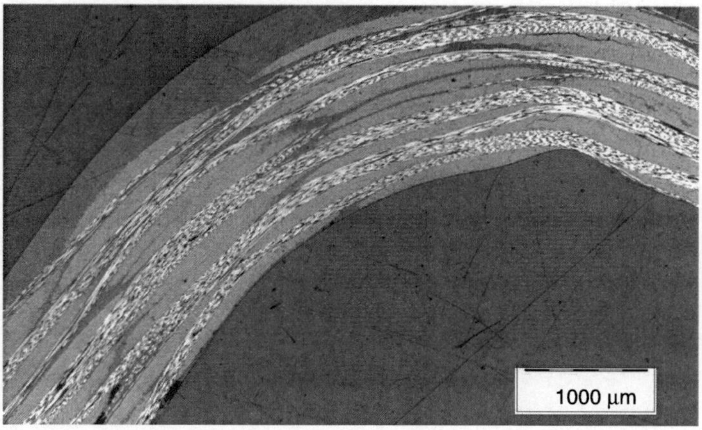

8.30 Fiber drawn in profile curves (IVW GmbH). (From reference 37.)

In contrast to the pultrusion process, the production speed and the maximum size of the profile are not limited by the required pulling forces. Due to the use of release films, there is no adhesion between the polymer and the surface of the mold. Furthermore, the friction between the release film and the tool is minimized because of the semi-continuous operating principle of the CCM process.

8.3.4 Process modeling

Due to the increasing complexity of manufacturing methods, process modeling is becoming more and more important. It enables the necessary number of cost-intensive experiments to be reduced. During the impregnation of the reinforcement fibers in the CCM process there is nearly only matrix flow in through thickness direction. Thus, an impregnation model developed by Mayer[44] is used for the process modeling. The so-called B-factor represents a non-dimensional process constant, which subsumes all factors influencing the impregnation. Thereby, it represents process conditions that enable the production of fully impregnated laminates. Thus, these process parameters have to be determined in advance by investigating different configurations, for example in a small static press. Using the model, it is possible to transfer these parameters to the CCM process afterward. If the determined B-factor is reached or exceeded with the CCM process with given process parameters too, the impregnation quality of the laminate will at least reflect the impregnation quality of the reference process. The general procedure for the determination of the B-factor is given in Fig. 8.31.

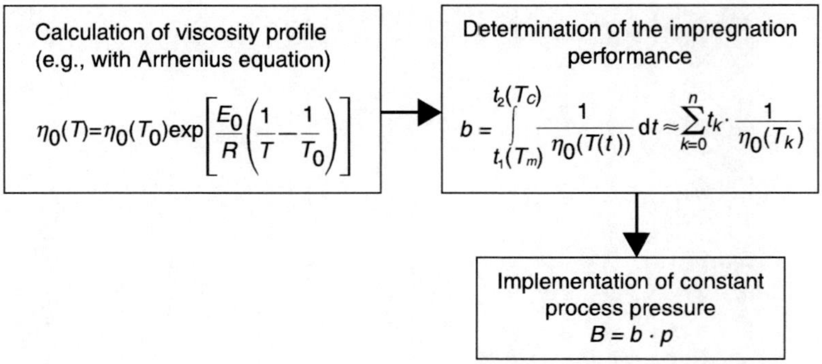

8.31 Approach for determination of the B-Factor. (From reference 44.)

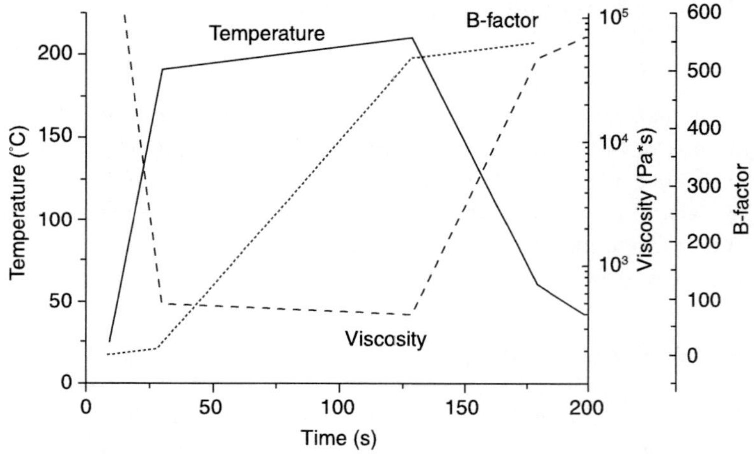

8.32 Example of B-Factor calculation.

The calculation starts with the determination of the viscosity profile of the matrix material. Because there are only very low shear rates during the impregnation process in a CCM machine, this can usually be done with the Arrhenius equation. In the second step the b-integral is determined by the integration of the reciprocal viscosity profile with the melting temperature (T_M) as lower limit and the crystallization temperature (T_C) as upper limit of integration. Thus, the temperature profile during the process is necessary too. Alternatively, if there are only discrete values of the temperature or the viscosity available, the b-integral can be determined with an empirical formula. Finally, the B-factor is determined by the multiplication of the b-integral with the applied process pressure during the process.[42–44] Figure 8.32 illustrates the calculation process.

8.3.5 Mechanical properties

In principle, every laminate structure is processable with a CCM machine so that the mechanical properties of profiles can be very different. Because of the working principle of the press, most profile designs will have no circumferential reinforcement fibers. For hollow profiles there is at least one joining zone, which has to transmit the load to the other part of the profile. The quality and the performance of this joining zone are mainly determined by the process conditions and by the used matrix materials, which makes it difficult to give a general assessment.

In order to demonstrate the mechanical properties, especially the performance of the *in situ* joined zone, results of a dynamic crash test are shown. The test profile is a closed cap profile made of glass fiber textile and polypropylene matrix material (TwinTex® – see Fig. 8.29). An impact energy of

8.33 Crash test behavior of a closed cap profile (IVW GmbH). (From reference 37.)

790 J is applied. A crash test sequence is shown in Fig. 8.33. The profile is bent up to the maximum deformation of about 107 mm. Subsequently the test sled was pushed backwards by the restoring force of the profile due to the elastic strain of the profile. This sequence clearly demonstrates the integrity of the profile and especially of the *in situ* joined zone, which is definitely not the weak point of the chosen profile geometry and material under the given load conditions.

8.4 Preferred application areas for the pultrusion process and continuous compression molding process

By means of the described technology properties, both processes are suitable for mass production. Nevertheless, there is a preferred field of application for both technologies. Unidirectional reinforcement fibers are best suited for the pultrusion process. In contrast, the CCM process is independent of the fiber orientation, thus it doesn't matter if unidirectional or multidirectional reinforcement materials are used. The assessment of the rates of production (= process speed) against the circumferences of the profiles, which are in contact with the tools' surface (see Fig. 8.34), defines relevant process parameters and emphasizes the preferred field of application. Thus the circumference characterizes the profile shape.

The rate of production subsumes the relevant process parameters (e.g., temperature, pressure, length of the respective tools, or the pulling resistance). Furthermore, the specific rate of production 'P' is shown in Fig. 8.35.[9,13,15,20–24,26,29,30,45–49] It is defined in terms of the profile geometry, in order to make the performance of the different processes comparable to each other (see Equation 8.1). For each material–process combination the values of the specific rate of production can be compared by means of the ISO-productivity curves.

$$P = circumference\,rate\,of\,production = \left[m \cdot \frac{m}{h}\right] = \frac{m^2}{h} \qquad [8.1]$$

(m: meter h: hour)

Figure 8.35 illustrates clearly that the pultrusion process is the preferred choice for the production of profiles with small circumference at high line speeds. In contrast, the reachable line speeds with a CCM press are lower, but independent of the profiles shape. That means that profiles with a huge circumference can be produced as fast as those having a smaller circumfer-

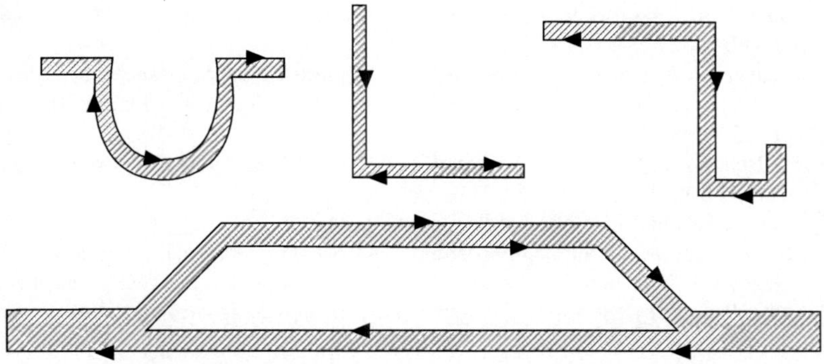

8.34 Circumference of different profiles contacting the die.

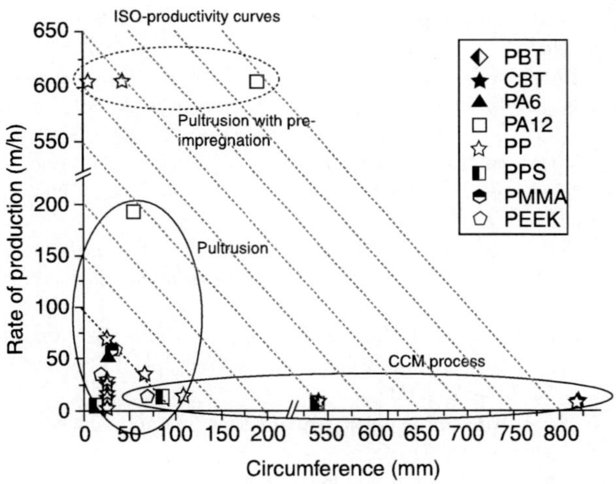

8.35 Specific production of the pultrusion and the continuous compression molding process.

ence. Comparing the performance 'P' of the two processes to each other, it is shown that both are able to work at a comparable level.

8.5 References

1. Schürmann, H. and Scholl, S., 'Continuous profiles with targeted orientation', *Kunststoffe International*, **98**(7), 55–58, 2008.
2. Starr, T. F., *Pultrusion for engineers*. Cambridge: Woodhead Publishing, 2000.
3. Meyer, R. W., *Handbook of pultrusion technology*. New York: Chapman & Hall, 1985.

4. Stewart, R., 'Pultrusion industry grows steadily in US', *Reinforced Plastics*, **46**(6), 36–39, 2002.
5. Fanucci, J. P., Nolet, S. and McCarthy, S., 'Pultrusion of composites'. In Gutowski, T. G. (ed.), *Advanced composites manufacturing*. New York: John Wiley, 1997, pp. 259–296.
6. Wilson, B. A., 'Pultrusion'. In Peters, S. T. (ed.), *Handbook of composites*, 2nd edn. Cambridge: Chapman & Hall, 1998, pp. 488–524.
7. Marine Products Inc.. http://pultrusiononline.com/
8. EPTA – European Pultrusion Technology Association. http://pultruders.com
9. Angelov, I., Wiedmer, S., Evstatiev, M., Friedrich, K. and Mennig, G., 'Pultrusion of a flax/polypropylene yarn', *Composites Part A: Applied Science and Manufacturing*, **38**, 1431–1438, 2007.
10. Mitschang, P. and Neitzel, M., *Handbuch Verbundwerkstoffe – Werkstoffe.* Verarbeitung, Anwendung, Munich: Carl Hanser Verlag, 2004.
11. Beever, W. H. and O'Conner, J. E., 'Pultruded thermoplastic composites structures'. In *International Sampe Symposium Proceedings*, **32**, 1309, 1987.
12. Goldsworthy, W. B., 'Thermoplastic composites: The new structurals', *Plastic World*, **42**(9), 56–58, 1984.
13. Dubé, M. G., Batch, G. L. and Vogel, J. H., 'Macosko, reaction injection pultrusion of thermoplastic and thermoset composites', *Polymer Composites*, **16**(5), 378–385, 1995.
14. Cho, B. G., 'Experimental studies of pultruded fiber reinforced nylon-6 composites', *47th Annual Conference, Composites Institute*, The Society of Plastics Industry, February, 3–6, 1992.
15. Ma, C.-C. M., Yn, M.-S., Chen, C.-H. and Chiang, C.-L., 'Processing and properties of pultrude thermoplastic composites (I)', *Composites Manufacturing*, **1**(3), 191–196, 1990.
16. Luisier, A., Bourban, P. E. and Manson, J.-A. E., 'Reaction injection pultrusion of PA12 composites: Process and modelling', *Composites Part A: Applied Science and Manufacturing*, **34**, 583–595, 2003.
17. Ehrenstein, G. W., *Polymer Werkstoffe. Struktur Eigenschaften Anwendung*, 2nd edn. Munich: Carl Hanser Verlag, 1999.
18. TEX MER GmbH&Co. KG. http://www.texmer.de
19. Martin, J., 'Pultrusion'. In Miller, E. (ed.), *Plastic products design handbook. Part B. Process and design for processes.* New York: Marcel Dekker, 1983, pp. 37–74.
20. Wiedmer, S., *Zur Pultrusion von thermoplastischen Halbzeugen: Prozessanalyse und Modellbildung*. Kaiserslautern: Schriftenreihe Institut für Verbundwerkstoffe GmbH, 2006.
21. Kerbiriou, V. and Friedrich, K., 'Pultrusion of thermoplastic composites: Process optimization and mathematical modeling', *Journal of Composites Materials*, **12**, 96–120, 1999.
22. Kerbiriou, V., *Imprägnieren und Pultrusion von thermoplastischen Verbundprofilen.* Düsseldorf: VDI-Fortschrittsbericht, 1997.
23. Bechthold, G., Sakaguchi, M., Friedrich, K. and Hamada, H., 'Pultrusion of micro-braided GF/PA6 yarn', *Advanced Composites Letters*, **8**(6), 305–314, 1999.
24. Miller, A. H., Dodds, N., Hale, J. M. and Gibson, A. G., 'High speed pultrusion of thermoplastic matrix composites', *Composites Part A: Applied Science and Manufacturing*, **29A**, 773–782, 1998.

25. Devlin, B. J., 'Pultrusion of unidirectional composites with thermoplastic matrices', *Composites Manufacturing*, **2**(3/4), 203–207, 1991.
26. Bechthold, G., *Pultrusion von geflochtenen und axial verstärkten Thermoplast-Halbzeugen und deren zerstörungsfreie Porengehaltsbestimmung*. Kaiserslautern: Schriftenreihe Institut für Verbundwerkstoffe GmbH, 2000.
27. Aström, G., Pipes, R. B. and Larsson, H., 'Development of a facility for pultrusion of thermoplastic-matrix composites', *Composites Manufacturing*, **1**, 114–123, 1991.
28. Lehmann, U., Blaurock, J. and Rau, S., 'Fertigung von Bauteilen und Halbzeugen aus endlosfaserverstärkten Verbundkunststoffen, Kombiniertes Flecht-Pultrus ions-verfahren'. In *18. IKV, Kunststofftechnisches Kolloquium Eurogress*, Aachen, 6–8 March, 8–20, 1996.
29. Bechthold, G., Wiedmer, S. and Friedrich, K., 'Pultrusion of thermoplastic composites: New developments and modeling studies', *Journal of Thermoplastic Composites Materials*, **15**(5), 443–465, 2002.
30. Larock, J. A. and Hahn, H. T., 'Pultrusion processes for thermoplastic composites', *Journal of Thermoplastic Composite Materials*, **2**, 216–228, 1989.
31. Dubé, M. G., Batch, G. L., Vogel, J. G. and Macosko, C. W., 'Reaction injection pultrusion of thermoplastic and thermoset composites', *Polymer Composites*, **16**(5), 378–385, 1995.
32. Aström, B. T. and Pipes, R. B., 'A modeling approach to thermoplastic pultrusion. I: Formulation of models', *Polymer Composites*, **14**(3), 173–183, 1993.
33. Sumerak, J. E. and Martin, J. D., 'Pultrusion process variables and their effect upon manufacturing capability', 39th Annual Conference, Composites Institute, SPI, 1984.
34. Bibbo, M. A. and Gutowski, T. G., 'An analysis of the pulling force in pultrusion', ANTEC'86, 1986.
35. Stavrov, V. P. and Narkevich, A. L., Applied for Patent in Belarus, No. a20081347, 24/10.2008.
36. Top Glass S.p.A.. http://www.topglass.it and http://www.fulcrumcomposites.com
37. IVW GmbH. http://www.ivw.uni-kl.de
38. xperion Aerospace GmbH. http://www.xperion-aerospace.de/
39. Spelz, U. and Schulze, V., 'Intervall-Heißpreßverfahren', *Kunststoffe*, **85**(5), 665–668, 1991.
40. Gardiner, G., 'Aerospace-grade, compression molding', *High Performance Composites*, July, 34–40, 2010.
41. Mayer, C. and Stadtfeld, H., Continuous production of prepregs from sized carbon fiber reinforcement, DE patent 19734417 (C1), 1998.
42. Steeg, M., *Prozesstechnologie für Cyclic Butylene Terephthalate im Faser-Kunststoff-Verbund*. Kaiserslautern: Schriftenreihe Institut für Verbundwerkstoffe GmbH, 2010.
43. Wöginger, A., *Prozesstechnologien zur Herstellung kontinuierlich faserverstärkter thermoplastischer Halbzeuge*. Kaiserslautern: Schriftenreihe Institut für Verbundwerkstoffe GmbH, 2004.
44. Mayer, C., *Prozeßanalyse und Modellbildung bei der Herstellung gewebeverstärkter, thermoplastischer Halbzeuge*. Kaiserslautern: Schriftenreihe Institut für Verbundwerkstoffe GmbH, 2000.
45. Ma, C.-C. M. and Chen, C.-H., 'Pultruded fiber reinforced thermoplastic poly(methyl methacrylate) composites. Part I: Correlation of processing parameters for optimizing the process', *Polymer Engineering and Science*, **31**(15), 1086–1093, 1991.

46. Bechthold, G. and Friedrich, K., 'Pultrusion of glass fiber/polypropylene composites'. In Karger-Kocsis, J. (ed.), *Polypropylene: An A–Z reference*. Dordrecht: Kluwer Academic Publishers, 1999, pp. 687–693.
47. Tomlinson, W. J. and Holland, J. R., 'Pultrusion and properties of unidirectional flass fibre-polypropylene matrix composites', *Journal of Materials Science Letters*, **13**, 675–677, 1994.
48. Aström, B. T., Larsson, P. H. and Hepola, P. H., 'Flexural properties of pultruded carbon/PEEK composites as a function of processing history', *Composites*, **25**(8), 814–882, 1994.
49. Aström, B. T. and Carlsson, A., 'Experimental investigation of pultrusion of glass fibre reinforced polypropylene composites', *Composites Part A: Applied Science and Manufacturing*, **29A**, 585–593, 1998.

Part III

Manufacturing of polymer matrix composites (PMCs): thermoset based processing

9 Resin transfer molding (RTM) in polymer matrix composites

E. M. SOZER, Koc University, Turkey and P. SIMACEK and S. G. ADVANI, University of Delaware, USA

Abstract: The resin transfer molding (RTM) process was adopted for composite manufacturing for high volume production net shape structural parts using low viscosity thermoset resins and continuous fibers. This chapter discusses how to overcome the challenges of RTM, which has led to many variations that have sprung over the last two decades to fill a niche need. The part quality manufactured using RTM varies due to the effects of inherent variations in the materials and process parameters. The following important issues that manifest themselves either during fiber preforming or mold filling stages of RTM have been identified and discussed: racetracking channels, deformation of fiber structure during draping, macrovoid formation, microvoid formation, transverse flow in the thickness direction and dual scale fiber structure in a preform. One can address and overcome them with process modeling, control and automation as discussed in this chapter.

Key words: issues of RTM, process modeling, resin flow, permeability, flow and cure monitoring.

9.1 Introduction

Resin transfer molding (RTM) was adopted for composite manufacturing in the mid-1980s. The driving force was the automotive industries that were looking for high volume production net shape structural parts. Injection and compression molding of discontinuous fibers could fabricate net shape structures at high volumes, but the structural performance could not be achieved by short fibers. Hence the idea emerged to have a woven or stitched fiber preform structure inside a net shaped mold and then inject the resin under high pressure to cover the empty spaces between the fibers. Only low viscosity resins were possible candidates due to the resistance to flow because of the micron level empty spaces between the fibers. Hence the resins of choice for this process are thermosets, although there has been some recent activity in bringing to market thermoplastic resins with low viscosity.

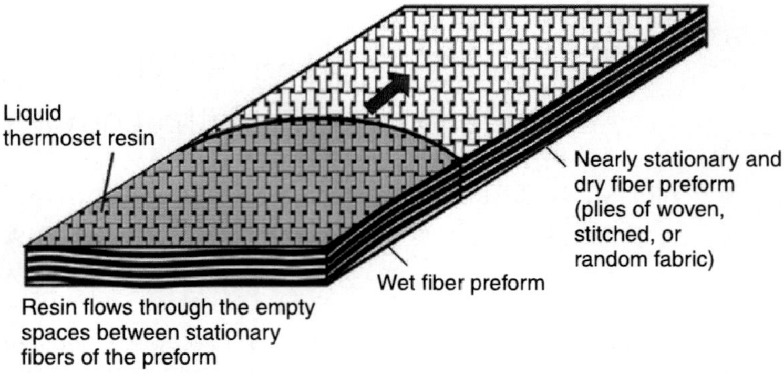

9.1 Flow of liquid thermoset polymer through porous fabric preform. (*Source*: Reprinted with permission from reference 1, copyright 2010 CRC Press, Taylor & Francis Group.)

The process development of RTM was slow and dogged with many challenges along the way which were gradually overcome. Interestingly, the challenges also led to many variations of the process that have sprung over the last two decades to fill a niche need. Today, the family of processes is broadly categorized as liquid composite molding (LCM) processes. In LCM processes, a dry or partially impregnated fibrous preform is placed in a closed mold cavity; and the empty spaces between the fibers are filled with resin by transferring the liquid resin from a reservoir under positive pressure or by drawing the resin into the mold by subjecting the mold to a vacuum. The mold filling must be completed before the gelation of thermoset resin or solidification of thermoplastic resin. Usually, thermoset resins are used due to their low viscosity (typically 0.1–0.5 Pa.s). Use of thermoplastic resins is not practical due to their high viscosity (typically 10^2–10^6 Pa.s) which requires very high injection pressures which translates into high cost equipment and it may also result in fiber washout or movement during injection creating resin-rich areas in the mold altering the final properties of the composite.

The common feature of all LCM processes is the injection of a liquid thermoset resin into a bed of stationary (or almost stationary) fibrous preform (see Fig. 9.1). Although this chapter is devoted to the RTM process only, we will point out advantages/disadvantages of RTM compared to the other processes also addressed in this book. First, we will describe the RTM process and briefly introduce other LCM processes in reference to RTM before highlighting the advantages and shortcomings of RTM as compared to the latter mutations of this process.

- **RTM**: Figure 9.2 shows a schematic of the various stages of the RTM process. The mold has a cavity that is in the shape of the part to be manufactured. The fiber preform is usually net shaped by draping the fabric

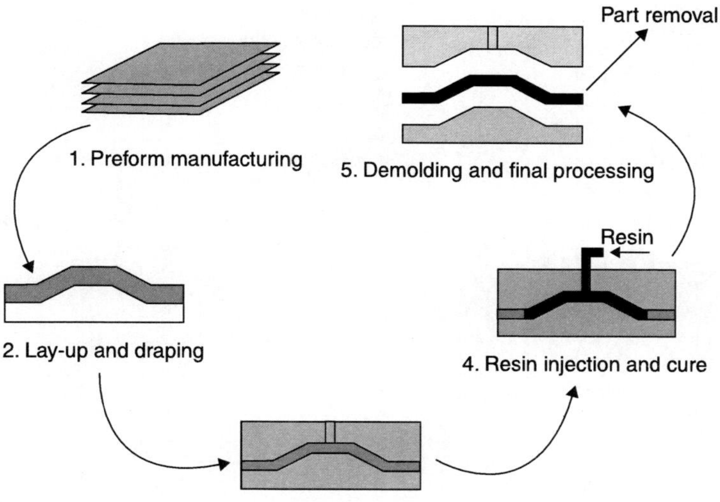

9.2 RTM process. (*Source*: Reprinted with permission from reference 1, copyright 2010 CRC Press, Taylor & Francis Group.)

plies on the tool surface of the mold. For complicated mold geometries, the plies may be partially wetted by brushing or spraying with tackifier and/or stitched together to prevent movement during the resin injection process. Once the fiber plies are placed in the mold and the edges are trimmed, the mold is closed and clamped. The resin is then injected into the mold cavity through one or more gates under positive pressure. The resin injection is continued until the resin is seen exiting from the vent ports. At this point, the resin injection is discontinued, vent ports are closed and the resin is allowed to cure. The cure may be initiated by heating the mold and/or by addition of inhibitors to the resin system initially. After the part is sufficiently hardened (usually referred to as green strength), the mold is opened and the part is demolded.

Other variations of this process will be described briefly to highlight the differences.

- **Vacuum-assisted resin transfer molding (VARTM)**: Figure 9.3 shows a schematic of 'vacuum bagged' VARTM process. In this process, a single sided rigid mold is used and the mold is sealed with a flexible bag by drawing a vacuum to remove the air from the mold and compacting the preform placed on the tool surface. The resin is drawn into the mold from a reservoir at atmospheric pressure. The two main differences between RTM and VARTM are (1) only one atmosphere is available for resin infusion so the filling can be very slow and (2) as the top surface is flexible, resin pressure

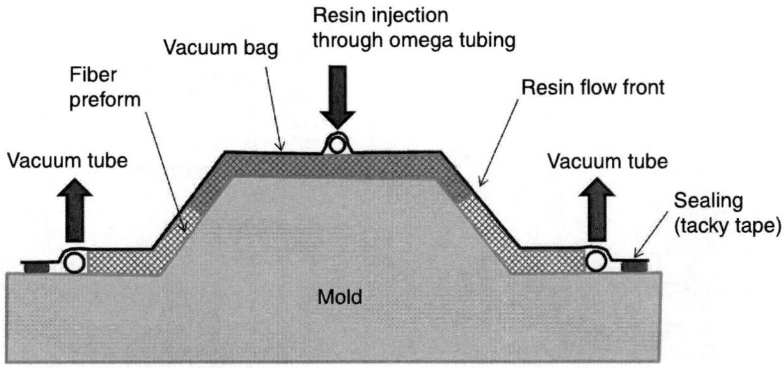

9.3 VARTM process. (*Source*: Adapted from reference 1.)

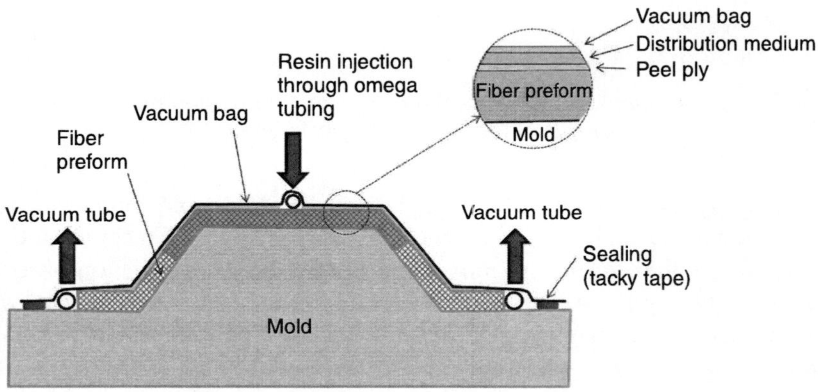

9.4 SCRIMP process. (*Source*: Adapted from reference 1.)

as it flows into the mold will change the compaction and hence the fiber volume fraction during impregnation resulting in non-uniform fiber volume fraction in the part unless certain steps are taken to prevent it.

- **Seemann's composite resin infusion molding process (SCRIMP)**: The major disadvantage of VARTM is slow filling due to the limited pressure differential between the inlet and exit which cannot exceed one atmospheric pressure. SCRIMP overcomes this by adding a peel-ply and a distribution medium of much higher porosity (and thus higher permeability) than the fiber preform and placing it between the fiber preform and the vacuum bag as shown in the schematic in Fig. 9.4. Resin initially races through the distribution medium due to its high permeability in the in-plane directions; it needs to penetrate only across the thickness direction which is usually of the order of 3 mm as compared to in-plane directions which are usually in 10–100 m long. The design of the distribution medium (i.e., its size and

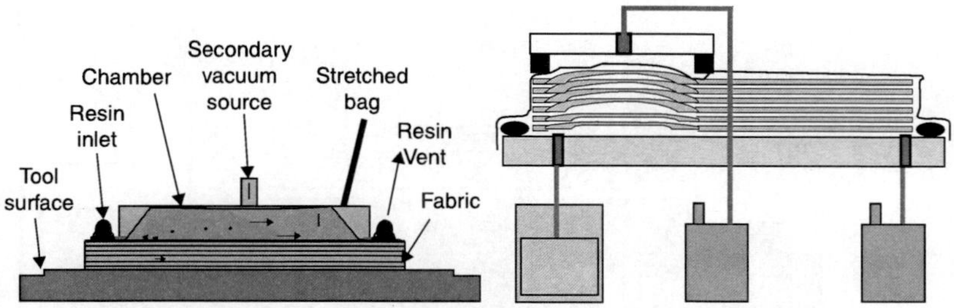

9.5 VIPR process. (Reprinted with permission from reference 1, copyright 2010 CRC Press, Taylor & Francis Group.)

location in the mold) must be carefully crafted to ensure complete mold filling, otherwise the resin could reach the vent before saturating the preform completely. As compared to RTM, the issue of gradient in the fiber volume fraction still remains due to the absence of a rigid top mold to prevent the preform from expanding as the resin enters the preform relieving some of the atmospheric pressure being applied on the flexible bag. The chapter that addresses VARTM in detail is presented in this book elsewhere.

- **Vacuum induced preform relaxation (VIPR)**: This is a modified version of VARTM and SCRIMP that assists in reducing the fill time. In this process, a vacuum chamber is placed on a section of the vacuum bag which locally reduces the compaction pressure on the fiber preform increasing its permeability and porosity temporarily and fiber preform enabling the resin to flow faster in the locally affected region as shown in Fig. 9.5. This modification also allows one to direct the flow to regions which are difficult to fill. This approach not only reduces the mold filling time but also provides the means to steer the flow in regions that may be difficult to fill as the increase in permeability and porosity lasts for only as long as the chamber is active. However, one must carefully design the location of external vacuum chamber and duration of the relaxation to achieve the desired modification in the flow pattern and ensure complete mold filling. In RTM one cannot modify the permeability or porosity during the infusion – hence the only way to redirect flow in RTM is by changing the flow rate or by introducing new gates in the mold.
- **Fast remotely actuated channeling (FASTRAC)**: Figure 9.6 shows a schematic of FASTRAC process which served as the inspiration for the VIPR process.[2] A primary vacuum bag is placed over the fiber preform; and vacuum is drawn as in a traditional VARTM process. The FASTRAC non-contacting tool is placed over the primary vacuum bag and covered with a secondary vacuum bag. The vacuum inside the primary vacuum bag is released which draws the primary vacuum bag into the channels of the FASTRAC non-contacting tool. These channels now serve as racetracking

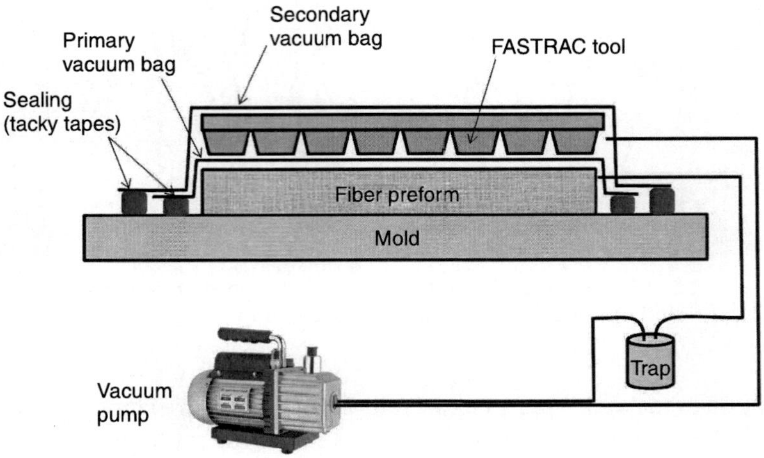

9.6 FASTRAC process.

distribution channels that distribute the resin over the preform surface quickly requiring it to penetrate only through the thickness after the vacuum is released on the secondary bag. Unlike RTM, in this process one can create temporary racetracking channels to assist flow which cannot be done in RTM.

- **Light RTM (LRTM)**: The upper mold half is made from semi-transparent composite shell which may be supported by a steel structure. LRTM was developed to reduce the tooling cost in RTM, and also shorten the labor time of vacuum-bagging in VARTM and thus shorten the cycle time.
- **Structural reaction injection molding (S-RIM)**: Multiple resin components (monomers and catalysts) are separately stored in the injection machine; and they are mixed just before transferring to the mold during the injection. This process requires one to address the cure kinetics and viscosity changes during the impregnation of resin into the rigid mold.
- **Co-injection resin transfer molding (CIRTM)**: Two or more resins can be simultaneously injected into a mold cavity. It allows for the manufacturing of multi-layer and multi-resin structures in a single manufacturing process. An impermeable separation layer is used to manufacture large parts or if the resins have significantly different viscosities. Unlike RTM the pressures across the separation layer may be different due to different viscosities and must be accounted for during infusion.
- **Compression resin transfer molding (CRTM)**: This process is also known as injection compression molding. To reduce the injection pressure and fill time as compared to RTM, the mold creates a gap between the mold surface and the fiber preform as shown in Fig. 9.7. The amount of resin needed to

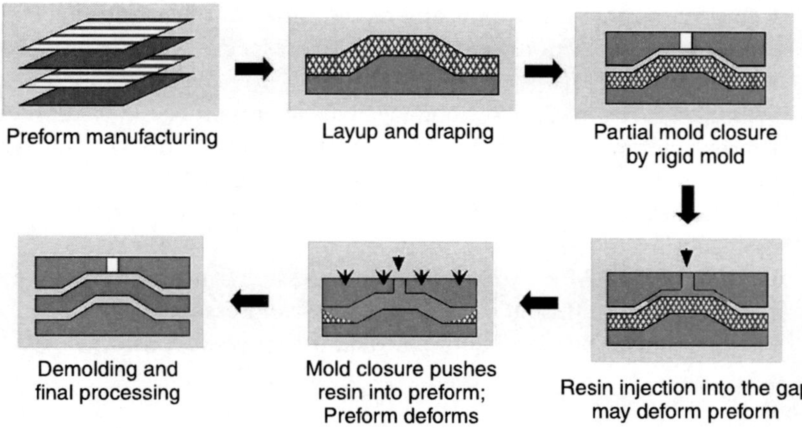

9.7 Compression RTM (CRTM) process. (*Source*: Reprinted with permissions from reference 1, copyright 2010 CRC Press, Taylor & Francis Group and reference 3, copyright 2009 Elsevier.)

fill the closed mold is carefully calculated and injected into this empty space on the top of the preform. Then, the mold is gradually closed as in compression molding to the final dimensions causing the resin to redistribute and flow in the thickness direction. This requires a press that can control the compression stage.[4] So unlike RTM, the fiber preform volume fraction and the permeability are changing during the compression phase. A chapter on compression RTM is presented elsewhere in this book.

- **Resin infusion between double flexible tooling (RIDFT)**: Resin is injected into a flat fiber preform that is placed between two flexible tools. The flexible tools with the saturated preform in them are transferred over a tool surface, and vacuum is drawn to conform the saturated preform to the final shape of the part. Unlike RTM, here the resin redistributes or squeezes out during the applied vacuum which dynamically changes the permeability and the fiber volume fraction.

To decide whether RTM is the most suitable manufacturing process or not the following pros and cons could help guide decision-making.

Pros:

- Near-net-shaped composite parts with a good surface finish and close dimensional tolerances can be produced since the reinforcing preform is compacted between the two sides of a rigid mold with gel-coated surfaces.
- Complicated shapes with ribs, channels and tapered thickness can be obtained; however, skilled preforming labor is needed.

- Compaction pressure is typically of the order of MPa, and this yields a fiber volume fraction as high as 60–70%, especially compared to wet hand layup, VARTM.
- With SCRIMP and LRTM, good mechanical properties can be achieved.
- Process automation and thus consistent reproducibility of composite parts compared to wet hand lay up, VARTM and SCRIMP.
- Fast manufacturing cycle due to (1) high resin pressure and (2) faster mold closing and demolding.
- Relative to CRTM, low mold clamping pressure is needed; thus manual clamps may be sufficient instead of expensive presses.
- Possible to automate and improve yield with process control.

Cons:

- Inconsistency in reproducibility: inherent variations in preforming due to inconsistent fabric rolls from vendors, and inconsistencies in fabric cutting, stacking, pre-impregnating and placement into the mold cavity.
- Compared to VARTM and SCRIMP, part dimensions are limited to a few meters typically due to the span limitations of typical milling machines during machining of a two-sided mold.
- Sensitivity to mold design: improper gating and venting can cause incomplete filling of the empty spaces and result in macroscale dry spots (resin-starved regions) in the composite.
- Long production cycles compared to parts made from aluminum or steel in the automotive industry using sheet metal forming processes.
- Fiber wash due to loose compaction and/or high resin pressure.

9.2 Resin transfer molding (RTM) process steps

The five steps of the process are shown in Fig. 9.2. These stages are briefly discussed below:

- **Fiber preform manufacturing:** Fiber preform is an assembly of the mat or fabric plies that are cut and placed over a tool surface. Depending on the design criteria such as cost limitations and required mechanical properties, the types of fiber (e.g., E-glass, S-glass, carbon, aramid, etc.), fabric (e.g., random, woven, non-crimp, etc.), fiber volume fraction (e.g., 30% or 65%) and orientations of the plies (e.g., [0/90°/0/90°], etc.) must be carefully decided using mechanics analysis models. Foam or wooden cores may be embedded to increase the stiffness of a part; and metal inserts may be embedded for assembly purpose.

- **Layup and draping:** While fiber preform is being placed in the mold cavity, it is draped on the mold surface due to the curvatures of the mold which causes the fiber bundles to undergo structural deformation. This may cause significant spatial variations in the fiber volume fraction, permeability and mechanical properties of the part. Excess deformations of the fiber structure may cause wrinkles of the preform. The dimensions of the fiber preform are crucial; for example, if the in-plane dimensions are shorter than the mold cavity dimensions, racetracking channels are formed along the preform–mold wall interface which causes a significant change in the resin flow pattern resulting in unwanted macroscale voids in the part. On the other hand, if the fiber preform is even slightly larger than the mold cavity, closing the mold and ensuring a good seal will be difficult. This usually causes wrinkles in the preform and leads to non-uniform compaction and racetracking channels. Thus, special care should be given to the preforming stage.
- **Mold closure:** The mold cavity is closed either manually using bolts and clamps, or under the force of a compression press. The former option is low cost; however, it is not suitable for high volume manufacturing and it may cause non-uniform fiber compaction and permeability and hence, non-repeatable resin propagation and incomplete mold filling. Sealing of the mold must not allow resin leakage under high resin pressure which is especially important near the inlet gates where the resin is under maximum pressure.
- **Resin injection and cure:** Liquid thermoset resin is transferred from the reservoir of an injection machine to the mold cavity under positive pressure either under (i) constant flow rate, or (ii) constant pressure. The injection location and whether it should be subjected to constant flow rate or constant pressure or a combination of the two can be designed by considering the following: the size and the geometry of the part, permeability of the preform, gel time of the resin, desired cycle time, and capability of the injection machine. After complete mold filling, the cross-linking of the thermoset polymer is usually achieved at elevated temperatures either by circulating hot water through the channels in the mold walls, or by placing the mold in an oven. Room temperature cure resins usually have an inhibitor which will initiate cure after a specific time interval in which the mold filling should be complete.
- **Demolding and final processing:** After the part reaches its 'green strength' so that it is sufficiently strong and stiff to be demolded, the mold is opened and the part is ejected from the mold.

9.3 Fibers, fabrics and preform manufacturing

9.3.1 Fibers

Fibers carry structural loads and provide stiffness and strength to the composite part. Short fibers and natural fibers (such as hemp, flax and jute) are usually used in composite parts not requiring high mechanical properties. Synthetic continuous fibers are made from glass, carbon or polymer for high performance applications:

- *Glass fibers* are manufactured in a furnace by melting sand and chemical additives and then drawing through small orifices at high speeds. These continuous fibers (filaments) are cooled with water and then sized, that is, coated with a chemical to improve the adhesion between the fibers and the polymer matrix during the manufacturing process. The sizings also minimize the abrasion of the fibers during the roving, stitching or weaving into a textile preform. A single fiber diameter is approximately 10–20 μm. Rovings are formed by bundling the continuous filaments together to manufacture tows or yarns. A tow has parallel filaments in a bundle, and a yarn has twisted filaments. A tow or yarn is generally elliptical in cross-section, and may contain hundreds to tens of thousands of single filaments. The cross-sectional width and thickness of the bundles are of the order of millimeters. Many woven fabrics are made from the twisted yarns. E-glass fibers are the most common low-cost reinforcing material. S-glass fibers have a different composition of silica (SiO_2) and metallic oxides. They are used in high performance parts requiring high strength and stiffness. The two common measures of a roving's linear density are yield and tex. For example, 450-yield roving means that 450 yards of the roving weighs one pound (where one yard is equal to 0.9144 m, and one pound is equal to 0.4536 kg approximately); and 750-tex roving means that 1 km of the roving weighs 750 g.
- *Carbon fibers* are made from a precursor polymer such as rayon, polyacrylonitrile (PAN) or petroleum pitch by drawing or spinning process and then subjected to a high temperature environment. During the thermal phase, the precursor is held under tension and heated in air at moderate temperatures, followed by heating in an oxygen-free inert gas at elevated temperatures. Typically, carbon fibers have lower density, smaller diameter and higher flexural strength than glass fibers. They can also be bundled into tows, and woven or stitched to form fabrics.
- *Aramid fibers* (also known by the trade name of Kevlar) have been developed and manufactured by DuPont since 1970s. The fibers have very long and longitudinally oriented linear polymer chains. Due to their higher specific strength (strength/density ratio) and impact toughness than glass and carbon fibers, aramid fibers are commonly used in ballistics applications. There are other aramid fibers being developed for impact performances as well but these are less common.

9.3.2 Mats and fabrics

A *mat* is a ply of chopped (short or long) yarns that are held together by a binder or stitches. A *fabric* is a ply of continuous yarns that are manufactured by using the laminating techniques of textile industry, such as weaving or knitting. Composites containing continuous synthetic yarns are known as *advanced composites*.

The major advantages of mats (compared to fabrics) are as follows: (1) high porosity and thus permeability to resin flow; (2) easy to handle (i.e., cut and place in the mold cavity) and (3) inexpensive. Their major disadvantages are as follows: (1) low strength and stiffness and (2) low fiber volume fraction even at high compaction pressures due to non-structured plies.

The common measure of a ply's density is its areal density (also known as superficial density). For example, a ply with 450 g/m^2 superficial density weighs 450 g/m^2.

Figure 9.8 shows common fiber mats and fabrics used in LCM. The first one is made from chopped (i.e., discontinuous) fiber yarns and the other three are made from continuous yarns.

- *Chopped strand mats (CSMs)* are made by randomly depositing chopped glass fiber bundles and binding agent, which is a polymer, on a conveyor belt. Due to the random orientation of the glass fibers, CSM has approximately isotropic in-plane material properties (both mechanical and flow such as strength and permeability). CSM is used in non-structural composite parts due to use of the discontinuous fibers and thus have limited load carrying capability. Low density CSM is used as an outer surface ply (known as surface tissue or veil) and thus a smooth resin-rich skin is formed.[6]

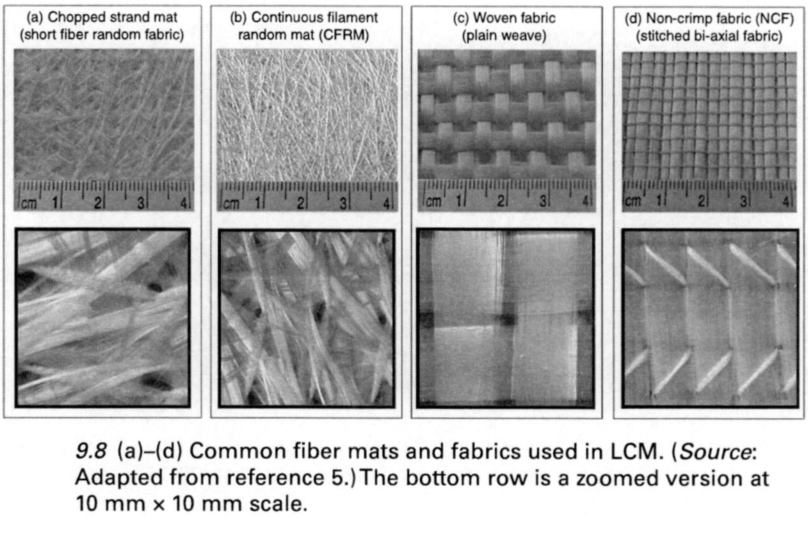

9.8 (a)–(d) Common fiber mats and fabrics used in LCM. (*Source*: Adapted from reference 5.) The bottom row is a zoomed version at 10 mm × 10 mm scale.

- *Continuous filament random mats (CFRMs)* are similar to CSM except that continuous filament yarns are swirled and deposited instead of chopped fibers while manufacturing the mats. Compared to CSM, CFRM has higher strength, higher formability during preform layup and higher resistance to fiber wash at high resin pressures. This is an attractive feature for high volume production parts such as automotive as one could spray the fibers on the mold surface and use a binder to hold the preform together before the resin injection thus eliminating labor and combining the preforming and the layup stage.
- *Woven fabrics* are manufactured by inserting horizontal weft yarns over and under the lengthwise warp yarns (a.k.a. filler) that are kept under tension. If the periodicity of the weaving is 1 over and 1 under, then the woven fabric is known as plain weave which is the simplest pattern, and it can be seen in Figs. 9.8c and 9.9. The fiber crimp (the undulations of the woven pattern) limits the in-plane strength of the fabric.[6] Woven fabrics with high twill and harness are manufactured to reduce the fiber crimp and also increase the drapability (ability to conform tool surface during preforming or layup stage). Their unit cell has a longer structural periodicity than the plain weave's periodicity of 1 over and 1 under. Skilled labor is needed when cutting the edges of the plies. Otherwise, some yarns may separate and no longer be part of the preform, thus increasing the porosity and the permeability along the edges. This will significantly change the resin flow pattern during resin injection due to the racetracking channels along these affected edges. Usually, the

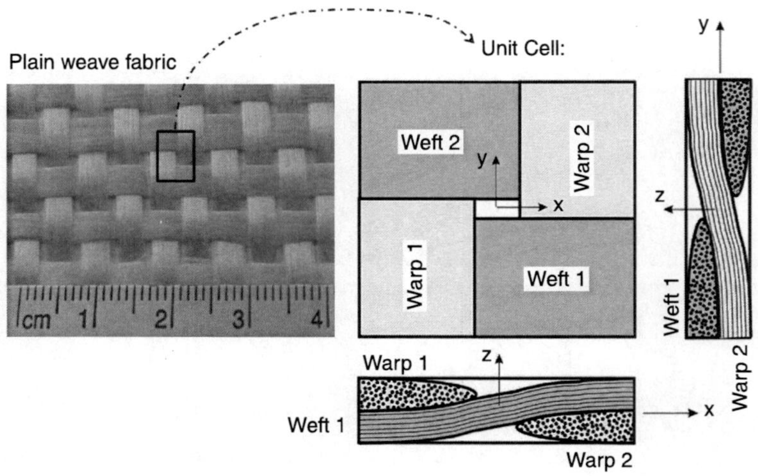

9.9 A periodic unit cell of a plain weave fabric ply. (*Source*: Reprinted with permission from reference 1, copyright 2010 CRC Press, Taylor & Francis Group.)

distances between adjacent warps and wefts is not equal, thus the fabric is not isotropic in terms of mechanical properties and permeability. The fabric cutting should be done consistently to have repeatable resin injections and flow patterns.

- *Unidirectional fabrics (UDFs)* are manufactured by stitching parallel yarns. They have much higher strength and stiffness in filament direction than all other fabric types. However, these mechanical properties are much lower in the transverse direction. UDFs are vulnerable to fiber wash under high resin pressure.
- *Non-crimp fabrics (NCFs)* are manufactured by stitching unidirectional yarns in a layer to another layer of unidirectional yarns. If the directions of the yarns in the two layers are perpendicular to each other, for example one layer is parallel to the roll direction and the other is perpendicular to it, then the NCF is referred as [0/90°]. There may be many other configurations such as [0/90° /0/90°] or [0/90°/+45° /−45° /90° /0]. NCF eliminates the fiber crimp, and thus high fiber volume fraction and in-plane mechanical properties can be obtained.

9.3.3 Preform manufacturing

Usually the reinforcing fibers are assembled on a tool surface by depositing chopped fibers, or draping mats or fabrics, and holding them together using thermoset or thermoplastic binder or with stitches. The tool surface is the inside surface of the mold cavity. Core materials such as foam and balsa wood may also be embedded in the preform to increase the stiffness of the finished composite part. Some metallic inserts may also be embedded to introduce multifunctionality. The overall cycle time is reduced by manufacturing the reinforcing fiber preform outside the mold and then transferring it to the mold cavity before the resin injection.

One should design the preform by selecting adequate fiber and fabric types, and fiber volume fraction (thus the number of plies in the preform) considering the required (1) mechanical performance, (2) permeability to resin flow, (3) fiber wet out, (4) formability and (5) cost. These items are briefly discussed below:

- *Mechanical performance*: To predict the mechanical properties of the composite part, there are several modeling approaches. The simplest one is the 'rule of mixtures' which is applicable for only continuous unidirectional filaments:

$$E_c = \left(1 - V_f\right) E_m + V_f E_f \qquad [9.1]$$

where E_c, E_m and E_f are the modulus of elasticity of the composite, matrix (cured resin) and fibers, respectively and V_f is the fiber volume fraction. One should not forget that E_c and E_f are the stiffnesses along the unidirectional filaments. For preforms made of woven and non-crimp plies, mechanical properties may be predicted using models such as classical laminate theory and finite element analysis.

- *Permeability to resin flow*: To achieve the highest mechanical properties, the designers would like to have high fiber volume fraction; but they should not forget that this decreases the permeability of the preform, and thus the resin flow will be more difficult. High preform permeability allows rapid mold filling under constant pressure injection, or lower injection pressure under constant flow rate injection. The permeability is dependent on the fiber structure (tex of the yarns, type of the fabric, orientations of the plies and fiber volume fraction which is dependent on the compaction of the preform in the mold cavity).
- *Fiber wet out*: To have a strong interface bond between the fibers and the resin matrix, the fibers must be fully wet out by the resin during resin injection. This can be achieved readily if the fiber type and resin system are selected such that the free surface energy of the fiber is higher than the surface energy of the resin.
- *Formability*: When the fiber preform is draped over the tool surface, the structure of the yarns is deformed. Depending on the surface curvature and fabric type(s) used, fibers may buckle and significant wrinkles may occur. To avoid or minimize this, one must carefully design the preform by selecting the twill and harness of the woven fabric.
- *Cost*: Although the fabrics made of E-glass fibers are the most inexpensive, they may not meet the mechanical performances required. In that case, fabrics made of S-glass, carbon or aramid fibers may be more suitable candidates.

Manual tailoring of mats and fabrics (i.e., cutting, binding and draping) is low cost, but it is time consuming and may not be repeatable. For high volume manufacturing, automated machines and processes should be used to reduce the cycle time and increase the repeatability. The common preforming processes are briefly described below:[6]

- *Spray-up of chopped fibers and binder*: Chopped fibers and binder are sprayed on a tool made of perforated screen. Vacuum is applied on the other side of the screen to help deposit the fibers on the screen. The main advantages of this preforming process are (1) trimming is not needed, thus there is no waste of raw material; (2) process automation can be achieved by using robots which allows consistency. The disadvantages are (1) low mechanical properties (strength and stiffness); (2) non-uniform and non-repeatable preform manufacturing if the spraying is done manually.

- *Filtration of chopped fiber slurry on a perforated screen*: A tool made of perforated screen is raised through a slurry of chopped fibers, binder and water. The deposited fibers and binder are compressed by pressing another screen on the perforated screen. The preform develops sufficient rigidity as it dries due to the binder between the fibers, and it can be transferred to the mold cavity readily.
- *Stamping of mats and fabrics in a two-sided mold*: Plies of mats or fabrics are cut and binder is added between them. The plies are stamped between a matched mold which is heated to activate the binder. Then the mold and the preform are cooled; and the edges of the preform are trimmed before transferring it to the mold cavity. This preforming process can be used with fabrics of continuous fibers. However, compared to chopped fiber preforming processes, the stamping process (1) results in higher waste when cutting the fabric plies; (2) has a higher possibility of forming wrinkles and tears during fiber draping and compaction.
- *Braiding*: Fiber bundles (yarns) from multiple carriers are wrapped around a mandrel to manufacture a preform. The wrapping pattern is designed to form an interlocked fiber structure that has (1) sufficient structural rigidity for handling during placement in the mold; (2) high resistance to fiber wash during resin injection.
- *3D weaving*: The weft yarns are inserted over and under the multiple layers of warp yarns. This process yields thicker woven structures than in 2D weaving where only one single layer of warp is used. 3D woven preforms have higher strength in the thickness direction than the 2D woven fabrics; however, their strength decreases in the in-plane directions due to the fiber crimp.
- *Knitting*: The yarn loops (also known as stitches) are formed using needles to manufacture 3D preforms. There are many patterns and types of knitting available. The major advantage of knitted preforms is that they can be draped on complicated mold surfaces with no significant wrinkles due to the enhanced formability of the fiber structure.

To ease the draping and stacking of mat and fabric plies on a complicated tool surface, pre-impregnated plies (prepregs) may be used instead of dry plies. Alternatively, tackifier may be used to bond the individual plies. In both cases, the resin flow will be affected during mold filling stage as binders will change the permeability of the fabric and previously wetted regions can cause blockage or redistribution of the resin during injection stage of the RTM process.

9.4 Resin system

A resin system is a liquid mixture of base thermoset monomers, promoters, fillers, mold release agents, pigments and catalysts. In RTM, the resin system is prepared in one of the following ways: (1) mixed manually and quickly stored in the tank of an injection machine just before the start of the resin injection or (2) the components are stored separately in different tanks of an injection machine with a mixing head. The resin is transferred to the mold cavity under positive pressure injection. Depending on the mixing ratio of the monomers and the curing agents, the resin system has a pot life which is also known as gelation time of the resin. The mold filling stage has to be complete before the end of the pot life because the viscosity of the resin starts to increase exponentially arresting the flow. The curing of resin (cross-linking of the polymers) can be initiated by heating the mold and/or activating the curing agents. The role of polymer matrix is to:

- bind the fibers together,
- provide the permanent shape and dimensions of the part,
- provide major contribution to the part's strength in compression and shear,
- provide overall durability by protecting the fibers against the harsh ambient conditions.

Nanoclays and nanotubes may be added to the resin system to improve mechanical and electrical properties of composite parts.[1]

When one selects a suitable resin system for RTM, the following features should be considered:

- *Low viscosity to ease resin injection.* A resin system with low viscosity (usually between 0.1 and 1 Pa.s) allows one to: (1) keep the resin injection pressure at low level under constant flow rate injection or (2) reduce the mold filling time under constant pressure injection. High resin viscosity not only increases the mold filling time but in addition may cause microvoids inside the fiber bundles since the resin may not penetrate the bundles readily. On the other hand, if one considers the other extreme case of having a resin system with very low viscosity (typically much lower than 0.1 Pa.s), then the resin races along the preform's least resistant regions, and some regions may not be completely filled.
- *Effective fiber wetting properties.* This creates a good bond between the fibers and the matrix by wetting of the fibers by the resin. Liquids with low surface energy effectively wet high energy solids. The surface energies for glass and thermoset polymers are 500 dynes/cm^2 and 30–40 dynes/cm^2, respectively. Hence, thermosets have good glass fiber wetting characteristics. Since carbon has a surface energy of around 50 dynes/cm^2, one must be careful while selecting the polymer to ensure that it can wet the fiber.

- *Uniform and consistent cure kinetics*. The cross-linking chemical reaction is an exothermic polymerization. The length of the cure cycle depends on the type of resin system and temperature. The degree of cure is proportional to the cross-linked material, and can be characterized by measuring the fractional exothermic heat of the chemical reaction. The cure kinetics are usually characterized separately and may not reflect the interaction with the sizings on the fiber surfaces which may change the cure kinetics.
- *Rapid cure* after gelation shortens the cycle time. The length of the cure cycle is adjusted with the fraction of catalyst and accelerator in the resin system. If one waits for the complete cure cycle for demolding the part, the cycle time will be unnecessarily long. The part may be demolded after the matrix reaches a 'green strength' such that the part will not deform or warp during demolding or afterwards.
- *High glass transition temperature*. This is required for resin systems used in structural composite parts. Otherwise, the part will soften above its T_g which is not desirable.
- *High tensile strength* to support the reinforcing fibers under load. Although the fibers carry the major fraction of the load, the matrix also helps to increase the overall strength of the composite part. Otherwise, the matrix may fail long before the fibers.
- *Low thermal expansion coefficient* keeps the dimensional changes and thermal stresses at low level.

9.5 RTM mold

Some useful guidelines are given in the literature about how to design and operate an RTM mold:[1,6]

- Steel, aluminum and copper alloys are the most commonly used metallic mold materials.
- Chromium plated tool steel is a commonly used mold material for high volume manufacturing due to its hardness and resistance to wear. It allows manufacturing of composite parts with high surface quality (low surface roughness).
- Tool steel is one of the most suitable mold materials since it has (1) high modulus of elasticity (and thus stiffness) to have dimensional control; (2) high thermal diffusivity to quickly transfer heat between the internal heating/cooling channels and the composite part during curing cycle; (3) high hardness for wear resistance and long tool life; (4) high corrosion resistance; (5) a lower thermal expansion coefficient than the other metallic alternatives (aluminum and copper) to control the part dimensions. Considering its high material and machining cost, tool steel is a feasible material only for high volume manufacturing (in the order of

hundreds of thousands of parts), or if the part quality and not the cost is the major concern.

- Aluminum offers an advantage of (1) quick machining of the mold (approximately half of the tool steel's machining time); (2) light weight (its density is approximately one third of steel) and thus easier mold operations; (3) quick heating time during cure cycle due to its high thermal diffusivity. The disadvantages include: (1) softer than tool steel and thus vulnerable to scratches and wear especially during demolding and (2) higher potential for surface contamination than steel mold surfaces.
- Although copper is another metallic alternative material for mold making due to its corrosion resistance and higher thermal conductivity than steel and aluminum, it is not as commonly used as steel and aluminum.
- The quality of the mold surface should be inspected regularly since the glass fibers are very abrasive, and the mold surface may get damaged when the mold is closed. Temperature fatigue, solvents, mold release agents and demolding tools such as wedges may also damage the mold surfaces.
- Although the temperature rise due to exothermic curing reaction and mold heating is not significant in RTM for typical resin systems (typically 100°C above the room temperature, or less), this may shorten the mold life especially if the mold is nonmetallic (such as composites).
- Low-cost and soft materials are used for low volume or prototype manufacturing, or when the dimensional accuracy and surface roughness are not critical. Shell composite molds (glass fiber reinforced polyester, vinly ester or epoxy) with an epoxy gel-coated inner surface are commonly used for these cases. To increase the mold stiffness, the shell composite may be supported by a steel frame.
- As the composite molds have much lower thermal conductivity than metallic molds, mold designers should consider having more and closer heating channels to the cavity in composite molds than in metallic molds if necessary.
- New epoxies with very low thermal expansion coefficient have been developed for fabricating composite molds which significantly reduce the shrinkage due to mold heating/cooling and due to the exothermic curing reaction.
- A matched mold has at least two rigid parts. For simple part geometries, the preform is placed between a two-sided mold (usually referred as male and female halves). If the part has large, intricate or complicated shape, the male and female halves might be constructed from smaller pieces (such as platens, frame and lids) to ease the preform placement and demolding. One such mold with multiple components is shown in Fig. 9.10. Another RTM mold is shown in Fig. 9.11 which is used to manufacture a complex structure.[1]

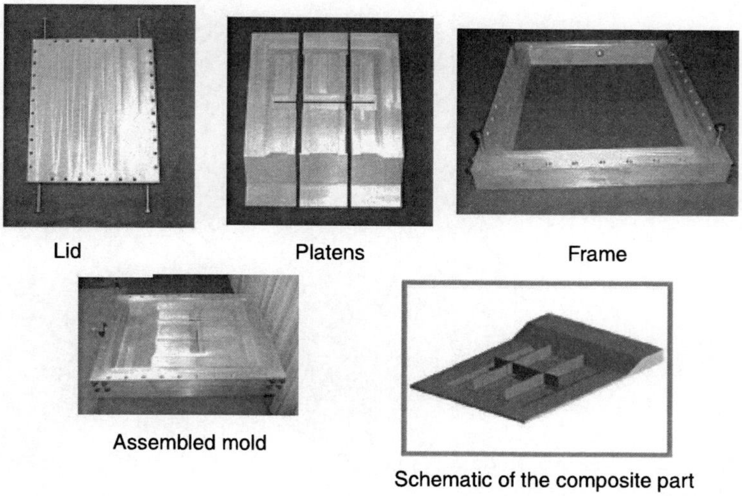

9.10 A matched mold constructed from smaller pieces to ease the preform placement and demolding. (*Source*: Reprinted with permission from reference 1, copyright 2010 CRC Press, Taylor & Francis Group.)

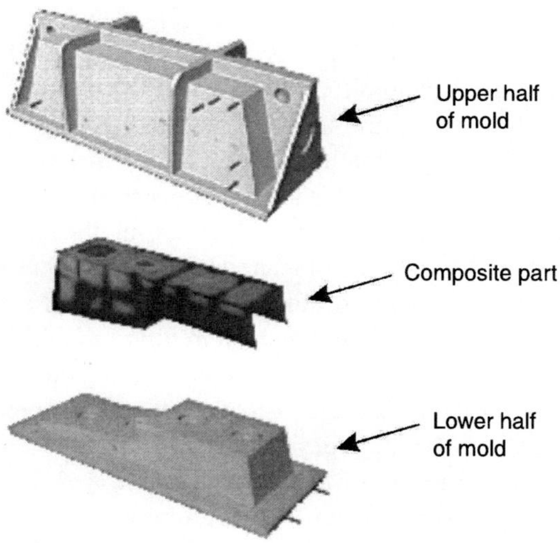

9.11 A matched (two-sided) mold for RTM to manufacture a complex composite structure. (*Source*: Reprinted with permission from reference 1, copyright 2010 CRC Press, Taylor & Francis Group.)

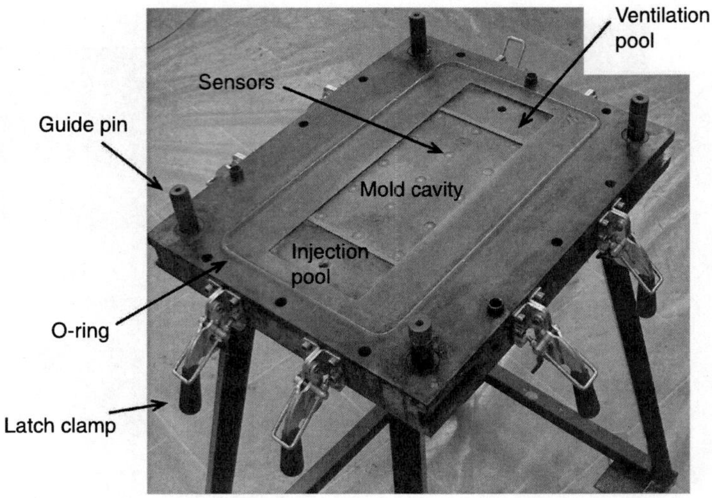

9.12 The lower plate of an RTM mold with latch clamps to close the upper and lower plates.

- The two sides of the mold are aligned using guide pins on one side (see Fig. 9.12) and guide bushes on the other side.
- A press is the ideal tool to close the mold and achieve an almost uniform fiber compaction. It also allows for high resin injection pressure in the mold cavity due to enhanced sealing. Instead of a press, if bolts or manual latch clamps are used around the periphery of the mold as shown in Fig. 9.12, the mold dimensions and its shape should be designed to carry sufficient stiffness especially if the in-plane mold dimensions are large.
- To shorten the cure cycle, the mold is heated by using one of the following alternatives: (1) placing the mold in an oven during the cure cycle, (2) circulating water or oil through a channel network that is either machined in thick mold plates or attached to the back side of a shell mold or (3) attaching an electrical resistance heating system to the mold. The second alternative has an advantage of using the channel as a cooling channel, as well as heating, after the cure cycle or to extract excessive exothermic heat of the cure reaction. The mold designer should carefully consider the heating rate, thermal diffusivity of the mold material, and mold dimensions to maintain temperature and cure distributions within the composite part to be as uniform as possible. High thermal gradients may (1) cause premature cure of some sections or result in long cycle times and (2) lead to shrinkage and warpage due to the residual stresses in the part.
- Excessive exothermic heat of the cure reaction may be harmful to thin mold plates (especially shell composite molds) as it may cause delamination.

- Before the preform is placed in the mold cavity, a release agent is sprayed or brushed on the mold cavity surface to prevent sticking of the composite part to the mold during the demolding.
- The design of mold and boundary conditions (locations of resin inlet (gate) and exit port (vent), and resin injection pressure/flow rate and ventilation pressure) is done either (1) by using experience and trial-and-error prototype manufacturing or (2) by using mold filling simulations such as LIMS.[1] The general approach is to use a line gate (usually at the center or along one edge of the part) instead of a point gate, and a line vent (usually around the periphery of the mold cavity, or along one edge of the part opposite to the line gate) to (a) reduce the filling time under constant injection pressure or (b) reduce the maximum injection pressure under constant flow rate injection. Usually, the mold cavity is vacuumed by evacuating the air from the vents before the start of the resin injection; otherwise macro size voids (dry regions) may remain if the resin reaches the vents before the mold cavity is filled completely.
- The resin injection is usually continued for a while (usually in the order of a few minutes or longer depending on the size of the part) even after the resin reaches the vents. The resin is allowed to bleed out of the vents to ensure complete mold filling, especially to fill the empty spaces between the fibers in bundles which have much lower porosity (and thus lower permeability) than the empty spaces between the bundles. Otherwise, microvoids (dry regions) may remain inside the fiber bundles.
- To decrease the microvoid content, the following practices may be applied: (1) vibrate the mold or (2) switch the inlet gate and exit vent, and thus reverse the direction of the resin flow.
- *Mold seal:* Heat-resistant silicone O-rings or tubes are commonly used around the periphery of the mold cavity to seal the mold and thus prevent the leakage of resin. Strip seals are used to increase the contact surface area between the seal and the mold plate. They are all used to prevent the resin outflow and air inflow. A mold designer should carefully set the width and depth of the groove with small tolerances (usually specified in the data sheet of the seal by the supplier company), and select the seal with appropriate hardness so that it can readily deform under the compaction pressure and create sufficient interference between the mold plates and itself when the mold is closed. The plate, groove and seal must be kept clean by removing the residual cured resin. In some applications,[6] the two sides of the metallic mold are closed together with a 'metal-to-metal' joint around the periphery of the mold cavity as shown in Fig. 9.13d. Although this seal type eliminates the need for an elastic seal, a groove on the mold plate and their cleaning and reduces maintenance, the dimensional tolerances and surface roughness of the

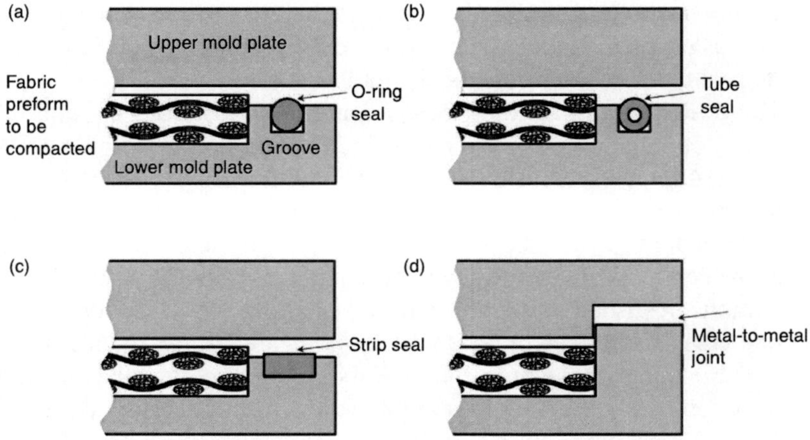

9.13 (a)–(c) Different silicone seals used in RTM molds and (d) a 'metal-to-metal' joint;[6] adapted from reference 6.

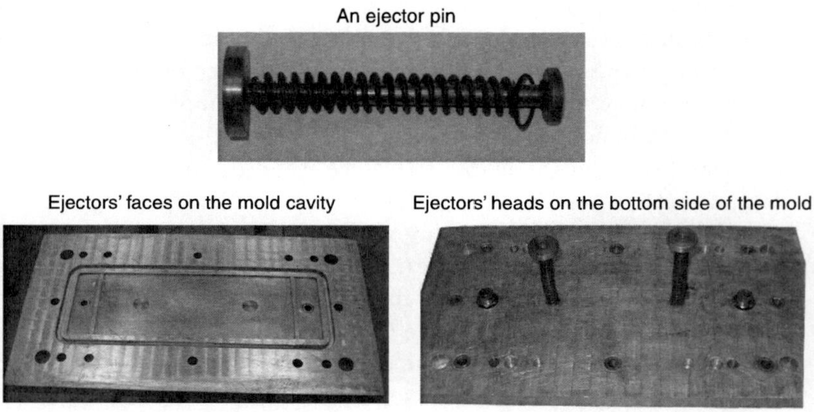

9.14 An RTM mold with two ejector pins mounted on the lower plate to ease the demolding of the composite part.

mold plates must be very small. Thus, the machining of the mold must be done with high accuracy so that the mold surfaces match all along the periphery.

- A large part with a complicated shape may not be demolded readily even when mold release agent is sprayed or brushed on the mold cavity; in this case, mechanical ejectors should be mounted on the mold to ease the demolding as shown in Fig. 9.14. However, the tip of the ejectors and the impact force should be designed so that the surface of the matrix of the composite is not damaged while demolding.

9.6 Resin injection equipment

Broadly, RTM resin injection equipment is categorized based on its mixing capability and whether the flow rate or pressure is controlled. Based on mixing capability, they can be classified as:

- *Simple injectors with no capability of mixing* have a pressurized reservoir that contains the premixed resin system ready to inject into the mold. They are preferred due to their low investment cost, but requires one to inject the system into the mold before the gelation which is not suitable for high volume production as it can only be done in small quantities to prevent a large amount of exothermic heat from being generated in the reservoir.
- *Injectors with capability of storing and mixing multiple components* store each resin and the catalyst and accelerators in separate tanks, and mix them in the mixer just before the resin is transferred from the injector to the mold. This way, one initiates the curing only after the mixing takes place and is desirable for high volume production and also for controlling the curing rate by changing the ratio of the catalyst to resin during the injection process.

The second type of classification depends on the injection boundary condition:

- *Pressure-controlled injectors* allow the user to set the injection pressure, P_{inlet} to a constant value. The corresponding flow rate, Q_{inlet} decreases with time as schematically shown in Fig. 9.15a. Notice that the flow rate is very high at the beginning of the mold filling; this may cause delayed fiber wetting inside the yarns (fiber bundles) of the preform. That means the flow front has a high velocity initially, but the yarns may not have been saturated. This is an important flow issue which may result in microvoids in the yarns; and it will be addressed in the next section.
- *Flow-rate controlled injectors* allow the user to set the injection flow rate, Q_{inlet} to a constant value. The corresponding resin pressure, P_{inlet} increases with time as schematically shown in Fig. 9.15b. Robust injectors such as flow-rate controlled RTM injector by Radius Engineering Inc.[7] also allow setting an upper bound of the resin pressure as shown in the same figure. When the resin pressure reaches a critical value, P_{limit}, then the injection boundary condition is switched to a pressure-controlled type so that $P_{\text{inlet}} = P_{\text{limit}}$, and the corresponding injection flow rate decreases with time. Setting an upper pressure bound allows the user to avoid (1) resin leakage, (2) fiber wash and (3) deformation of the mold walls.

An ideal RTM resin injector should have the following features:

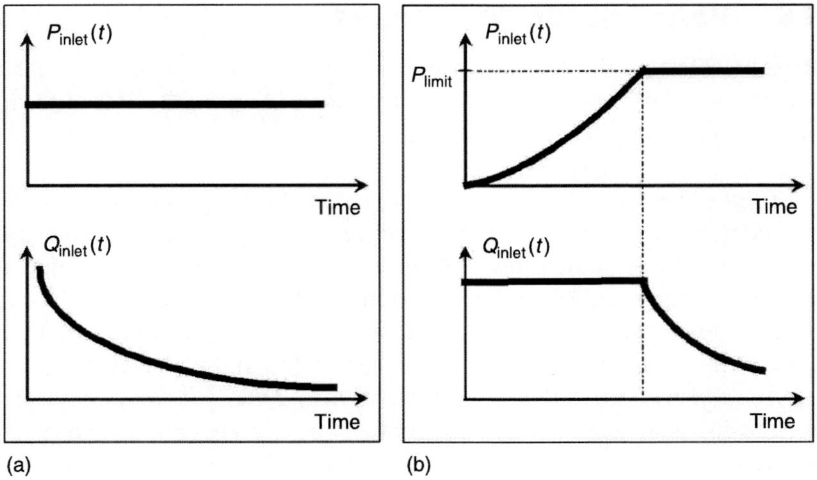

9.15 Inlet pressure $P_{inlet}(t)$ and flow rate $Q_{inlet}(t)$ for two common types of injection boundary conditions: (a) constant inlet pressure, and (b) constant inlet flow rate with an upper pressure limit. (*Source*: Reprinted with permission from reference 1, copyright 2010 CRC Press, Taylor & Francis Group.)

- Multiple components of the polymer should be mixed accurately at the user's defined ratio. The user may want to adjust the mixing ratio to achieve the desired gelation time (a.k.a. pot life) of the resin at different temperatures.
- The boundary conditions (resin injection pressure or flow rate) should be adjusted *in situ* readily (e.g., the injector in Fig. 9.16a allows one to control the boundary conditions using either one of the two alternatives: a control console or via a PC).
- Process variables such as resin pressure, flow rate and temperature, and amount of remaining resin should be monitored on the user interface window.
- In usual RTM applications, the resin is kept in the reservoir of the injector at room temperature, and the mold is heated for the cross-linking chemical reaction of the resin. In an ideal injector, the resin storage tank and the connection tubes should be heated to keep the resin at a desired temperature (usually at the same temperature of the mold), otherwise the resin temperature may vary due to the heat transfer between the mold and the resin as it propagates. For example, the heater jacket around the resin cylinder in Fig. 9.16a allows one to control the resin temperature up to 175°C. Additional heater jackets or insulator jackets are placed around the inlet tubes between the injector and the mold

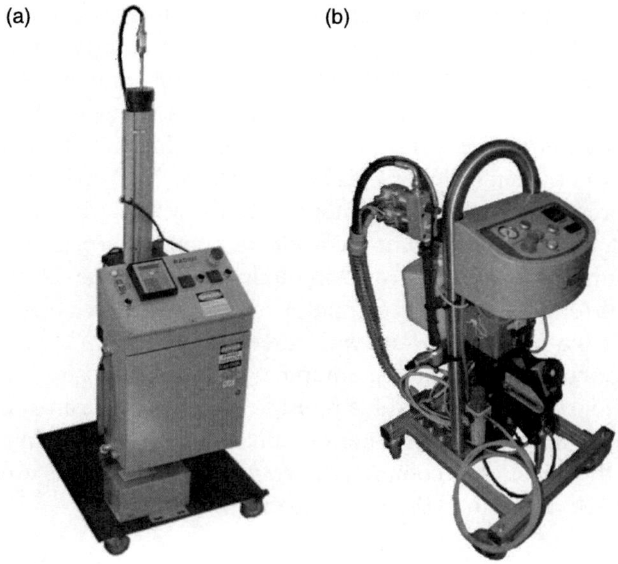

9.16 Two RTM injection equipments: (a) flow-rate controlled injector for single component or pre-mixed multiple component resin systems from Radius Engineering Inc. [reference 7] and (b) pressure-controlled injector for multiple component systems from Composite Integration [reference 8]. (*Source*: Reprinted with permission from reference 1, copyright 2010 CRC Press, Taylor & Francis Group.)

to reduce the heat loss to the ambient and keep the resin temperature constant.

- No back flow should be allowed. That means the resin should flow from the injector's outlet to the mold inlets (gates), but not in the opposite direction. This is especially important when multiple injection gates are used with constant flow rate injection boundary conditions. If the resin pressure inside the mold exceeds the instantaneous injection pressure at any of the gates, an injector with a one-way flow feature does not allow the resin to flow backward by closing the valve at the exit of the injector.
- It should have an emergency stop button.
- It should be easy to clean the resin storage tank/cylinder, and maintain the components such as seals, pistons, switches and sensors.
- Either the reservoir volume should be large enough, or reloading the resin reservoir should be possible during the injection with no interruption.

Well-known injection equipment manufacturers for RTM are Radius Engineering Inc.,[7] Composite Integration,[8] Magnum Venus Plastech Limited[9] and Liquid Control Corp.[10]

Depending on the needs and expectations, one should decide what type and capacity of the injection equipment to purchase. Usually, the initial investment cost of a flow rate-controlled injection equipment is higher than a pressure-controlled equipment. Figure 9.16a shows a flow-rate controlled injection equipment from Radius Engineering Inc.[7] Its major components are (1) a resin cylinder (with a capacity of 2100 cc for this model) with a heater jacket, (2) an electric stepper motor which drives a piston at a desired resin flow rate, (3) a control console to input the process parameters and monitor process variables, (4) communication interfaces to control the equipment through a personal computer instead of the control console if desired and (5) a frame and stand with wheels.

A pressure-controlled injection equipment from Composite Integration[8] is shown in Fig. 9.16b. It is used for multiple component resin systems and stores the resin and catalyst in separate tanks and mixes them in the mixer head. After the injection is completed, a solvent flush system is used to prevent the cure of the resin in the mixer head.

9.7 Issues that influence manufacturing with RTM

Compared to consistent and repeatable manufacturing in metal forming and shaping processes, the part quality in composite manufacturing processes suffers from the effects of inherent variations in the materials and process parameters which results in variations in the mechanical properties of the parts. Important issues have been identified that manifest themselves either during fiber preforming or mold filling stages of RTM. One can address or overcome them with process modeling, control and automation as discussed later. These issues are schematically shown in Fig. 9.17, and will be briefly discussed in this section.

One needs to design the mold and process parameters to achieve two key goals: (1) to fill the mold cavity completely without the presence of either macro or micro voids and (2) to reduce the cycle time and total cost.

Usually a process simulation can be used to design the injection and vent locations with the input parameters of geometry of the mold and the permeability of the fabric. The vents are usually placed at the locations where the resin arrives last so voids can be prevented. This will produce void-free parts if these conditions are replicated from one part to the next. However, material placement and variability will change the permeability of the fabric in certain locations from one part to the next altering the resin flow pattern which will not ensure that the resin will arrive at the vents for all the parts. Hence it is important for the designer to anticipate disturbances in the flow due to material placement and variability and develop optimization and control approaches to address them. On the next page we identify various micro and macro issues that may cause variation in the flow pattern.

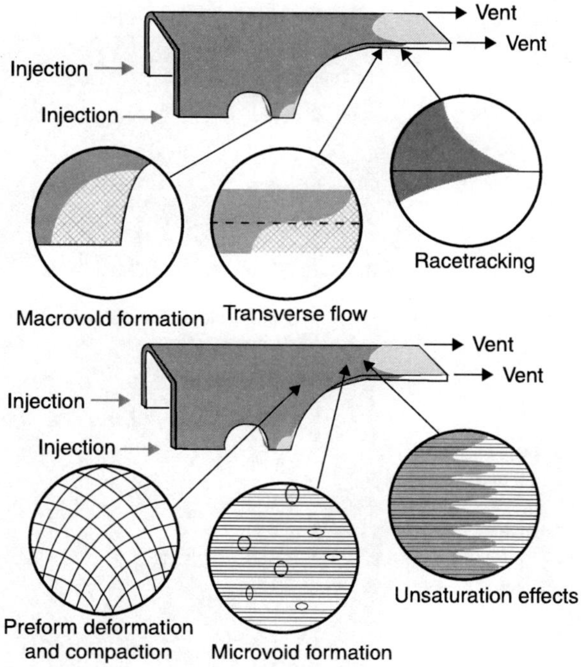

9.17 Material and process variations that can alter flow in RTM. (*Source*: Reprinted with permission from reference 1, copyright 2010 CRC Press, Taylor & Francis Group.)

9.7.1 Racetracking channels

When a fiber preform is placed and compacted in the mold cavity, regions in contact with the mold walls or inserts placed in the mold usually will have lower fiber volume fraction than the bulk. These regions have lower resistance (thus higher permeability) to resin flow than the bulk preform, and the resin races along the path of the highest permeability. This phenomenon is known as 'racetracking', and the path is called 'racetracking channel'.[11–18] Usually, the *racetracking channels are formed*

- *Along the mold edges*: The fiber volume fraction may be lower along the mold edges due to (1) missing fiber bundles which may have dislodged from the plies during cutting and placement and (2) smaller in-plane preform dimensions than the mold dimensions. Both of these material handling and placement issues can form racetracking channels as illustrated in Fig. 9.18b; whereas no racetracking channel is present in Fig. 9.18a as a result of careful fiber preforming and placement. Figure 9.19 shows how

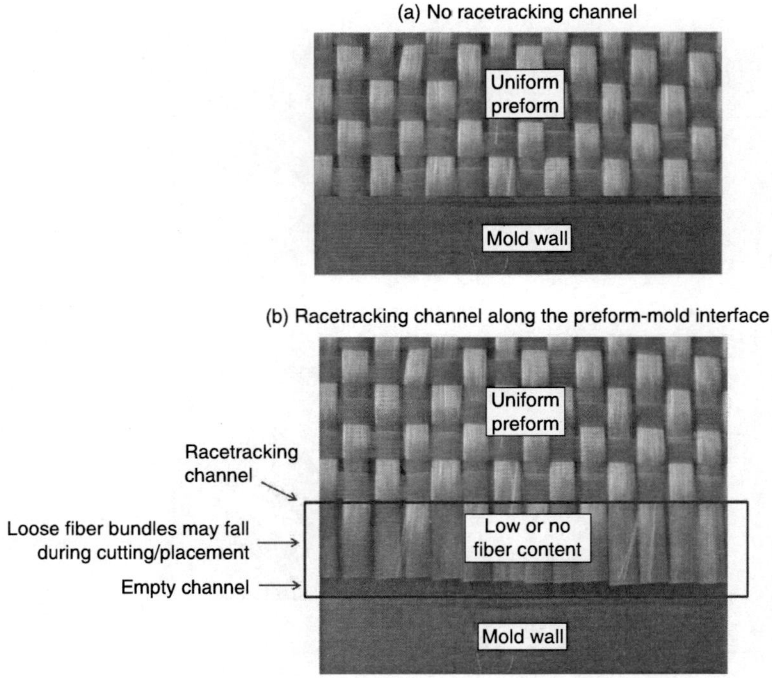

9.18 An edge of mold cavity (a) without and (b) with a racetracking channel along the preform-mold interface. Racetracking is caused by air channels and other regions with high porosity (due to lower fiber content than the bulk fiber preform) which provide paths of low resistance to fluid flow. The resin flows faster through these paths than through the bulk preform. (*Source*: Reprinted with permission from reference 1, copyright 2010 CRC Press, Taylor & Francis Group.)

low fiber volume along the edge as compared to the bulk alters the flow pattern and results in resin racing along that edge.

- *Along the ribs or bend sections*: If the inner and outer bend radii of the mold are not carefully machined, the fiber volume fraction will be lower there as compared to the bulk regions, providing resin with low resistance to flow paths which will alter the flow. On the other hand, if the bend radii create smaller mold gap than the bulk regions, the fiber bundles may be highly compacted in in-plane and thickness directions and fiber volume fraction may be higher here than the bulk regions. In this case, the preform will have higher resistance (lower permeability) to resin flow than the bulk regions. The effect of these cases on the mold filling has been investigated.[19,20]
- *Along line injection gates*: Although the above racetracking channels are formed unintentionally, racetracking channels may be formed intentionally

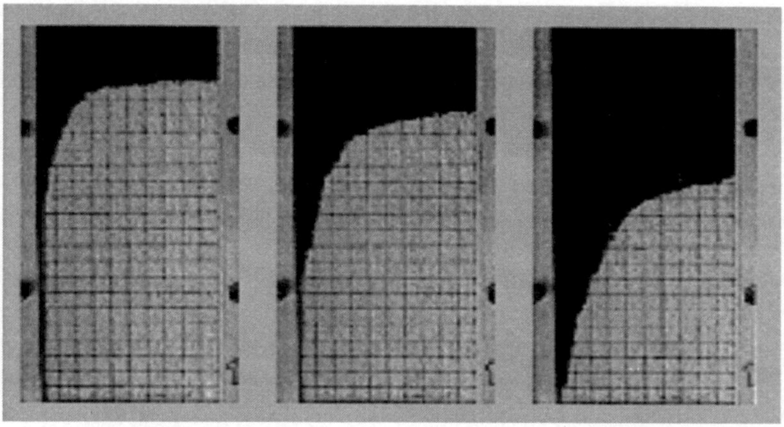

9.19 Effect of racetracking channel on the flow patterns. Intentional racetracking channel was generated along the left edge; and the three photographs were taken from the top of a transparent mold at different times. (*Source*: Reprinted with permissions from reference 1, copyright 2010 CRC Press, Taylor & Francis Group and reference 11, copyright 1999 Elsevier.)

to create line gates and vents, and also along other paths to intentionally race the resin flow and thus reduce the mold filling time and/or reduce the required maximum resin pressure. This is usually accomplished by machining in the mold plates. However, the designer should be careful that it does not lead to multiple flow fronts that can entrap air between them as they approach each other.

9.7.2 Deformation of fiber structure during draping

When a fiber preform is draped over a tool surface, the orientation of the fibers in the preform will also change. This will change the fiber volume fraction and hence the permeability to resin flow. Depending on the type of fabric and radius of the mold curvature, the deformation and permeability may vary spatially. The designer should consider the effect of permeability variations on the resin propagation when the fabric is placed over a compound curvature surface. Rudd *et al.*[6] discussed four different *mechanisms of fiber deformations*:

- *Inter-fiber (intraply) shear*: occurs when the fibers rotate about the stitches or weave centers.
- *Inter-fiber slip*: occurs when the fibers move relative to each other.
- *Fiber buckling*: occurs due to the in-plane compression of the fibers which causes fiber wrinkling.

- *Fiber extension*: may occur under high tensile stress during draping. However, this mechanism is not as common as the other three listed previously due to the high stiffness (modulus of elasticity) of fiber materials.

Rudd *et al.*[6] developed a kinematic drape model. Its numerical solution with constrained fiber paths predicts fiber shear deformations and their effect on the fiber volume fraction. Bickerton *et al.*[21] modeled the draping of a compound curved preform and its effect on the resin flow. They validated the model results with mold filling experiments.

There are commercial programs available such as FiberSIM[22] that will provide the draping angles due to the layup. One can use that information to update the permeability and fiber volume fraction values to access the changes in the flow patterns and the time to fill due to the draping of the fabric.

9.7.3 Macrovoid formation

Macrovoids (macro size dry regions in the preform) are formed under the following conditions:

- if the resin flow front reaches the vents before impregnating the preform completely,
- if air is present in the dry region (i.e., no perfect vacuum is applied before the resin injection),
- if the resin pressure around the dry region is not sufficiently high to shrink and collapse the void or move the void toward the vents.

Mold filling (i.e., resin flow) simulations help the design engineers to determine the last point(s) of resin arrival under vacuum, and then place the vents at those points to avoid macrovoid formation. There are two approaches to eliminate the macrovoids:

1. Allow the resin to bleed out of the vents for sufficient time (usually of the order of minutes) so that any potential macrovoid (along with the air inside) is pushed toward the vents and/or it shrinks.
2. Apply process control by (a) monitoring the flow front position by using sensors, (b) predicting if any macrovoid is likely to be formed at the end of the resin injection and (c) adjusting process parameters (resin pressure or flow rate, opening/closing inlet gates) if necessary.

The disadvantage of item (1) is that flushing of the extra resin increases both waste and cycle time. Item (2) is powerful, but it requires process modeling and control and embedding of robust sensors in the mold.

9.7.4 Transverse flow in the thickness direction

In typical RTM applications, in-plane part dimensions are much larger than its thickness. Thus, as will be studied in this chapter, many resin flow models in the literature are based on the assumption that there is no significant transverse flow which simplifies the modeling to a 2D flow in the in-plane directions, and only a shell mold cavity is used for the solution domain. *2D flow assumption is violated when:*

- the part thickness varies significantly,
- the in-plane permeabilities of the multiple plies of the fiber preform change by orders of magnitude or
- core materials such as foam are embedded between the plies.

In that case, significant transverse flow develops, and macrovoids may be entrapped as illustrated in Fig. 9.20. From the solution procedure point of view, fully three-dimensional flow modeling and simulations are not much more difficult than the two-dimensional version; however, one needs (1) to discretize the solution domain in 3D instead of 2D (which takes much longer CPU time to solve the pressure distributions and advance the flow front in a time marching scheme) and (2) to measure the transverse permeability of the preform which is much more difficult than measuring in-plane permeability components.[1,23–25]

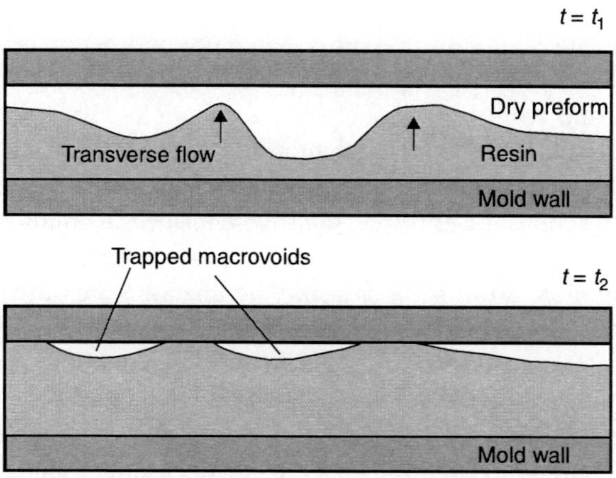

9.20 Macrovoids trapped in the composite part due to the transverse resin flow in the thickness direction. (*Source*: Reprinted with permission from reference 1, copyright 2010 CRC Press, Taylor & Francis Group.)

9.7.5 Dual scale fiber structure in a preform

A typical RTM fiber preform has *two scales of permeabilities* to resin flow:

- *Inside a fiber bundle*, the porosity (volume fraction of empty spaces) is low. Thus, its permeability to resin flow is low.
- *The empty spaces between the woven or stitched fiber bundles* are relatively larger than the empty spaces in the fiber bundles. Thus its permeability to resin flow is high.

Typically, a fiber bundle has hundreds to thousands of fibers in an elliptical cross-section with a width of a few millimeters. Considering that a glass or carbon fiber has a diameter of 10 microns approximately, and if all the fibers are densely packed in a bundle, the gap between the fibers is only in the order of microns (i.e., 10^{-6} m). This is much smaller empty space than the empty space between the fiber bundles which is typically of the order of millimeters (i.e., 10^{-3} m). These two types of empty spaces give rise to two scales of permeabilities encountered by the resin flow, and they may result in microvoid entrapment inside the fiber bundles.

9.7.6 Microvoid formation

Due to the dual scale permeabilities in a fiber preform (as explained previously), two types of microvoids may be entrapped in a composite part during resin injection:

- *Intra-bundle microvoid* is the most common microvoid type which occurs due to the lower permeability of the fiber bundles than the permeability of the empty spaces between the bundles. The resin flow is faster between the fiber bundles than inside the bundles, and the resin encircles itself when it reaches a stitch or another bundle perpendicular to the flow direction, and entraps a microvoid inside the bundle as shown in Fig. 9.21. To avoid this type of microvoid, the common practice is to slow the resin flow down by decreasing the injection pressure/flow rate boundary condition. This allows sufficient time for the encircled microvoids to shrink and collapse due to the higher resin pressure around the microvoids than inside the voids.
- *Inter-bundle microvoid* is formed if the dual scale permeabilities are formed such that the resin flows faster inside the bundles along the fiber direction due to the capillary forces than in between them. Microvoids are entrapped between fiber bundles when it gets drawn due to the capillary action across a stitch[1,26] as shown in Fig. 9.21.

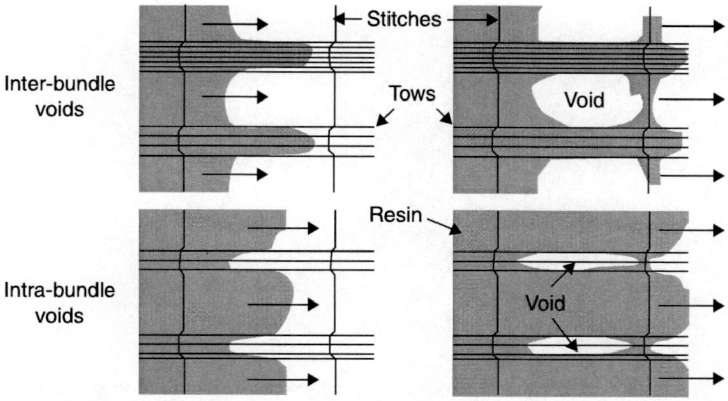

9.21 Microvoids trapped in the composite part [reference 1]. (Top): Inter-bundle voids are trapped when the resin races along the fiber bundles and encircles itself when it reaches a stitch. (Bottom): Intra-bundle voids are trapped in the fiber bundles when the resin races in between the bundles due to the high permeability, and the impregnation of the bundles is delayed due to the low permeability of bundles. (*Source*: Reprinted with permission from reference 1, copyright 2010 CRC Press, Taylor & Francis Group.)

Figure 9.22 illustrates how intra-bundle microvoids are formed when the resin is injected into a fiber preform. Delayed impregnation of the fiber bundles is clearly observed when one examines the dark (completely saturated) and light (partially saturated) regions in the preform, and compares them with the actual flow front position.

9.8 The need for process modeling

The main objectives of RTM are listed below:

- Manufacture composite parts with no significant macro or microvoids.
- Reduce the cycle time (length of time from the mold preparation and fiber preform placement to part demolding after the part reaches a green strength) for high volume manufacturing.
- Enhance the specific strength and stiffness for high performance applications.
- Reduce the cost.

Trial-and-error injections are still commonly used to achieve these goals in the industry. The locations of the injection gates and ventilation ports and injection boundary conditions are decided after these time-consuming trials. These injections require experience for iterating the boundary locations and conditions, and may not result in the optimal design.

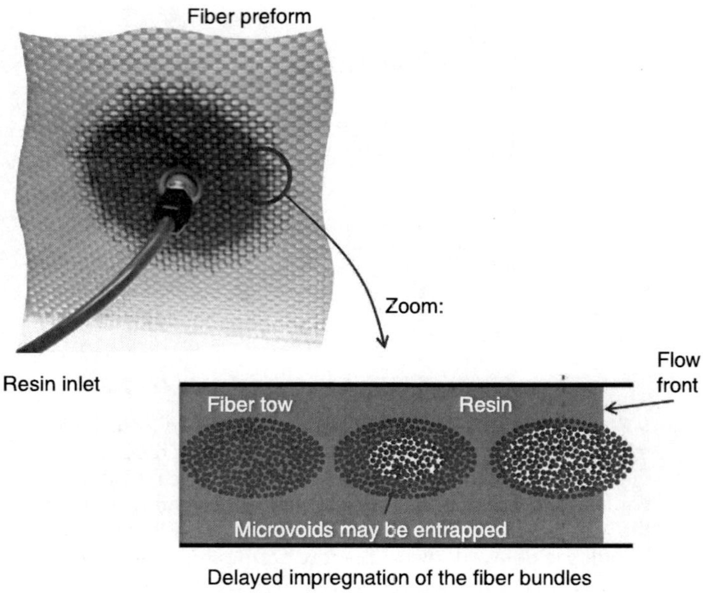

9.22 Delayed impregnation of the fiber bundles may cause entrapping microvoids in the bundles. (*Source*: Reprinted with permission from reference 1, copyright 2010 CRC Press, Taylor & Francis Group.)

Mathematical modeling of different stages of the RTM process has been applied for the last several decades to design the process. Flow, cure and mechanics models were developed to achieve the first three objectives listed earlier, respectively. In this chapter, flow modeling is discussed which allows the user to control the process parameters (resin injection pressure/flow rate) and design the mold (locations of the gates and vents). The details of the process modeling have been studied in many other books and book chapters.[1,27–30]

The cycle time is the total of the three stages: mold preparation and preform preparation, mold filling and resin cure. Mold filling time is predicted by using the flow models. Cure model is used to determine the degree of resin cure; and if a database is available which relates the degree of cure to the mechanical properties, the user can decide when to demold the part.

Flow and cure models require important material properties as input. These are permeability of the fiber preform and reaction kinetics of the resin, respectively. The accuracy of the models depends on the accuracy in characterizing these properties.

9.9 Resin flow models for RTM

In fluid mechanics, the velocity field is solved by using the differential equations for mass and momentum conservation with boundary conditions around the problem domain. One can follow the same procedure to solve the resin flow through a porous fiber preform in RTM, but the solution will be impractical because of the difficulty in representing the flow domain which is a very complicated network of empty channels between the fibers in bundles and also between the bundles. The accepted approach has been to replace the momentum conservation equation with an empirical macroscopic relationship between the velocity and the pressure gradient which was originally proposed by Henry Darcy in 1856 for flow of underground water through a porous sand column.[31] This macroscopic equation is solved in the entire mold domain (i.e., not only in the empty spaces, but in the total space occupied by the empty spaces and also the solid fibers). Thus, the representation of the mold domain is much easier than the empty spaces alone; and the numerical discretization of the mold geometry is done readily using commercial programs such as PATRAN, SolidWorks or I-DEAS.

Darcy's law describes the overall relationship between the velocity and the pressure gradient, and not the details of the velocity and pressure at each point inside the microscale fiber structure. Usually, this is sufficient to model the resin flow in RTM while designing the mold as we are interested in macroscopic variables such as mold fill time, maximum value of the injection pressure required and locations of gates/vents to ensure complete mold filling.

One could also address micro flow using Darcy's law for flow within the fiber bundles but characterized by a different permeability as will be shown later, but to rigorously describe the flow within fiber bundles, one must also account for surface tension and couple this flow with the macro flow in between the fiber bundles. More details can be found in the references[32–37] that model the microvoid formation in the fiber bundles by adding a sink term to the equation of mass conservation.

9.9.1 1D Darcy flow through a porous medium

In 1856, Henry Darcy, a French engineer, conducted 1D water flow experiments through sand columns as schematically shown in Fig. 9.23. He measured the flow rate of the water, Q by varying the height of the water column and thus varying the pressure drop, ΔP between the top and bottom of the sand column. He observed a linear relationship between the 1D average water velocity $\bar{u}$ and the pressure gradient $\Delta P/L$ in the porous region:[31]

$$\bar{u} = -\frac{K}{\mu}\frac{\Delta P}{L} \qquad [9.2]$$

where K is the isotropic permeability of the porous sand column, μ is the viscosity of the water, ΔP is the pressure drop and L is the height of the sand column.

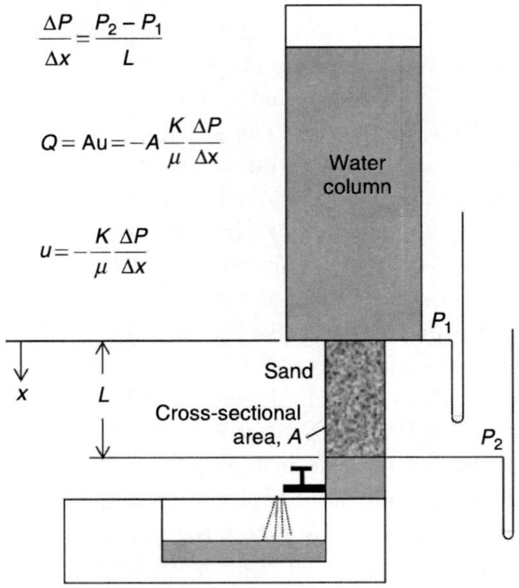

9.23 The experimental setup used by Darcy in 1856 (reference 31) to measure the permeability of porous sand column. (*Source*: Reprinted with permission from reference 1, copyright 2010 CRC Press, Taylor & Francis Group.)

Permeability is dependent on the porosity of the sand column which is a function of the size distribution and compaction of the sand particles and rocks.

9.9.2 Anisotropic Darcy flow through a porous medium

Darcy's law adequately models typical RTM applications in which the Reynolds number (the ratio of inertial force over viscous force on the liquid resin) is small. This allows one to ignore the inertial forces and assume a quasi-steady resin flow.[1]

In RTM, the porous medium is a fiber preform, and its permeability is dependent on the fiber structure, fiber volume fraction (which is dependent on the fiber compaction) and fiber sizing. Considering flow through an anisotropic medium in general (i.e., having different permeabilities in different directions), Darcy's law can be generalized as follows:

$$\bar{\mathbf{U}} = -\frac{\mathbf{K}}{\mu}\cdot\nabla P \qquad [9.3]$$

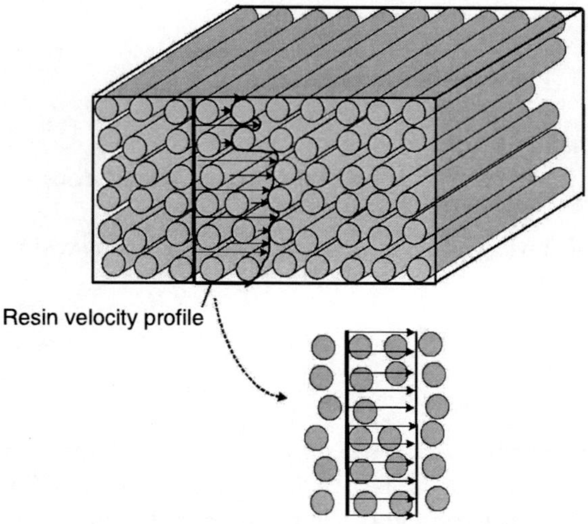

9.24 Volume-averaged Darcy's velocity through a porous fiber preform. (*Source*: Reprinted with permission from reference 1, copyright 2010 CRC Press, Taylor & Francis Group.)

where $\overline{\mathbf{U}}$ is the volume-averaged 'Darcy's velocity' (see Fig. 9.24), μ is the viscosity of the fluid, ∇P is the pressure gradient and $\mathbf{K}$ is the permeability tensor of the preform.

The three-dimensional Darcy's law takes the following form:

$$\overline{\mathbf{U}} = \begin{pmatrix} \bar{u} \\ \bar{v} \\ \bar{w} \end{pmatrix} = -\frac{1}{\mu} \begin{pmatrix} K_{xx} & K_{xy} & K_{xz} \\ K_{yx} & K_{yy} & K_{yz} \\ K_{zx} & K_{zy} & K_{zz} \end{pmatrix} \begin{pmatrix} \partial P/\partial x \\ \partial P/\partial y \\ \partial P/\partial z \end{pmatrix} \qquad [9.4]$$

where $\bar{u}$, $\bar{v}$, and $\bar{w}$ are the Darcy's volume-averaged resin velocity components in x, y and z directions, respectively.

If the principal directions of the preform permeability coincide with the x, y and z directions (thus, $K_{ij} = K_{ji} = 0$ for $i \neq j$), then Darcy's law simplifies to:

$$\bar{u} = -\frac{K_{xx}}{\mu}\frac{\partial P}{\partial x} \qquad [9.5a]$$

$$\bar{v} = -\frac{K_{yy}}{\mu}\frac{\partial P}{\partial y} \qquad [9.5b]$$

$$\bar{w} = -\frac{K_{zz}}{\mu}\frac{\partial P}{\partial z} \qquad [9.5c]$$

In most RTM applications, the resin flow is two-dimensional approximately, thus one can ignore $\bar{w}$ compared to $\bar{u}$ and $\bar{v}$, and does not have to characterize K_{zz}.

The equation of conservation of mass for 2D quasi-steady resin flow is the divergence of velocity:

$$\frac{\partial \bar{u}}{\partial x} + \frac{\partial \bar{v}}{\partial y} = 0 \qquad [9.6]$$

In Eq. [9.6], the fiber bundles are assumed to be completely impregnated instantaneously as the flow front passes through. If one would like to investigate the microvoid formation in the bundles, usually a sink term is added to Eq. [9.6].[32–37] In that approach, the microvoid dimensions and strength of the sink are related to the viscosity, pressures of the void-free surface and capillary, and the permeability of the fiber bundle.

Substitution of the Darcy's velocities in Eq. [9.6] results in the following second order partial differential equation on the resin pressure, P:

$$\frac{\partial}{\partial x}\left(\frac{K_{xx}}{\mu}\frac{\partial P}{\partial x} + \frac{K_{xy}}{\mu}\frac{\partial P}{\partial y}\right) + \frac{\partial}{\partial y}\left(\frac{K_{yx}}{\mu}\frac{\partial P}{\partial x} + \frac{K_{yy}}{\mu}\frac{\partial P}{\partial y}\right) = 0 \qquad [9.7]$$

for 2D flow. Equation [9.7] simplifies to

$$\frac{\partial}{\partial x}\left(\frac{K_{xx}}{\mu}\frac{\partial P}{\partial x}\right) + \frac{\partial}{\partial y}\left(\frac{K_{yy}}{\mu}\frac{\partial P}{\partial y}\right) = 0 \qquad [9.8]$$

for $K_{xy} = K_{yx} = 0$. Equation [9.7] is solved to calculate the resin pressure, P in the impregnated preform domain which enlarges with time as the flow front propagates with time. At any instant of time, the pressure distribution $P(x, y, t)$ is unique for a given set of boundary conditions around the impregnated domain. This is possible because a quasi-steady state assumption is made. This assumption allows one to solve for the pressure at each time step as a steady state problem. This assumption is justifiable for viscous resins due to their fast momentum transfer and they can achieve steady state quite rapidly. All possible types of boundary conditions are tabulated in Table 9.1, and two examples of boundary condition sets are shown in Figs. 9.25 and 9.26.

Table 9.1 Boundary conditions for resin flow

Type of boundary	Boundary condition
Specified pressure injection gate	$P = P_{\text{inj}}$
Specified flow-rate injection gate with an inlet cross-sectional area of A and with an outward unit normal vector, **n**	$Q = Q_{\text{inj}}$ $A\bar{U}_n = A\bar{\mathbf{U}} \cdot \mathbf{n} = -Q_{\text{inj}}$ $A\left(-\dfrac{\mathbf{K}}{\mu} \cdot \nabla P\right) \cdot \mathbf{n} = -Q_{\text{inj}}$ for example, $\dfrac{A}{\mu}\left(K_{xx}\dfrac{\partial P}{\partial x} + K_{xy}\dfrac{\partial P}{\partial y}\right) = Q_{\text{inj}} \quad \text{if} \quad \mathbf{n} = +\text{i}$ $\dfrac{A}{\mu}\left(K_{yx}\dfrac{\partial P}{\partial x} + K_{yy}\dfrac{\partial P}{\partial y}\right) = Q_{\text{inj}} \quad \text{if} \quad \mathbf{n} = +\text{j}$
No outflow along an impermeable mold edge with an outward unit normal vector, **n**	$\bar{U}_n = \bar{\mathbf{U}} \cdot \mathbf{n} = 0$ (normal velocity component is zero) for example, $\bar{u} = 0 \;\rightarrow\; K_{xx}\dfrac{\partial P}{\partial x} + K_{xy}\dfrac{\partial P}{\partial y} = 0 \quad \text{if} \quad \mathbf{n} = \pm\text{i}$ $\bar{v} = 0 \;\rightarrow\; K_{yx}\dfrac{\partial P}{\partial x} + K_{yy}\dfrac{\partial P}{\partial y} = 0 \quad \text{if} \quad \mathbf{n} = \pm\text{j}$
Vent (specified pressure)	$P = P_{\text{vent}}$ if P is the absolute pressure $P = 0$ if P is the gage pressure
Pressure at the free surface which is moving toward a vent	$P = P_{\text{vent}}$ if P is the absolute pressure $P = 0$ if P is the gage pressure
Pressure at the free surface of an entrapped macrovoid	$\dfrac{PV_{\text{void}}}{T_{\text{void}}} = \text{constant}$ where V_{void} and T_{void} are the volume and temperature of the air inside the macrovoid

9.9.3 Experimental characterization of permeability

Experimental procedures for measuring the in-plane permeability of a porous preform are summarized in this section. The schematics of the experimental setups are shown in Figs. 9.27–9.30 for four different approaches. Table 9.2 summarizes the boundary conditions, required data to be collected during the experiments and the formulas used to calculate the permeability.

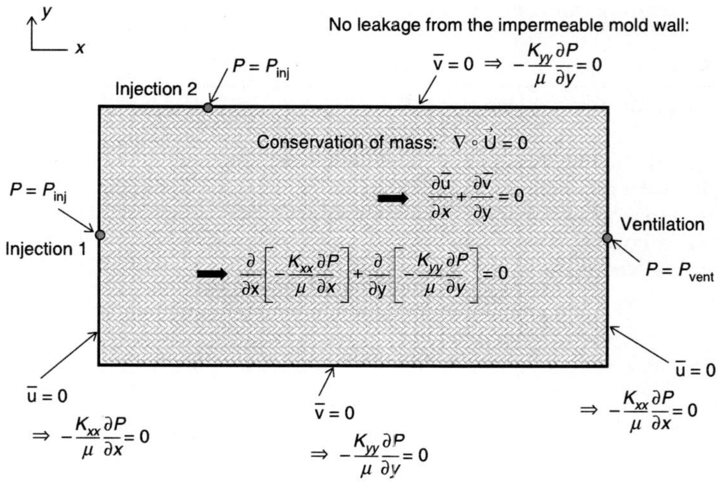

9.25 An example of boundary conditions for resin flow. Here, K_{xy} was assumed zero for simplicity.

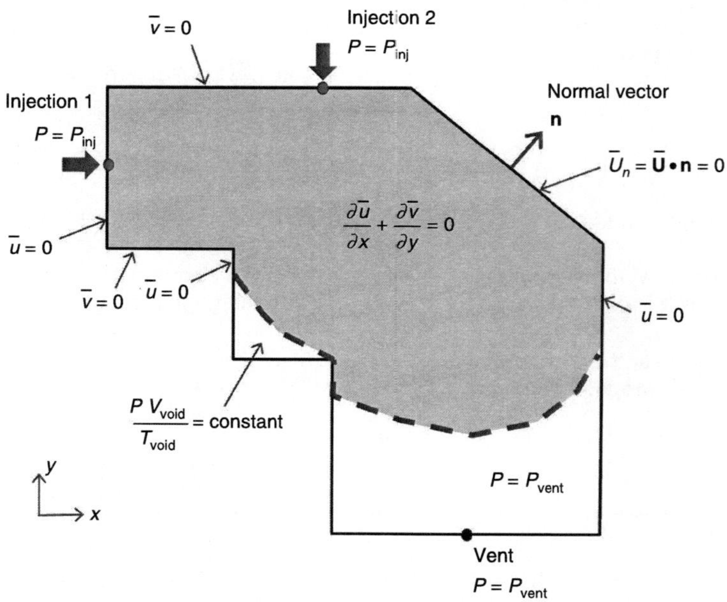

9.26 Another example of boundary conditions for resin flow. Notice that two different boundary conditions should be used at the two free surfaces since one is directly connected to the vent, and the other is an entrapped macrovoid in which the void pressure increases as its volume decreases (i.e., as it shrinks).

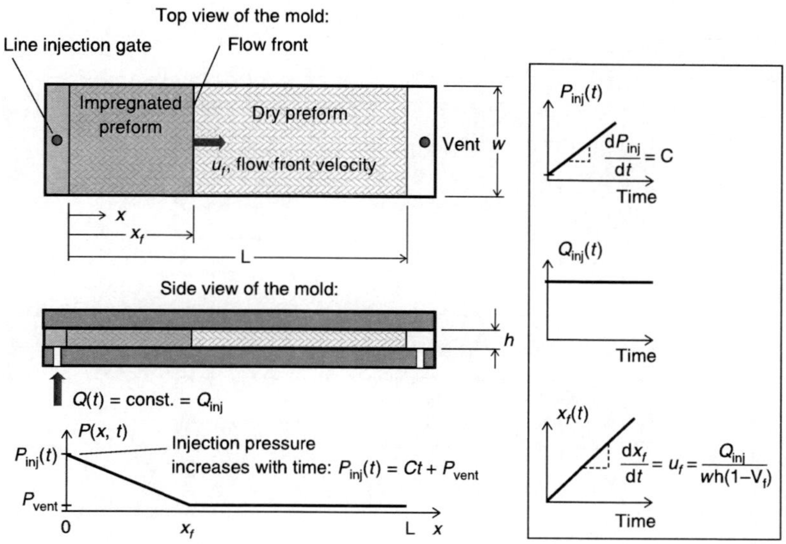

9.27 Permeability measurement experimental setup for transient filling 1D constant-flow rate injection. x_f is the position of the flow front. The flow front velocity, u_f and the volume-averaged velocity $\bar{u}$ are related with $u_f = \bar{u}/\phi = \bar{u}/(1-V_f)$) where ϕ and V_f are the porosity and the fiber volume fraction of the preform.

The following is a list of important assumptions and guidelines while performing permeability measurement experiments:

- In all experiments, Darcy's law is assumed to be satisfied. Thus, the Reynolds number should be low; and almost instantaneous complete saturation of the fiber bundles is assumed when the flow front arrives there.
- One-dimensional linear injections (unsteady or steady versions) are performed to measure the permeability along the flow direction. To measure the permeability along the transverse in-plane direction, one has to perform another experiment by using a different specimen which is rotated by 90° and then placed in the mold cavity accordingly.
- In one-dimensional experiments, care must be taken to ensure that the resin flow is uniform and truly 1D. Thus, no significant racetracking channel should exist between the mold edges and the preform. Preforming (cutting and placement) should be done carefully so that the preform fits snugly with the edges of the mold cavity. The gate should be a line gate which can be a racetracking channel in the width direction near the inlet hole (see Figs. 9.27—9.29), thus the resin fills this region before entering as a uniform flow front across the width in the region containing the preform. A transparent mold lid (usually made of acrylic or glass) is commonly used

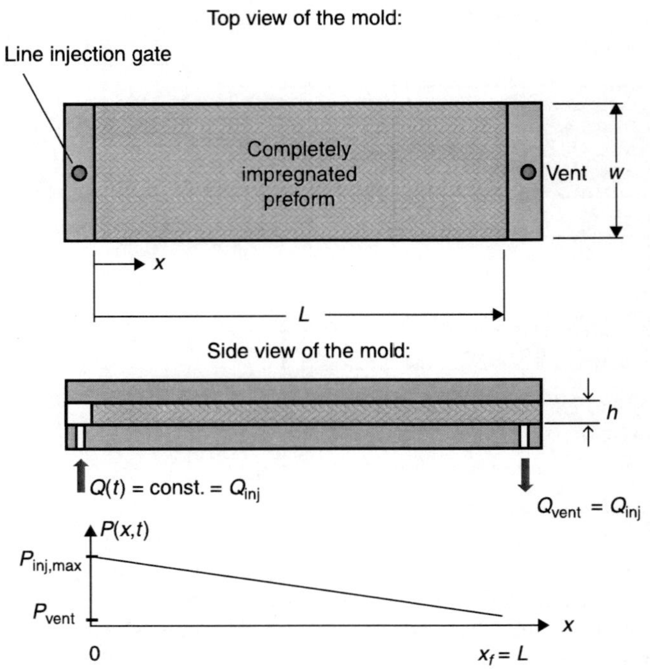

9.28 Permeability measurement experimental setup for steady 1D constant-flow rate injection.

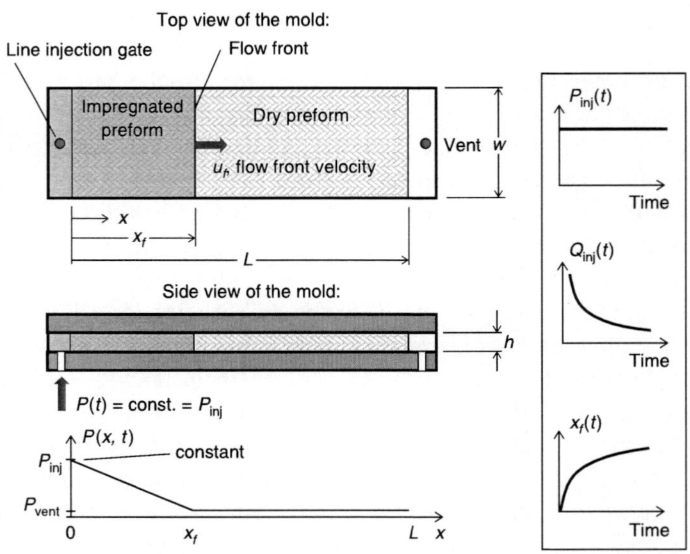

9.29 Permeability measurement experimental setup for transient filling 1D constant-pressure injection.

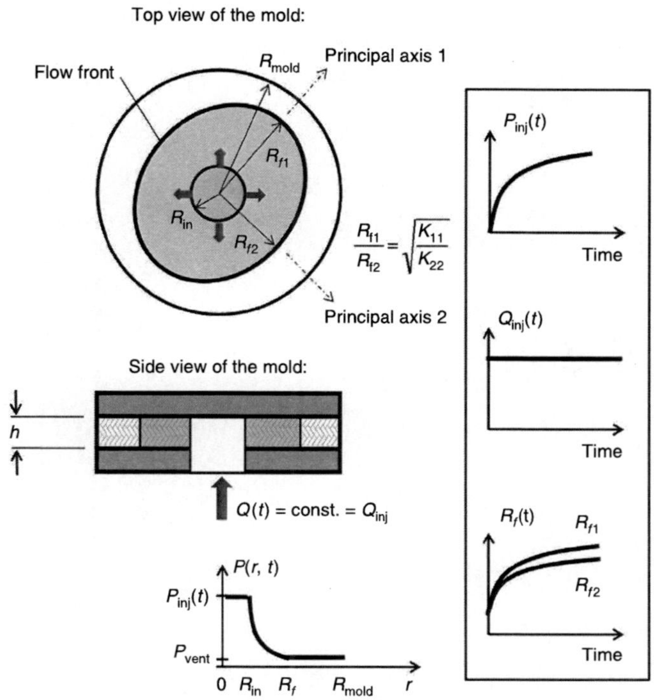

9.30 Permeability measurement experimental setup for transient filling radial constant-flow rate injection. Here, an elliptical flow front propagates with major and minor dimensions of $R_{f1}(t)$ and $R_{f2}(t)$ for an anisotropic preform. The flow front would be circular ($R_{f1}(t) = R_{f2}(t) \equiv R_f(t)$) when the preform is isotropic ($K_{11} = K_{22} \equiv K$). Note the injection hole should be through the preform as well to ensure two-dimensional flow.

to visually verify that the flow is 1D. One must ensure that the plexiglass plate is sufficiently thick or is reinforced with cross bars to prevent bending of the plate. Alternatively, one may use a steel plate to ensure the mold is stiff and introduce sensors to monitor the flow front position and shape for verification purposes. However, these sensors must be carefully embedded in the mold walls so as not to cause any disturbance to the flow.

- The thickness of the mold cavity, h, should be uniform everywhere. Thus, the mold should be clamped appropriately to prevent mold bending under high fiber compaction and resin pressures.
- In the unsteady 1D linear experiment with constant flow rate injection, the injection pressure, P_{inj} is expected to increase linearly with time (see Fig. 9.27) if the flow is 1D and fiber tows are filled at the instant the resin arrives there. Since the slope, dP_{inj}/dt is used to calculate the permeability, K (see Table 9.2), one must carefully check if $P_{inj}(t)$ is almost a first order

Table 9.2 Permeability measurement experiments

Experiment	Figure	Boundary condition	Required data to be collected	Formula to calculate K along the flow direction
Transient filling 1D linear with constant flow rate injection	9.27	$Q = Q_{inj}$	$P_{inj}(t)$	$K = \left(\frac{Q_{inj}}{wh}\right)^2 \frac{\mu}{\phi} \frac{1}{(dP_{inj}/dt)}$ where dP_{inj}/dt is the slope of the 1st order curve-fit to the data.
Steady 1D linear with constant flow rate injection when the mold is completely filled (both the gate and vent are kept open to achieve steady flow)	9.28	$Q = Q_{inj}$	$P_{inj}(t_{fill})$ and P_{vent}	$K = -\frac{Q_{inj}\mu}{wh}\left(\frac{dP}{dx}\right)^{-1}$ where $\frac{dP}{dx} = \frac{\Delta P}{L} = \frac{P_{vent} - P_{inj}(t_{fill})}{L}$ is assumed to be constant (i.e., not varying with x or t).
Transient filling 1D linear with constant pressure injection	9.29	$P = P_{inj}$	$x_f(t)$	$x_f(t) = \sqrt{\frac{2KP_{inj}t}{\mu\phi}}$ i.e., $K = \frac{C\mu\phi}{2P_{inj}}$ where C is to be determined by using $x_f(t) = \sqrt{Ct}$ curve-fit to the data.

Transient filling radial with constant flow rate injection	9.30	$Q = Q_{inj}$	$P_{inj}(t)$ $R_{f1}(t)$ and $R_{f2}(t)$	$P_{inj}(t) = \frac{\mu Q_{inj}}{4\pi h\sqrt{K_{11}K_{22}}}\ln\left(1+\frac{Q_{inj}t}{\pi h R_{in}^2\phi}\right)$ where $\frac{K_{11}}{K_{22}} = \left(\frac{R_{f1}}{R_{f2}}\right)^2$ $K \equiv \sqrt{K_{xx}K_{yy}} = \frac{\mu Q_{inj}}{4\pi h D}$ where D is to be determined by using $P_{inj}(t) = D\ln\left(1+\frac{Q_{inj}t}{\pi h R_{in}^2\phi}\right)$ curve-fit to the data.

Process parameters: μ: resin viscosity [Pa.s]; ϕ: preform porosity (note that $\phi = 1 - V_f$ where V_f is the fiber volume fraction); Q_{inj}: injection flow rate [m³/s]; P_{inj}: injection pressure [Pa]; w: mold width [m]; h: mold thickness [m]; L: mold length [m]; K: preform permeability along the flow direction [m²]; t: time [s]; R_{in}: inlet hole radius [m]; R_{f1} and R_{f2}: major and minor principal dimensions of the elliptical flow front [m] (see Fig. 9.30).

Sources: From references 1,12, 27–30.

function by curve fitting. If the residual (the error between the experimental and curve-fit data) of the first order polynomial least-square curve fit is not small, one needs to investigate whether there was racetracking during the flow in which case the permeability values will not be valid. Even if racetracking is not present and the increase in pressure at the injection gate does not show linear behavior, it suggests that the fabric is dual scale in which the fiber tows are being filled at a slower rate than the regions in between the fibers. The issue of characterizing the dual scale permeability has been addressed by Zhou *et al.*[38,39] and Kuentzer *et al.*[40]

- A single radial injection experiment allows one to measure both principal in-plane permeability components (K_{11} and K_{22}). One does not have to address the issue of racetracking during radial injection. These are the major advantages of the radial injection compared to the 1D linear injection.
- The permeability of the preform specimens that are prepared from the same fabric roll by the same person may have significant variations. This is mainly due to the inconsistency in the fiber structure and racetracking channels which may vary from one experiment to the next. Slight variations in the preform ply layup and nesting effects may cause significant variations in the permeability. Hence it is necessary to perform experiments on a large number of specimens, and document the mean and standard deviation of the results.
- Very low or high flow rates can cause deviations as assumptions are violated. At very low flow rates, capillary forces become more dominant than the applied pressure leading to a fingering effect in which the flow may be faster within the fiber bundles. On the other hand, very fast flow rates may not allow sufficient time for fiber tows to be saturated.
- The results of transient and steady 1D experiments may have very significant variations mainly due to the delayed saturation of the fiber bundles during the transient experiments. It is more accurate to use the permeability values of the transient one-dimensional filling experiment when simulating the flow than the steady state filled values especially for dual scale fabrics.
- The simulated resin used for permeability measurements should not change viscosity during the filling stage as it will influence the permeability calculations as viscosity is assumed to be constant during the experiment. The same is true for ensuring that the fiber volume fraction of the preform used in the mold is uniform and constant in the entire mold.
- Considering the tedious and time-consuming conventional permeability measurement experiments mentioned earlier, 'continuous permeability measurement experimental setups'[41–43] allow performing one transient filling experiment and a set of steady experiments on a single preform specimen at different fiber volume fractions by adjusting the gap

between the two mold parts. The sealing system of these molds should be designed carefully so as not to leak resin or buckle the fibers.

Figure 9.31 shows some samples of experimentally measured permeability values for mats and fabrics published in the literature. Eleven institutions and companies participated in calculating in-plane permeabilities of two fabrics using different experiment types (1D linear and radial),

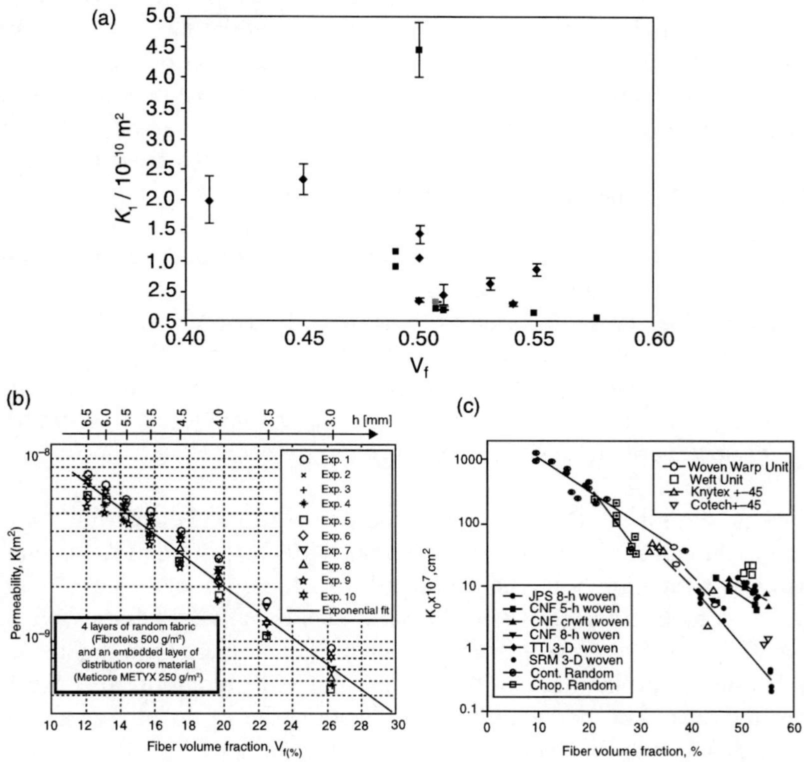

9.31 Some samples of experimentally measured permeability values for reinforcing materials: (a) for an E-glass woven (2 × 2 twill) by HEXCEL with 390 g/m^2 superficial density (dark diamond: transient filling radial, light diamond: steady radial, dark square: transient filling linear, light square: steady linear). (*Source*: Reprinted with permission from reference 44, copyright by 2011 Elsevier); (b) for a preform made of four layers of stitched random fabric (Fibroteks 500 g/m^2) and an embedded layer of distribution core medium (Meticore 250 polypropylene by METYX 250 g/m^2) at the center. (*Source*: Reprinted with permission from reference 43, copyright 2009 Elsevier); (c) for a variety of woven and random fabrics. (*Source*: Reprinted with permission from reference 45, copyright 2001 Elsevier.)

boundary conditions (constant pressure or flow rate) and filling state (transient or steady) in an international benchmark exercise by Arbter *et al.*[44] A very significant variation was seen in their results (see Fig. 9.31a) which was speculated to be due to human factors mainly. Some guidelines were given to reduce the scatter in the results.

9.10 Heat transfer and cure model

The cure of the resin (cross-linking of the monomers) is usually initiated by one of the following approaches: (1) by addition of the second component (curing agents) just prior to resin injection or (2) initiating cure by UV, heat or other prompters. Even if the resin type does not require being externally heated, exothermic heat evolved from the chemical reaction may cause the resin temperature to increase. Due to the external heating and/or the exothermic heating of the resin, the heat is transferred between the resin, fibers and mold with time. As the resin temperature increases:

- the resin viscosity, μ decreases (which is desired since the required injection pressure or fill time decreases) and
- the degree of cure, α increases (it helps to shorten the cycle time by first gelling the resin and then achieving the green strength to demold the part; but it also increases the viscosity which is undesired during mold filling as it may even cause incomplete mold filling if the flow and heating stages are not designed appropriately).

For situations in which the temperature of the resin and the curing changes during the filling, one requires additional characterization of the resin viscosity as a function of temperature and cure. In addition, the cure kinetics equation for resin needs to be characterized. In RTM, the increase in the viscosity due to the cure is usually much more dominant than the decrease in it due to the increase in the temperature.[1] By coupling the non-isothermal resin flow model along with the equations of the energy and cure kinetics and viscosity model as a function of temperature and cure, one can design the two stages (mold filling and resin cure) of the RTM process so that premature resin gelation is prevented, and also the cycle time is reduced.

Due to its simplicity, the most common model employed for the heat transfer between the resin, fibers and mold is called the 'local equilibrium model'. The average temperature of a control volume containing the resin and the fibers is used instead of the individual temperatures by assuming that respective temperatures equalize at the instant the fibers were covered with the resin.[46] For low Reynolds number flows, this assumption was supported with experimental results. The heat equation is given as follows:[1,47]

$$\left[\left(1-V_f\right)\rho_r c_{p_r}+V_f\rho_f c_{p_f}\right]\frac{\partial T}{\partial t}+\rho_r c_{p_r}\mathrm{U}\cdot\nabla T$$
$$=\nabla\cdot\left[\left(\mathrm{k}_e+\mathrm{K}_D\right)\cdot\nabla T\right]+\left(1-V_f\right)\dot{s}+\mu\mathrm{U}\cdot\mathrm{K}^{-1}\cdot\mathrm{U} \qquad [9.9]$$

where T, t, ρ and c_p are the averaged temperature of resin and fibers, the time, the density, and the specific heat capacity, respectively. The subscripts r and f denote the resin and fibers, respectively. V_f is the fiber volume fraction, and $\dot{s}$ is the source term that accounts for heat released during curing. The effective thermal conductivity, $\mathbf{k}_e$ is dependent on the fiber and resin thermal conductivities κ_r and κ_f.[48] The heat dispersion coefficient, $\mathbf{K}_D$ represents the differences in heat dispersion not accounted for by the convection of heat in the transverse direction.[1,47]

The general equation can be further simplified depending on the extent of non-isothermal nature of resin during flow and its impact on resin cure kinetics. Some general guidelines are listed below:

- Although one can easily justify solving for flow in two dimensions for RTM as the through thickness flow velocity is small as compared to in-plane velocity, one cannot ignore the heat transfer in the through thickness direction as it may be higher or of the same order as plane dimensions. The through thickness conduction lumps the convective effects of the heat due to resin movement in that direction in the heat dispersion term and would require one to characterize it as a function of in-plane flow velocity.
- In-plane heat conduction and heat convection in the transverse direction are negligible for thin shell-like domains[47]; however, heat dispersion may be significant.[49,50]
- Viscous dissipation is negligible for resin flows with low Brinkman number (viscous dissipation over conductive heat transfer) and this could be ignored.
- The heat dispersion term, $\mathbf{K}_D$ (mainly in the thickness direction) should not be neglected especially for high Peclet number flows (heat dispersion over heat conduction).[50,51] However, $\mathbf{K}_D$ is usually neglected in the heat models since the evaluation of it is difficult.[1,51–53]
- $\dot{s}$ represents the exothermic heat release during the resin cure. A constitutive equation is needed to characterize the resin cure as a function of temperature and degree of cure.
- Equation [9.9] needs to be solved with boundary conditions on the entire spatial domain and an initial condition must be supplied for the temperature. The reader may refer to the references[1,29,54] for the common boundary conditions used to solve for temperatures.

The equations of the flow, energy and cure kinetics are coupled due to the resin viscosity which is dependent on T and α. One can account for changing viscosity if the resin initiates cure during injection and if the thermal heat evolved is negligible by modifying the viscosity at each time step during the filling process. However, if the mold is being heated and/or if the heat evolved is significantly changing the temperature of the resin, one should couple the cure kinetics and viscosity model with the flow and heat transfer governing equations.

9.10.1 Resin cure models

Empirical resin cure models have been formulated for various resin systems.[1,46,54–56] The heat generation due to the cure reaction, $\dot{s}$ may be linearly related to the rate of the reaction, R_α as follows:

$$\dot{s} = R_\alpha E_\alpha \tag{9.10}$$

Here, E_α is the heat of reaction, and R_α is dependent on temperature T and degree of cure α which may be expressed by a semi-empirical equation as follows:

$$R_\alpha = \left(k_1 + k_2\alpha^m\right)(1-\alpha)^n \tag{9.11}$$

where

$$k_1 = A_1 e^{E_1/RT} \qquad k_2 = A_2 e^{E_2/RT} \tag{9.12}$$

Here, R is the universal gas constant, and the constants m, n, A_1, A_2, E_1 and E_2 must be determined by conducting characterization experiments.[1]

Due to the convection of the resin during flow in the mold cavity, one should account for convection of cure as well:

$$R_\alpha = \frac{\partial \alpha}{\partial t} + \bar{u}\frac{\partial \alpha}{\partial x} + \bar{v}\frac{\partial \alpha}{\partial y}. \tag{9.13}$$

9.10.2 Temperature and cure dependent viscosity

Chemorheological models are commonly used to characterize the viscosity μ of the thermoset resins used in the RTM process which is dependent on its temperature T and degree of cure α.[1] One of the examples of the constitutive equation for viscosity is listed below and requires characterization of constants used to describe the viscosity[1,27,29]:

$$\mu = a_0 e^{-b_0 \alpha} e^{(a+b\alpha)/(RT)} \qquad [9.14]$$

where R is the universal gas constant, $a + b\alpha$ is the activation energy, and a, b, a_0 and b_0 are determined experimentally. The user of these models should not exceed the temperature ranges used for characterization.

If the resin flow is modeled in the shell-like 2D domain for thin mold cavities, one calculates the Darcy's volume-averaged velocity components, $\bar{u}$ and $\bar{v}$ using Eq. [9.7] and boundary conditions. One can notice that the resin viscosity μ in the same equation would vary in the thickness direction besides the in-plane directions due to the temperature changes in all three directions, and the viscosity is dependent on the temperature. Therefore, the common approach is to use a local average viscosity through the thickness μ as follows[1]:

$$\frac{1}{\bar{\mu}} = \frac{1}{h}\int_0^h \frac{1}{\mu}\,\mathrm{d}z \qquad [9.15]$$

where h is the mold cavity thickness and z is the axis in the thickness direction.

Thermal, kinetic and chemorheological properties of a typical polyester resin system and a typical glass fiber are given in Table 9.3.[27]

9.11 Numerical simulation of resin flow

For process designers, sophisticated and validated numerical simulations are now commercially available to describe flow in RTM such as LIMS,[1,57] PAM-RTM[58] and RTM-Worx.[59] These simulations can be very useful and help in cost mitigation by reducing expensive trial-and-error approaches to finalize the process design for successful and reliable manufacturing, especially if the composite part has a complicated geometry, many inserts and varying thickness and permeability.

If the simulations are used to design the mold and process parameters (injection pressure or flow rate, locations of the gates and vents, and layup of the preform) in which the manufacturing engineer anticipates very small variation in material and layup process, the computational efficiency is not important. For such cases, simulation times of the order of hours or days may be acceptable. However, if the design and process engineer anticipates local variations and would like to address them in the process design, this

Table 9.3 Properties of a typical polyester resin system with inhibitor and a typical glass fiber

	Parameter	Value	Unit
Thermal properties:	ρ_r	1280	kg/m^3
	ρ_f	2560	kg/m^3
	c_{pr}	1900	J/(kg. °K)
	c_{pf}	880	J/(kg. °K)
	κ_r	0.2	W/(m. °K)
	κ_f	1.0	W/(m. °K)
Kinetic properties:	A_1	1.36×10^{28}	s^{-1}
	A_2	1.25×10^{28}	s^{-1}
	E_1	224.4	kJ/mol
	E_2	223.0	kJ/mol
	m	0.226	
	n	1.856	
	E_α	347	J/g
Chemorheological properties:	a_0	6.41×10^{-5}	N.s/m^2
	b_0	23.1	
	a	32.6	kJ/mol
	b	18.3	kJ/mol

Note: The subscripts r and f denote the resin and fiber, respectively.
Source: From reference 27.

would require the engineer to run hundreds of simulations to guide the most robust design by utilizing optimization algorithms and developing process control scenarios that can counteract flow disturbances due to unpredictable racetracking created by the non-repeatable layup procedure. For such cases, one must run each simulation within seconds requiring reasonable approximations such as 2D flow instead of 3D flow, coarse meshes and fast solution algorithms.[57,60] Currently, non-isothermal solutions with full coupling with cure and energy cannot be run in a few seconds, but isothermal versions with 3000–5000 elements can be executed on a PC in 2–3 s in LIMS.[57]

In the numerical solution of the flow model, Eq. [9.7] and the solution domain are discretized. The part geometry, which is equivalent to the mold cavity, is discretized as a shell mesh in 3D using usually triangular and/or quadrilateral elements in 2D. The most common numerical method is to use a finite element/control volume (FE/CV) approach[1] although boundary element and finite difference methods have been used.[61–65]

1. *Finite element (FE)* is used to solve the pressure distribution in the resin-covered subdomain. The pressure within an element e, P^e is expressed in terms of the unknown nodal pressures, P_i of that element as follows:

$$P^e = \sum_{i=1}^{n} N_i P_i \qquad [9.16]$$

where n is the number of nodes of the element e (three for a triangular element, four for a quadrilateral element and so on), and N_i is the interpolation function. The Galerkin finite element method converts the partial differential equation on resin pressure, Eq. [9.7], into a system of algebraic equations:

$$\left[[S^e]\right][P] = [f] \qquad [9.17]$$

where the components of the element stiffness matrix, $[[S^e]]$ are given by

$$S_{ij}^e = \int_\Omega \left(\frac{K_{xx}}{\mu} \frac{\partial N_j}{\partial x} \frac{\partial N_i}{\partial x} + \frac{K_{yy}}{\mu} \frac{\partial N_j}{\partial y} \frac{\partial N_i}{\partial y} + \frac{K_{xy}}{\mu} \frac{\partial N_j}{\partial x} \frac{\partial N_i}{\partial y} + \frac{K_{xy}}{\mu} \frac{\partial N_j}{\partial y} \frac{\partial N_i}{\partial x} \right) \mathrm{d}\Omega. \qquad [9.18]$$

Equation [9.17] is solved in the resin-covered subdomain (i.e., fully impregnated preform) at each time step of the mold filling. The ith component of $[[S^e]][P]$ vector is the amount of resin volume generated per unit time at the ith node of the mesh. Thus, the corresponding component of the forcing vector, $[f]_i$ is zero for all nodes except for the injection and ventilation nodes. $[f]_i$ is equal to the flow rate at the injection node, which is either a specified boundary condition, or can be calculated by using Darcy's law and the known injection pressures at those nodes.

Material properties (permeability, fiber volume fraction and part thickness) are assigned for each element individually. This allows for simulating the resin flow through spatially non-uniform preform with varying material properties and part thickness.

If the filling is non-isothermal or the viscosity is affected significantly due to the degree of cure during resin flow, the energy equation is also solved, and the volume-averaged resin viscosity $\mu(T, \alpha)$ is updated in the flow equation. The details of the solution for non-isothermal case is given in reference 1.

2. *Control volume (CV)* is used to calculate the Darcy's velocity components and fluxes between the adjacent control volumes to advance the flow front position for one time step as follows. The mesh is divided into small control volumes (i.e., control area times the local part thickness) such that each node has its own control volume bounded by the neighbor elements'

centeroids. The flow rates between the adjacent control volumes are calculated as follows:[1]

$$q_{ij} = -\int_{s_{ij}} h\left(-\frac{1}{\mu}\mathbf{K}\cdot\nabla P\right)\cdot\mathbf{n}\,\mathrm{d}s \qquad [9.19]$$

where q_{ij} is the flow rate from the control volume associated with node i to the control volume associated with node j, s_{ij} is the boundary between the two control volumes, $\left(-\frac{1}{\mu}\mathbf{K}\cdot\nabla P\right)\cdot\mathbf{n}$ is the normal component of the volume-averaged Darcy's velocity at the boundary of the control volume, h is the part thickness and $\mathbf{n}$ is the outward normal unit vector on the boundary.

Time to fill the control volume associated with node j, $\Delta t_{j,\text{fill}}$ is calculated as follows:

$$\Delta t_{j,\text{fill}} = \frac{A_j h\left(1-f_j\right)}{\sum_{i=1}^{n} q_{ij}} \qquad [9.20]$$

for all node j's at or near the flow fronts. Here, A_j is the in-plane surface area of the control volume associated with node j, f_j is the fill factor of that control volume just before this current time step. Nodal fill factors are used to monitor the position of the resin flow front. The fill factor f_j is the fraction of the control volume associated with the node j occupied by the resin. $f_j = 0$ indicates that the preform is completely dry, and $f_j = 1$ implies that the preform is completely impregnated with the resin.[27,29,66]

In this approach, the time increment Δt is not constant. At an instant of the resin flow, Δt is taken as the smallest of the $\Delta t_{j,\text{fill}}$ for node j's at or near the flow fronts:

$$\Delta t = \left(\Delta t_{j,\text{fill}}\right)_{\text{minimum}} \qquad [9.21]$$

so that at least one new control volume is completely filled at this current time step.

This approach ('FE' + 'CV') is repeated in a loop until the mold cavity is filled completely. This approach is based on the marker and cell method.[65,67,68] One of the major advantages of this approach is that it does not require remeshing the solution domain after each time step although the resin-covered subdomain expands with time. That means the mesh is fixed, but the unknown pressures are solved in a subdomain (the impregnated subdomain) only.

9.11.1 Important features of mold filling simulations

A robust resin flow simulation should be able to address the important issues in the RTM process and allow the designer and process engineer to implement a robust manufacturing process in a cost effective manner. These features should include:

- *Design of the mold*: Be able to determine the locations of the injection gates and ventilation ports such that no dry regions (voids) remain at the end of the mold filling. The ventilation ports are placed where resin arrives last when the mold is perfectly vacuumed.
- *Design of the process parameter window*: Should be able to conduct a parametric study of the effect of various disturbances on the flow and determine a process window which will ensure complete filling.
- *Design of multiple gates with timed injectors*: To reduce the injection pressure and/or mold filling time, the gates may be opened/closed sequentially. This approach can be explored manually with multiple simulation runs or coupled with optimization routines such as neural networks or genetic algorithms.[69–71]
- *Resin flow through non-uniform preform* can be handled assigning non-uniform material properties (preform fiber volume fraction and permeability which may vary spatially and resin viscosity which may vary with time and spatially).
- Effect of the *racetracking* channels on the resin flow can be integrated by assigning high permeabilities to the elements along the edges or corners of the mold and/or inserts. This may be done by scaling the permeabilities of those elements by one to three orders of magnitude compared to the bulk permeability of the inner regions depending on the level of racetracking (such as weak or strong). For more accurate simulations, the reader is referred to the literature to calculate the permeability of a racetracking channel with a rectangular cross-section.[1,12,72–74] The permeability $K_{\text{racetracking}}$ of a channel with a width w and thickness of h is calculated by using the following analytical solution:

$$K_{\text{racetracking}} = \frac{h^2}{12}\left[1 - \frac{192h}{\pi^5 w}\sum_{n=1,3,5,\ldots}^{\infty}\frac{\tanh(n\pi w/2h)}{n^5}\right] \qquad [9.22]$$

For example, for channel dimensions of $h = w = 2$ mm, $K_{racetracking} = 1.41 \times 10^{-7}$ m^2, which is much higher than typical bulk permeability values (10^{-9}–10^{-12} m^2) in the RTM applications.

- Simulations should be able to address the slow filling of the tows in textile like fabrics and predict the *delayed impregnation of fiber bundles* as compared to the impregnation of the empty spaces between the tows due to the dual scale permeabilities in the fiber preform. This feature allows one to investigate whether microvoids remain in the preform after the resin injection is discontinued at the end of the mold filling.[57,75]
- Simulation should be able to account for drapability of the fiber preform to investigate any wrinkle formation and also to predict the permeability of the sheared fiber bundles. The permeabilities of the elements should be modified accordingly to accurately reflect the flow physics.[21,76–78]
- If the simulation is to be used to optimize gate and vent locations and for process control, the simulation should be completed in less than a few seconds to allow for execution of hundreds of simulations within reasonable time without the use of parallel processing or super computers. Hence development of fast fill algorithms is essential to further this approach.[57,60]

9.12 Process control

In RTM, process control is necessary and applied by monitoring the resin flow and cure, and making some adjustments in the process parameters to correct the resin flow and/or heating if the inherent variations in the preforming causes significant disturbance on the flow pattern, otherwise uncontrolled injection may result in incomplete mold filling due to void entrapment or exceeding the gelation time before the mold is filled. A robust process control should integrate the following:[79]

- Mold with additional injection gates and vent ports that can be activated if necessary.
- Sensors embedded flush with the mold walls to monitor the resin flow and cure and provide feed-back on the state of flow front location and cure.
- Data acquisition (DAQ) card and computer program to evaluate the *in situ* sensor data and implement a control algorithm based on the state of filling.
- Injection equipment connected to the DAQ system so that the actuators can be controlled automatically implementing the desired controlled action dictated by the state of filling and cure.

- Ability to turn the valves on and off which are connected with injection gates and vents as dictated by the control design formulated with the simulations.
- Ability to implement appropriate control actions as dictated by a decision tree created based on offline simulations that have considered all the possible disturbances.[79–81]
- Ability to integrate online resin flow simulations to decide on appropriate control actions if the *in situ* sensor data correspond to an unexpected disturbance 'scenario'.[57]

9.12.1 Sensor systems for flow and cure monitoring

The following sensor systems can be used for monitoring resin flow and/or resin cure.[82,83]

- *Thermocouples*[84] are used to track the temperature change in the mold cavity which occurs due to the conductive and convective heat transfer between the mold walls and resin for non-isothermal mold filling applications, and also due to the exothermic heat release of the resin cure. If the temperature of the resin is considerably different than the mold walls, the sensor readings will change when the resin wets the sensor, and the data are used to monitor the flow front position by embedding many sensors in the mold cavity. The response rate of the thermocouples is dependent on the temperature differential between the walls and the resin inlet, and thermal diffusivities of the mold material and resin. These sensors cannot come in contact with the resin if they are to be re-used, thus one needs to remove the cured resin on top of the sensor with acetone or some other process.
- *Pressure sensors*[83] measure the pressure in the mold cavity which is the static compaction pressure initially. When the resin flow front reaches a sensor location, the reading changes to static compaction pressure plus the resin pressure. Thus, the increase in the pressure is evaluated as the resin arrival to the sensor location. These data are used to monitor the flow front position by embedding many pressure sensors in the mold cavity.
- Instead of using many point pressure sensors, thin and flexible pressure mapping systems such as Tekscan products[85] can be placed on the mold cavity surface to monitor the resin pressure in a planar domain. Its electrical wire mesh is sandwiched between impermeable films. However, the films should be protected against degradation due to the contact with the resin and its exothermic reaction by using special demolding agents.
- *Resistive sensors*[81,86,87] measure the electrical resistance of the medium between two point or line probes. The medium is dry fiber preform and air initially. The voltage reading of the circuit changes when the resin

arrives at the sensor location since the electrical resistance of the typical RTM resins is relatively lower than the dry glass fibers and air. This sudden change in the resistivity can be detected to signal the arrival of the resin at that location and serves as an inexpensive way to monitor the arrival of the resin. Besides being used for flow monitoring, resistive sensors can also be used for cure monitoring since the resistance of the resin increases as it cures. However, the voltage change due to the resin arrival may be low depending on the resin. The user should be careful about amplifying the output voltage and reducing the electrical noise in the system.

- *SMARTweave*[88–90] uses a mesh of conductive thin wires embedded in between the non-conductive dry plies of the fiber preform. The conductivity of the medium at the junction of two orthogonal wires increases when the resin flow front reaches that location. The increase in the effective conductivity of the junction will change the readings of the electrical circuit. The major three disadvantages of this type of sensor are (1) tedious and time-consuming labor for sensor placement, (2) disturbance to the resin flow and (3) being a permanent insert in the finished composite part which may cause stress concentrations. The shortcomings are overcome by employing the same concept for a point sensor which is embedded in the mold walls.
- *Ultrasonic sensors*[91,92] send out sound waves from one side of the mold. The phase and magnitude of the reflected waves change when the dry fiber preform is impregnated with resin, thus the flow front position can be monitored.
- *Fiberoptic sensors*[93,94] are placed between the plies of the fiber preform, and they transmit light from one side of the mold. The intensity and delay of the light is measured on the other side. The intensity decreases when the dry fiber preform is impregnated with resin, thus the flow front position can be monitored by using many fiberoptic sensors embedded in the preform. These sensors can be later used for 'structural health-monitoring' of the composite part during the service time to track any changes in the structural loading and damage detection.
- *Electric time-domain reflectometry (E-TDR) wire sensors*[95,96] apply high speed pulses from one side of the wire. The delay in the reflection signals is dependent on the impedance of the sensing medium around the wire, thus the recorded data are used to monitor the regions of the dry and impregnated fiber preform. This method allows sensing multiple and separate wetted regions on a sensor, thus multiple flow fronts can be monitored.
- *Dielectric sensors*[82,97–104] measure the capacitance of the medium between the sensor plates. A 'parallel-plate' type sensor is placed on two sides of the mold (i.e., on the top and bottom of the fiber preform), and 'coplanar' type sensor is placed on only one side of the mold. The effective

capacitance of the dry fiber and air is different than the impregnated fibers with resin, thus the sensor data are used to monitor the resin flow. Additionally, the capacitance of the resin changes as it cures. Thus, this sensor can also be used for cure monitoring.

9.13 Conclusions and future trends

In this chapter we have addressed various issues that when addressed can make RTM more competitive with compression and injection molding for high volume applications. Various guidelines and process modeling tools have been introduced to tackle and overcome these issues. In the last two decades, progress has been noticed in the industry where the manufacturing and processing engineers are comfortable using science-based tools such as flow simulations to design robust processes with fewer trials which has mitigated the cost and moved the industry toward lean manufacturing by spending less time and money from design to prototype development.

The next step we see in RTM is to introduce passive and active control which can account for part to part variability and reduce the scrap rate and make the process more reliable and attract manufacturers to consider lightweight composites as a serious alternative to stamped aluminum parts. There are control algorithms available which in unison with flow simulations (offline in which one anticipates the differences and creates a decision tree based on the disturbance that can be identified with sensors, and online simulations which try to correct the flow during injection) can be integrated in an automated environment which will make RTM competitive for high volume production. This methodology has been demonstrated in a laboratory environment and hopefully will be adopted by some industries to reap the benefits. Until then RTM will serve as prime candidate for low volume specific design components.

9.14 References

1. Advani, S. G. and Sozer, E. M., *Process modeling in composite manufacturing*, 2nd edn. Boca Raton: CRC Press, Taylor and Francis Group, 2011.
2. Walsh, S. M., Rigas, E. J., Spurgeon, W. A., Roy, W. N., Heider, D. and Gillespie, J., 'A non-contact distribution scheme for promoting and controlling resin flow for VARTM processes'. In *International SAMPE Technical Conference*, Boston, MA, 5–9 November 2000, pp. 284–293.
3. Bhat, P., Merotte, J., Simacek, P. and Advani, S. G., 'Process analysis of compression resin transfer molding', *Composites Part A: Applied Science and Manufacturing*, **40**(4), 431–441, 2009.
4. Pillai, K. M., Tucker, C. L. and Phelan, F. R., 'Numerical simulation of injection/compression liquid composite molding. Part 2: Preform compression', *Composites Part A: Applied Science and Manufacturing*, **32**(2), 207–220, 2001.

5. Yenilmez, B. and Sozer, E. M., 'Compaction of E-glass fabric preforms in the vacuum infusion process. A: Characterization experiments', *Composites Part A: Applied Science and Manufacturing*, **40**(4), 499–510, 2009.
6. Rudd, C. D., Long, A. C., Kendall, K. N. and Mangin, C. G. E., *Liquid molding technologies*. Cambridge: SAE International, Woodhead Publishing, 1997.
7. Radius Engineering, Inc., 3474 So. East, Salt Lake City, Utah 84109. Web page: www.radiusengineering.com. *Radius FloWare TM 2500cc RTM Injector Operation & Maintenance Manual Version: 1.4.99*, 1999.
8. Composite Integration. http://www.composite-integration.co.uk
9. Magnum Venus Plastech Limited. http://www.plastech.co.uk
10. Liquid Control Corp., 7576 Freedom Ave. N.W., P.O. Box 2747, North Canton, Ohio 44720. Web page: www.liquidcontrol.com. *Meter, Mix and Dispense Equipment Engineering Handbook*, 1996.
11. Bickerton, S. and Advani, S. G., 'Characterization of racetracking in liquid composite molding processes', *Composites Science and Technology*, **59**, 2215–2229, 1999.
12. Bickerton, S., 'Modeling and control of flow during impregnation of heterogeneous porous media, with application to composite mold filling processes', PhD thesis, University of Delaware, Newark, DE, 1999.
13. Bickerton, S. and Advani, S. G., 'Characterization of corner and edge permeabilities during mold filling in resin transfer molding'. In *Proceedings of the ASME AMD-MD Summer Meeting*, 56, Los Angeles, CA, June 1995, pp. 143–150.
14. Bickerton, S. and Advani, S. G., 'Experimental investigation and flow visualization of the resin transfer molding process in a non-planar geometry', *Composites Science and Technology*, **57**(1), 23–33, 1997.
15. Bickerton, S., Advani, S. G., Mohan, R. V. and Shires, D. R., 'Experimental analysis and numerical modeling of flow channel effects in resin transfer molding', *Polymer Composites*, **21**, 134–153, 2000.
16. Bickerton, S., Simacek, P., Pillai, K. M., Mogavero, J. and Advani, S. G., 'Important mold filling issues in liquid composite molding processes: Modeling and experiments'. In *Annual Technical Conference – ANTEC, Materials, Proceedings of the 55th Annual Technical Conference*, volume 2, 27 April—2 May 1997, pp. 2411–2423.
17. Sozer, E. M., 'Effect of preform non-uniformity on mold filling in RTM process'. In *33rd International SAMPE Technical Conference – Advancing Affordable Materials Technology*, Seattle, WA, 5–8 November 2001, pp. 176–189.
18. Han, K., Lee, B. and Rice, B., 'Permeability measurements of fiber preforms and applications'. In *Mechanical Behavior of Advanced Materials, Proceedings of the 1998 ASME International Mechanical Engineering Congress and Exposition*, Anaheim, CA, 15–20 November 1998, p. 275.
19. Bickerton, S., Sozer, E. M., Graham, P. J. and Advani, S. G., 'Fabric structure and mold curvature effects on preform permeability and mold filling in the RTM process. Part I: Experiments', *Composites Part A: Applied Science and Manufacturing*, **31**(5), 423–438, 2000.
20. Bickerton, S., Sozer, E. M., Simacek, P. and Advani, S. G., 'Fabric structure and mold curvature effects on preform permeability and mold filling in the RTM process. Part II: Predictions and comparisons with experiments', *Composites Part A: Applied Science and Manufacturing*, **31**(5), 439–458, 2000.
21. Bickerton, S., Simacek, P., Guglielmi, S. E. and Advani, S. G., 'Investigation of draping and its effects on the mold filling process during manufacturing of a compound

curved composite part', *Composites Part A: Applied Science and Manufacturing*, **28A**, 801–816, 1997.

22. http://www.vistagy.com/products/fibersim-composite environments.aspx
23. Parnas, R. S., Howard, J. G., Luce, T. L. and Advani, S. G., 'Permeability characterization. Part 1: A proposed standard reference material for permeability', *Polymer Composites*, **16**(6), 430–446, 1996.
24. Ahn, S. H., Lee, W. I. and Springer, G. S., 'Measurement of the three-dimensional permeability of fiber preforms using embedded fiber optic sensors', *Journal of Composite Materials*, **29**(6), 714–733, 1995.
25. Ballata, B., Walsh, S. and Advani, S. G., 'Measurement of the transverse permeability of fiber preforms', *Journal of Reinforced Plastics and Composites*, **18**, 1450–1464, 1999.
26. Patel, N., 'Micro scale flow behavior, fiber wetting and void formation in liquid composite molding', PhD thesis, Ohio State University, 1997.
27. Advani, S. G., Bruschke, M. V. and Parnas, R., 'Resin transfer molding'. In S. G. Advani (ed.), *Flow and rheology in polymeric composites manufacturing*. Amsterdam: Elsevier Publishers, 1994, pp. 465–516.
28. Lee, L. J., 'Liquid composite molding'. In T. G. Gutowski (ed.), *Advanced composites manufacturing*. New York: John Wiley, 1997, Chapter 10.
29. Advani, S. G. and Simacek, P., 'Modeling and simulation of flow, heat transfer and cure'. In T. Krukenberg and R. Paton (eds.), *Resin transfer molding for aerospace structures*. Dordrecht: Kluwer Academic Publishers, 1999, pp. 225–281.
30. Parnas, R. S., *Liquid composite molding*. Munich: Hanser Publishers, 2000.
31. Darcy, H., *Les fontaines publiques de la ville de Dijon*. Paris: Dalmont, 1856.
32. Binetruy, C., Hilaire, V. and Pabiot, J., 'The interactions between flows occurring inside and outside fabric tows during RTM', *Composites Science and Technology*, **57**, 587–596, 1997.
33. Patel, N., Rohatgi, V. and Lee, J. L., 'Micro scale flow behavior and void formation mechanism during impregnation through a unidirectional stitched fiberglass mat', *Polymer Engineering and Science*, **35**(10), 837–851, May 1995.
34. Phelan, Jr., F. R., Leung, Y. and Parnas, R. S., 'Modeling of microscale flow in unidirectional fibrous porous media', *Journal of Thermoplastic Composite Materials*, **7**(3), 208–218, 1994.
35. Amico, S. and Lekakou, C., 'Mathematical modeling of capillary micro-flow through woven fabrics', *Composites Part A: Applied Science and Manufacturing*, **31**(12), 1331–1344, 2000.
36. Pillai, K. M. and Advani, S. G., 'Numerical simulation of unsaturated flow in woven fiber preforms during resin transfer molding process', *Polymer Composites*, **19**(1), 71–80, 1998.
37. Slade, J., Pillai, K. and Advani, S. G., 'Investigation of unsaturated flow in woven, braided and stitched fiber mats during mold-filling in resin transfer molding', *Polymer Composites*, **22**, 491–505, 2001.
38. Zhou, F., Alms, J. and Advani, S. G., 'A closed form solution for flow in dual scale fibrous porous media under constant injection pressure conditions', *Composites Science and Technology*, **68**(3–4), 699–708, 2008.
39. Zhou, F., Kuentzer, N., Simacek, P., Advani, S. G. and Walsh, S., 'Analytic characterization of the permeability of dual-scale fibrous porous media', *Composites Science and Technology*, **66**(15), 2795–2803, 2006.
40. Kuentzer, N., Simacek, P., Advani, S. G. and Walsh, S., 'Permeability characterization of dual scale fibrous porous media', *Composites Part A: Applied Science and Manufacturing*, **37**, 2057–2068, 2006.

41. Sozer, E. M., Chen, B., Graham, P. J., Chou, T.-W. and Advani, S. G., 'Characterization and prediction of compaction force and preform permeability of woven fabrics during the resin transfer molding process'. In *Proceedings of 5th International Conference on Flow Processes in Composite Materials*, July 1999, pp. 25–36.
42. Stadtfeld, H. C., Erninger, M., Bickerton, S. and Advani, S. G., 'An experimental method to continuously measure permeability of fiber preforms as a function of fiber volume fraction', *Journal of Reinforced Plastics Composites*, **21**, 879–899, 2002.
43. Yenilmez, B., Senan, M. and Sozer, E. M., 'Variation of part thickness and compaction pressure in vacuum infusion process', *Composites Science and Technology*, **69**(11–12), 1710–1719, 2009.
44. Arbter, R., Beraud, J. M., Binetruy, C., *et al.* 'Experimental determination of the permeability of textiles: A benchmark exercise', *Composites: Part A: Applied Science and Manufacturing*, **42**, 1157–1168, 2011.
45. Lomov, S. V., Huysmans, G., Luo, Y., Parnas, R. S., Prodromou, A., Verpoest, I. and Phelan, F. R., 'Textile composites: modelling strategies', *Composites Part A: Applied Science and Manufacturing*, **32**(10), 1379–1394, 2001.
46. Chiu, H.-T., Yu, B., Chen, S. C. and Lee, L. J., 'Heat transfer during flow and resin reaction through fiber reinforcement', *Chemical Engineering Science*, **55**(17), 3365–3376, 2000.
47. Tucker III, C. L. and Dessenberger, R. B., 'Governing equations for flow and heat transfer in stationary fiber beds', In S. G. Advani (ed.), *Flow and rheology in polymer composites manufacturing*. Amsterdam: Elsevier, 1994, pp. 257–324
48. Kaviany, M., *Principles of heat transfer in porous media*. New York: Springer, 1995.
49. Advani, S. G. and Hsiao, K.-T., 'Heat transfer during mold filling in liquid composite manufacturing'. In K. Vafai (ed), *Handbook of porous media*. New York: Marcel Dekker, 2000, pp. 845–891.
50. Hsiao, K.-T., Laudorn, H. and Advani, S. G., 'Experimental investigation of heat dispersion due to impregnation of viscous fluids in heated fibrous porous during composites processing', *ASME Journal of Heat Transfer*, **123**, 178–187, 2001.
51. Dessenberger, R. B. and Tucker III, C. L., ' Thermal dispersion in resin transfer molding', *Polymer Composites*, **16**(6), 495–506, 1995.
52. Hsiao, K.-T. and Advani, S. G., 'A method to predict microscopic temperature distribution inside a periodic unit cell of non-isothermal flow in porous media', *Journal of Porous Media*, **5**, 69–86, 2002.
53. Chiu, H.-T., Chen, S.-C. and Lee, L. J. 'Analysis of heat transfer and resin reaction'. In *Proceedings of the 55th Annual Technical Conference, ANTEC, Part 2*. Toronto, Canada, 27 April–2 May 1997, pp. 2424–2429.
54. Bruschke, M. V. and Advani, S. G., 'A numerical approach to model non-isothermal viscous flow through fibrous media with free surfaces', *International Journal for Numerical Methods in Fluids*, **19**, 575–603, 1994.
55. Kamal, M. R., Sourour, S. and M. Ryan. Integrated thermo-rheological analysis of the cure of thermosets. *SPE Technical Paper*, 18:187–191, 1973.
56. Kamal, M. R. and Sourour, S., 'Kinetics and thermal characterization of thermoset cure', *Polymer Engineering and Science*, **13**, 59–64, 1973.
57. Simacek, P. and Advani, S. G., 'Desirable features in mold filling simulations for liquid molding processes', *Polymer Composites*, **25**, 355–367, 2004.
58. PAM-RTM. http://www.esi-group.com/products/composites-plastics/pam-rtm
59. RTM-Worx. Polyworx, http://www.polyworx.com/apz/

60. Maier, R. S., Rohaly, T. F., Advani, S. G. and Fickie, K. D., 'A fast numerical method for isothermal resin transfer mold filling', *International Journal of Numerical Methods in Engineering*, **39**, 1405–1422, 1996.
61. Guceri, S. I., 'Finite difference solution of field problems'. In C.L. Tucker, III (ed.), *Fundamentals of computer modeling for polymer processing.* Munich: Hanser Publishers, 1989, Chapter 5.
62. Barone, M. R. and Osswald, T. A., 'Boundary element solution for field problems'. In C. L. Tucker, III (ed.), *Fundamentals of computer modeling for polymer processing.* Munich, Germany: Hanser Publishers, 1989, Chapter 7.
63. Kwang, U. M. and Il, L. W., 'Numerical simulation of the resin transfer molding process using the boundary element method'. In *35th International SAMPE Symposium*, 1992.
64. Trochu, F. and Gauvin, R., 'Limitations of a boundary-fitted finite difference method for the simulation of the resin transfer molding process', *Journal of Reinforced Plastics and Composites*, **11**, 772–786, 1992.
65. Hieber, C. A. and Shen, S. F., 'A finite element/finite difference simulation of the injection mold filling process', *Journal of Non-Newtonian Fluid Mechanics,* **7**(1), 1–31, 1980.
66. Bruschke, M. V. and Advani, S. G., 'A finite element/control volume approach to mold filling in anisotropic porous media', *Polymer Composites*, **11**(6), 398–405, 1990.
67. Osswald, T. A. and Tucker, C. L., 'An automated simulation of compression mold filling for complex parts'. In *Proceedings of 43rd SPE ANTEC*, Washington, D.C., 1985, pp. 169–172.
68. Shen, S. F., 'Simulation of polymer flows in the injection molding process', *International Journal for Numerical Methods in Fluids*, **4**, 171–183, 1984.
69. Mathur, R., Advani, S. G. and Fink, B. K., 'Use of genetic algorithms to optimize gate and vent locations for the resin transfer molding process', *Polymer Composites*, **20**, 167–178, 1999.
70. Mathur, R., Advani, S. G. and Fink, B. K., 'A sensitivity-based gate location algorithm for optimal mold filling during the resin transfer molding process', *Advances in Computational Engineering and Sciences*, **1**, 138–144, 2000.
71. Mathur, R., Advani, S. G. and Fink, B. K., 'A real-coded hybrid genetic algorithm to determine optimal resin injection locations in the resin transfer molding process', *Computer Modeling in Engineering and Sciences*, **4**(5), 587–602, 2003.
72. Advani, S. G. and Sozer, E. M., 'Resin impregnation in liquid molding processes'. In A. Kelly (ed.), *Comprehensive composite materials.* Amsterdam: Elsevier Science, 2000, Chapter 24.
73. Fong, L. L. and Advani, S. G., 'Preforming analysis of thermoformable fiber mats – Preforming effect on mold filling', *Journal of Reinforced Plastics and Composites*, **13**(7), 637–663, 1994.
74. Tadmor, Z. and Gogos, C. G., *Principles of polymer processing*. New York: John Wiley, 1979.
75. P. Simacek, Neacsu, V. and S. G. Advani. A phenomenological model for fiber tow saturation of dual scale fabrics in liquid composite molding. *Polymer Composites*, **31**(11), 1881–1889, 2010.
76. Bickerton, S., Advani, S. G., Fickie, K. and Fong, L. H., 'Effect of draping of fiber preforms on process parameters during manufacturing with resin transfer molding'. In *11th Annual ESD Advanced Composites Conference and Exposition*, Dearborn, MI, November 1995, pp. 213–220.

77. Sozer, E. M., Simacek, P. and Advani, S. G., 'Draping and filling of long axisymmetric molds in RTM process'. In *Proceedings of the 8th Japan–U.S. Conference on Composite Materials*, Baltimore, MD, September 1998, pp. 73–84.
78. Indermaur, M., Simacek, P. and Advani, S. G., 'The influence of draping on the filling of doubly curved molds in the resin transfer molding process', *International Journal of Forming Processes*, **2**, 321–352, 2000.
79. Sozer, E. M., Bickerton, S. and Advani, S. G., 'On-line strategic control of liquid composite mold filling process', *Composites Part A: Applied Science and Manufacturing*, **31**(12),1383–1394, 2000.
80. Lawrence, J. M., Sozer, E. M., Stadtfeld, H., Simacek, P., Estrada, G., Don, R. and Advani, S. G., 'Use of sensors and simulations for process control to manufacture a feature based composite with resin transfer molding process'. In *Proceedings of the Society for the Advancement of the Material and Process Engineering (SAMPE)*, Long Beach, CA, 2000, pp. 146–155.
81. Lawrence, J. M., Hsiao, K. T., Don, R. C., Simacek, P., Estrada, G., Sozer, E. M., Stadtfeld, H. C. and Advani, S. G., 'An approach to couple mold design and online control to manufacture complex composite parts by resin transfer molding', *Composites Part A: Applied Science and Manufacturing*, **33**(7), 981–990, 2002.
82. Yenilmez, B. and Sozer, E. M., 'A grid of dielectric sensors to monitor mold filling and resin cure in resin transfer molding', *Composites Part A: Applied Science and Manufacturing*, **40**, 476–489, 2009.
83. Heider, D., Don, R., Thostensen, E. T., Tackitt, K., Belk, J. H. and Munns, T., 'Cure monitoring and control'. In *ASM Handbook, Volume 21, Composites*, ASM International, Materials Park, OH, 2001.
84. Tuncol, G., Danisman, M., Kaynar, A. and Sozer, E. M., 'Constraints on monitoring resin flow in the resin transfer molding (RTM) process by using thermocouple sensors', *Composites Part A: Applied Science and Manufacturing*, **38**(5), 1363–1386, 2007.
85. Products for pressure mapping and force measurement. Tekscan. http://www.tekscan.com/products
86. Danisman, M., Tuncol, G., Kaynar, A. and Sozer, E. M., 'Monitoring of resin flow in the resin transfer molding (RTM) process using point-voltage sensors', *Composites Science and Technology*, **67**(3–4), 367–379, 2007.
87. Barooah, P., Berker, B. and Sun, J. Q., 'Lineal sensors for liquid injection molding of advanced composite materials', *Journal of Materials Processing and Manufacturing Science*, **6**(3), 169–184, 1999.
88. S. Walsh. *In situ* sensors method and device. US Patent 5,210,499, May 1993.
89. Fink, B. K., Walsh, S. M., DeSchepper, Jr., D. C., Gillespie, J. W., McCullough, R. L., Don, R. C. *et al.*, 'Advances in resin transfer molding flow monitoring using SMARTweave sensors'. In *Proceedings of ASME, International Mechanical Engineering Congress and Exposition*, San Francisco, California, 1995, pp. 999–1015.
90. Bradley, J. E., Diaz-Perez, Jr., J., Gillespie, J. W. and Fink, B. K., 'On-line process monitoring and analysis of thick-section composite parts utilizing smartweave *in situ* sensing technology'. In *Proceedings of International SAMPE Symposium and Exhibition*, Long Beach, CA, 1998, pp. 254–267.
91. Rath, M., Doring, J., Stark, W. and Hinrichsen, G., 'Process monitoring of moulding compounds by ultrasonic measurements in a compression mould', *NDT&E International*, **33**(2), 123–130, 2000.

92. Schmachtenberg, E., zur Heide Schulte, J. and Topker, J., 'Application of ultrasonics for the process control of resin transfer moulding (RTM)', *Polymer Testing*, **24**(3), 330–338, 2005.
93. S.T. Lim and W. Lee. An analysis of the three-dimensional resin-transfer mold filling process. *Composites Science and Technology*, **60**(7), 961–975, 2000.
94. Dunkers, J. P., Lenhart, J. L., Kueh, S. R., van Zanten, J. H., Advani, S. G. and Parnas, R. S., 'Fiberoptic flow and cure sensing for liquid composite molding', *Optics and Lasers in Engineering*, **35**(2), 91–104, 2001.
95. Dominauskasa, A., Heider, D. and Gillespie, J. W., 'Electric time-domain reflectometry sensor for online flow sensing in liquid composite molding processing', *Composites Part A: Applied Science and Manufacturing*, **34**(1), 67–74, 2003.
96. Dominauskasa, A., Heider, D. and Gillespie, J. W., 'Electric time-domain reflectometry distributed flow sensor', *Composites Part A: Applied Science and Manufacturing*, **38**(1), 138–146, 2003.
97. McIlhagger, A., Brown, D. and Hill, B., 'The development of a dielectric system for the on-line cure monitoring of the resin transfer molding process', *Composites Part A: Applied Science and Manufacturing*, **31**(12), 1373–1381, 2000.
98. Hegg, M. C., Ogale, A., Mescher, A., Mamishev, A. V. and Minaie, B., 'Remote monitoring of resin transfer molding processes by distributed dielectric sensors', *Journal of Composite Materials*, **39**(17), 1519–1539, 2005.
99. Kim, H. G. and Lee, D. G., 'Dielectric cure monitoring for glass/polyester prepreg composites', *Composite Structures*, **57**(1–4), 91–99, 2002.
100. Bang, K. G., Kwon, J. W., Lee, D. G. and Lee, J. W., 'Measurement of the degree of cure of glass fiber epoxy composites using dielectrometry', *Journal of Materials Processing Technology*, **113**(1–3), 209–214, 2001.
101. Vaidya, U. K., Jadhav, N. C., Hosur, M. V., Gillespie, J. W. and Fink, B. K., 'Assessment of flow and cure monitoring using direct current and alternating current sensing in vacuum-assisted resin transfer molding', *Smart Materials and Structures*, **9**, 727–736, 2000.
102. Rowe, G. I., Yi, J. H., Chiu, K. G., Tan, J., Mamishev, A. V. and Minaie, B., 'Fill-front and cure progress monitoring for VARTM with auto-calibrating dielectric sensors'. In *SAMPE 2005 Symposium and Exhibition*, Long Beach, CA, May 2005.
103. Skordos, A. A., Karkanas, P. I. and Partridge, I. K., 'A dielectric sensor for measuring flow in resin transfer moulding', *Measurement Science and Technology*, **11**(1), 25–31, 2000.
104. Mounier, A. L., Binetruy, C. and Krawczak, P., 'Multipurpose carbon fiber sensor design for analysis and monitoring of the resin transfer molding of polymer composites', *Polymer Composites*, **26**(5), 717–730, 2005.

10
Vacuum assisted resin transfer molding (VARTM) in polymer matrix composites

K.-T. HSIAO, University of South Alabama, USA and
D. HEIDER, University of Delaware, USA

Abstract: The vacuum assisted resin transfer molding (VARTM) process is a closed-mold process that is capable of manufacturing high-performance and large-scale fiber reinforced polymer (FRP) parts with low tooling cost. This chapter is divided into two parts: the first providing an overview of the VARTM process and the second highlighting the membrane-based VARTM process variation used in aerospace applications. In particular, the chapter explains the physics including mold filling, cure kinetics, fiber preform compaction, spring-in, dry spot and microvoid formation. Reported studies to address the related processing issues and a collection of useful analytical models are reviewed. The development trends and emerging uses of VARTM in nanocomposites and the green FRP are also covered. The second part highlights the membrane-based infusion process that is a promising variation of VARTM. The additional membrane, which is permeable to gas and impermeable to the resin, facilitates the improved means to avoid dry spot formation and to continuously degas the resin during the infusion and curing stages compared to conventional VARTM.

Key words: vacuum assisted resin transfer molding (VARTM), vacuum-assisted process (VAP), modeling, defects, compaction.

10.1 Vacuum assisted resin transfer molding (VARTM) processing

The vacuum assisted resin transfer molding (VARTM) process, which has been developed during the past two decades and has many process variants (e.g., SCRIMP process[1]), is now a widely used process for manufacturing fiber reinforced polymer (FRP) composite laminates. The VARTM process, which is a closed-mold process with reduced volatile organic compounds (VOC) emission, combines the benefits of high quality, repeatability and clean handling of the resin transfer molding (RTM) process with the advantages of flexibility and scalability of open-mold hand layup processing. The VARTM process plays many important roles in promoting the quality, affordability and part complexity of large closed-mold FRP composite structures. VARTM processes are widely used in the marine, energy, infrastructure building, aerospace and defense industries. Many variations of VARTM have also been developed recently for manufacturing more complex composite parts with improved quality and lower cost.

10.1.1 Setup and procedure of a typical VARTM process

Figure 10.1 illustrates a commonly used VARTM setup. For a VARTM process, the environmental pressure (e.g., the atmospheric pressure) is typically utilized to provide the compressive pressure against the fiber preform

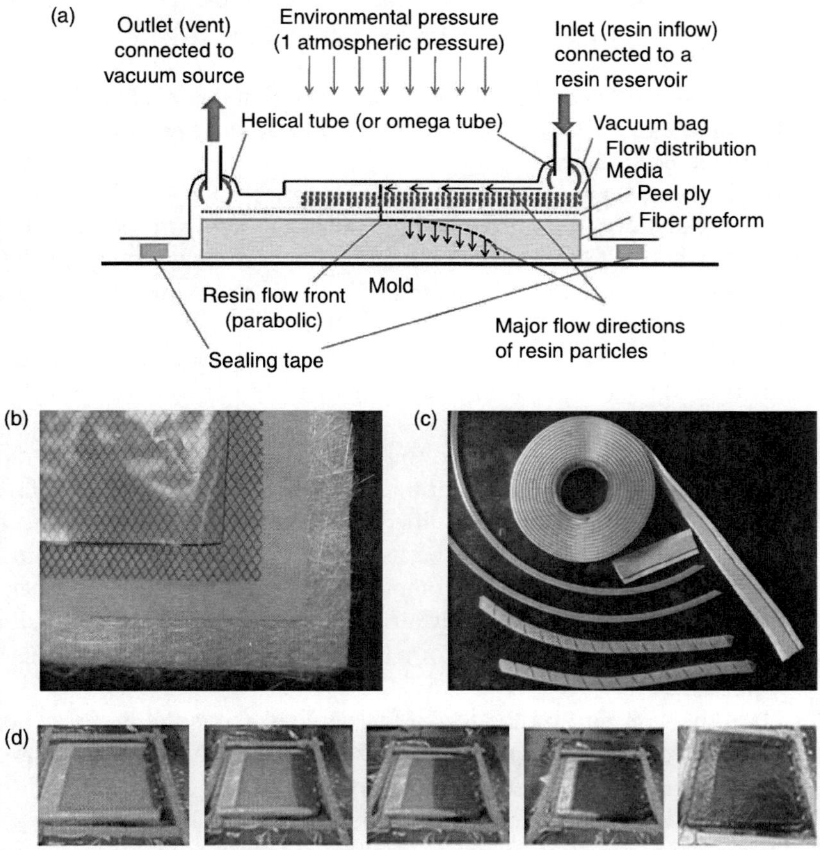

10.1 A typical VARTM process: (a) VARTM setup; (b) a picture showing a random glass fiber mat, a peel ply, a flow distribution layer and a vacuum bag (from bottom to top); (c) a picture showing the sealing tap roll, sealing tap, resin tubes and helical tubes (from top to bottom); (d) photos recorded during the resin infusion (from left to right).

assembly (i.e., fiber preform/peel ply/flow distribution media/helical tubes) that is sealed and sandwiched between the vacuum bag and the VARTM mold as shown in Fig. 10.1. As a result, the VARTM mold, which looks identical to the open mold of a hand layup process, can be constructed with much larger dimensions than an RTM mold. Note that the RTM process utilizes a positive injection pressure that is significantly higher than the environmental pressure and its mold thickness (or the rigidity) must be significantly increased to overcome the mold deflection for manufacturing a large composite part. Different to the RTM process, the VARTM process uses the pressure difference between the vacuum pressure and the environmental pressure to compress the preform, secure the preform against the mold and draw the resin into the preform.

The major processing steps of a typical VARTM process (also see Fig. 10.1) include:

1. Clean the mold and apply mold release (wax) on the mold surface.
2. Lay up the dry fiber preform, which could also be layers of dry fabrics, on the mold surface.
3. Apply the peel ply to cover the fiber preform.
4. Apply the flow distribution medium layer on the top of the peel ply as necessary. The flow distribution medium layer can help to enhance the resin infusion speed and is commonly used for manufacturing large composite laminate parts. Note that the flow distribution medium layer will later be connected to the resin injection port; and the flow distribution medium layer must not directly contact with the vent port.
5. Place the resin injection port on one end of the flow distribution medium layer (or on the preform if the flow distribution medium layer is not used). A helical open tube (like a helical spring shape) or an omega-shaped tube can also be used as a resin injection line source, which serves the purpose of promoting fast resin supply in the helical or omega tube and simultaneously infusing the resin into the flow distribution medium layer (or the preform if the flow distribution medium layer is not used).
6. Place the vent port on the top of the peel ply above the preform; note that the vent must be placed slightly away from the flow distribution medium layer to avoid direct competition against the dry fiber preform for the resin supplied from the distribution medium layer.
7. Apply the sealing tape, which is a double-sided tacky tape for adhering to the mold surface and the vacuum bag together, surrounding the preform assembly.
8. Carefully lay up the vacuum bag on the assembly and secure it against the sealing tape on the mold.

9. Connect the vent tube and the injection tube to the vent port and the injection port, respectively. Connect the vent tube and the injection tube to a vacuum source and a resin reservoir, respectively. Do not fill the resin into the reservoir at this time.
10. Close the injection port and open the vacuum port to apply the vacuum inside the bagged preform assembly. Carefully check for and fix any air leakage.
11. Apply debulking process (optional) by cyclically compressing and relaxing the preform to better compact the fiber preform.
12. Fill the resin into the resin reservoir. Keep the vacuum port on. Open the resin injection port to allow the resin to be drawn into the vacuum bagged fiber preform assembly.
13. One should notice that the resin flows quickly through the flow distribution medium layer and gradually infuses into the fiber preform in the thickness direction (see Fig. 10.1). As the length of a VARTM part is usually much larger than its thickness, the flow distribution medium layer will greatly accelerate the resin infusion process.
14. Once the resin reaches the vent, allow some extra resin to be bled out for a few more minutes to remove the tiny air bubbles in the resin flow front.
15. Close the injection port and keep the vacuum port open until the resin cures into the solid phase. The vacuum will keep the preform assembly tightly pressed against the mold and will also maintain the uniform compressive pressure on the preform to create a composite part with a uniform thickness (i.e., a uniform compression ratio or a uniform fiber volume fraction).
16. Once the resin fully cures into the solid phase, one can turn off the vacuum and demold the composite part from the mold.

Note that depending on the resin system used in VARTM, the mold temperature may need to be elevated during the curing cycle of the VARTM process. For a large or complex composite part (with inserts, hybrid fabric systems, co-cured parts, etc.), multiple injection lines and vents could be used to improve the resin infusion. The flow distribution medium layer could also be placed in different patterns to create versatile resin infusion paths that can promote the resin infusion quality of a large or complex composite part.

10.1.2 Major advantages and disadvantages of VARTM

The VARTM process was originally developed for manufacturing high quality and large composite parts such as ship structures. It utilizes the pressure difference between the environmental pressure and the vacuum pressure to achieve the unique requirements. The advantages and disadvantages of the VARTM process are summarized below.

Advantages:

- Flexible mold tooling design and selection of mold materials.
- Able to manufacture large and complex composite parts with good quality.
- A VARTM mold, which is similar to the open mold of a hand layup process, can be easily modified for manufacturing different part geometries.
- The resin and the catalyst can be stored separately and mixed just before the resin infusion.
- With a transparent plastic vacuum bag, a visible dry spot occurring during the resin infusion process can be removed by inserting a vacuum needle at the dry spot and drawing the air out.
- Low VOC (i.e., VOC) emission. The resin mixing process is the only step with major VOC emission.

Disadvantages:

- Vacuum bag, flow distribution medium, peel ply, sealing tape and resin tubing may not be reusable. These consumables will need to be prepared for each individual VARTM process every time.
- Chance of air leakage is high and this strongly depends on the worker's skill, experience and the consumable (sealing tape, vacuum bag, etc.) quality of each VARTM process. The air leakage can cause dry spot and incomplete resin infusion. A careful and frequent inspection for the air leakage is necessary before the resin infusion, during the infusion resin and during the curing cycle as a leakage can be initiated at any time during these three processing stages and ruin the composite part.
- The resin injection pressure is limited between the environmental pressure (e.g., the atmospheric pressure) and the vacuum. The resin injection pressure of a VARTM process is much less than the pressure applied during a typical RTM process or an autoclave/vacuum bagging process, and can limit the air void compressibility.
- The compressive pressure on the preform is limited between the environmental pressure (e.g., 1 atmospheric pressure) and the vacuum. A lower compressive pressure on the fiber preform can limit the fiber volume fraction of the composite part. Typical fiber volume fraction achieved by VATRM is within the low 40% to high 50% range and mainly depends on the fiber preform used.

Overall, the VARTM process has its strengths in flexibility and can be used for manufacturing complex and large composite parts with good quality.

However, it requires more preparation and is less robust than the RTM process and the autoclave/vacuum bagging process.

10.2 Fundamentals of VARTM

10.2.1 Resin flow phenomenon

The resin flow within the distribution medium layer and fiber preform can be treated as flow through anisotropic porous media[2] and described by the generalized Darcy's law:[3]

$$\vec{u}_{\mathrm{D}} = -\frac{\underline{\underline{K}}}{\mu} \cdot \nabla P \qquad [10.1]$$

where $\vec{u}_{\mathrm{D}}$ is the Darcy velocity (which is the volume averaged velocity with respect to a small control volume containing both the solid phase porous medium and the fluid inside it), μ is the dynamic viscosity of the fluid (i.e., the liquid resin or air), P is the fluid pressure and $\underline{\underline{K}}$ is the permeability tensor for the stationary porous media. Equation [10.1] can precisely quantify the relation between the Darcy velocity and the pressure for a resin saturated porous medium. For the liquid resin in the solid porous medium, it further requires the mass flow continuity for the incompressible fluid and solid system as:

$$\nabla \cdot \vec{u}_{\mathrm{D}} = 0 \qquad [10.2]$$

Combining Eqs. [10.1] and [10.2] for a resin saturated porous medium, one obtains:

$$0 = \nabla \cdot \left(\frac{\underline{\underline{K}}}{\mu} \cdot \nabla P \right) \qquad [10.3]$$

By specifying the boundary conditions of the pressure for the resin filled porous medium domain, the pressure distribution inside the resin filled porous medium domain can be solved by using Eq. [10.3]. Then the Darcy velocity ($\vec{u}_{\mathrm{D}}$) distribution in the resin saturated porous medium domain can be solved by using Eq. [10.1].

During the mold filling process, the resin flow front is indeed a moving boundary. At the resin flow front, there are two significant velocities, that is, the Darcy velocity $\vec{u}_{\mathrm{D}}$ and the flow front velocity $\vec{u}_{\mathrm{F}}$. The flow front velocity is the phase-volume averaged velocity of the liquid resin (with respect to the control volume of the liquid phase only) and is related to the Darcy velocity as:

$$\vec{u}_{\mathrm{F}} = \frac{\vec{u}_{\mathrm{D}}}{\varphi} \qquad [10.4]$$

where φ is the porosity of the porous medium. The porosity can be related to the fiber volume fraction V_f of the fiber preform as:

$$V_f + \varphi = 1 \qquad [10.5]$$

Once one obtains the Darcy velocity at the flow front from Eq. [10.1] for a given time step, one can use Eq. [10.4] to project the expansion of the resin saturated porous medium domain for the next time step and reiterate the pressure, Darcy velocity and flow front velocity solutions for the next time step by using Eqs [10.3], [10.1] and [10.4], respectively. For the numerical modeling, one can also perform the finite element-control volume (FE-CV) numerical simulation to fill the control volume of the mesh of a numerical finite element model and update the resin flow front and the pressure distribution for each time step until the resin front reaches the vent of the VARTM mold.[2] More details and improvements regarding the numerical solution methods have been reported by many researchers.[4–8] Since the VARTM parts are usually very large (long and wide) and relatively thin, the numerical simulation method such as the FE-CV usually encounters the element aspect ratio limitation issue and the numerical compatibility issue between the flow distribution medium layer and the fiber preform. As a result, a numerical simulation of a large and thin VARTM part can be very costly and challenging in terms of mesh generation, number of elements used and mold filling computation.

In the beginning stage of a VARTM process design, the analytical solution proposed by Hsiao *et al.*[9] for predicting the VARTM mold filling process has been proven useful in determining the major manufacturing parameters and the design window. The two-dimensional (the longitudinal direction and the through-thickness direction) analytical solution employs a dimensionless analysis to divide the resin saturated porous medium domain into two regions, namely the saturated region and the flow front region (see Fig. 10.2).

In the saturated region, the resin flow is simply along the longitudinal direction in both the flow distribution medium layer and the fiber preform. In the flow front region, the resin flow is more complex and couples the effects from the through-thickness resin flow (the major resin flow), the in-plane resin flow supplied from the saturated region (the minor resin flow) and the fact that the resin flow front shape actually remains unchanged

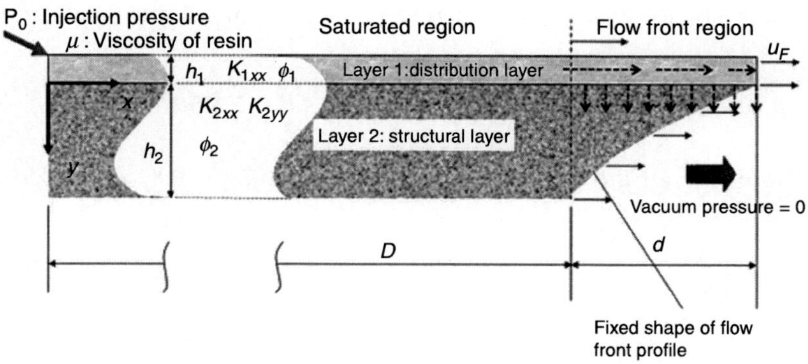

10.2 Geometry and parameters of the analytical VARTM flow model by Hsiao *et al.* (Redrawn based on reference 9.)

during a VARTM mold filling process. Hsiao *et al.*[9] used a dimensionless analysis to argue that the ratio between in-plane resin flow rate (Q_x) and the through-thickness resin flow rate (Q_y) is close to:

$$\frac{Q_x}{Q_y} \sim \frac{K_{2xx}h_2^2}{K_{2yy}d^2} << 1 \quad [10.6]$$

where K_{2xx}, K_{2yy}, h_2 and d (as also shown in Fig. 10.2) are the permeability of the preform in the x-direction, the permeability of the preform in the y-direction, the thickness of the preform and the flow front region length, respectively. A closed form solution of a VARTM mold filling process can then be derived if the assumption of Eq. [10.6] is reasonably approximated. By combining all the arguments and equations, the time required for the resin flow front to travel from an initial location D_0 (at the initial time t_0) to a new location D (at the time t) can be calculated as:

$$t = t_0 + C_1\left(D^2 - D_0^2\right) + C_2\left(D - D_0\right) \quad [10.7]$$

Alternatively, the distance of a resin flow front travels during a period from the initial time t_0 till an arbitrary time t can be calculated as:

$$D(t) = \frac{\sqrt{(\Lambda h_2)^2 + (2\Gamma K_{1xx}P_0/\mu)\left(t - t_0 + C_1 D_0^2 + C_2 D_0\right)} - \Lambda h_2}{\Gamma} \quad [10.8]$$

The resin flow front region length (lag length) can also be calculated as:

$$d = h_2 \cdot 3h_1^* \cdot \sqrt{\frac{K_{1xx}}{K_{2yy}}} \left(\sqrt{\frac{\Phi_1}{\Phi_2 - U_2^*} + \frac{2}{3h_1^*}} - \sqrt{\frac{\Phi_1}{\Phi_2 - U_2^*}} \right) \qquad [10.9]$$

Note that all the coefficients used in Eqs. [10.7], [10.8] and [10.9] are defined as following:

$$C_1 = \frac{\Gamma\mu}{2K_{1xx}P_0}$$

$$C_2 = \frac{\mu\Lambda h_2}{K_{1xx}P_0}$$

$$\Lambda = \frac{K_{1xx}\left(\Phi_2 - U_2^*\right)}{K_{2yy}} \sqrt{\frac{K_{2yy}\Phi_1}{K_{1xx}\left(\Phi_2 - U_2^*\right)} + \frac{2K_{2yy}}{3K_{1xx}h_1^*}}$$

$$\Gamma = \frac{K_{1xx}U_2^*}{K_{2xx}}$$

$$U_2^* = \frac{\Phi_1 h_1^* + \Phi_2}{\left(K_{1xx}h_1^*/K_{2xx}\right) + 1}$$

$h_1^* = \frac{h_1}{h_2}$ thickness ratio between the flow distribution medium layer and the fiber preform

h_1, h_2 thicknesses of the flow distribution medium layer and the fiber preform, respectively

Φ_1, Φ_2 porosities of the flow distribution medium layer and the fiber preform, respectively

P_0 pressure difference between the resin inlet and the resin outlet (vacuumed)

K_{1xx}, K_{2xx} x-direction permeabilities of the flow distribution medium layer and the fiber preform, respectively

K_{2yy} y-direction (i.e., through-thickness direction) permeability of the fiber preform

Figure 10.2 shows all the geometric parameters used in the VARTM flow model.[9] The example results of using the analytical model are illustrated in Fig. 10.3. It shows that the VARTM flow front velocity (U_F) decreases significantly as the saturated region length (D) increases. A series of parametric case studies in reference 9 further show that the VARTM mold filling slowing down issue can be mitigated by: (1) using a thicker flow distribution medium layer, (2) using a flow distribution medium with a higher

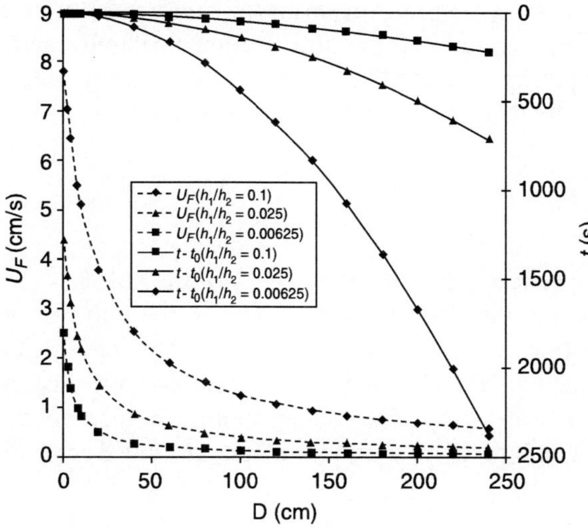

10.3 Results of the analytical VARTM flow model [9]; the flow front velocity (U_F) and the fill time (t) are functions of the length (D) of the saturated region and the thickness ratio (h_1/h_2) between the distribution medium layer and preform. (*Source*: Reprinted from reference 9, with the permission from ASME.)

permeability, (3) using a fiber preform with a higher through-thickness permeability. Equations [10.7], [10.8] and [10.9] are very useful for the initial design of a VARTM mold filling process in determining: (1) locations of gates and vents, (2) arrangement and selection of the flow distribution medium layer(s), (3) mold filling time (or acceptable resin gel time if one can tune the resin gel time) and (4) flow front region profile. After choosing the initial design parameters for the VARTM mold filling process, one can further use a numerical mold filling simulation tool to refine the design.

10.2.2 Fiber preform compaction

During a VARTM process, the pressure differences between the vacuum and the environmental pressure and the flexible vacuum bag are utilized to compact the fiber preform. The relation between the compacted fiber volume fraction (V_f) and the compaction pressure (P_{comp}) (i.e., the pressure difference between the environmental pressure and the local pressure inside the fiber preform) has been studied by Gutowski *et al.*[10] and is given as:

$$P_{comp} = A \frac{\left(\left(V_f / V_{f0}\right) - 1\right)}{\left(\left(1/V_f\right) - \left(1/V_{f\infty}\right)\right)^4} \qquad [10.10]$$

where A, V_{f0} and $V_{f\infty}$ are the preform spring constant, fiber volume fraction at zero compaction pressure and the ultimate fiber volume fraction at infinite compaction pressure, respectively. Another commonly used empirical model introduced by Robitaille and Gauvin[11] is given as:

$$V_f = V_{f1} P_{\text{comp}}^{B} \tag{10.11}$$

where the two empirical constants B and V_{f1} are the stiffness exponent of the fiber preform and the fiber volume fraction at the unit reference compaction pressure (e.g., 1 atm, 1 Pa or 1 psi, etc.). This empirical model is widely used because it requires only two empirical constants and can be easily curve-fitted from compaction experiments. The composite preform thickness (h_2) can be directly related to the fiber volume fraction[12] as:

$$h_2 = \frac{n_p W_A}{\rho_f V_f} \tag{10.12}$$

where n_p, W_A, ρ_f are the number of plies of fiber mats, the weight per unit area of a fiber mat and the density of the fiber material. The change in the fiber volume fraction also has a direct influence on the permeability. The relationship between the permeability K and the compacted fiber volume fraction V_f can be described by the Kozeny-Carman equation[13] as:

$$K = k\frac{\left(1-V_f\right)^3}{V_f^2} \tag{10.13}$$

where k is a constant determined from experiments. Note that the compaction variation of a fiber preform is more obvious on the random fiber mat system and is less obvious on the unidirectional fiber mat system.

Li *et al.*[12] conducted a numerical study about the relaxation process of fiber preform compaction during VARTM process. The fiber preform compaction has some influence on the resin infusion process due to the change of preform thickness, porosity of preform and permeability of preform; however, its influence on the final cured VARTM part thickness variation (i.e., thickness non-uniformity) may not be too significant if one gives enough relaxation time to allow the compaction pressure to be evenly distributed inside the vacuum bag by closing all the resin injection gates. As there is no resin entering the porous medium system (i.e., the fiber preform), the pressure gradient inside the porous medium system will slowly relax and become uniform according to Darcy's law (i.e., Eq. [10.1]). The compaction relaxation process is related to the permeability and dimensions of fiber preform, the vent (vacuum port) locations, the mold geometry, the resin viscosity and the mold temperature that affects the resin viscosity and curing reaction. Apparently, a longer time between the completion of resin filling and the resin gelation point will permit a more complete relaxation process and a more uniform VARTM part. Using more vents or switching the injection gates into vents during the post-filling compaction relaxation stage can also accelerate the compaction relaxation process.[14]

10.2.3 Resin viscosity

Resin viscosity is an important factor for the mold filling, the fiber preform compaction and the curing process during a VARTM process. During the mold filling stage of a well-controlled VARTM process, the resin has negligible change in its degree of cure and hence the viscosity is typically assumed as a function of temperature as:

$$\mu = \mu_0 \exp\left(\frac{E}{RT}\right) \qquad [10.14]$$

where E, R, T and μ_0 are the flow activation energy (J/mol), the universal gas constant (8.314 J/mol·°K), the absolute temperature (°K) and the viscosity constant (Pa·s), respectively. If one would like to include the degree of cure effect into the viscosity, one can consider the combined viscosity model proposed by Lee *et al.*[15] as:

$$\mu = \mu_0 \exp\left(\frac{E}{RT} + a_c c\right) \qquad [10.15]$$

where a_c and c are a constant and the degree of cure of the resin, respectively. For instance, Lee *et al.*[15] used this model to fit the viscosity of a commercial epoxy and found $\mu_0 = 7.93 \times 10^{-14}$ Pa·s, $E = 9.08 \times 10^4$ J/mol and $a_c = 14.1 \pm 1.2$.

Based on the two viscosity models, one can find that the viscosity of resin can be effectively controlled by the mold temperature. For many high-performance resin systems, the mold has to be heated during the VARTM mold filling process in order to reduce the resin viscosity to a manageable range.

10.2.4 Composite curing behavior

Thermosetting resins are typically used in VARTM processes. Due to the geometry of a typical VARTM panel, the heat transfer analysis of the curing process only needs to be performed in the thickness direction (i.e., the z-direction). The one-dimensional energy balance equation in the thickness direction is given as:[16]

$$\rho_c c_{pc} \frac{\partial T}{\partial t} = \frac{\partial}{\partial z}\left(k_{czz} \frac{\partial T}{\partial z}\right) + \rho_r \varepsilon_r H_r \frac{\partial c}{\partial t} \qquad [10.16]$$

where ρ_c, c_{pc}, k_{czz} and T are the density, the specific heat capacity, the z-direction thermal conductivity and the temperature of the composite, respectively. The heat generation rate due to the resin cure is related to the resin density ρ_r, the porosity ε_r, the reaction heat of resin H_r and the resin cure rate $\partial c/\partial t$. The reaction model[17] of the resin is given as:

$$\frac{\partial c}{\partial t} = A \cdot \exp\left(-\frac{E}{RT}\right) \cdot c^m \cdot (1-c)^n \qquad [10.17]$$

where c is the degree of cure (or conversion) and is ranged between zero and unity, A is the pre-exponential factor, E is the activation energy and R is the universal gas constant. The exponents m and n are ranged between zero and two, and $m + n = a$ & $a \cong 2$.[18] Table 10.1 lists some example values of the cure kinetic parameters for an epoxy resin and an unsaturated polyester resin. The cure kinetic parameters can be characterized by using Differential Scanning Calorimeter (DSC)[17] or being directly fitted from the temperature history of a VARTM experiment.[16] Table 10.2 shows some thermal property examples of the polyester/random E-glass composite panel made by VARTM. By imposing the initial conditions of T, c and the boundary conditions of T, one can easily solve the solutions of T and c for a VARTM curing process. The direct cure kinetics characterization technique developed by

Table 10.1 Cure kinetic parameters for an epoxy resin (epon862/w curing agent) and for an unsaturated polyester resin with 2 wt% MEKP

	H_r(kJ/kg)	A(s^{-1})	E/R(°K)	m	n
Epoxy (epon862/W curing agent)[17]	399	6879	6480	0.32	1.66
Unsaturated Polyester (with 2 wt% MEKP)[16]	385.8	4800	4811	0.54	1.36

Table 10.2 Thermal properties of random E-glass fiber/polyester composites manufactured by VARTM

E-glass fiber/polyester composite	Polyester resin
ρ_c = 1690 kg/m^3	ρ_y = 1080 kg/m^3
c_{pc} = 1160 J/kg·°K	ε_y = 0.58
k_{czz} = 0.27 W/m·°K	

Source: From reference 16.

Hsiao *et al.*[16] further utilizes the curing simulations with various sets of cure kinetic parameters generated by an artificial genetic algorithm optimizer to minimize the difference between the experimentally measured VARTM temperature history and the numerically simulated temperature history. By using a powerful artificial genetic algorithm optimizer, one can directly fit the cure kinetic parameters from a VARTM curing process. Figure 10.4 shows the simulated temperature and curing history of a VARTM experiment based on the direct cure kinetic characterization method.

10.2.5 Critical elements of VARTM process design

After understanding the fundamentals of a VARTM process, one can summarize the design elements of a successful VARTM process as follows:

- Mold temperature selection: the mold temperature control is very critical for (1) resin curing management, (2) resin gel time control, (3) resin viscosity control, (4) material selection of vacuum bag, sealing tapes, flow distribution medium layer, flow distribution tubes, resin flow inlet and outlet tubes, peel ply, mold release agent and the construction material of the mold itself.
- Flow process design: after the mold temperature has been decided, the resin viscosity and the resin gel time can be determined or measured. Then one can work on designing the flow process parameters: (1) locations of vacuum ports (vents) and injection gates, (2) locations and sizes of flow distribution lines, (3) type, number of layers and locations of flow distribution medium, (4) timing to open and close gates and vents, (5) in some cases, one may like to control the vacuum pressures of vents to steer the resin flow during the resin infusion stage.[19]
- Fiber preform compaction and fiber volume fraction control: the fiber preform compaction is very important for the part thickness control. To have a more uniform part thickness, it is preferred to close all injection gates and leave the vents on after filling the mold. The pressure gradient as well as the non-uniformity of fiber preform compaction will gradually relax as the liquid resin redistributes itself inside the fiber preform. The relaxation process is related to the resin viscosity, the mold temperature, the pressures and locations of vents, the flow distribution medium and the fiber preform.

10.3 Defects and challenges of VARTM

10.3.1 Air entrapment

A dry spot is an area of the composite part that is not saturated with the matrix material (e.g., thermoset resin). In a VARTM process, the vacuum pressure inside the fiber preform can never be the true zero pressure. As a result, there is a considerable chance of air entrapment inside the final composite parts if the air cannot be completely displaced by the resin during the mold filling.

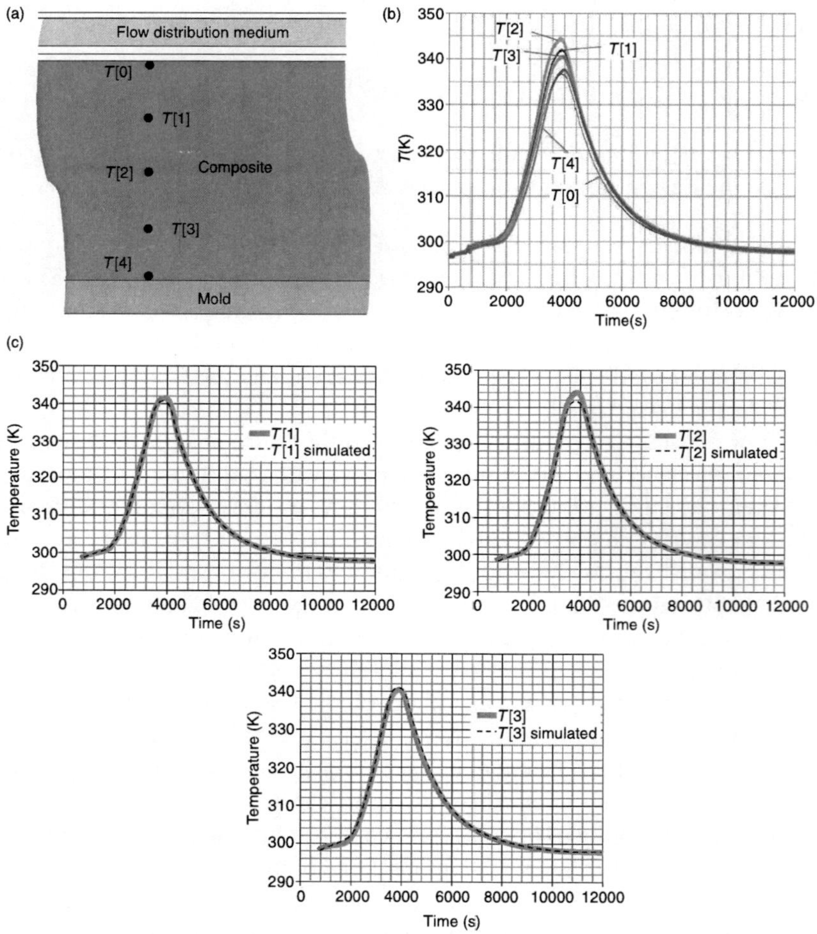

10.4 Direct cure kinetics characterization for determining the resin cure kinetic parameters by best matching the simulated temperatures and the measured temperature during a VARTM curing stage. (a) Thermocouple locations. (b) Experimentally measured temperature history. (c) Comparison between the experimentally measured temperature history and the simulated temperature history based on the optimized resin cure kinetic parameters. (d) Predicted degree of cure evolution during the VARTM process based on the simulation. (Figures are redrawn based on the data of reference 16.)

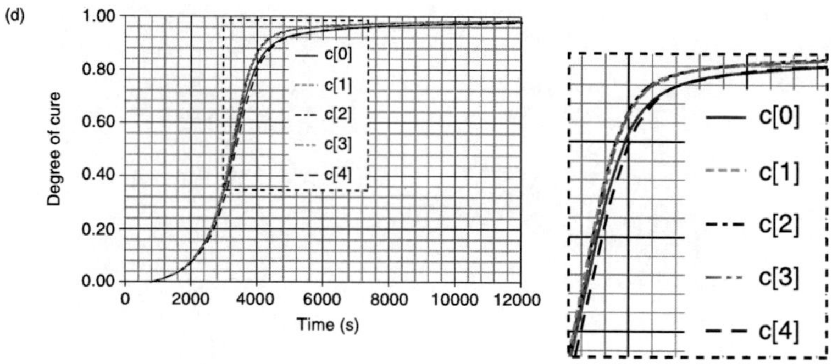

10.4 Continued

One of the major reasons for dry spot formation is an improper mold filling design that causes the resin flow reaches the vent before all the air inside the fiber preform is fully displaced by the resin. In this scenario, although the entrapped air can still be slowly washed out by a continuously resin flow with the inlet and vent opened, the limited resin gelation time and the increased cost to supply the extra resin for the air washout process make this approach unpopular in general VARTM processes. On the other hand, an optimized mold filling design or an active mold filling control technology can mitigate the dry spot issue effectively. For a complex mold geometry, one can use the optimization method such as the artificial genetic algorithm and the molding filling simulations to optimize the arrangement of flow distribution medium layers, flow lines and vent locations or even control the on/off timing of the gates and vents.[20–22] Figure 10.5 proves the feasibility of the automated VARTM design process. It shows that a successful molding filling design can be independently generated by a computer for a complex VARTM part, which includes a co-cured rib structure, by coupling the artificial genetic algorithm optimizer with flow simulation software.[20] Johnson and Pitchumani[23] have reported an active VARTM flow control method by heating the resin locally during the VARTM mold filling stage to reduce the resin viscosity and accelerate the VARTM flow locally. The local flow acceleration can be utilized to control the resin flow front motion during the mold filling process to minimize the dry spot in a VARTM part. However, they also point out that the heating may accelerate the curing reaction of the resin locally and one has to take this factor into consideration while using this local heating flow control technique.

The second cause for dry spot formation is that the filling process could be too slow to completely fill the mold before the resin becomes too viscous to flow. To accelerate the infusion speed, one may consider using more flow distribution layers, injection ports and vents. Alternatively, one may also consider slowing the resin curing process by using fewer curing accelerators or

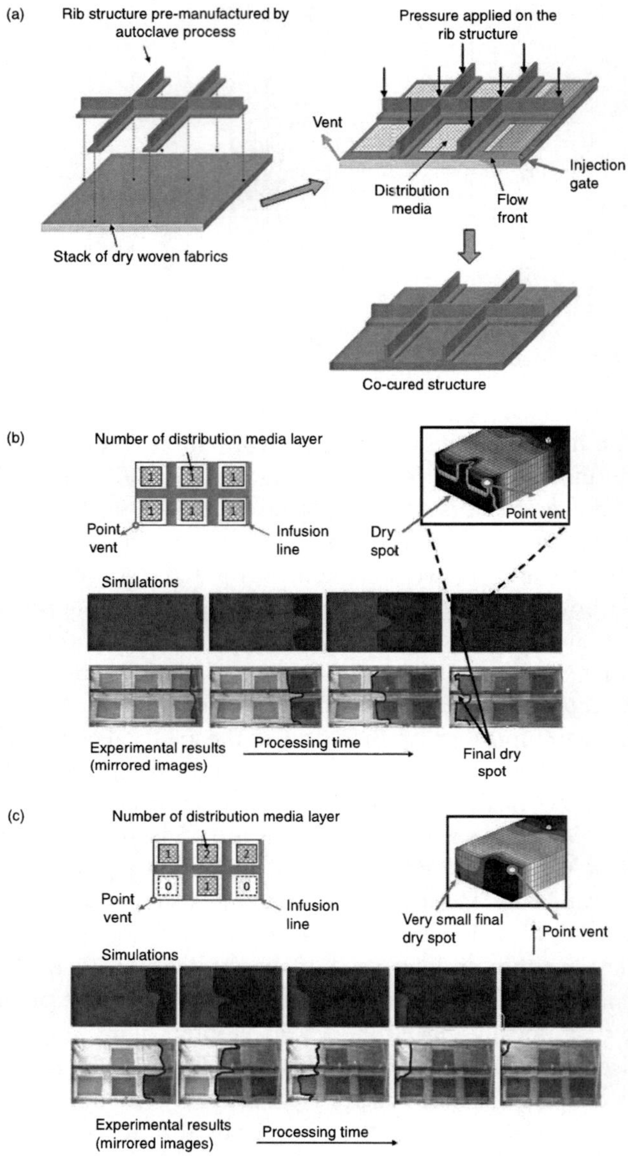

10.5 An artificial genetic algorithm optimizer is coupled with flow simulation software for delivering a successful mold filling design for a complex VARTM part. (a) The geometry of a VARTM part co-cured with a rib structure. (b) Comparison of the experimental results and the flow simulation of the mold filling process for an intuitive VARTM design. (c) Comparison of the experimental results and the flow simulation of the mold filling process for a VARTM design by the artificial genetic algorithm optimizer coupled with flow simulation software. (*Source*: Reprinted from reference 20, with permission from IOP Publishing Ltd.)

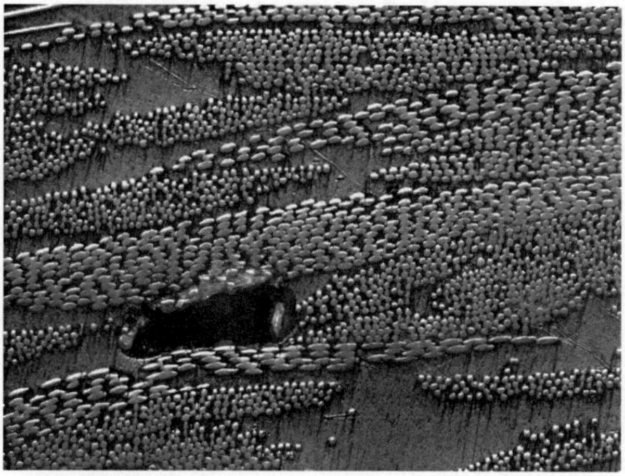

10.6 The cross-section microscope picture shows a microvoid in a random E-glass fiber/unsaturated polyester FRP sample manufactured with VARTM. Also note that there are two different scales of flow pathways for the resin infusion (i.e., the large flow channel between fiber tows and the small flow channel between individual fibers in a fiber tow).

by changing the mold temperature, etc. A modeling based mold filling analysis can also be helpful to prevent early resin gelation before the VARTM mold is completely filled.

The leakage in the vacuum bagging system is also a common cause of dry spots. The leakage may include but not is limited to: (1) vacuum bag damage, (2) leakage in tubing, connectors or resin supply lines, (3) leakage near the sealing tapes and (4) newly formed leakage due to vacuum bag shrinkage or the composite part deformation during the VARTM process. The leakage may be prevented by carefully selecting the bagging and related materials and paying attention to mold tool cleaning and layup process.

Besides the visible dry spots, microvoids (see Fig. 10.6) are another type of air entrapment in VARTM processes. The cause of microvoids is different from that of dry spots. The microvoids are formed due to the incompatible dual scale flow behavior of the wetting process inside a fiber tow (or fiber bundle), which is used to form the fiber mat, and the resin flow process in the gap between fiber tows. The resin flow in the gap between two fiber tows is governed by the Darcy's law (i.e., Eq. [10.1]). However, the resin filling inside a fiber tow is driven by the capillary effect if air/resin interface is involved. The relationship between the microvoid formation in RTM processes and the dual scale effects of the Darcy's flow and the capillary flow has been investigated by many researchers.[24–30] Generally, at the resin flow front, small amounts of air could be trapped and form microvoids if the capillary flow front velocity is significantly faster or slower than the Darcy's flow front velocity during a mold filling process. For a given resin viscosity and flow velocity, the microvoids larger than a certain critical size can be mobilized and washed away by the resin flow according to the void mobilization model proposed by Chen *et al.*[28] Since the resin viscosity, the fiber preform compaction, the Darcy's flow behavior and the capillary flow behavior (resin surface tension and the contact angle between resin and fiber) strongly depend on the temperature and the pressure used in a VARTM process, Kedari *et al.*[31] demonstrated that it is possible to reduce the microvoid content as well as to achieve high fiber volume fraction by optimally controlling the mold temperature and the resin inlet pressure (assuming the environmental pressure is fixed and the vent vacuum is maintained at a constant level) during a VARTM mold filling process. Their experimental results suggest that a higher mold temperature and a stronger vacuum level (i.e., lower absolute pressure) at the vent are useful in increasing the fiber volume fraction and enhancing the fiber volume fraction consistency due to the reduction in resin viscosity and the increase in thickness-direction compression. However, their experimental data also show that a higher mold temperature may sometimes increase the void content if the injection pressure is not modified accordingly. Since the capillary pressure of polyester/E-glass system is reduced at a higher temperature, they utilized a compatibility model to predict that a reduced pressure difference between the inlet and the vent of a heated VARTM process must be used so one can obtain the same low void content of the part infused at a room temperature. By using a dual pressure control VARTM configuration with a heated mold, they experimentally validated this prediction by applying a reduced inlet pressure and an elevated mold temperature to fabricate a VARTM part with enhanced fiber volume fraction and reduced void content. In their study, Kedari *et al.*[31] concluded that the flow compatibility and the thermal-pressure coupling effects have significant influence on both the microvoid formation and the fiber volume fraction control and should be considered for optimizing the flow and thermal control of a heated VARTM process.

10.3.2 Thickness and fiber volume fraction uniformity

As mentioned earlier, the fiber volume fraction control (or laminate thickness control) is important in both the mold filling stage and the post-mold filling stage. Furthermore, the fiber volume fraction distribution of the final composite part is determined and locked during the post-filling compaction relaxation process. The post-filling compaction relaxation process depends on many parameters such as: (1) the preform and the fiber system, (2) the resin viscosity and the cure kinetics, (3) the mold temperature and (4) the type and the arrangement of the flow distribution network. Currently, the compaction relaxation can only either be measured from experiments or predicted by numerical simulations.[11–14] For a practical VARTM process, one should try to have the resin stay at a low viscosity for enough time and close all injection gates during the post-filling stage. Generally, increasing the VARTM mold filling speed and using a resin with longer gel time will benefit the part thickness control.

10.3.3 Curing and thermal management

The VARTM process has been recognized as a preferred method for manufacturing large composite parts in many industries. However, some thick composite parts may not be manufactured by only one single VARTM process due to the potential thermal degradation and thermal-mechanical management issues caused by the fast cure reaction of the thermoset resin in a thick VARTM part. Due to the low thermal conductivity and high reaction heat of the thermoset resin such as polyester or epoxy, the center temperature of a thick composite panel can rise to a high temperature in a short period (i.e., thermal spiking). Out of control thermal spiking can cause the thermal degradation of resin, cracks due to thermal expansion, the deformation of part due to non-uniform thermal-mechanical property evolutions or even fire. Furthermore, the reaction rate of a thermoset resin will be exponentially increased by a high temperature (see Eq. [10.17]) if the curing of the resin is in its early development stage.

To avoid the thermal spiking of a thick composite part, a multi-stage curing (MSC) technique, which helps to avoid the degradation regarding the interlaminar fracture toughness and the interlaminar shear strength of the composite parts, was studied by White and Kim.[32] In the MSC process, a thick polymer matrix composite part is manufactured through several sequential VARTM processes. Each 'stage VARTM process' can only cure a set of manageable layers of composite laminars. Once the composite laminars cure, the vacuum bag system is removed and a new set of preform plies are placed on the top of the existing laminar stack. A subsequent VARTM process is performed on the newly stacked preform plies. By repeating the 'stage VARTM processes', one can achieve the desired laminate thickness and minimize the risk associated with thermal spiking.

10.3.4 Spring-in

For a large composite structure, the dimensional tolerance control is important for the structure assembling process and the residual stress control. For a VARTM process, the vacuum bag side of the part during the curing process is solely subject to the environmental pressure (e.g., the atmospheric

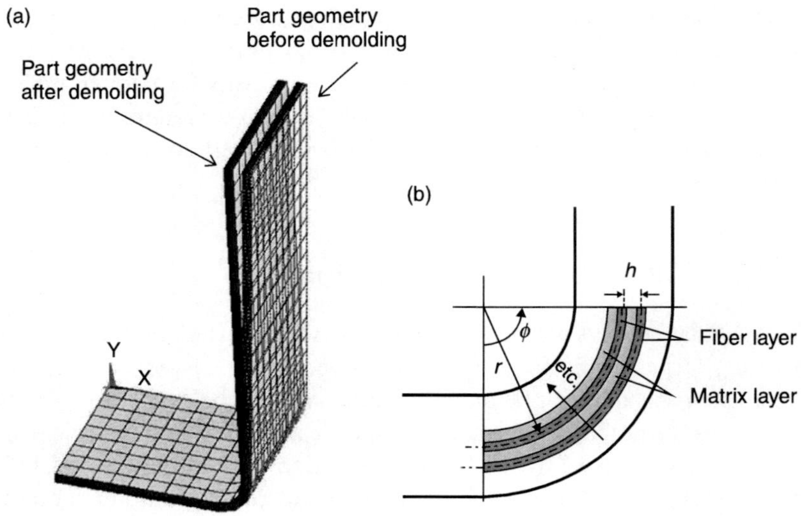

10.7 Spring-in problem defined in reference 33. (a) An L-shaped composite laminate part tends to spring-in after demolding. (b) The spring-in is mainly caused by the matrix layer shrinkage in the laminate thickness direction of a curve-shaped section; note that the fiber system has much less shrinkage or expansion than the matrix system during the curing process. (*Source*: Reprinted from reference 33, with permission from Elsevier.)

pressure). The final composite laminate thickness after the curing process can be different from the thickness right after the compaction relaxation process (i.e., the beginning stage of the curing process) due to resin thermal contraction and resin volume shrinkage (i.e., resin cross-linking shrinkage) during the curing cycle. The residual stress or residual strain is consequently generated inside the composite laminate by the resin cross-linking shrinkage and the mismatched thermal contractions between the fibers and the matrix. The residual stress or strain can further trigger a dimensional infidelity problem called spring-in (see Fig. 10.7) that is commonly encountered for curve-shaped VARTM parts. During the curing process, the significant thermal contraction and the cross-linking shrinkage cause the composite laminate to shrink in its thickness direction, which usually strongly depends on the matrix rather than on the fiber. On the other hand, the fiber system, which has less thermal contraction and a much higher elastic modulus than the resin, helps to maintain the laminate dimensions in the length and width directions (i.e., in-plane directions) during the curing process. For a flat laminate part, this will not cause any issue; however, for a curve-shaped laminate part, this kind of non-isotropic dimensional changes will cause the curve-shaped composite laminate part to be further bent inward after being demolded. The inward bending of a curve-shaped laminate part caused by the curing process is called 'spring-in'. The change in the angle of the curve-shaped composite part is called 'spring-in angle' that is about 1°–4° for a 90° curve-shaped part.

An analytical model proposed by Hsiao and Gangireddy[33] for predicting the spring-in angle (θ) of a curve-shaped laminate is summarized as:

$$\theta = \phi(T)\left\{\left(\frac{h(T_{\text{ref}})}{h(T)}\right)^{-1}\left[1+\alpha_{c,//}(T_{\text{ref}}-T)\right]-1\right\} \quad [10.18]$$

$$\frac{h(T_{\text{ref}})}{h(T)} = (1-V_f)\left\{1+\left[\alpha_m^e - 2\nu_m\left(\alpha_{c,//}-\alpha_m^e\right)\right](T_{\text{ref}}-T)\right\} + V_f\left\{1+\alpha_f(T_{\text{ref}}-T)\right\} \quad [10.19]$$

$$\alpha_{c,//} = \frac{V_f E_f \alpha_f + (1-V_f)E_m \alpha_m^e}{V_f E_f + (1-V_f)E_m} \quad [10.20]$$

where,

$\phi(T)$ = mold arc angle at the peak cure temperature

$\frac{h(T_{\text{ref}})}{h(T)}$ = thickness expansion ratio

$\alpha_{c,//}$ = in-plane thermal expansion coefficient of composite laminate

α_f = thermal expansion coefficient of fiber

α_m^e = the equivalent thermal expansion coefficients of matrix

T = peak cure temperature of laminate (average temperature)

T_{ref} = reference temperature (room temperature in this study)

V_f = fiber volume fraction

ν_m = Poisson's ratio of the matrix

E_f, E_m = tensile moduli of fiber and matrix, respectively

The equivalent thermal expansion coefficients of an isotropic matrix is modeled as,

$$\alpha_m^e = \alpha_r + \frac{1}{3}\frac{V_{\text{SH},r}}{T_{\text{ref}}-T} \quad [10.21]$$

where α_r and $V_{\text{SH},r}$ are the thermal expansion coefficient of resin and the resin volumetric cross-linking shrinkage, respectively. In the experiments done by Hsiao and Gangireddy, they further used nanofibers (CNFs) to reduce the equivalent thermal expansion coefficients of the matrix; for the CNF-enhanced matrix, the equivalent thermal expansion coefficient is given as:

$$\alpha_m^e = \frac{V_{f,\mathrm{CNF},m} E_{\mathrm{CNF}} \alpha_{\mathrm{CNF}} + \left(1 - V_{f,\mathrm{CNF},m}\right) E_r \left[\alpha_r + \varepsilon_{\mathrm{SH},m} / (T_{\mathrm{ref}} - T)\right]}{V_{f,\mathrm{CNF},m} E_{\mathrm{CNF}} + \left(1 - V_{f,\mathrm{CNF},m}\right) E_r} \qquad [10.22]$$

where E_{CNF} and E_r are the elastic moduli of the CNF and the resin, respectively. The volume fraction of CNFs in the matrix is denoted as $V_{f\mathrm{CNF},m}$. The volume fraction of CNFs in the matrix can be calculated from the weight fraction information by:

$$V_{f\,\mathrm{CNF},m} = \frac{W_{f\,\mathrm{CNF},m} / \rho_{\mathrm{CNF}}}{\left(1 - W_{f\,\mathrm{CNF},m}\right) / \rho_r + W_{f\,\mathrm{CNF},m} / \rho_{\mathrm{CNF}}} \qquad [10.23]$$

where ρ_r, ρ_{CNF} and $W_{f\mathrm{CNF},m}$ are the density of resin, the density of CNF, and the weight fraction of CNFs in the CNF/resin matrix, respectively. For a CNF-enhanced matrix used in the VARTM process, the linear cross-linkage shrinkage of the resin in the laminate thickness direction is substantially inhibited by the presence of CNFs during the resin curing process and is approximated as:

$$\frac{\varepsilon_{\mathrm{SH},m}}{V_{\mathrm{SH},r}} = \frac{1 - V_{f,\mathrm{CNF},m}}{3\left[\left(1 - V_{f,\mathrm{CNF},m}\right) + V_{f,\mathrm{CNF},m}\left(E_{\mathrm{CNF}} / E_r\right)\right]} \qquad [10.24]$$

The elastic modulus of the CNF-enhanced matrix is given as:

$$E_m = V_{f,\mathrm{CNF},m} E_{\mathrm{CNF}} + \left(1 - V_{f,\mathrm{CNF},m}\right) E_r \qquad [10.25]$$

With the material properties listed in Table 10.3, the model successfully predicts the spring-in angles for 90° curve-shaped E-glass/polyester composite laminate parts without and with different CNF contents as shown in Fig. 10.8. As shown in Fig. 10.8, indicated by both the experimental data and the modeled results, the spring-in angle is reduced from 2.4° to 0.6° by adding 1.5 wt% CNFs into the polyester matrix. Therefore, it is concluded that the addition of CNFs into the polymer matrix of a curve-shaped VARTM part can reduce the spring-in angle of the part thus enhanced the dimensional accuracy. It one adds more CNFs in a VARTM part, the spring-in angle can be further reduced. According to the spring-in model, with 10 wt% CNFs in the polymer matrix, the model trendline suggests the spring-in angle can be reduced to 0.03°, which could be negligible in many applications.

Table 10.3 Material properties of polyester resin, E-glass fiber and CNFs used in the spring-in study

Material property	Polyester resin	E-glass fiber	CNF
Density (g/c.c.)	1.09	2.58	2.1
Young's modulus (GPa)	6.55	74	600
Poisson's ratio	0.38	0.22	—
CTE (μm/m-°C)	100	5.74	−1
Volumetric shrinkage (%)	Between −5 and −12	—	—
Length (μm)	—	—	30–100
Diameter (nm)	—	~20 000	50–200
Tensile strength (GPa)	0.075	3.44	7

Source: From reference 33.

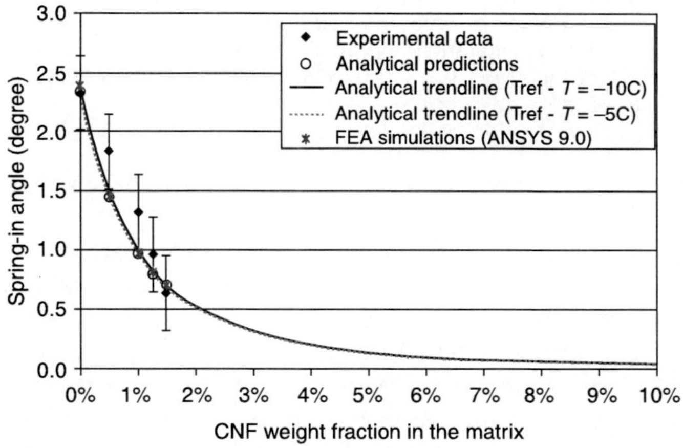

10.8 Comparison of the spring-in angle data from the experiments, the analytical predictions and the finite element analysis (FEA) simulations by Hsiao and Gangireddy. (*Source*: Reprinted from reference 33, with permission from Elsevier.)

For a thick VARTM part, as mentioned before, the MSC technique is recommended for better managing the thermal spiking issue during the curing process. However, in addition to the thermal management, the MSC technique further offers the advantage in mitigating the spring-in problem according to Teoh and Hsiao.[34] Their experimental data and modeled results both indicated that the spring-in angles of 90° VARTM parts can be reduced as the number of curing stages increases (see Fig. 10.9). As explained by their analytical model, the MSC method provides the opportunity for the freshly added laminars of every curing stage to partially slide against the already cured laminas. The partially sliding of stage laminars helps to reduce the bending moment (with respect to the neutral plane of the curve-shaped composite laminate) that causes the final spring-in of the composite

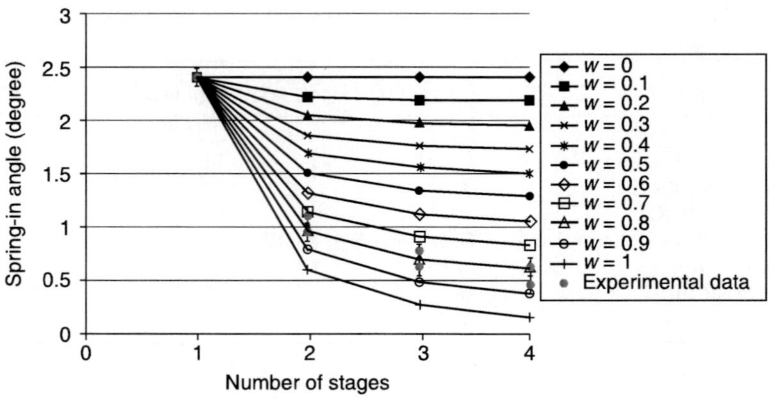

10.9 Comparison of experimental spring-in angles and the modeling results provided by Teoh and Hsiao. Note that the best fit between the experimental data and the prediction was found as the partial slipping interface factor $w = 0.8$. (*Source*: Reprinted from reference 34, with permission from Elsevier.)

laminate part. The sliding effect strength is indicated by a partial slipping interface factor w.

As shown in Fig. 10.9, the partial slipping interface factor w is found to be about 0.8 by curve-fitting the model with the experimental results. Teoh and Hsiao have performed the study on the both concave mold and convex mold and concluded the same spring-in mitigation effect by applying the MSC technique in curve-shaped VARTM parts. The analytical MSC spring-in model is further developed based on the analytical spring-in model discussed in Eq. [10.18] by balancing the bending moment due to the manufacturing process and the bending moment caused by spring-in after demolding. The spring-in angle of a MSC curved part with total 'k' curing stages can be calculated by:

$$\theta^{\text{lam}} = \frac{\sum_{i=1}^{k} \int_{r_{i,i}^{\text{st}}}^{r_{o,i}^{\text{st}}} E\left(\left\{\rho - \left[w.r_{n,i}^{\text{st}} + (1-w)r_n^{\text{lam}}\right]\right\}\left(\rho - r_n^{\text{lam}}\right)\left(\theta_i^{\text{st}}\right)/\rho\phi\right) \mathrm{d}\rho}{\sum_{i=1}^{k} \int_{r_{i,i}^{\text{st}}}^{r_{o,i}^{\text{st}}} E\left(\left(\rho - r_n^{\text{lam}}\right)^2/\rho\phi\right) \mathrm{d}\rho} \quad [10.26]$$

where i, $r_{i,i}^{\text{st}}$ and $r_{o,i}^{\text{st}}$ are the index of the stage-laminate, the inner radius of the '*i-th*' stage-laminate and the outer radius of the '*i-th*' stage-laminate, respectively. In general, the elastic modulus E is a function of the radial coordinate ρ. The radial coordinates of the neutral plane of the '*i-th*' stage-laminate and the complete laminate are denoted as $r_{n,i}^{\text{st}}$ and r_n^{lam}, respectively.

θ_i^{st} is the free-standing spring-in angle calculated from Eq. [10.18] for the '*i-th*' stage-laminate. For a 90° VARTM mold, the model predicts the spring-in angle of the composite laminate part will be reduced from 2.4° to 0.6° when one replaces the single stage VARTM process with a 4-stage MSC VARTM process (see Fig. 10.9). The analytical MSC spring-in model agrees with the experimental results well when the partial slipping interface factor w is assumed to be 0.8.

In conclusion, there are three different approaches to manage the spring-in problem: (1) predict the spring-in angle and then modify the mold to compensate the spring-in angle; (2) use the MSC technique to reduce the spring-in angle if the part is thick; (3) reduce the resin shrinkage by adding non-shrinking additives such as the CNFs into the matrix system. One may have to consider all these approaches when the spring-in phenomenon becomes a problem in a composite structural assembling process.

10.4 Recent advances in VARTM

Recent developments in sustainable green materials have an increasing impact on FRP composites technologies. The VARTM process, due to its flexibility and low tooling cost for large composite part manufacturing, has been selected and successfully used for manufacturing the green composite structures. Bio-based composites that consist of cellulous fiber mats and soil oil-based resins were successfully manufactured using the VARTM process.[35,36] Note that due to the facts that most cellulous fibers are hygroscopic and the water absorbed by the cellulous fibers can inhibit the free radical effect during the curing process, the fiber drying is necessary and critical for using cellulous fibers in the VARTM process.

The VARTM process is also a very important method for manufacturing nano-enhanced FRP composites. Nanoparticles such as carbon nanofibers (CNFs), single-walled carbon nanotubes (SWCNTs), double-walled carbon nanotubes (DWCNTs), multi-walled carbon nanotubes (MWCNTs) and nanoclays have been reported as the nano-reinforcement additives in the polymer matrix. The nano-enhanced polymer matrices can be infused into the fiber mats consisting of microfibers such as carbon fibers or glass fibers to form the nano-enhanced FRP. From many processing methods being investigated for manufacturing the nano-enhanced FRP, the VARTM process has been identified as a promising method due to its unique through-thickness mold filing flow pattern as we discussed in this chapter. The through-thickness flow can reduce the traveling distance of the nano-resin flowing through the fiber preform and reduce the change of nanoparticles being filtered by the fiber preform.[37] The nanoparticle filtration is related to the porosity of the fiber preform. To temporarily create a higher porosity of the fiber preform and permit the nano-enhanced resin flowing through the preform with fewer nanoparticles being filtered during the resin infusion, Fan *et al.*[38] introduced an injection and double vacuum-assisted resin transfer molding (IDVARTM) process. The IDVARTM process adds a rigid vacuum chamber outside the vacuum bag of a regular VARTM setup and utilizes the chamber vacuum pressure to increase the fiber preform porosity (see Fig. 10.10). The IDVARTM process further employs the capillary flow feature in fiber tows to achieve a good nano-resin infusion process. The IDVARTM includes six steps: (1) close the resin inlet and vacuum the preform through the vent, (2) close the vent and inject the nano-enhanced resin, (3) close the resin inlet and apply the chamber vacuum pressure to increase the porosity and help nano-enhanced resin flowing into the preform, (4) allow the nano-enhanced resin to further saturate the fiber tows with the capillary pressure, (5) remove the chamber vacuum pressure so the atmosphere pressure will help to push the nano-enhanced resin everywhere inside the preform, (6) open the vent and reapply the vacuum to remove the excessive resin as the preform is compressed by the atmospheric pressure. Fan *et al.*[38] reported that IDVARTM can help to infuse 2 wt% of MWCNTs through a glass fiber preform in comparison with only 0.5 wt% of MWCNTs by regular VARTM. However, they also reported that a part made with typical VARTM shows aligned MWCNTs in the thickness direction but a part by IDVARTM does not have the preferential MWCNT alignment and is not as strong as a typical VARTM part of the same MWCNT concentration. Despite the difference between the VARTM and IDVARTM, a VARTM or IDVARTM composite part with nanoparticles in the matrix system typically

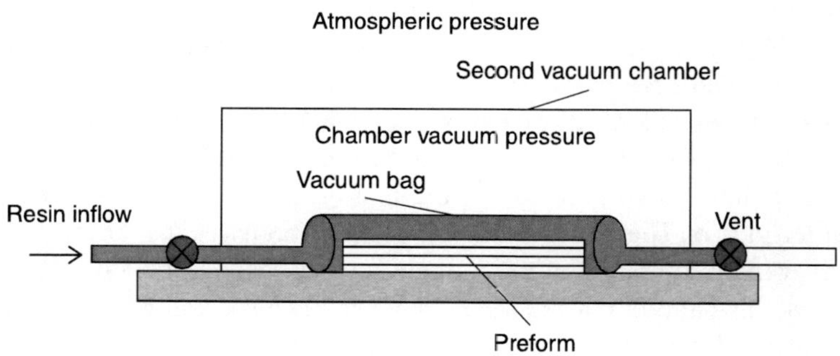

10.10 Setup of the injection and double vacuum-assisted resin transfer Molding (IDVARTM) process proposed by Fan *et al.* The IDVARTM setup controls the preform porosity, compaction and the pressure inside the preform during the resin infusion process by changing the chamber vacuum pressure. The capillary flow effect and the Darcy's flow effect are both utilized in the IDVARTM process to infuse the MWCNTs enhanced epoxy into the glass fiber preform. (This original schematic is created based on the work reported in reference 38.)

shows improved performance in terms of shear strength and delamination strength in comparison with a VARTM part without nanoparticles in the matrix.

Movva *et al.*[39] studied another VARTM based approach to add CNFs into FRP. They pre-bonded the CNFs on the glass fiber mats with a solvent assisted spray process. The CNFs were dispersed with a solvent (i.e., acetone in their study) and the CNF/solvent mixture was sprayed on dry glass fiber mats. The solvent was removed with a vacuum drying process thereafter. The vacuum dried CNF/glass fiber mats were used for manufacturing the CNF/glass fiber/polyester FRP with the SCRIMP process. This method may reduce the chance of CNF filtration during the resin infusion; but the safe handling of the CNF/glass fiber mats and the dry spot related problems could cause additional issues for a large-scale production process.

In addition to using the VARTM process for emerging new composite material systems, many variations from the basic VARTM process have been developed to help manufacture larger, thicker and sophisticated composite parts with improved capability, reliability and cost-effectiveness. Focused areas can be (1) optimal design and fabrication of the fiber preform to achieve good permeability control and compaction control, (2) co-curable interlayer flow channel and interlayer distribution medium layer for successfully infusing a thick composite laminate part or a composite part with inserts, (3) new resin system with the viscosity and cure kinetics customized for VARTM, (4) improved reliability in detecting and fixing any leakage during VARTM process, (5) reusable bagging systems, (6) advanced flow and curing control of the VARTM process and (7) an additional selective membrane (i.e., permeable to gas and impermeable to liquid) sandwiched between the vacuum bag and the flow distribution layer to prevent the dry spot formed on the surface of a VARTM part (see 10.6 for details).

10.5 Conclusion and future trends in VARTM

The VARTM process is a closed-mold liquid composite molding process that is especially suitable for manufacturing large composite laminate parts with low tooling cost and reduced VOC emission. The mold tooling system typically consists of a solid open mold, flow distribution network (i.e., resin inflow gates, resin distribution medium layer, flow channels and vents (or vacuum ports)), a peel ply layer, and a vacuum bag. By utilizing the pressure difference between the environmental pressure and the vacuum pressure applied through the vent, the fiber preform encapsulated between the vacuum bag and the solid mold is compressed against the mold surface. Then the resin is infused into the fiber preform due to the relatively higher pressure in the resin reservoir. A flow distribution medium layer is commonly applied on the top of fiber preform to assist the resin for fast travel across the flow distribution medium and simultaneously infusing into the preform in the thickness direction. Once the resin infusion is completed, one may close the resin inflow gate to allow the excess resin be removed though the vent before the resin becomes too viscous to flow. This preform compaction relaxation helps to uniformly compact the composite part into a higher fiber volume fraction. The full cure cycle is then completed after the preform compaction relaxation; and the final part can be demolded once the cure cycle is finished. Quality composite parts can be obtained from a carefully executed VARTM process. A well designed VARTM process usually has a good repeatability if there is no air leakage in the bagging system.

As discussed in this chapter, a VARTM process is affected by many fundamental factors such as resin viscosity, fiber preform compaction, fiber preform permeability, resin cure kinetics and temperature control. Flow modeling and curing process modeling can help one to avoid these potential problems during the resin infusion and the curing cycle.

Due to the flexible vacuum bagging system used in VARTM, a composite part manufactured by VARTM could have a thickness non-uniformity issue. Generally, better thickness control (i.e., preform compaction relaxation) can be achieved during the post-mold filling process if one allows enough time for the excess resin to be completely vented out from the resin saturated fiber preform with all resin inflow gates closed. Under such a scenario, the whole fiber preform could be uniformly compacted to its compaction limit (or fiber volume fraction limit) under the pressure difference between the environmental pressure and the vacuum pressure. The preform compaction relaxation can either be measured from experiments or predicted by numerical simulation tools.

Another dimension tolerance issue of a curve-shaped VARTM part is the spring-in problem. The spring-in angle of a curve-shaped part is the difference between the bending angles of the mold and of the demolded composite laminate part. The spring-in angle can be reduced either by adding fillers such as the carbon nanofibers (CNFs) or by using a MSC technique. One can model the spring-in angle analytically or numerically.

The VARTM process has a great potential to be used in many new applications. It has been recognized as one of the compatible processes for manufacturing green FRP and nano-enhanced FRP. The versatile VARTM process has also been well accepted in many emerging applications that use FRP parts; many new variations of VARTM have been developed for cost-effectively manufacturing larger and more sophisticated composite parts with improved success rate and quality.

The advantages of the VARTM process including its flexibility, versatility, scalability and cost-effectiveness have been well aligned with the future trend of FRP composite structural applications in aerospace, marine, automotive, infrastructure, wind energy and defense industries. Many process modeling methods and characterization techniques have been established for better understanding and designing the VARTM process. To further enhance the capability and the success rate of the VARTM process, several advanced sensing and control technologies have been developed and become available. The advances in the new bagging materials, sealing tape materials, flow distribution materials and selective membranes have further enabled the VARTM process to be used for manufacturing larger, thicker, more sophisticated or even high temperature FRP parts. It is foreseen that, with the aforementioned advantages along with the supporting knowledge and advanced technologies of VARTM, new manufacturing processes could be further spun off from the VARTM process in the future when particular needs are identified. It is hoped that through the ongoing effort in research, the composites industry will enjoy the further enhanced reliability and versatility of the VARTM process.

10.6 Membrane-based infusion processing

The vacuum-assisted process (VAP) is a VARTM variation patented[40] by EADS using an air-permeable membrane on top of the distribution media to allow continuous and areal venting reducing void content and creating a robust process variant. The membrane is permeable to gas and impermeable to the resin at typical vacuum pressure and can be used at temperatures up to at least 200°C. During the infusion, the resin cannot penetrate through the membrane resulting in uniform vacuum on the surface. The membrane provides also continuous degassing of the resin during infusion and staging.

10.6.1 Process description and infusion behavior

A typical setup for a VARTM infusion process with membrane is illustrated in Fig. 10.11. Here, the layup includes the dry reinforcing fibers as a preform on the mold. A peel ply separates the high-permeability layer (i.e., distribution media) from the preform. The membrane is placed on the distribution media and sealed to the tooling surface. Often, an additional breather material is placed between the membrane and bag to facilitate fast volatile flow to the vent.

The infusion behavior is identical to standard VARTM processes where the vents are placed at the last location of fill. The membrane layer enables any point of the fabric surface to be connected to the vacuum port thus reducing the need for an optimized placement of the venting system. Nevertheless, the post-infusion behavior can be significantly different compared to standard VARTM processing. No resin bleeding occurs as the vent is placed on the impermeable membrane surface. Hence, the pressure behavior and consequently the overall thickness gradient and fiber volume

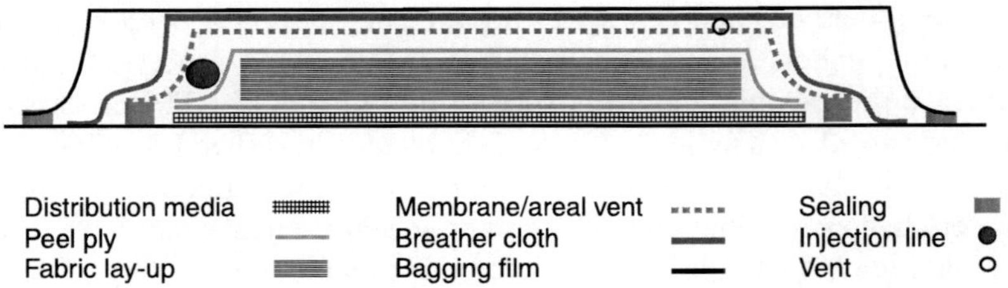

10.11 Layup of the VAP process. (Redrawn based on reference 42.)

fraction development after infusion are significantly different as compared to the VARTM process.[41]

Figure 10.12 illustrates the difference of the pressure and thickness development between the VARTM (a and c) and VAP process (b and d) after infusion in a 70 cm long tool. Figures 10.12a (pressure) and 10.12c (thickness) shows the behavior when the resin is allowed to bleed after infusion (VARTM condition). The pressure drops significantly after the vent line is closed (after approximately 600 s) and reaches ultimately the vacuum pressure of the vent bucket. Resin bleeds through the vent line from the component and reduces the overall thickness of the system. On the other side, the membrane does not allow any resin bleeding. Thus, the pressure behavior during VAP processing is significantly different as seen in Fig. 10.12b. The pressure increases significantly after the infusion line is closed until the pressure reaches equilibrium depending on the total resin amount infused. This is verified by the thickness measurement where the actual steady-state thickness is reached quickly (after around 800 s) but is significant thicker compared to the VARTM approach. Optimum closing of the injection prior to full fill of the preform would allow resin transfer from the resin-rich area near the infusion line into the dry area close to the vent location fully filling the preform. The advantage would be no resin bleeding and no resin waste while minimizing thickness and maximizing fiber volume fraction.

10.7 Membrane evaluation

The success of the membrane process is directly related to the membrane material morphology and to its interactions with the fluid (resin in the case of membrane-based VARTM). The original membrane materials developed by Gore consist of two layers, one being the membrane itself, often made of polytetrafluoroethylene (PTFE) and the other being the support (Fig. 10.13). The support is attached to provide structural support during handling, placement in the preform over complex geometries and to retain integrity (limit stretching and tearing) during compaction and infusion during manufacture.

The barrier properties can be directly related to the membrane pore size distribution measured by a standard porometer, the contact angle of the resin to the membrane surface and surface tension of the resin. A model has been developed[44] to relate these properties to the permeability of the membrane as a function of applied pressure. Figure 10.14 shows the permeability of the Gore membrane for a fluid system with a surface tension of 0.072 N/m for various contact angles. The membrane permeability changes as a function of pressure for any non-wetting fluids where the capillary pressure of each pore contributes to the barrier characteristics. Pressure increases and more pores are opened resulting in increasing permeability. For a standard VAP process (100 kPa) and the fluid system under investigation, it can be

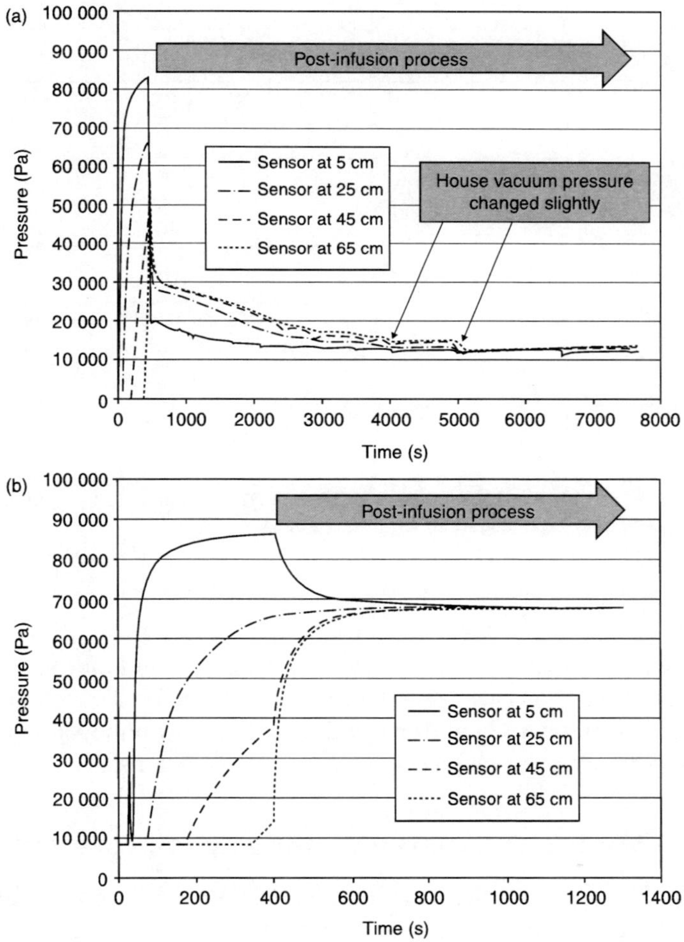

10.12 Thickness and pressure behavior comparison between VARTM (a, c) and VAP (b, d) processing. (*Source*: Reprinted based on reference 42.)

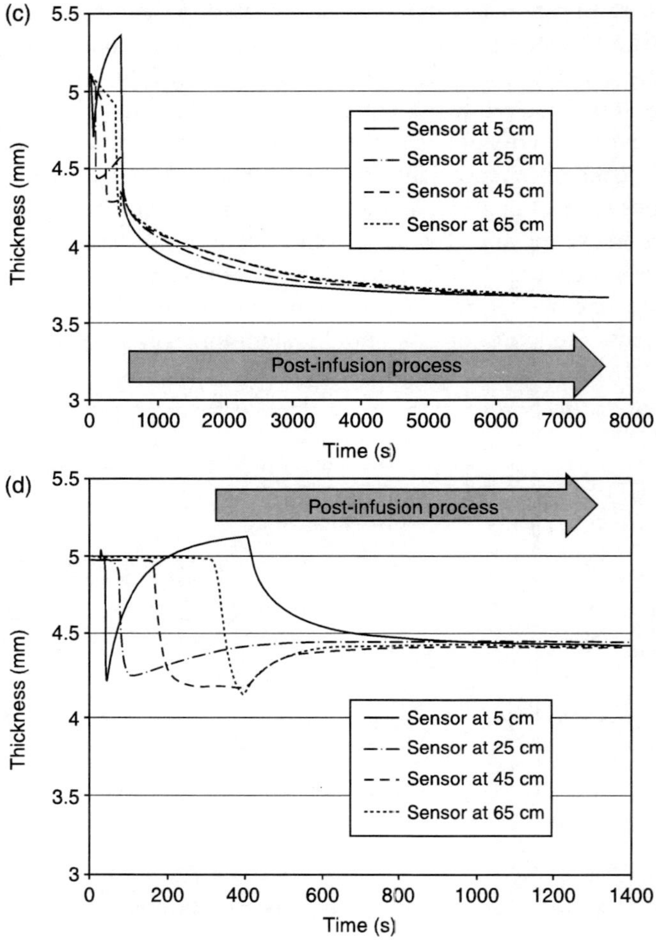

10.12 Continued

observed that the permeability is approaching zero and thus provides near perfect barrier properties. Nevertheless, the particular resin/membrane combination has pressure limitations and reaches a maximum value at around 1000 kPa for even very high contact angle combinations.

10.8 Process and material property improvement

The VAP process results in a more robust VARTM process that minimizes the potential for dry spot formation as well as lower void content and improved dimensional tolerances. The complete surface is connected to the vent reducing the need for overall vent location optimization. Also the

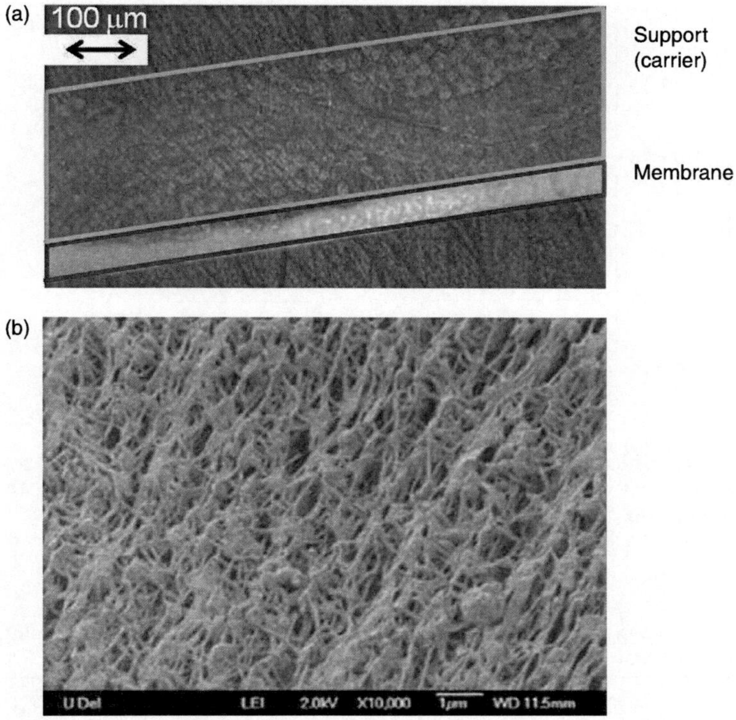

10.13 Micrographs of 'Albatros' membrane by W. L. Gore and Associates: (a) Optical picture showing membrane and support and (b) SEM of nano-porous membrane system. (*Source*: Reprinted based on reference 43.)

VAP process allows continuous degassing during infusion and cure reducing overall void content levels. Volatiles generated during processing can escape through the membrane layer. Table 10.4 summarizes the void content and fiber volume fraction of panels fabricated with the VAP and VARTM process. None of the VAP specimens had more than 1% void content. The standard deviations for VAP panels without degassing and with degassing the resin are constant, which shows the degassing effect of the membrane. The SCRIMP part without degassing had a void content of 1.64% ± 1.2%, which was reduced by the degassing step to 1.07% ± 0.7%. The VAP panels showed a large reduction of void content to autoclave process levels. Nevertheless, fiber volume fractions were lower in the VAP panels as the resin inlet was closed after the resin filled out the complete preform. As outlined in the previous section, this closing approach overfilled the fabric and resulted in lower fiber volume fraction. An optimized VAP process would shut down the resin infusion prior to complete fill maximizing the fiber volume fraction while minimizing the void fraction in the panel.

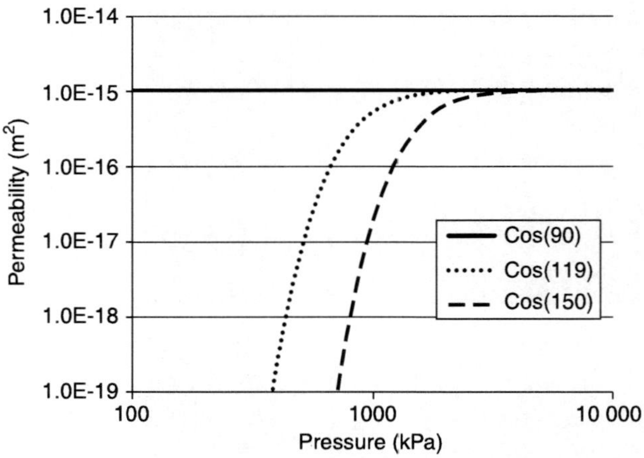

10.14 Permeability behavior for various contact angles properties of the Albatros membrane/fluid system. (*Source*: Reprinted from reference 44, with permission from Begell House, Inc.)

Table 10.4 Fiber volume fraction and void content for panels with no degassing and degassing of resin

	Fiber volume fraction (%)/standard deviation (%)		Void content (%)/standard deviation (%)	
	No degassing	Degassing	No degassing	Degassing
VAP	50.9/0.5	52.8/0.7	0.37/0.3	0.23/0.2
SCRIMP	56.0/1.0	56.21/1.2	1.64/1.2	1.07/0.7

Source: From reference 45.

10.9 Summary of membrane-based infusion processing

This study focuses on a new liquid molding process called VAP. The process adds a membrane to the SCRIMP process allowing uniform venting on the complete preform surface. It evaluates the flow behavior, membrane material and final material properties resulting from the process. Overall, the membrane approach provides a more robust process with the correct selection of the membrane material.

The membrane layer has to be carefully selected to be compatible with the resin and process cycle times and pressure. Models have been developed to evaluate the effective permeability of the membrane as a function of its nanoporous structure and the resin–membrane interaction. The models can be used to evaluate the penetration time of the membrane or the allowable maximum pressure of the process. A wider selection of membrane materials has recently being made commercially available.

Overall, the VAP process is a more robust filling process able to manufacture composite parts with complex geometries and/or inserts with low risk of dry spot formation. The flow behavior during resin infusion is identical to standard VARTM processing while significant differences can be seen for the post-infusion process as no resin bleeding occurs. Optimum closing of the resin inlet would allow full resin transfer into the fabric while minimizing resin waste. Also the VAP process allows continuous degassing during infusion and cure reducing overall void content levels.

10.10 References

1. Seeman II, W. (1990), 'Plastic transfer molding techniques for the production of fiber reinforced plastic structures', United States Patent 4,902,215, February.
2. Bruschke, M. V. and Advani, S. G. (1990), 'A finite-element control volume approach to mold filling in anisotropic porous-media', *Polymer Composites*, **11**(6), 398–405.
3. Darcy, H. (1856), *Les Fontaines Publiques de la Ville de Dijon*. Paris: Victor Dalmont.
4. Chen, Y. F., Stelson, K. A. and Voller, V. R. (1997), 'Prediction of filling time and vent locations for resin transfer molds', *SPE ANTEC Conference, Toronto*, 2424–2429.
5. Lin, R., Lee, L. J. and Liou, M. (1991), 'Non-isothermal mold filling and curing simulation in thin cavities with preplaced fiber mats', *International Polymer Processing*, **6**, 356–369.
6. Trochu, F., Boudreault, J. F., Gau, D. M. and Gauvin, R. (1995), 'Three-dimensional flow simulations for the resin transfer molding process', *Materials and Manufacturing Processes*, **10**, 21–26.
7. Phelan Jr., F. R. (1997), 'Simulation of the injection process in resin transfer molding', *Polymer Composites*, **18**, 460–476.
8. Maier, R. S., Rohaly, T. F., Advani, S. G. and Fickie, K. D. (1996), 'A fast numerical method for isothermal resin transfer mold filling', *International Journal of Numerical Methods in Engineering*, **39**, 1405–1422.
9. Hsiao, K.-T., Mathur, R., Advani, S. G., Gillespie Jr., J. W. and Fink, B. K. (2000), 'A closed form solution for flow during the vacuum assisted resin transfer molding process', ASME *Journal of Manufacturing Science and Engineering*, **122**(3), 463–475.
10. Gutowski, T. G., Morigaki, T. and Cai, Z. (1987), 'The consolidation of laminate composites', *Journal of Composite Materials*, **21**, 172–188.
11. Robitaille, F. and Gauvin, R. (1998), 'Compaction of textile reinforcements for composites manufacturing. I: Review of experimental results', *Polymer Composites*, **19**, 198–216.
12. Li, J., Zhang, C., Liang, R., Wang B. and Walsh, S. (2008), 'Modeling and analysis of thickness gradient and variation in vacuum-assisted resin transfer molding process', *Polymer Composites*, **29**(5), 473–482.
13. Correia, N. C., Robitaille, F., Long, A. C., Rudd, C. D., Simacek, P. and Advani S. G. (2004), 'Use of resin transfer molding simulation to predict flow, saturation, and compaction in the VARTM process', *Journal of Fluids Engineering, Transactions of the ASME*, **126**(2), 210–215.

14. Simacek, P., Heider, D., Gillespie Jr., J. W. and Advani, S. G. (2009), 'Post-filling flow in vacuum assisted resin transfer molding processes: Theoretical analysis', *Composites: Part A: Applied Science and Manufacturing*, **40**, 913–924.
15. Lee, W. I., Loos, A. C. and Springer, G. S. (1982), 'Heat of reaction, degree of cure, and viscosity of Hercules 3501-6 resin', *Journal of Composite Materials*, **16**, 510–520.
16. Hsiao, K.-T., Little, R., Restrepo, O. and Minaie, B. (2006), 'A study of direct cure kinetics characterization during liquid composite molding', *Composites Part A: Applied Science and Manufacturing*, **37**(6), 925–933.
17. O'Brien, D. J. and White, S. R. (2003), 'Cure kinetics, gelation, and glass transition of a Bisphenol F Epoxide', *Polymer Engineering Science*, **43**(4), 863–873.
18. Michaud, D. J., Beris, A. N. and Dhurjati, P. S. (2002), 'Thick-sectioned RTM composite manufacturing. Part I: *In situ* cure model parameter identification and sensing', *Journal of Composite Materials*, **36**(10), 1175–1199.
19. Hsiao, K.-T., Gillespie Jr., J. W., Advani, S. G. and Fink, B. K. (2001), 'Role of vacuum pressure and port locations on flow front control for liquid composites molding processes', *Polymer Composites*, **22**(5), 660–667.
20. Hsiao, K.-T., Devillard, M. and Advani, S. G. (2004), 'Simulation based flow distribution network optimization for vacuum assisted resin transfer molding process', *Modeling and Simulation in Materials Science and Engineering*, **12**(3), S175–S190. DOI: 10.1088/0965-0393/12/3/S08.
21. Hsiao, K.-T. and Advani, S. G. (2004), 'Flow sensing and control strategies to address race-tracking disturbances in resin transfer molding. Part I: Design and algorithm development', *Composites Part A: Applied Science and Manufacturing*, **35**(10), 1149–1159.
22. Devillard, M., Hsiao, K.-T. and Advani, S. G. (2005), 'Flow sensing and control strategies to address race-tracking disturbances in resin transfer molding. Part II: Automation and validation', *Composites Part A: Applied Science and Manufacturing*, **36**(11), 1581–1589.
23. Johnson, R. J. and Pitchumani, R. (2008), 'Active control of reactive resin flow in a vacuum assisted resin transfer molding (VARTM) process', *Journal of Composite Materials*, **48**(12), 1205–1229.
24. Parnas, R. S. and Phelan Jr., F. R. (1991), 'The effect of heterogenous porous media on mold filling in resin transfer molding', *SAMPE Quarterly*, **22**(2), 53–60.
25. Deparseval, Y., Pillai, K. M. and Advani, S. G. (1997), 'A simple model for the variation of permeability due to partial saturation in dual scale porous media', *Transport in Porous Media*, **27**(3), 243–264.
26. Kang, M. K., Lee, W. I. and Hahn, H. T. (2000), 'Formation of microvoids during resin-transfer molding process', *Composites Science and Technology*, **60**, 2427–2434.
27. Rohatgi,V., Patel, N. and Lee, L. J. (1996), 'Experimental investigation of flow-induced microvoids during impregnation of unidirectional stitched fiberglass mat', *Polymer Composites*, **17**(2), 161–170.
28. Chen, Y.-T., Macosko, C. W. and Davis, H. T. (1995), 'Wetting of fiber mats for composites manufacturing: II. Air entrapment model', *AIChE Journal*, **41**(10), 2274–2281.
29. Pillai, K. M. (2004), 'Unsaturated flow in liquid composite molding processes: A review and some thoughts', *Journal of Composite Materials*, **38**(23), 2097–2118.
30. Tan, H. and Pillai, K. M. (2012), 'Multiscale modeling of unsaturated flow in dual-scale fiber preforms of liquid composite molding. I: Isothermal flows', *Composites Part A: Applied Science and Manufacturing*, **43**(1), 1–13.

31. Kedari, V. R., Farah, B. and Hsiao, K.-T. (2011), 'Effect of vacuum pressure, inlet pressure, and mold temperature on the void content, volume fraction of polyester/E-glass fibre composites manufactured with VARTM process', *Journal of Composite Materials*, **45**(26), 2727–2742.
32. White, S. R. and Kim, Y. K. (1996), 'Staged curing of composite materials', *Composites Part A: Applied Science and Manufacturing*, **27**(3), 219–227.
33. Hsiao, K.-T. and Gangireddy, S. (2008), 'Investigation on the spring-in phenomenon of carbon nanofiber-glass fiber/polyester composites manufactured with vacuum assisted resin transfer molding', *Composites Part A: Applied Science and Manufacturing*, **39**(5), 834–842.
34. Teoh, K. J. and Hsiao, K.-T. (2011), 'Improved dimensional infidelity of curve-shaped VARTM composite laminates using a multi-stage curing technique: Experiments and modeling', *Composites: Part A: Applied Science and Manufacturing*, **42**(7), 762–771.
35. O'Donnell, A., Dweib, M. A. and Wool, R. P. (2004), 'Natural fiber composites with plant oil-based resin', *Composites Science and Technology*, **64**, 1135–1145.
36. Dweib, M. A., Hu, B., Shenton III, H. W. and Wool, R. P. (2006), 'Bio-based composite roof structure: Manufacturing and processing issues', *Composite Structures*, **74**, 379–388.
37. Sadeghian, R., Gangireddy, S., Minaie, B. and Hsiao, K.-T. (2006), 'Manufacturing carbon nanofibers toughened polyester/glass fiber composites using vacuum assisted resin transfer molding for enhancing the mode-I delamination resistance', *Composites Part A: Applied Science and Manufacturing*, **37**(10), 1787–1795.
38. Fan, Z., Santare, M. H. and Advani, S. G. (2008), 'Interlaminar shear strength of glass fiber reinforced epoxy composites enhanced with multi-walled carbon nanotubes', *Composites Part A: Applied Science and Manufacturing*, **39**(3), 540–554.
39. Movva, S., Zhou, G., Guerra, D. and Lee, L. J. (2009), 'Effect of carbon nanofibers on mold filling in a vacuum assisted resin transfer molding system', *Journal of Composite Materials*, **43**, 611–620.
40. Filsinger, J., Lorenz, T., Stadler, F. and Utecht, S. (2001), 'Method and device for producing fiber-reinforced components using an injection method', German Patent WO 01/68353 A1.
41. Simacek, P., Heider, D., Gillespie Jr., J. W. and Advani, S. G. (2009), 'Post-filling flow in vacuum assisted resin transfer molding processes: Theoretical analysis', *Composites Part A: Applied Science and Manufacturing*, **40**(6–7), 913–924.
42. Simacek, P., Eksik, O., Heider, D., Gillespie Jr., J. W. and Advani, S. G. (2012), 'Experimental validation of post-filling flow in vacuum assisted resin transfer molding processes', *Composites Part A: Applied Science and Manufacturing*, **43**, 370–380.
43. Amouroux, S. C., Heider, D. and Gillespie Jr., J. W. (2010), 'Characterization of membranes used in pressure driven composite processing', *Composites Part A: Applied Science and Manufacturing*, **41**(2), 207–214.
44. Amouroux, S. C., Heider, D. and Gillespie Jr., J. W. (2010), 'Permeability estimation of nano-porous membranes for nonwetting fluids', *Journal of Porous Media*, **13**(4):, 19–329.
45. Li, W., Krehl, J., Gillespie Jr., J. W., Heider, D., Endrulat, M., Hochrein, K., Dunham, M. G. and Dubois, C. J. (2004), 'Process and performance evaluation of the vacuum-assisted process', *Journal of Composite Materials*, **38**, 1803–1814.

11

Compression resin transfer moulding (CRTM) in polymer matrix composites

S. BICKERTON and P. A. KELLY,
The University of Auckland, New Zealand

Abstract: This chapter discusses the compression resin transfer moulding (CRTM) process for the manufacture of composite materials. The process is particularly suited to the rapid production of high-quality composite parts. CRTM grew out of the need to reduce production cycle-times, and this is achieved by reducing resistance to resin flow in high fibre content reinforcing materials. The reduction in fill time is offset by the necessity for larger mould-closing forces. The benefits of CRTM are highlighted and the impediments facing a more widespread use are also discussed, including the careful process design required. A number of different variants of CRTM currently in use are examined. Process dependence on the material properties of the fibrous reinforcement are discussed, including material permeability and viscoelasticity. Modelling and analysis, based on the work of Terzaghi and Darcy, is carried out and experimental validation of numerical models as applied to planar and non-planar geometries is presented. Optimisation of the CRTM process is discussed and it is shown how the results of such analyses can be used for design purposes by the manufacturer. The chapter ends with a short review of current challenges facing CRTM and possible future trends.

Key words: compression resin transfer moulding, compaction response, permeability, numerical analysis, optimisation.

11.1 Introduction

Closed mould processes for the manufacture of fibre reinforced plastics continue to grow in popularity, being applied within an increasing range of industries. The term liquid composite moulding (LCM) denotes those processes that involve infiltration of a dry fibrous preform with a liquid polymeric resin. The preform can be composed of short discontinuous, or long continuous fibres. Some form of mould is used to conform the fibre reinforcement to the intended shape of the part, and to withstand local forces due to resin pressures generated internally, and the resistance to deformation offered by the preform. Required mould construction and supporting clamping equipment depend on the composition of the part (i.e., type of fibre reinforcements, desired fibre volume fraction), and intended scale of production. For low production rates single sided moulds are often employed, in conjunction with resin infusion under a flexible bag (i.e., VARTM, SCRIMP). Multiple tool pieces are used with increasing production numbers and the requirement for better part tolerances and improved surface finish. For moderate production levels, fibre composite tooling can be employed, as is commonly used for the RTM Light process. RTM Light is characterised by significant mould deflections during processing. As mould rigidity is increased, mould deflections are limited, and part thickness is assumed to remain constant during filling. The resin transfer moulding (RTM) process provides for excellent part quality, the relatively large cost of tooling being justified for medium to high production numbers. However, for high fibre content parts from continuous fibre reinforcements, the high resistance to resin flow has proven to be a barrier to the rapid processing required in some industries, for example automotive. The Compression RTM (CRTM, a.k.a. I/CM, RTCM, RI/CM and CTM) process has evolved from RTM to fill this need, by enabling very short mould filling times.

To extend the application of high fibre volume fraction composites into high-level production runs, factors limiting the cycle times for RTM need to be addressed. An RTM cycle is primarily composed of preforming, mould filling, resin cure, part demoulding and mould cleaning and preparation. Accurate and repeatable preforming is required for the manufacture of quality components, and this stage can be very time-consuming. However, preforming can be carried out independently of moulds used for injection and cure, preform units being prepared in advance using separate equipment. Injection and cure must occur within the mould imparting the final component shape, mould heating and cooling being desirable for the reduction of cycle times. Injection times are governed by the flow resistance offered by the preform, and the capacity of the available resin injection equipment. The time required for resin cure depends on the reactivity of the resin system, the thermal properties of the laminate (fibres and resin), and mould heating/cooling capacity. Injection and resin cure are commonly completed sequentially, while additional time savings can be made if some amount of cure occurs during filling. Faster curing resin systems are also advantageous, although careful mould temperature control is required to counter the risk of premature resin gel. By working in combination with methods to significantly reduce mould filling times, faster curing resins can be used with more confidence and large improvements made in total cycle time.

The CRTM process is described schematically in Fig. 11.1. The filling stage comprises a combination of injection and compression driven flows. Unlike RTM, the mould is not completely closed prior to the initiation of filling. The mould is closed to some predetermined position, which results in lower overall fibre volume fraction, and hence a lower global resin flow resistance offered by the preform. It is also possible to leave a small empty cavity on

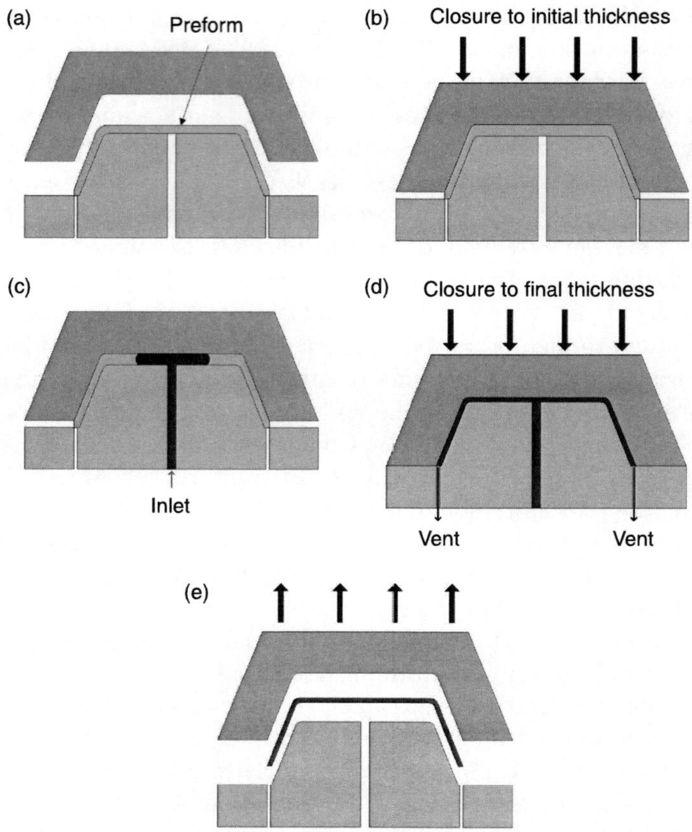

11.1 Schematic description of CRTM. (a) Placement of preform, (b) closure to cavity thickness for injection, (c) injection, (d) compression driven flow and (e) cure and demoulding.

one side of the preform, further reducing flow resistance during injection. Once the required volume of resin is injected, injection gates are closed, and the mould is closed in a controlled manner. This secondary mould-closing phase drives the resin front through the preform to the vents, and advances the laminate to the final part dimensions and composition. With careful selection of processing parameters, CRTM can realise significant reductions in mould filling times, relative to RTM. However, faster mould filling is typically gained at the expense of larger required mould clamping forces, process design therefore being a complex optimisation problem. CRTM is therefore suited to rapid production of high-quality fibre composite parts, and has to date attracted significant attention from the automotive industry. For any application in which reductions in cycle time are advantageous, the CRTM process is a promising candidate.

The CRTM process has many of the advantages provided by RTM, including containment of gases hazardous to an operator, good surface finish on all faces of the product, capability to tailor the arrangement of fibre reinforcement within the preform and the potential for automated manufacture of parts with high volume content of continuous fibre reinforcement. In addition to significantly faster mould filling, there is some evidence that the compression flow phase of CRTM leads to reduced void content. These benefits are accompanied by several challenges, including:

- Mould filling is inherently more complex, due to introduction of compression driven flows. Preform flow resistance, or permeability, continuously changes through the compression flow stage. The balance of flow and preform related forces is complex, and is influential to processing times. How can these effects be efficiently modelled?
- A larger quantity of parameters must be defined to specify a filling process. Along with the selection of where to inject and vent, a decision is required on how far open the mould will be during injection. How should secondary mould closure be controlled, specifying a constant closing speed or application of a specified force?
- Mould sealing arrangements are more complex. Sealing must contain resin and maintain pressure levels within the mould, as the mould pieces move together during the secondary mould-closing phase.

The earliest publicly reported research on the CRTM process was presented in the mid-1990s, this work primarily focusing on the extension of RTM mould filling simulations. Subsequent research has addressed a wider range of processing conditions, detailed force analyses and process optimisation, these topics being summarised in this chapter. The knowledge developed is helping to overcome the challenges that hinder a broader adoption of the CRTM process. By the successful and repeatable application of low filling times, coupled with rapidly curing resin systems, this technology provides the potential to manufacture structural automotive parts at cycle times to rival stamping of steel. With improved modelling and process optimisation, the benefits of compression driven flows may be realised with lower required tooling forces, and therefore smaller presses.

11.2 Process description

The combination of injection and compression driven flows results in the relatively large number of parameters required to define a CRTM mould filling stage. Figure 11.2 presents schematically a mould, defines several key process parameters and illustrates two simple mould filling scenarios. Mould filling is initiated with the loading of the preform into the mould.

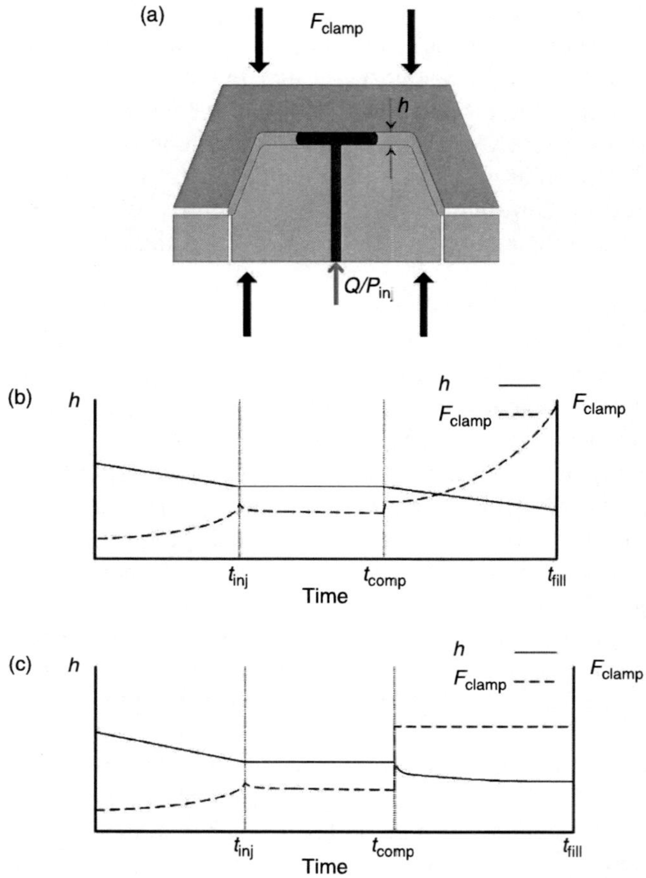

11.2 (a) Key process parameters. Variation of cavity thickness and clamping force with time, (b) applying constant speed and (c) constant force during compression filling.

h is considered to be the cavity thickness at an arbitrary datum point, measured in the closing direction of the mould. The preform is compressed until $h = h_{inj}$, being the cavity thickness at which resin injection is initiated. As for RTM, resin is injected at the gates, typically using a controlled flow rate (Q) or controlled gate pressure (P_{inj}). A predetermined mass of resin is injected, with a small overflow specified to ensure complete mould filling. Compression can be completed using either constant mould-closing speed ($\dot{h}$), or constant applied clamping force (F_{clamp}). Figure 11.2b illustrates the evolution of h and F_{clamp}, during a CRTM process utilising constant $\dot{h}$ during the compression flow phase. Until t_{inj}, cavity thickness reduces at a constant rate, and the force required to compress the preform rises as the average fibre volume fraction increases. At t_{inj}, h is maintained constant, and injection occurs. F_{clamp} can rise or fall during this period, due to the competing effects of stress relaxation within the preform, and the increasing force applied to the mould due to internal resin pressure. At t_{comp} the injection gates are closed, and h is reduced at a constant rate to the final part thickness. F_{clamp} rises at an increasing rate, due to the increasing fibre volume fraction of the part. Both the resistance to preform compaction and the resistance to resin flow are increasing rapidly, resulting in a rapid increase in F_{clamp}. Alternatively, the compression phase of filling can be performed using force control, as depicted in Fig. 11.2c. The total applied force is balanced by the internally generated fluid pressure, and the compaction resistance of the preform. As both increase with decreasing h, h drops rapidly after t_{comp}, the rate of reduction slowing with time until the final part thickness is reached.

Of the two process scenarios described in Fig. 11.2, application of constant force is most efficient, potentially utilising the full capacity of the available press during the compression flow stage.[1] Further reductions in fill time may be achieved by performing the injection and compression phases concurrently. Figure 11.3 depicts one scenario, in which compression is initiated part way through injection. However, many combinations are possible, further complicating the task of designing an optimal CRTM process. Such an approach may be complicated to implement in practice, additional hardware being required to control resin injection, and prevent backflow of resin out of gates during compression.

There are two widely used variations of the CRTM process. As depicted in Fig. 11.4, a mould can be closed prior to injection, such that some compaction is applied to all surfaces of the preform. The reduced average fibre

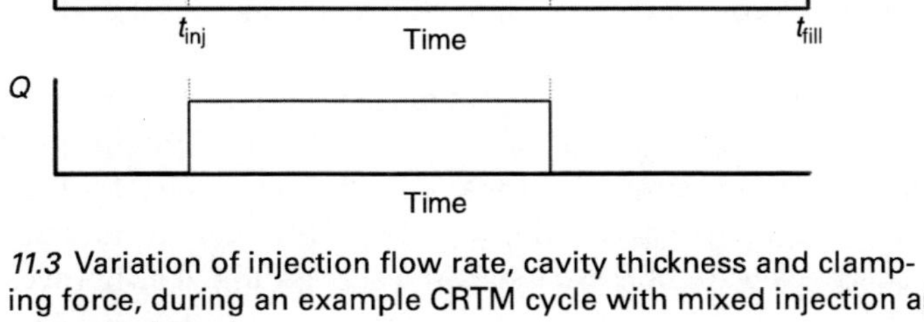

11.3 Variation of injection flow rate, cavity thickness and clamping force, during an example CRTM cycle with mixed injection and compression.

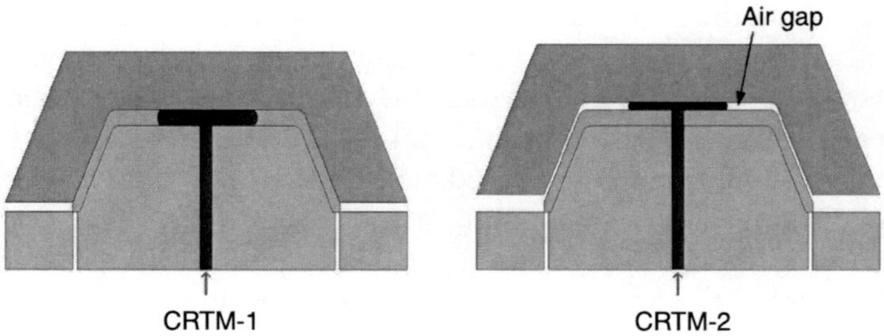

11.4 Definition of CRTM-1 and CRTM-2 process variants, based on conditions at the initiation of resin injection.

volume fraction presents a significantly lower flow resistance, as compared to RTM. This process is denoted here as CRTM-1, and has been addressed by a number of authors. As an approach to further reduce flow resistance within the mould, injection can be initiated when an air gap still remains adjacent to the preform. This variation is denoted here as CRTM-2, and is shown schematically in Fig. 11.4. The air gap can have a permeability several orders of magnitude higher than typical fibre reinforcements, significantly reducing time required for injection. While this approach can lead to a very rapid mould filling process, the physical processes are significantly complicated, and part to part filling may be very sensitive to variations in preform thickness.

CRTM moulds can be constructed in a very similar manner to RTM tools. As for RTM, capacity for heating and cooling provides a significant advantage. Filling can be performed at raised temperatures, reducing resin viscosity. Control of mould temperature allows for better management of resin cure, and the potential for significant reductions in time to demould. One point of difference to RTM is the form of the resin pressure field generated during compression. During injection, local resin pressure drops rapidly with increasing radius (r) from the gate, the analytical pressure solution being proportional to $\ln(1/r)$. During the compression flow phase pressure varies proportional to r^2, with high resin pressures acting over a larger effective area of the mould. Peak resin pressures can grow rapidly during compression flow periods, resulting in large local and total forces exerted on tools.

Another considerable practical challenge is posed by the requirement for effective mould sealing during CRTM filling. Sealing systems must contain the resin within the mould, and maintain resin pressure levels, while the mould is progressively closed during filling. Designers of tooling and process need to be aware of these issues, if a successful CRTM process is to be specified.

11.3 Material properties and characterisation

As for any composites processing technique, the physical properties of the constituent materials strongly influence the progression of a CRTM cycle. Low resin viscosity is desirable for the reduction of mould filling times, and a good understanding of the cure kinetics is required in order to manage the cure process. The flow resistance offered by the preform, or permeability, governs the speed and spatial distribution of resin. Lower permeabilities result in large in-mould resin pressures, contributing significantly to the forces exerted on tooling. The other significant contribution to tooling forces is due to the resistance to deformation offered by the preform. Fibre reinforcement compaction response can be very influential, being dependent on local fibre volume fraction, speed of compaction and the deformation history. This section briefly considers aspects of reinforcement permeability and compaction response specific to the CRTM process.

As part thickness remains constant during RTM filling, it is sufficient to characterise the permeability of the preform at the fibre volume fractions defining the final part. Primarily in-plane permeability is required, unless thick parts are considered for which significant through-thickness flow is required. For CRTM-1, in-plane resin flow occurs during both injection and compression. During injection, part thickness remains constant, but fibre volume fraction may vary widely due to part complexity. During compression, fibre volume fraction, and hence permeability is changing continuously, and potentially across a significant range. For CRTM-2, the situation is further complicated by the presence of through-thickness flow, requiring an additional permeability component to model the flow. Therefore, permeability characterisation must cover a wide fibre volume fraction range, to meet the requirements for an accurate numerical model of CRTM.

In general, the thickness-permeability relation is stated in the following form

$$K_{ij} = K_{ij}\left(h\left(V_f\right)\right) \qquad [11.1]$$

where V_f is the fibre volume fraction. Apart from special cases such as aligned fibre beds, none but purely empirical relations are used for Eq. [11.1]. The standard Carman-Kozeny relation (e.g., reference 2), exponential relations[3] and polynomials[4] have been used in CRTM studies. A very useful relation in this context is the modified Carman-Kozeny model,

$$K\left(V_f\right) = \frac{1}{C}\frac{\left(1-V_f\right)^{n+1}}{V_f^n} \qquad [11.2]$$

where C and n are material parameters defining the material, since it can be used to fit $K - V_f$ curves over a wide range of volume fraction and for a wide range of different reinforcing materials.[5]

CRTM involves periods of compression ($\dot{h} \neq 0$), and periods in which the mould components remain stationary ($\dot{h} \neq 0$). Fibre reinforcement will experience compression at different speeds, depending on the position in a mould for a complex part, or the type of control maintained during compression flow (i.e., force control results in continuously varying $\dot{h}$ in time). During periods in which $\dot{h} = 0$ stress relaxation is also exhibited, which can be influential during injection. Figure 11.5 presents a series of compaction tests on dry samples of a glass fibre chopped strand mat. Each test comprises compaction to a target V_f at constant $\dot{h}$, followed by a period with thickness held constant. The data demonstrate the influence of compaction speed (increasing peak stress with increasing $\dot{h}$), and show the effect of stress relaxation.

The required constitutive equation for the fibrous material is:

$$\sigma_s = \sigma_s\left(V_f, \dot{V}_f\right) \tag{11.3}$$

This is most often taken to be a nonlinear elastic relation; again, a wide range of expressions have been used in CRTM studies, from power-laws[6] to relations involving the hyperbolic tangent.[7] Equation [11.3] can also express the viscoelastic properties of fibrous materials[8–10] (and hence the possible dependence on the rate of change of volume fraction, $\dot{V}_f$, in Eq. [11.3]) and

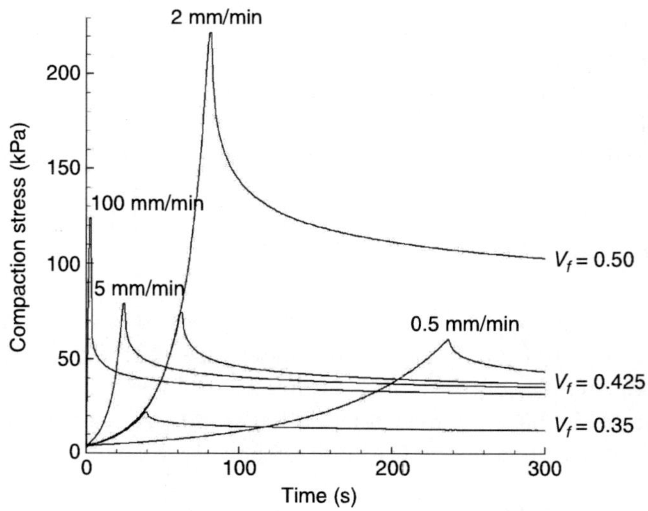

11.5 'Compress and hold' compaction tests performed on samples of a glass fibre chopped strand mat.

permanent deformations which are known to occur.[11] Equation [11.3] should also in general incorporate the differences which occur when the preform is dry and when it is saturated – in the latter case giving lower stress values for a given fibre volume fraction.[12] Figure 11.6 demonstrates the performance of a viscoelastic compaction model,[13] compared against compaction and hold experiments for the same chopped strand mat addressed in Fig. 11.5. The model compares well at several V_f levels and compaction speeds, and provides good reproduction of stress relaxation.

It should be emphasised here that a careful characterisation of the materials under study, captured through the material relations (Eqs. [11.1] and [11.3]), is essential in order to obtain an accurate and reliable mathematical model of the CRTM process. While such models are often proven at the laboratory scale, several additional challenges are posed by the complex shape and composition of preforms for industrial application. Preforms can be composed of multiple fibre reinforcements, combined in a myriad of possible stacking arrangements. Application of these reinforcements to doubly curved part geometry requires reinforcement draping, altering local flow and deformation behaviour. Effective permeability and compaction response need to be computed, which has been addressed in the RTM literature. As discussed previously, numerical models of CRTM processes will require response to be calculated across a wide range of V_f. A balance must be struck between the effort required to specify the numerical model, and the expected accuracy and outcomes from the analysis.

11.4 Modelling and analysis of the CRTM-1 process

Initial reinforcement layup, followed by compaction and resin injection, are phases common to RTM. The volume of literature devoted to RTM can thus be applied directly to these phases (e.g., references 14,15). Once the resin has been injected and the gates have been closed, wet compression is initiated. The first theoretical analyses on the compression of saturated fibrous materials were carried out by Davé *et al.*[16] and Gutowski *et al.*,[17] in the context of autoclave processing and bleeder ply moulding. Therein it was noted the similarity between these processes and the classic geomechanics consolidation problem, that is, the compression of a water-saturated soil or clay.[18,19] The theory can be formulated in terms of Lagrangian coordinates,[17,20] but is more conveniently presented for the present purposes, as in what follows, in terms of an Eulerian description.[21]

For composite parts which are relatively thin and saturated through the thickness, one can reasonably neglect fluid-flow in the thickness direction, particularly when the transverse permeability is very much smaller than the in-plane permeability.[22] For this case, one can derive the fluid mass conservation equation in general three-dimensional form and then specialise it to

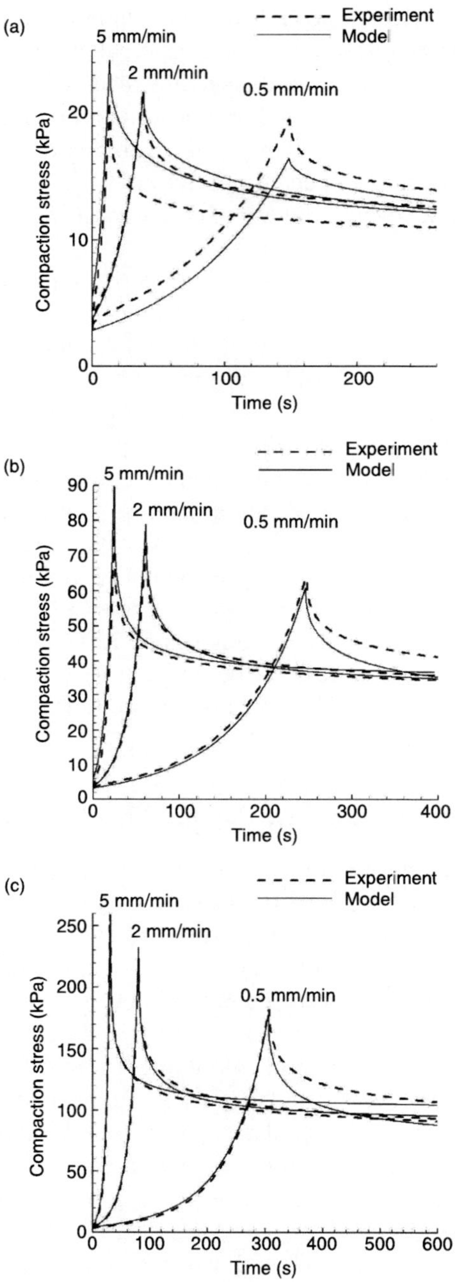

11.6 Application of a viscoelastic compaction model to a series of 'compress and hold' tests on a glass fibre chopped strand mat. (a) Target $V_f = 0.35$, (b) 0.425 and (c) 0.50.

the case of compression in the thickness direction only, with flow in the lateral directions only.[22–25] However, in order to allow for thickness gradients, one can alternatively consider two-dimensional flow at the outset and use a thickness-averaged fluid mass conservation, leading to:[26,27]

$$-\frac{\partial(h\phi\rho S)}{\partial t}=\frac{\partial}{\partial x}(h\rho S\,\mathbf{q}) \qquad [11.4]$$

where ρ is the resin density, φ the porosity of the fibre reinforcement, S is the saturation (volume of fluid divided by volume of pore-space) and $\mathbf{q}$ is the Darcy velocity vector. In general, the Darcy velocity is related to the relative fluid velocity: $\mathbf{q} = \phi(\mathbf{v}^{\mathrm{fl}} - \mathbf{v}^{\mathrm{sol}})$, where $\mathbf{v}^{\mathrm{fl}}$ and $\mathbf{v}^{\mathrm{sol}}$ are, respectively, the fluid velocity and the velocity of the solid skeleton; however, it is assumed that the solid moves only in the thickness direction.

It is generally accepted that Darcy's law is valid for the low-Reynold's number flow of resin through a fibrous material:

$$\mathbf{q}=-\frac{\mathbf{K}}{\mu}\nabla p \qquad [11.5]$$

where $\mathbf{K}$ is the permeability tensor and μ is the fluid viscosity. Combining Eqs. [11.4] and [11.5], and assuming a constant resin density and fully saturated fabric, leads to

$$\frac{\partial(h\phi)}{\partial t}=\nabla\cdot\left(h\frac{\mathbf{K}}{\mu}\nabla p\right) \qquad [11.6]$$

From conservation of mass for the solid component, hV_f is constant, equal to $A_mN/\rho s$, where A_m is the areal mass, ρ_s is the fibre density and N is the number of layers. Thus, with $\phi = 1 - V_f$,

$$\nabla\cdot\left(h\frac{\mathbf{K}}{\mu}\nabla p\right)=\frac{\mathrm{d}h}{\mathrm{d}t} \qquad [11.7]$$

In this equation, the thickness h, pressure p and permeability $\mathbf{K}$ are in general coupled. To relate these, one must consider equilibrium of forces within the preform. At the foundation of such an analysis is Terzaghi's principle of effective stress,[19] which states that the effective stress σ_s carried by the solid skeleton is the difference between the total applied stress σ and the pore fluid pressure p:

$$\sigma_s=\sigma-p \qquad [11.8]$$

Any deformation of the solid skeleton is due to this effective stress. Being a principle, Terzaghi's Law does not rely on any 'proof', and indeed was formulated through experimentation and 'intuition'.[28] Nevertheless, because of its centrality to the theory, it is instructive to see how it relates to the underlying physics: let the total force applied to an element of area A of saturated material be $F = F_s + F_f$, where F_s and F_f are, respectively, the forces taken up by the solid fibrous skeleton and the fluid. Then $\sigma = \bar{\sigma}_s (A_s/A) + p(A_f/A)$, where A_s and A_f are the respective areas within the element consisting of solid and fluid, and $\bar{\sigma}_s$ is the average stress in the solid component. From the Delesseian Law,[29] that is, that the areal porosities are the same as the volume porosities, $V_f\bar{\sigma}_s = \sigma - p + pV_f$. From this and Eq. [11.8] it can be seen that the effective stress is not equal to the actual stress in the solid (averaged over the total area) $V_f\bar{\sigma}_s$, but is the excess stress, that is, the solid stress $\bar{\sigma}_s$ minus the pore pressure, since $V_f(\bar{\sigma}_s - p) = \sigma - p$.[30] Terzaghi's Law can be modified to account for swelling or compressible constituents and materials which are not fully saturated.[31,32]

11.4.1 Solution methods

If one knows the compaction velocity $\dot{h}$, and the permeability as a function of thickness (Eq. [11.1]), one can solve Eq. [11.7] directly for the fluid pressure p, at any given instantaneous thickness h (without recourse to Eqs. [11.8] or [11.3]). For complex geometries, one can use a numerical scheme such as the finite element method (FEM) with control volumes (CV/FEM), using low-order conforming (e.g., reference 23) or non-conforming (e.g., reference 33) elements, or higher-order elements.[34] Once the pressure is known, the fluid flow-front can be updated using Darcy's law, Eq. [11.5]; an increment in thickness (reduction) is next carried out according to $\dot{h}$, $\mathbf{K}$ is updated and the process repeated until compression is complete. Deleglise *et al.*[4] (see also reference 35), used a simple, rapid and novel solution method, whereby the $\dot{h}$ term was neglected from Eq. [11.7]; accurate results for pressure were obtained, but the accuracy dropped closer to the flow front. Once the pressures are evaluated, the local mould stresses can be evaluated from Eqs. [11.8] and [11.3]; the total force acting on the mould is then obtained by integration of the total mould stress σ.

On the other hand, if the process is force controlled, so that $\dot{h}$ is unknown at any instant, Eqs [11.1], [11.3], [11.7] and [11.8] are coupled. This is similar to the situation in resin infusion (RI) and the vacuum-assisted resin transfer moulding (VARTM) process. A convenient way to solve the system in this case is to use an iterative procedure,[36] where $\dot{h}$ is first estimated, then adjusted until the calculated clamping force converges to the actual value. In the one-dimensional case, the problem can be reduced to a single ODE which can be solved numerically.[37]

11.4.2 Validation

Figures 11.7 and 11.8 depict normal stress distributions exerted on a mould during a CRTM process, considering circular preforms of glass-fibre chopped strand mat (CSM), A_m = 450 g/m^2.[1,13] In Fig. 11.7, seven layers of material were dry compacted at 5 mm/min to V_f = 0.3, the compaction was halted and resin was injected from the preform centre at 400 kPa; finally a wet compaction occurred at 10 mm/min until the final volume fraction of 0.35 was reached. In Fig. 11.8, ten layers are compacted at 5 mm/min to V_f = 0.28, resin was injected at 400 kPa, but in this case the wet compaction occurred at 10 mm/min to V_f = 0.5.

Figures 11.7a–11.7d show experimental results at four different instances, obtained using the Tekscan distributed pressure measurement system. Figure 11.7e and 11.7f show the corresponding numerical results obtained using SimLCM, the LCM simulation code developed at the University of Auckland.[38] The comparison is excellent, with the numerical simulation slightly underestimating the peak stresses. Since the final volume fraction here is quite low, the total compaction stresses is dominated by the fluid pressures. Figure 11.8a–11.8d and 11.8e and 11.8f are analogous to those of Fig. 11.7. Here, the stresses are very large because of the relatively large fibre volume fractions reached at the end of wet compression. For this case the total stresses are now dominated by the reinforcement compaction stress.

The total clamping force traces corresponding to the results shown in Figs. 11.7 and 11.8 are shown in Fig. 11.9, with Fig. 11.9a corresponding to Figs. 11.7 and 11.9b corresponding to Fig. 11.8. The clamping force traces for a force-controlled experiment are shown in Fig. 11.10. Comparisons are made to predictions using SimLCM, demonstrating good agreement with regards to process timings, and magnitude of predicted force through periods of injection and compression flow. Prediction of time to complete force-controlled wet compression is challenging; the analysis whose results are presented here involves not only the iterative procedure mentioned at the end of Section 11.4.1, but also a fully speed-dependent viscoelastic reinforcement compaction model.[36]

Extension to non-planar part geometry introduces complexity to the flow and force analyses, and requires consideration of friction between laminate and mould surfaces.[39] A truncated pyramid geometry has been studied as verification of SimLCM, having a constant thickness of 4.0 mm (see Fig. 11.11). Chopped Strand Mat was again studied, care being taken to produce consistent preforms resulting in target V_f of 0.36 and 0.49. Experimental and numerically predicted clamping force traces are presented for constant speed (Fig. 11.12) and constant force wet compression cases (Fig. 11.13). Comparisons are very good, while the low V_f constant speed case is influenced by an overprediction of injection time. The simulated force traces have been decomposed into contributions due to internal fluid pressure, normal reinforcement compaction, and shear force between mould and preform, providing valuable insight into their relative contributions.[39] In summary, good experimental validation has been provided for numerical

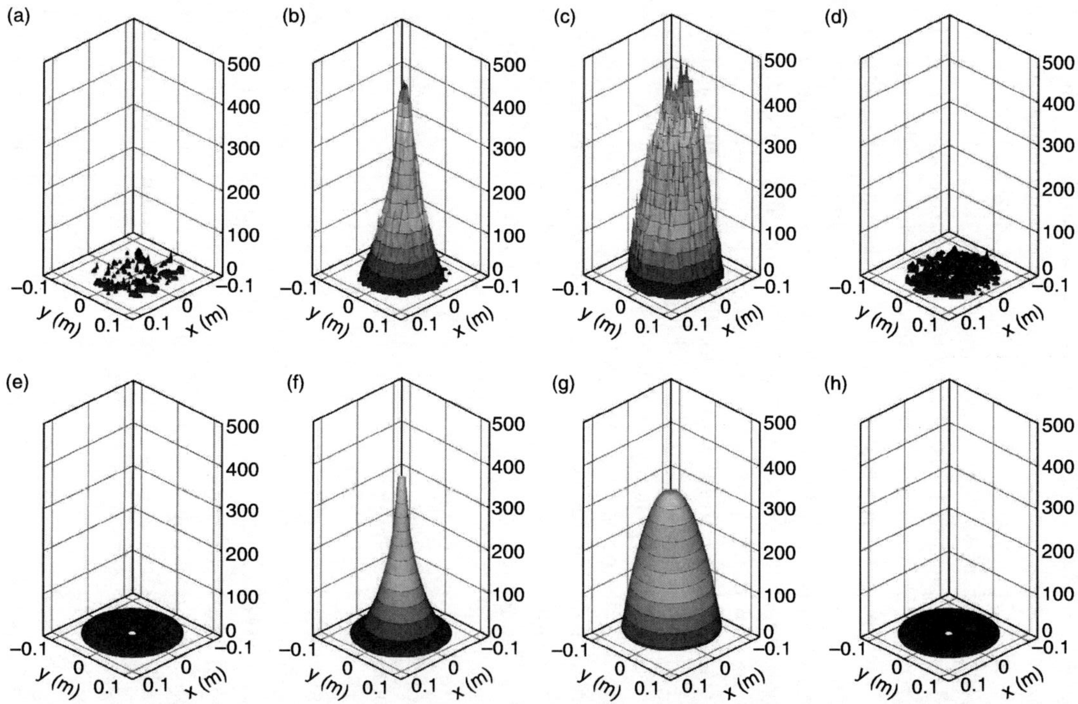

11.7 Comparison of normal stress distributions exerted during CRTM of a 200 mm part to $V_f = 0.35$. Experimental distributions at (a) end of primary compaction, (b) end of injection, (c) end of secondary compaction and (d) long-term relaxed states. Predicted distributions at (e) end of primary compaction, (f) end of injection, (g) end of secondary compaction and (h) long-term relaxed states. Contour levels represent 31 kPa increments.

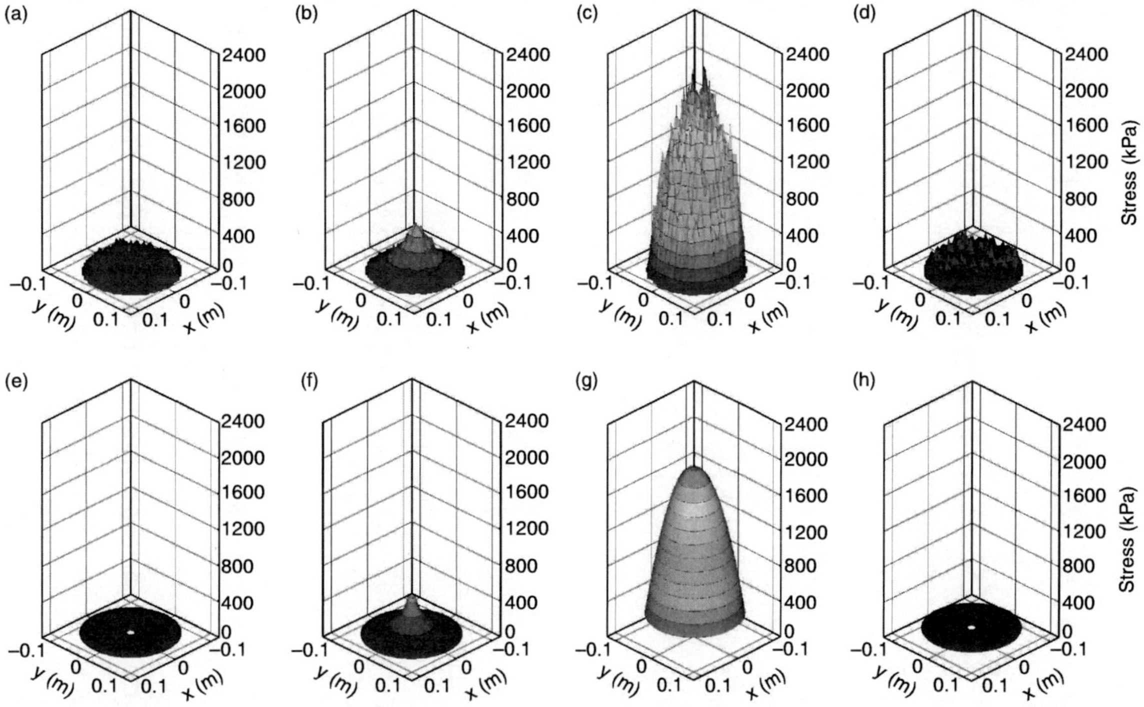

11.8 Comparison of normal stress distributions exerted during CRTM of a 200 mm part to $V_f = 0.50$. Experimental distributions at (a) end of primary compaction, (b) end of injection, (c) end of secondary compaction and (d) long-term relaxed states. Predicted distributions at (e) end of primary compaction, (f) end of injection, (g) end of secondary compaction and (h) long-term relaxed states. Contour levels represent 136 kPa increments.

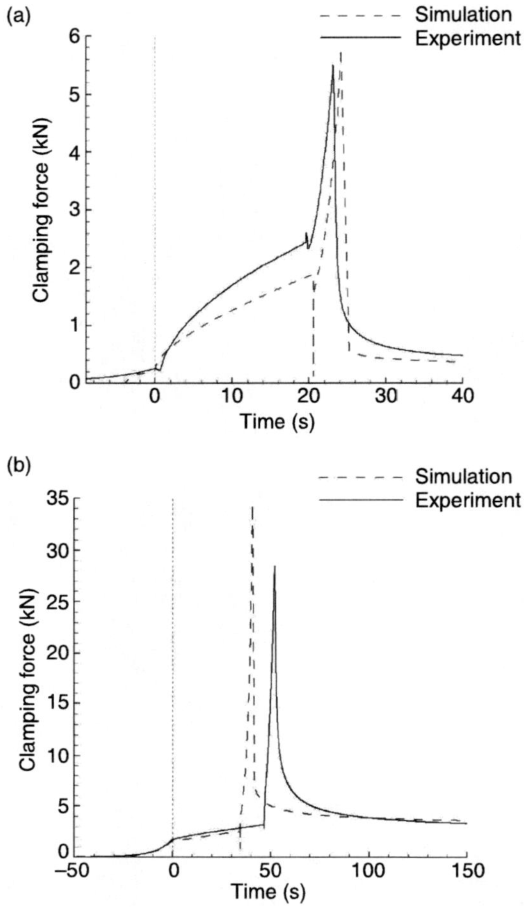

11.9 Total clamping force trace comparisons, for a 200 diameter flat part. Constant speed compression phase with $\dot{h}$ = 10.0 mm/min, (a) target V_f = 0.35 and (b) V_f = 0.50.

models of CRTM-1, including complex geometry and reinforcement deformation behaviour.

11.5 Modelling and analysis of the CRTM-2 process

The initial phase of the CRTM-2 process involves resin injection into a gap between the reinforcement material and the mould; this situation is closely related to racetracking[40] and to the rapid flow through high-permeable distribution media in the VARTM, SCRIMP and similar LCM processes.[41–43] In fact, numerical codes for the VARTM process have been

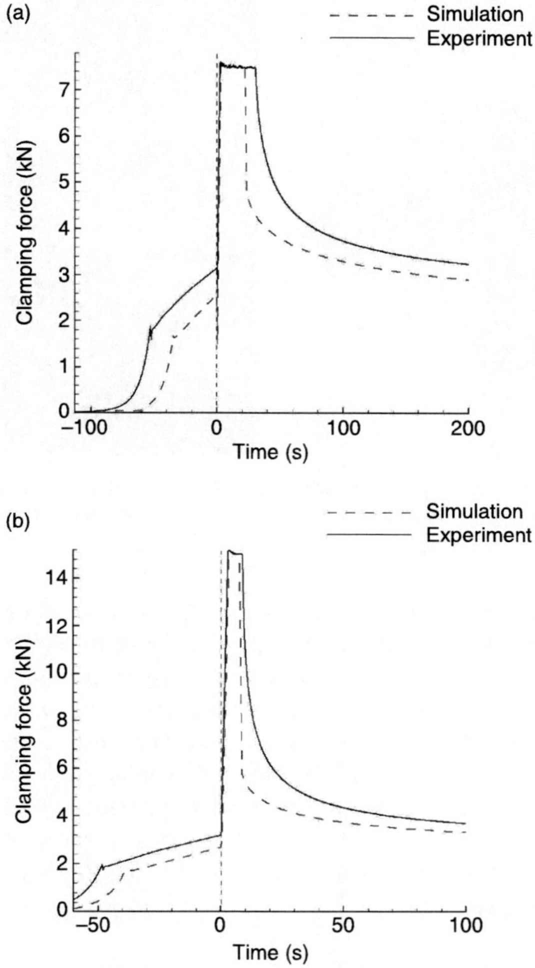

11.10 Total clamping force trace comparisons, for a 200 diameter flat part. Constant force compression phase to $V_f = 0.50$, (a) target F_{clamp} = 7.5 kN and (b) F_{clamp} = 15.0 kN.

utilised for this purpose.[44] The simplest means of modelling the phenomenon is to assume a Hele-Shaw creeping flow, that is, flow in an open channel between two parallel plates with small separation H. The gap can in this case be treated like a virtual porous medium with an equivalent permeability.[23,35,45]

$$K_{\text{eq}} = \frac{H^2}{12} \tag{11.9}$$

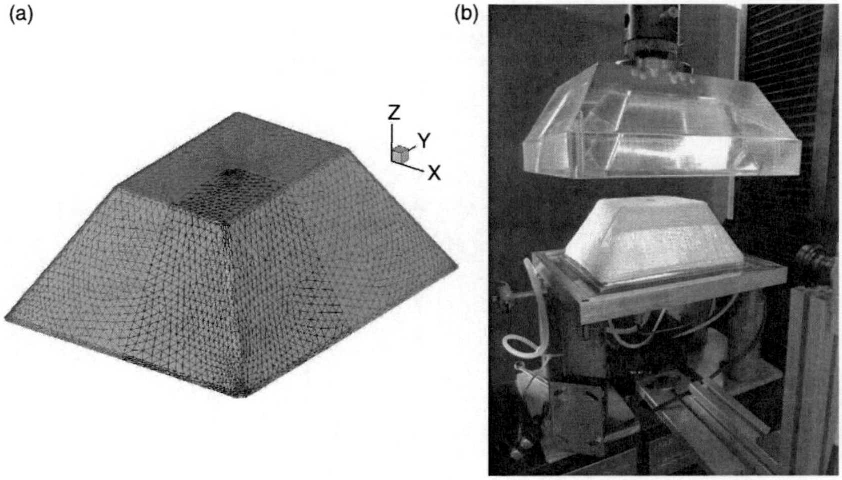

11.11 Four millimetre thick truncated pyramid part geometry. (a) Finite element mesh and (b) mould installed in a universal testing machine.

With the gap on the order of a few millimetres, K_{eq} is several orders of magnitude greater than the permeability of typical composite reinforcements (and orders of magnitude greater than typical distribution media). More involved expressions can be derived to account for the shape of the gap, but Eq. [11.9] should be satisfactory so long as the width of the channel is about three times or more the thickness of the gap. One can account for the fact that one side of the gap is a permeable medium by considering a flow with velocity matched at the boundary between the gap and the porous reinforcement, a situation first examined by Beavers and Joseph.[46] The effective permeability is then given by:[47]

$$K_{eq} = \frac{H^2 + 4H\sqrt{K}/\alpha + 6K}{12\left(1 + \sqrt{K}/\alpha H\right)} \qquad [11.10]$$

where K is the permeability of the preform and α is an empirical constant which needs to be found from experiment; however it is found to be close to unity (indeed, Han *et al.*[48] and others use a value of $\alpha = 1$). Nevertheless, for typical values of K, the output from Eq. [11.10] differs little from Eq. [11.9] and the simpler expression is perfectly adequate.

With the equivalent permeability of the gap known, the flow in the gap can be calculated using the standard solution techniques for the RTM process, with the flow velocity now given by Darcy's law with K given by this equivalent permeability. For less computational effort, one can obtain fairly

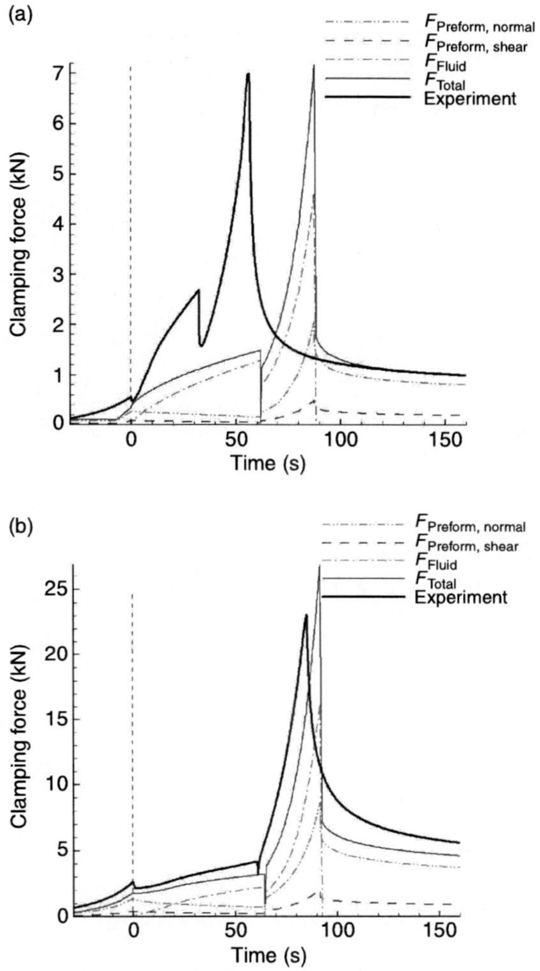

11.12 Total clamping force trace comparisons, for a truncated pyramid part. Constant speed compression phase with $\dot{h}$ = 5.0 mm/min, (a) target $V_f = 0.36$ and (b) $V_f = 0.49$.

accurate results and useful information using simple closed form solutions in conjunction with one-dimensional elements.[49]

During injection into the gap, there will be some seepage of resin into the underlying reinforcement. Seepage will be more significant for materials with relatively large transverse permeability, such as random mats, than for those with smaller permeabilities such as woven or stitched continuous fibre reinforcements. This transverse flow has been modelled a number of times in the context of flow through distribution media, for example

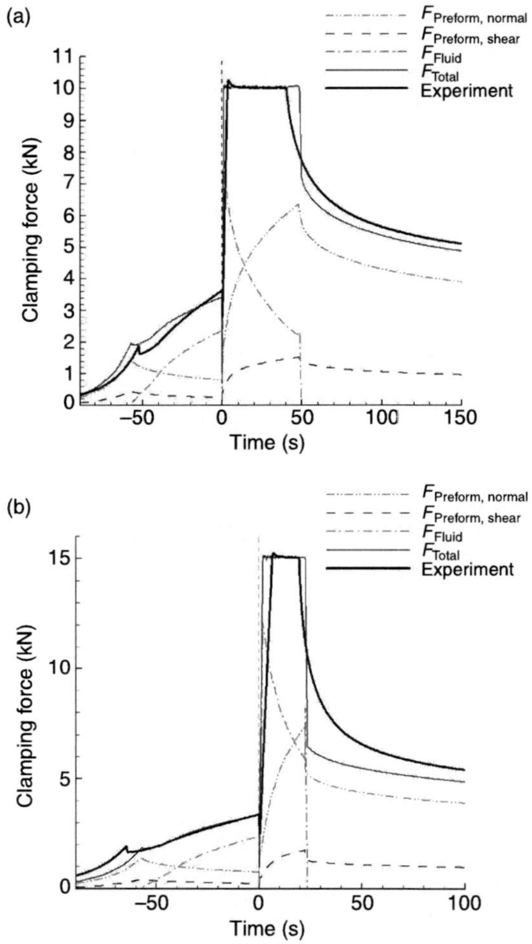

11.13 Total clamping force trace comparisons, for a truncated pyramid part. Constant force compression phase to $V_f = 0.49$, (a) target F_{clamp} = 10.0 kN and (b) F_{clamp} = 15.0 kN.

analytically[50] and using control volume finite elements (CVFEM).[51,52] Feather[53] showed that, since the injection times are much shorter than the compression times (see, for example, the study of Wirth and Gauvin[54]), transverse seepage during injection can often be safely neglected (as done by Chang[45]). However, seepage may be significant in the vicinity of an injection port[49] and consideration of seepage will of course be necessary if the volume of the gap is smaller than the volume of resin injected. Another complication which might arise here is possible deformation of the reinforcement during resin injection; however, this is likely to be small, if not insignificant, as the injection pressures for injection into a gap will be relatively low.

Once resin injection has been completed, compression is initiated. The resin is now forced down into the preform. If the gap is large, and due to the relatively high transverse permeability of preforms (as compared to their in-plane permeability) the resin will also move through any remaining gaps between the mould and perform; in this case, as the gap reduces, the gap effective permeability reduces also. One can carry out one-dimensional analyses of this state of affairs (through the thickness of the preform), by using various approximations to actual events.[53] For example, one can assume a uniform preform deformation and hence permeability through the thickness, such that the dry portion of the preform does not deform[35] or that a constant fluid velocity pertains throughout the fluid filled region.[45] Once the gap is closed, preform deformation occurs. Here again, one-dimensional analyses can be useful. For example, Merotte *et al.*[7] show that the transverse permeability and stiffness characteristics of the preform play a dominant role in the complete process; this is as opposed to other parameters which come into play such as gap thickness.

For a comprehensive analysis of CRTM-2, one must carry out a fully three-dimensional analysis. The governing equations and solution methods are similar to those used in CRTM-1. The presence of the gap is accommodated by using Eq. [11.9] and treating the problem as one of flow through a multi-layered preform stack. During gap-closure, one can either ignore perform deformations in the thickness direction[49] or one can treat the gap and perform as a stack of two materials with different compaction characteristics, the gap being given a nominal insignificant elastic stiffness of, for example, 100 Pa.[25] Once the mould makes contact with the preform, deformation is tracked by allowing the mesh-elements in the thickness direction to reduce in thickness at each time step. One can achieve accurate solutions to the problem posed, but the computational cost can be very high. Analyses of typical geometries can take many hours.[35]

Verifying models and simulations of CRTM-2 is not a trivial matter and only some preliminary work has been done in this area. For example, Merotte *et al.*[7] carried out one-dimensional (through thickness) experiments on E-glass woven material with a 4 mm gap. Results for normalised fill time were presented and there was found to be a reasonable match between (one-dimensional) simulation and experiment. A major factor in the discrepancy between results was attributed to viscous effects in the preform, which were neglected in the model. Simacek *et al.*[49] analysed a complex automotive part. Although it proved difficult to determine some of the process parameters, for example the gap thickness during resin injection (which varied over the mould), good qualitative agreement was observed between numerical and experimental results.

11.6 Optimisation of CRTM

One of the impediments to a wider usage of composites products is the relatively high cost of fabrication, and it is important that composites manufacturing processes are designed with the aim to keep cost to a minimum. This can be achieved by maintaining low production times, capital investment, material wastage, labour costs, and so on. Using an iterative trial-and-error approach to achieve this, based mainly on the experience and intuition of process engineers, can be time-consuming and expensive, particularly for the complex parts which are so well served by the LCM processes. Fortunately, the development of process simulation software over the last few decades, and the increase in power of computers during that time, now allow one to carry out cost-effective and timely optimisation analyses of manufacturing processes. Such virtual, computational, experiments can help the designer select appropriate processing parameters in order to increase overall efficiency and reduce manufacturing cost.

In general, an optimisation problem will consist of a set of design (decision) variables and process objectives. The design variables are process parameters which can be controlled and varied by the designer, for example the number and location of the gates and vents, mould temperature, the materials used and the placement and positioning of high-permeable layers. Process objectives are quantifiable performance measures which are sought to be minimised, for example the process cycle time, curing stresses, the amount of racetracking, mould deformation and the number of part defects. Physical and practical limits are also set on the design variables, for example an upper bound on allowable injection pressure or flow rate.

The optimal solution to a problem can be found by using an exhaustive search (ES) which is where the objective function is evaluated for all possible sets of process variables. Process simulation software is used for this purpose, for example LIMS, SimLCM, PAM-RTM or any other tool which can evaluate the numerical value of a process objective function for a given set of design variables. ES is a useful procedure when the solution space is small. However, although guaranteeing an optimal solution, such a method is simply impractical for most problems, because of the huge computational cost; the computational time for a single process simulation is often many minutes or even hours, and the many thousands of runs typically demanded by ES makes it unviable.

More generally, the process optimisation problem is solved using an optimisation solver, which interacts with the process simulator to minimise the objective function. The optimisation solver can be based on any one of many optimisation techniques. One group of possible optimisation techniques are the gradient-search methods. These methods rely on analytic or semi-analytic expressions for the objective function and objective function gradients (with respect to the design variables). However, referring to Eqs [11.1] to [11.8], such expressions are not easily available. Even when they are (see for example the method of Henz *et al.*[55]), gradient schemes cannot be guaranteed to find the optimal solution. This is because they start from an initial objective function value, associated with an initial set of design variables, and move the solution incrementally towards a nearby minimum. However, the CRTM problem can be shown to involve multiple minima, that is, it is non-convex,[56] and the minimum so found is unlikely to be the global minimum. This leaves one with the heuristic methods, in which the simulation software is treated as a 'black box';[57] the details of how the black box evaluates the objective function given a set of design variables are inconsequential to the optimisation solver. These heuristic methods solve problems by exploring various possible solutions (objective function values) and then evaluating feedback in order to improve upon these solutions. They find a 'good' solution, which might not be the actual optimal solution, but is likely to be at least very close to it. Some of the more popular methods include the Greedy Sequential Algorithm, Tabu Search, Ant Colony Optimisation and Genetic Algorithms (GA). It is worth pointing out that, given that the simulation software is but an approximation of a real process, rapid and robust methods for finding 'good' solutions are very desirable.

The bulk of the reported work on optimisation of LCM processes has focused on the RTM process. In particular, a large amount of research has been conducted on evaluating the most suitable methods for determining the optimal placement of injection gates and vents such that the fill time is minimised.[55,58,59] Another area of keen interest has been the minimisation of dry-spot formation during filling, through the analysis of flow-front smoothness and merging[57,60] and through flow-front capillarity.[61] Lin *et al.*[62] have looked to minimise resin wastage through the vents by controlling material permeability. The curing phase of RTM has also been examined;[63,64] here, the design variable is principally the time-temperature profile of the mould and a number of objectives can be examined, for example the maximum exothermic temperature, the curing time and the non-uniformity in degree of cure. One of the few studies to consider an LCM process other than RTM was that carried out by Trochu *et al.*,[65] who examined void-formation minimisation in resin infusion (RI). Almost all of these studies use heuristic optimisation algorithms, most often GA, although it is sometimes possible to use more mathematically precise methods for specific RTM scenarios,[62] or to use a combination of heuristic methods and gradient methods.[55,63] Further, it is sometimes possible to reduce the work required of the process simulator by rapidly calculating approximations of the objective function[66] or by tailoring the objective function so that a complete process simulation is not required.[67,68]

The CRTM process clearly offers more challenges than RTM to efficient optimisation, principally due to the changing preform fibre volume fraction, permeability and compaction response during processing. Furthermore, the designer is faced with a larger number of parameters which can be selected as design variables and objective functions as compared to RTM, and this increases the difficulty in formulating an optimisation problem whose results are useful and informative to the designer.

Very little work has been conducted on the optimisation of the CRTM process. That said, much of the large body of work on the optimisation of RTM, some of which was mentioned earlier, is directly relevant to the CRTM process also. This includes the work on, for example, void minimisation and the optimisation of the cure process. The new optimisation issues which arise with CRTM are in the main related to the increased complexity of designing the mould filling stage. Le Riche *et al.*[24] examined rectilinear flow CRTM in a flat plate, for which analytic expressions are available. They also examined the coupled problem of optimising both the process and the final composite structure. In previous work at the University of Auckland, Buntain[69] carried out an initial study optimising CRTM with respect to the fill time, tooling forces and maximum fluid pressure, using an Adaptive Response Surface Method (ARSM). Na[56] minimised fill times using a heuristic pattern-analysis. Hsu *et al.*,[70] based on the work of Kam *et al.*,[71] carried out a comprehensive study of the optimisation of CRTM-1, and this study is discussed in the text that follows.

In order to keep matters manageable, Hsu *et al.*[70] selected three important but easily controlled processing parameters as the design variables: the injection pressure P_{inj} (which was held constant during filling), the mould thickness at which fluid is injected h_{inj} and the mould-closing velocity during wet compression v. The two objective functions which were deemed to be the most important were the fill time t_{fill} and the maximum tooling force F_{clamp}, the latter chosen with the minimisation of equipment cost in mind:

$$\text{Minimise: Objective } t_{\text{fill}}\left(P_{\text{inj}}, h_{\text{inj}}, v\right)$$

$$\text{Minimise: Objective } F_{\text{clamp}}\left(P_{\text{inj}}, h_{\text{inj}}, v\right)$$

Note that these two objective functions oppose one another, because shorter process times can be achieved by increasing the tooling force. The requirement to optimise more than one objective function simultaneously arises commonly. There are two approaches one can take. First, one can form a weighted sum of the individual sub-objective functions to form a single overall performance objective parameter.[56,57,64,69] Numerical values can then be assigned to the weights by the designer. However, the results for such an analysis only apply to very particular manufacturing scenarios. In order to maintain greater applicability of results, the problem can be treated as a multi-objective optimisation problem, that is, with more than one objective function,[24,66,72] and this approach is followed here.

The SimLCM software is used to evaluate the objective functions. The Genetic Algorithm (GA) approach is taken, since it has been shown to perform well for other LCM optimisation problems, as mentioned above, and since it is known to perform well for generally non-convex multi-objective problems.[73] The popular NSGA-II (Non-dominated Sorting Genetic Algorithm), as used by Ratle *et al.*[66] in their study of optimal gate location, is used to optimise the problem. In brief, the GA methodology requires one to randomly generate an initial 'population' (of objective function solutions) from randomly generated sets of design variables (see Fig. 11.14). Next, 'inefficient' solutions are eliminated; in the present example, this means that solutions which have both a higher fill time and a higher force than other solutions (Fig. 11.14a). Surviving efficient solutions are then used to create new solutions, through 'reproduction' (using a weighted sum of two sets of design variables to generate a new set of design variables and associated solution) and 'mutation' (randomly adjusting a set of design variables to generate a new set) (Fig. 11.14b). The GA process is then continued by repeatedly removing inefficient solutions and generating new solutions until sufficient convergence is achieved.

As a precursor to the optimisation, an optimum mesh density for SimLCM needs to be found for the particular problem (geometry, materials, etc.) under study. This involves finding a mesh which is fine enough to ensure an adequately converged objective function for a wide range of sets of the design variables $\{P_{inj}, h_{inj}, v\}$, but coarse enough not to increase computational expense unnecessarily.

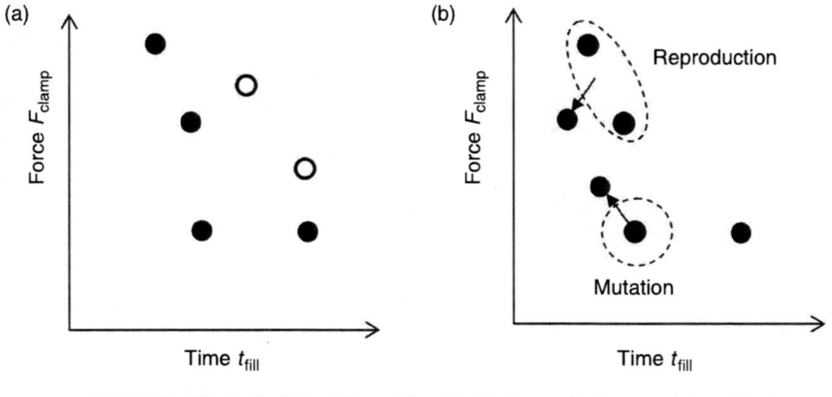

11.14 The Genetic Algorithm (a) an initial population and the elimination of inefficient solutions (hollow circles), (b) generation of a new population through reproduction and mutation.

This optimisation methodology was applied to a number of simple test cases, including filling of rectangular plates with a number of different fibre reinforcements under constant injection pressure. The GA code was seen to converge satisfactorily in less than 20 iterations, that is, generations of new populations. A further study was carried out on the fireman's helmet shown in Fig. 11.15. The dimensions of the helmet are 0.42 m (long), 0.32 m (wide) and 0.17 m (height), and it is considered to be manufactured using a continuous filament mat (CFM) reinforcement for the analysis presented here. The mesh used is shown in Fig. 11.15a. Output from SimLCM showing local normal stress distributions during closure of the mould is shown in Fig. 11.15b.[38]

In Fig. 11.16 are shown the first three generations output by the GA for the helmet problem. It can be seen how each successive generation dominates the previous generation, in terms of lower process time and lower

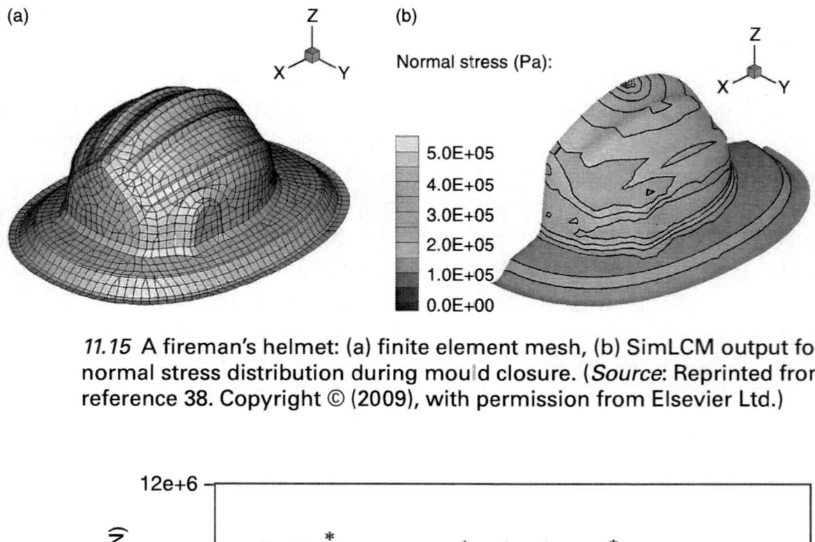

11.15 A fireman's helmet: (a) finite element mesh, (b) SimLCM output for normal stress distribution during mould closure. (*Source*: Reprinted from reference 38. Copyright © (2009), with permission from Elsevier Ltd.)

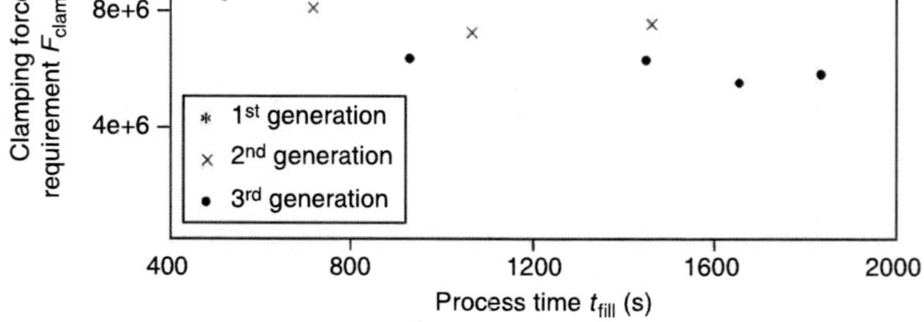

11.16 The first three generations output from the GA for the helmet geometry.

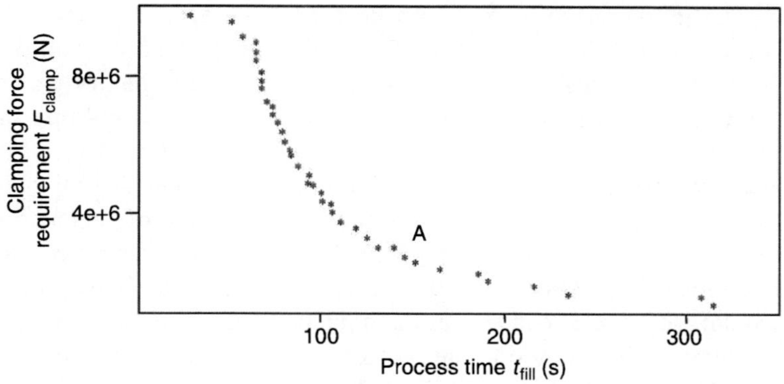

11.17 Final result for optimisation of the helmet geometry.

clamping force. The situation after 45 generations, by which time the procedure has converged satisfactorily, is shown in Fig. 11.17. The large clamping forces and fill times are due to the fibre reinforcement used (CFM), and the large target fibre volume fraction.

The final results of a GA CRTM analysis, of which Fig. 11.17 is an example, can be interpreted and used for design purposes by the manufacturer. For example, if one has a press with a certain force capability, one can discard the solutions for which F_{clamp} is greater. On the other hand, the results can be used to inform manufacturers whether it is advisable to invest in more powerful presses in order to reduce the fill time. Examining the particular data from Fig. 11.17, the solution marked 'A' shows a good trade-off between clamping force and process time. The clamping force is only about 20% of that of the leftmost solution, which is achieved with fill times that are approximately three times longer.

11.7 Future trends

CRTM filling is inherently more complicated than RTM, and requires careful process design if the full benefits of the technique are to be realised. Existing numerical models are able to address mould filling and resin cure, for simple preform shape and composition. Significant industrial and academic effort is currently being placed on development of efficient preforming techniques for complex shapes. Complex industrial preforms include multiple reinforcement types, stacked in many possible combinations, and these may be draped over doubly curved part surfaces. If numerical CRTM process models are to be adopted by industry, efficient techniques are required to characterise such complex preforms, across a much wider fibre volume fraction range than that considered for RTM.

Continuous fibre textile reinforcements are typically built with nominally repeating unit cells. However, significant in-plane variability can exist in the reinforcement architecture (i.e., tow width, direction and spacing). Additional geometric variability is introduced when multiple layers are stacked together, introducing the issue of layer nesting. This variability in preform architecture influences manufacturing (permeability and compaction response), and the structural performance of the manufactured laminate. The need has been acknowledged in the research community, to examine such variability to generate predictions of both average, and typical spread of permeability and compaction response. Addressing this, and the complexities of industrial preforms, may require application of predictive methods based on textile modelling, in conjunction with a limited set of characterisation experiments.

The primary advantage of the CRTM process is the potential to significantly reduce mould filling, and hence total cycle times. To date, the focus has been placed on rigid tool processes, for high production rates. Compression driven flows could be explored further in the context of semi-rigid tooling, typically used for RTM Light. This may serve to reduce cycle times, and extend the range of fibre reinforcements that can be used in such processes.

11.8 References

1. Walbran, W. A., Bickerton, S. and Kelly, P. A., 2009. 'Measurements of normal stress distributions experienced by rigid Liquid Composite Moulding tools', *Composites Part A: Applied Science and Manufacturing*, **40**, 1119–1133.
2. Han, K., Ni, J., Toth, J. and Lee, L. J., 1998. 'Analysis of an injection/compression liquid composite molding process', *Polymer Composites*, **19**, 487–496.
3. Shojaei, A. and Spah, A., 2007. 'A theoretical analysis of resin injection/compression molding', *Key Engineering Materials*, **334–335**, 209–212.
4. Deleglise, M., Binetruy, C. and Krawczak, P., 2006. 'Simulation of LCM processes involving induced or forced deformations', *Composites Part A: Applied Science and Manufacturing*, **37**, 874–880.
5. Gutowski, T. G., 1997. *Advanced composites manufacturing*. New York: John Wiley.
6. Shojaei, A., 2006. 'A numerical study of filling process through multilayer performs in resin injection/compression molding', *Composites Science and Technology*, **66**, 1546–1557.
7. Merotte, J., Simacek, P. and Advani, S. G., 2010. 'Resin flow analysis with fiber perform deformation in through thickness direction during compression resin transfer molding', *Composites Part A: Applied Science and Manufacturing*, **41**, 881–887.
8. Cai, Z., 1995. 'A nonlinear viscoelastic model for describing the deformation behaviour of braided fiber seals', *Textile Research Journal*, **65**(8), 461–470.
9. Bréard, J., Henzel, Y., Trochu, F. and Gauvin, R., 2003. 'Analysis of dynamic flows through porous media. Part II: Deformation of a double-scale fibrous reinforcement', *Polymer Composites*, **24**(3), 409–421.

10. Kelly, P. A., Umer, R. and Bickerton, S., 2006. 'Viscoelastic response of dry and wet fibrous materials during infusion processes', *Composites Part A: Applied Science and Manufacturing*, **37**(6), 868–873.
11. Somashekar, A. A., Bickerton, S. and Bhattacharyya, D., 2007. 'Exploring the non-elastic compression deformation of dry glass fibre reinforcements', *Composites Science and Technology*, **67**(2), 183–200.
12. Saunders, R. A., Lekakou, C. and Bader, M. G., 1999. 'Compression in the processing of polymer composites 1. A mechanical and microstructural study for different glass fabrics and resins', *Composites Science and Technology*, **59**(7), 983–993.
13. Walbran, W. A., Verleye, B., Bickerton, S. and Kelly, P. A., 2012. 'Prediction and experimental verification of normal stress distributions on mould tools during liquid composite moulding', *Composites Part A: Applied Science and Manufacturing*, **43**(1), 138–149.
14. Phelan, F. R., 1997. 'Simulation of the injection process in resin transfer molding', *Polymer Composites*, **19**, 460–476.
15. Shojaei, A. and Ghaffarian, S. R., 2003. 'Modelling and simulation approaches in the resin transfer molding process: A review', *Polymer Composites*, **24**, 525–544.
16. Davé, R., Kardos, J. L. and Dudukovic, M. P., 1987. 'A model for resin flow during composite processing. Part 1: General mathematical development', *Polymer Composites*, **8**, 29–38.
17. Gutowski, T. G., Morigaki, T. and Cai, Z., 1987. 'The consolidation of laminate composites', *Journal of Composite Materials*, **21**, 172–188.
18. Biot, M. A., 1941, 'General theory of three-dimensional consolidation', *Journal of Applied Physics*, **12**, 155–164.
19. Terzaghi, K., 1936. 'The shear resistance of saturated soils', *Proceedings of the 1st International Conference on Soil Mechanics and Foundation Engineering*, Cambridge, MA, **1**, 54–56.
20. Gibson, R. E., England, G. L. and Hussey, M. J. L., 1967. 'The theory of one-dimensional consolidation of saturated clays', *Géotechnique*, **17**, 261–273.
21. Davé, R., 1990. 'A unified approach to modelling resin flow during composite processing', *Journal of Composite Materials*, **24**, 22–41.
22. Pham, X.-T. and Trochu, F., 1999. 'Simulation of compression resin transfer molding to manufacture thin composite shells', *Polymer Composites*, **20**, 436–459.
23. Kang, M. K. and Lee, W. I., 1999. 'Analysis of resin transfer/compression molding process', *Polymer Composites*, **20**, 293–304.
24. Le Riche, R., Saouab, A. and Bréard, J., 2003. 'Coupled compression RTM and composite layup optimisation', *Composites Science and Technology*, **63**, 2277–2287.
25. Shojaei, A., 2006. 'Numerical simulation of the three-dimensional flow and analysis of filling process in compression resin transfer moulding', *Composites Part A: Applied Science and Manufacturing*, **37**, 1434–1450.
26. Bickerton, S. and Abdullah, M. Z., 2003. 'Modelling and evaluation of the filling stage of injection/compression moulding', *Composites Science and Technology*, **63**, 1359–1375.
27. Chang, C.-Y., Hourng, L.-W. and Chou, T.-Y., 2006. 'Effect of process variables on the quality of compression resin transfer molding', *Journal of Reinforced Plastics and Composites*, **25**, 1027–1037.
28. De Boer, R., 2000. *Theory of porous media: Highlights in historical development and current state*. Dordrecht: Springer-Verlag.

29. Delesse, M., 1848. 'Pour determiner la composition des roches', *Annals Mines*, **4**, 379–388.
30. Bishop, A. W., 1959. 'The principle of effective stress', *Teknisk Ukeblad*, **106**, 859–863.
31. Nuth, M. and Laloui, L., 2008. 'Effective stress concept in unsaturated soils: Clarification and validation of a unified framework', *International Journal for Numerical and Analytical Methods in Geomechanics*, **32**, 771–801.
32. Vlahini , I., Jennings, H. M., Andrade, J. E. and Thomas, J. J., 2011. 'A novel and general form of effective stress in a partially saturated porous material: The influence of microstructure', *Mechanics of Materials*, **43**, 25–35.
33. Pham, X.-T., Trochu, F. and Gauvin, R., 1998. 'Simulation of compression resin transfer molding with displacement control', *Journal of Reinforced Plastics and Composites*, **17**, 1525–1556.
34. Chang, C.-Y., 2008. 'A numerical study of filling process in resin injection/compression molding', *Journal of Reinforced Plastics and Composites*, **27**, 781–795.
35. Bhat, P., Merotte, J., Simacek, P. and Advani S. G., 2009. 'Process analysis of compression resin transfer molding', *Composites Part A: Applied Science and Manufacturing*, **40**, 431–441.
36. Verleye, B., Walbran, W.A., Bickerton, S. and Kelly, P. A., 2011. 'Simulation and experimental validation of force controlled compression resin transfer moulding', *Journal of Composite Materials*, **45**(7), 815–829.
37. Merotte, J., Simacek, P. and Advani, S. G., 2010, 'Flow analysis during compression of partially impregnated fiber perform under controlled force', *Composites Science and Technology*, **70**, 725–733.
38. Kelly, P. A. and Bickerton, S., 2009. 'A comprehensive filling and tooling force analysis for rigid mould LCM processes', *Composites Part A: Applied Science and Manufacturing*, **40**, 1685–1697.
39. Walbran, W. A., 2011. 'Modelling and development of rigid tool liquid composite moulding processes', PhD thesis, University of Auckland.
40. Hammami, A., Gauvin, R. and Trochu, F., 1998. 'Modelling the edge effect in liquid composites molding', *Composites Part A: Applied Science and Manufacturing*, **29**, 603–609.
41. Mohan, R. V., Shires, D. R., Tamma, K. K. and Ngo, N. D., 1998. 'Flow channels/fiber impregnation studies for the process modelling/analysis of complex engineering structures by resin transfer molding', *Polymer Composites*, **19**, 527–542.
42. Hsiao, K.-T., Mathur, R., Advani, S. G., Gillespie, J. W. and Fink, B. K., 2000. 'A closed-form solution for flow during the vacuum assisted resin transfer molding process', *Journal of Manufacturing Science and Engineering*, **122**, 463–475.
43. Chen, R., Dong, C., Liang, Z., Zhang, C. and Wang, B., 2004. 'Flow modelling and simulation for vacuum assisted resin transfer molding process with the equivalent permeability method', *Polymer Composites*, **25**, 146–164.
44. Modi, D., Simacek, P. and Advani, S. G., 2002. 'Numerical issues in mold filling simulations of liquid composites processing', *Proceedings of the 10th US–Japan conference on composite materials*, Stanford University, California.
45. Chang, C.-Y., 2006. 'Simulation of mold filling in simultaneous resin injection/compression molding', *Journal of Reinforced Plastics and Composites*, **25**, 1255–1268.
46. Beavers, G. S. and Joseph, D. D., 1967. 'Boundary conditions at a naturally permeable wall', *Journal of Fluid Mechanics*, **30**, 197–207.

47. Ni, J., Zhao, Y., Lee, L. J. and Nakamura, S., 1997. 'Analysis of two-regional flow in liquid composite molding', *Polymer Composites*, **18**, 254–269.
48. Han, K., Lee, L. J. and Liou, M., 1998. 'Fiber mat deformation in liquid composite molding II: Modelling', *Polymer Composites*, **14**, 151–160.
49. Simacek, P., Advani, S. G. and Iobst, S. A., 2008. 'Modelling flow in compression resin transfer molding for manufacturing of complex lightweight high-performance automotive parts', *Journal of Composite Materials*, **42**, 2523–2545.
50. Tari, M. J., Imbert, J. P., Lin, M. Y., Lavine, A. S. and Hahn, H. T., 1998. 'Analysis of resin transfer molding with high permeable layers', *Journal of Manufacturing Science and Engineering*, **120**, 609–616.
51. Sun, X., Shoujie, L. and Lee, L. J., 1998. 'Mold filling analysis in vacuum assisted resin transfer molding. Part 1: SCRIMP based on a high-permeable medium', *Polymer Composites*, **19**, 807–817.
52. Han, K., Jiang, S., Zhang, C. and Wang, B., 2000. 'Flow modelling and simulation of SCRIMP for composites manufacturing', *Composites Part A: Applied Science and Manufacturing*, **31**, 79–86.
53. Feather, B., 2008. 'Modelling of the GAICM Composite Materials Manufacturing Process', Part IV Report, Dept. Engineering Science, University of Auckland.
54. Wirth, S. and Gauvin, R., 1998. 'Experimental analysis of mold filling in compression resin transfer molding', *Journal of Reinforced Plastics and Composites*, **17**, 1414–1430.
55. Henz, B. J., Mohan, R. V. and Shires, D. R., 2007. 'A hybrid global–local approach for optimisation of injection gate locations in liquid composite molding process simulations', *Composites Part A: Applied Science and Manufacturing*, **38**, 1932–1946.
56. Na, S., 2008. 'Global optimisation of mould filling parameters during the constant speed injection compression moulding process', Master's of Engineering thesis, University of Auckland.
57. Ye, X., Zhang, C., Liang, Z. and Wang, W., 2004. 'Heuristic algorithm for determining optimal gate and vent locations for RTM process design', *Journal of Manufacturing Systems*, **23**, 267–277.
58. Kim, B., Nam, G., Ryo, H. and Lee, J., 2000. 'Optimisation of filling process in RTM using genetic algorithm', *Korea-Australia Rheology Journal*, **12**, 83–92.
59. Luo, J., Liang, Z., Zhang, C. and Wang, B., 2001. 'Optimum tooling design for resin transfer molding with virtual manufacturing and artificial intelligence', *Composites Part A: Applied Science and Manufacturing*, **32**, 877–888.
60. Gokce, A., Hsiao, K.-T. and Advani, S. G., 2002. 'Branch and bound search to optimise injection gate locations in liquid composite molding processes', *Composites Part A: Applied Science and Manufacturing*, **33**, 1263–1272.
61. Ruiz, E., Achim, V., Soukane, S., Trochu, F. and Bréard, J., 2006. 'Optimisation of injection flow rate to minimise micro/macro-voids formation in resin transfer molded composites', *Composites Science and Technology*, **66**, 475–486.
62. Lin, M. Y., Murphy, M. J. and Hahn, H. T., 2000. 'Resin transfer molding process optimisation', *Composites Part A: Applied Science and Manufacturing*, **31**, 361–371.
63. Pantelelis, N. G., 2003. 'Optimised cure cycles for resin transfer moulding', *Composites Science and Technology*, **63**, 249–264.
64. Ruiz, E. and Trochu, F., 2005. 'Comprehensive thermal optimisation of liquid composite molding to reduce cycle time and processing stresses', *Polymer Composites*, **26**, 209–230.

65. Trochu, F., Ruiz, E., Achim, V. and Soukane, S., 2006. 'Advanced numerical simulation of liquid composite molding for process analysis and optimisation', *Composites Part A: Applied Science and Manufacturing*, **37**, 890–902.
66. Ratle, F., Achim, V. and Trochu, F., 2009. 'Evolutionary operators for optimal gate location in liquid composite moulding', *Applied Soft Computing*, **9**, 817–823.
67. Jiang, S., Zhang, C. and Wang, B., 2002. 'Optimum arrangement of gate and vent locations for RTM process design using a mesh distance-based approach', *Composites Part A: Applied Science and Manufacturing*, **33**, 471–481.
68. Sánchez, F., García, J.A., Chinesta, F., Gascón, L. L., Zhang, C., Liang, Z. and Wang, B., 2006. 'A process performance index based on gate-distance and incubation time for the optimisation of gate locations in liquid composite molding processes', *Composites Part A: Applied Science and Manufacturing*, **37**, 903–912.
69. Buntain, M. J., 2003. 'Optimisation of mould filling during the injection compression moulding of fibre reinforced plastics', Master's of Engineering thesis, University of Auckland.
70. Hsu, S., Ehrgott, M. and Kelly, P. A., 2010. 'Optimisation of mould filling parameters of the compression resin transfer moulding process', *45th Annual ORSNZ Conference*, Auckland, 29–30 November 2010.
71. Kam, W., Kelly, P. A., Ehrgott, M., Verleye, B. and Bickerton, S., 2010. 'Optimisation of mould filling parameters during the compression resin transfer moulding process', *14th European Conference on Composite Materials (ECCM)*, Budapest, Hungary, 7–10 June 2010.
72. Ehrgott, M., 2005. *Multicriteria optimisation*, 2nd edn. Berlin: Springer Verlag.
73. Deb, K., 2002. *Introduction to evolutionary multiobjective optimisation.* New York: John Wiley.

12
The pultrusion process for polymer matrix composites

S. C. JOSHI,
Nanyang Technological University, Singapore

Abstract: Pultrusion is a continuous process for manufacturing composites with constant cross-sections or structural profiles having significantly long length. It is widely employed in the composites industry due to its continuous, automated and highly productive nature. This chapter presents a comprehensive study that covers the description, market, industry, innovations, variants and literature review of the pultrusion process technology. The first section gives an overview of the pultrusion process, its market and its history. The second section provides a detailed description on each component, the materials and the operations of pultrusion. Section three covers the literature review on various experimental and computational analyses of pultrusion. The fourth section encompasses the industrial standards, applications, patents, global markets and future trends of pultrusion process and pultruded products. The fifth section closes the chapter with concluding remarks.

Key words: process optimization, polymer, fibre reinforcement, innovations, future applications.

12.1 Introduction

With the passing of the first decade in the twenty-first century, advances in the field of materials have led to a need for revisiting the processes technology and science, and taking note of the latest developments and improvements. Current trends are that newer materials are being implemented in various applications as transitions from the conventional materials. For example, metals and plastics are making way for advanced composite materials that offer superior quality and performance. Composites are materials engineered from at least two distinguishable materials having different properties such that the resultant composites have the combined advantages and performance.

Fibre-reinforced polymer (FRP) composites are among the increasingly used composites in the industry. FRPs are composite materials made of a polymer matrix reinforced with fibres. The polymer matrix is either a thermosetting or a thermoplastic resin while the fibres are usually fibreglass,

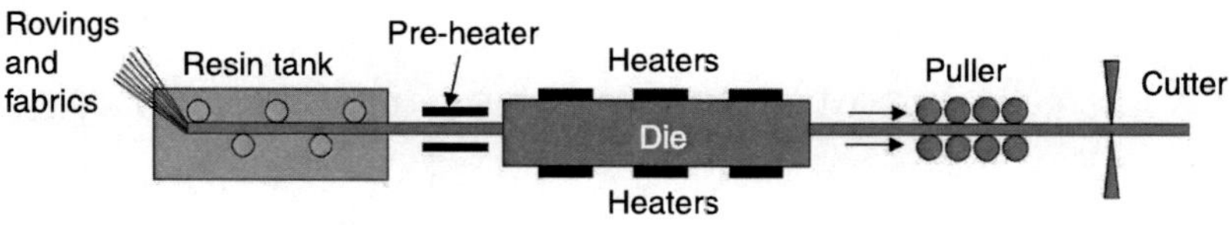

12.1 Schematic of pultrusion process.

carbon or aramid. Pultrusion is an effective process for manufacturing FRP composites parts with uniform cross-section. It is widely employed in the composite manufacturing industry due to its continuous, automated and highly productive nature.

12.1.1 Traditional pultrusion: an overview

The term pultrusion comes from the words 'pull' and 'extrusion'. It is the process where continuous fibres soaked in a polymer are pulled through a heated die and extruded (Hota *et al.*, 2009).

From Fig. 12.1, the pultrusion process starts with the pulling of reinforcing fibres in the forms of continuous rovings or mats from a series of creels. The fibrous material is fed continuously through a guiding system into a resin bath to get it impregnated. The resin-soaked fibres are guided to a heater assembly to preheat the fibre–resin mix while shaping it close to the desired finished profile. The same is then taken into a heated die where curing of thermosetting resin is initiated by the heat. The cured part subsequently leaves the die in the desired shape where it is cut into the required lengths using a cut-off saw without interrupting the continuous pultrusion process (Sharma *et al.*, 1998).

12.1.2 Historical background

Pultrusion is relatively an old process of manufacturing long FRP composites. Goldsworthy is believed to have pioneered the pultrusion process in the early 1950s. In those days the technique was mainly adopted to fabricate parts requiring uniaxial performance such as rods, poles, handles, etc.

The process has undergone considerable changes subsequently over the years. The variety and quality of structural profiles has been increasing. In 1960s, there were about 20 manufacturers, primarily in the United States. The number has been continuously growing, and in 2006 it was tallied to 300 pultruders in the world. In addition, the number of users of pultrusion in Asia and Eastern Europe is growing (Jacob, 2006).

12.1.3 Pultrusion applications

Pultrusion can be used to fabricate a wide range of solid and hollow structures with constant cross-sections (Mazumdar, 2009). In the beginning, around 1976, pultrusion products were more for the recreational, sporting and electrical sectors. As new players entered the market, the quantity and innovation of the applications increased. During the period 1976–1981, the number of corrosion resistant products by pultrusion went up at such a rapid rate that by 1981, these were at second place behind the electrical market (Meyer, 1985).

To date, the process has penetrated the application areas such as land transport, construction, infrastructure, electrical, marine and offshore, and aerospace. With the introduction of newer technologies, pultrusion can be used to produce nearly any constant cross-sectional shape (Meyer, 1985) such as channels, tubes, beams, bars and many other profiles for handrails, ladders, cable ducts, antennae, rail covers, tent poles, hockey sticks, fishing rods, pipes, slide guides, luggage racks, leaf springs, window frames, farm wagons, bridge sections, roof profiles, etc.

Advantages

Like other FRP composites, pultrusion products exhibit a high strength-to-weight ratio, excellent corrosion resistance, good electrical insulation and better dimensional stability. In addition, pultrusion offers additional advantages over other composite manufacturing processes (Sumerak and Martin, 1987):

- *Customizable product length*: The continuous nature of pultrusion allows any transportable length to be produced, ranging from cm to km (Sumerak and Martin, 1987).
- *High fibre content*: Pultrudates are under sufficient tension while they cure. This aligns the fibres well and results in a compact, high fibre content and strong product.
- *Low production cost*: Studies on products requiring optimum fibre placement and part performance have shown that pultrusion cost is about 41% of the filament winding and 26% of the prepeg hand lay-up (Krolewski and Gutowski, 1986; Strong, 2002). About 80–90% of the cost of pultruded composites comes from the raw materials it uses. The amortized machine cost and the labour to run pultrusion machines make up only small portion of the total factory costs.
- *Good production rate*: With pultrusion being a highly automatable process, the production rate can be quite high.
- *Consistent quality*: More automation with little human interference makes it possible to achieve a uniform quality in pultrusion.

Limitations

However, pultrusion also has some limitations as below (Mazumdar, 2009; Strong, 2002):

- *Tapered and complex shapes*: Pultrudates having constant cross-section can be easily produced, but production of tapered and complex shapes is unlikely.
- *Dimensional accuracy*: Part dimensions and their tolerances are not as good as those of other manufacturing methods.
- *Thin wall parts*: Producing thin parts is a challenge with pultrusion.
- *Processing problems*: A few processing difficulties such as resin buildup on the side of the die, voids and blister formation in the composites, non-uniform curing, etc., do arise from time-to-time in pultrusion.

12.2 Process description

As discussed earlier, the pultrusion process consists of pulling continuous rovings or mats of typical carbon or glass fibre reinforcement through a bath of thermosetting resin impregnating them, then into a fixture that shapes the composite partially by removing excess resin and air. Finally the shaped reinforcement is pulled into a temperature- and shape-controlled die where the composite is cured as it moves through the die from one end to the other (Meyer, 1985). The process components, the materials and the operations are discussed below in detail.

12.2.1 Components of pultrusion process operation

The basic pultrusion operation consists of six main and one optional components (Sumerak and Martin, 1987):

- reinforcement dispenser,
- resin impregnator,
- forming guides,
- Temperature controlled die,
- puller and clamp system,
- cut-off saw,
- mandrel (only for hollow sections) (see Fig. 12.2).

Reinforcement dispenser

The pultrusion process begins with this unit where either fibre or mat-shaped reinforcement is supplied from creels. Special reinforcements such

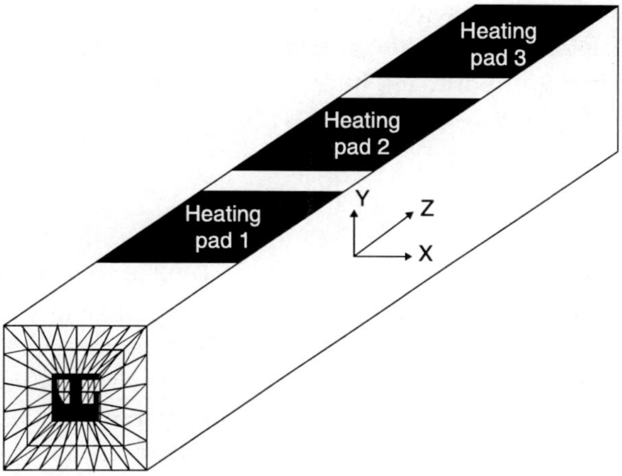

12.2 A typical hollow section pultrusion requiring mandrels.

as knitted, woven and braided fabrics, veils, fibre tows, etc., are mixed and dispensed as per the product design. Reinforcements are provided continuously from the mounted reels or bobbins generally with the help of ceramic eyes or pulleys to avoid entanglement and rubbing among themselves. Breakage and static charge buildup may occur due to any contacts or friction between dry fibrous reinforcements before they enter the resin.

Resin impregnator

The dry reinforcement is then allowed to pass through this unit to get impregnated by the resin. There are three types of impregnation systems currently used: (1) dip bath, (2) straight through bath and (3) resin injection system (Owens Corning, 2003).

The dip bath, also called an open bath, involves the use of a resin tank where polymer and hardener are premixed and maintained at desirable viscosity. The fibrous reinforcement travelling from the dispenser generally comes through a guiding comb, which keeps it aligned. The bar-guides then direct the reinforcement into and out of the resin dip. Another comb is sometimes used at the exit to squeeze out the dripping resin of the reinforcement. Although this allows good impregnation, it is a messy process.

In a straight through resin bath, liquid resin is allowed to leak through a resin trough with forming cards from open areas. Any excess resin is collected and fed back into the resin trough. The use of this unit avoids unnecessary bending of reinforcement.

A relatively new concept is to inject resin into the dry fibrous preform just before it enters the die. A resin–hardener mix is stored and supplied through a steel chamber attached to the front end of the die. An optimized combination of the mould cavity, resin injection pressure and pulling speed of the reinforcement into the die generates enough pressure to force the resin into the fibrous preform and impregnates it fully. This technique not only allows good fibre wet-out, but also reduces process time and avoids resin waste.

Forming guides

These guides, when provided, are meant to consolidate the resin-impregnated reinforcements, densifying them into a required shape. The sizing of the forming guide slots and holes as well as the clearances between the forming plates should be designed to prevent excessive tension on the wet reinforcements while allowing sufficient resin removal (Sumerak and Martin, 1987). The construction material used could be steel or ultrahigh molecular weight polyethylene (UHMWPE) (Owens Corning, 2003). Steel is less expensive, but is hard, corrosive and more difficult to machine. The UHMWPE is lighter, chemically resistant, gentle on fibre reinforcement, easy to fabricate, but not as long lasting for it wears out faster than steel.

Temperature controlled die

The die is the heart of the pultrusion process. The resin-impregnated reinforcement is heavily compressed while it enters the die, resulting in squeezing off of all extra resin. The component gets its density and final shape in the front region of the die, before the composites starts to heat up and cure. The temperature at the entrance of the die is generally kept low to avoid premature curing of the resin thereby causing blockage in the mould cavity.

The die heating may be facilitated using electrical cartridge heaters, strip heaters or by circulating oil. Even though the main heat transfer is affected by conduction, in some cases radio frequency radiations are used. The dies are mainly fabricated of steel with their interior surfaces generally chrome plated for increased die life and improved surface finish (Meyer, 1985).

A die can either be one-piece (see Fig. 12.3a) or split cavity die (see Fig. 12.3b). A one-piece die provides good uniform finished part without parting lines, yet is usually costly to maintain in the long run. Repairs are complex due to poor accessibility. Split cavity dies can be easily opened, thus allowing easy cleaning and maintenance. More care is needed in designing these dies so that the parting line between the mould halves is compatible with the component being fabricated (Owens Corning, 2003).

Puller and clamp system

There are two types of puller/clamp systems that are currently in use. The method that utilizes hydraulic reciprocating puller consists of two identical

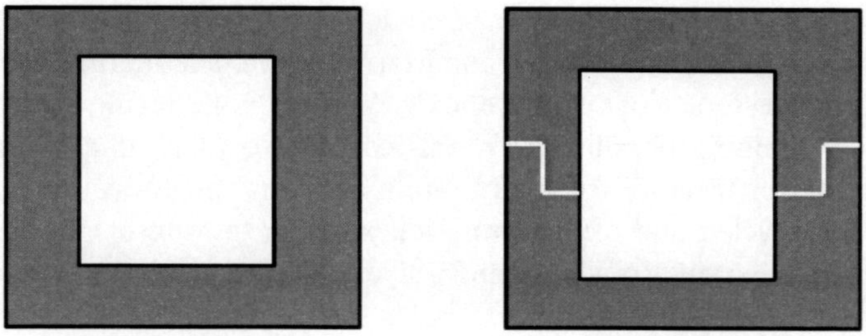

12.3 Schematic showing type of die design.

units that operate to grip and pull the profile alternately during pultrusion. The continuous caterpillar-tractor type system is used for mat/roving type composites in which grips clamp to the part and the puller drags the part through the die.

Sufficient distance (of about 3 m) between the die exit and the pulling device should be maintained to cool the hot pultruded product coming out of the die in the open atmosphere or in an air-cooling stream. This allows the pultrudate to develop adequate strength so as to withstand clamp and puller forces.

Cut-off saw

This is the last station where a flying cut-off saw is installed. The saw action can be synchronized with the puller movement so that sections of required lengths are cut automatically. The saw usually consists of an abrasive or a continuous rim diamond wheel. A coolant spray is sometimes employed to cool the cut-off wheel as well as to minimize dust.

12.2.2 Materials

The pultruded composites mainly consist of continuous fibrous reinforcement, polymeric matrix, fillers and additives.

Reinforcements are the primary load carrying constituents and thus determine the strength and rigidity of the resulting pultrudate. The process that requires pulling is only possible because the reinforcement allows the part to be pulled through the die while it cures at the same time. The common reinforcements include glass (E, S or A type), carbon, aramid, boron and several new thermoplastic (polyesters, nylon) fibres.

The matrix can be a thermosetting or thermoplastic polymer. The matrix resin determines the actual level of effective bonding and load transfer between reinforcing units. It also imparts properties such as high-temperature performance, corrosion resistance, dielectric properties, flammability and thermal conductivity (Sumerak and Martin, 1987). The thermosetting resins include polyesters, vinyl esters, epoxies and phenolic resins. Thermoplastic resins are gaining popularity as matrices due to their unique advantages such as better fracture toughness, ability to join and form thermally, ability to be recycled and offer lower weights than thermosets. Polyethylenes, polyurethanes, polypropylenes and polyamides are primarily used as matrix materials.

Fillers and additives are specifically added to enhance the specific performance, reduce cost, change viscosity or improve processibility of resin systems (Owens Corning, 2003). Dry fillers usually make up the largest proportion (up to 50 wt%) of a resin formulation. Commonly used fillers in pultrusion include calcium carbonate as a volume extender, alumina silicate or clay to build corrosion resistance and electrical insulation, and alumina trihydrate for better flame or smoke retardation and electrical arc resistance. Additives are meant to tailor specific performance or properties. These typically include initiators to influence resin curing, mould release compounds such as metallic stearates or organic phosphate ester, antimony oxide for flame retardance, pigments for coloration and agents for surface smoothness and crack suppression.

12.2.3 Technological challenges

Material composition

Pultrusion is a versatile process that allows the use of selective type of reinforcement made of different materials. Resin formulation and reinforcement selection determine the properties of the final product. A trade-off is mostly required to accomplish desired properties while keeping the product cost low. However, to have the trade-off done optimally, some groundwork and preliminary studies are required. The material composition also decides the thermal mass in the system that plays a part in its rate of conversion to the final form.

Pultrudate shaping

As discussed earlier, resin impregnated fibrous reinforcement is given the required form before entering the heated die. This is done either completely by passing the wet lay-up through forming guides or partially in the guides and the rest within the entrance section of the die.

The forming guides should be designed to ensure consistent and continuous placement of the different reinforcement types in the required configuration. Especially for large and complicated sections, the guide design can become very complex which requires a good knowledge of the reinforcements, their location within the pultrudate, the resin system and its characteristics, as well as the pultrusion process itself.

The pultrudate shaping is the key to enhancing the quality of the final product.

Heat transfer and pull speed

Thermosetting resins require a well-designed cure cycle to develop the necessary properties and thus both the heat ramp-up rate and the cooling-down rate should be maintained such that the curing progresses gradually and uniformly. It is necessary to understand the coupled effect of the die heating and the pull speed. Too slow a movement of the pultrudate through the hot die may result in over-cured component and vice versa. Also, the composites should not experience temperature overshoot while within the die due to its exothermic curing process. Thermoplastics should also be given enough time to melt and impregnate the fibres before final consolidation and cooling. Gentle cooling rates should be adopted to avoid any residual stresses, warping or cracking in the component.

12.2.4 Process controlling equations and formulae

Governing equations

With the assumption that a fibrous medium impregnated with incompressible resin forms a macroscopically homogeneous material system for heat transfer analysis, the three-dimensional energy equation for a moving composites lay-up in a pultrusion die may be written as:

$$\rho_c C_{pc}\left(\frac{\partial T}{\partial t}+w\frac{\partial T}{\partial z}\right)=\nabla\left(\bar{k}_c\nabla T\right)+\frac{\partial H}{\partial t} \qquad [12.1]$$

where T is the temperature, H the heat energy and t the time. The ∇ denotes differential operator and Z the pull direction. The term $\partial H/\partial t$ defines the rate of internal heat generation in the composites, which for the case of thermosetting resin represents the heat of exothermic cure reaction. By substituting $w = 0$, Eq. [12.1] can be used to express heat transfer in a non-moving die section, where $\partial H/\partial t$ represents the rate at which heating power is supplied to or lost from the die.

When a composites mass containing reactive resin moves through a die, the conservation of chemical species can be expressed as (Lam *et al.*, 1998):

$$\frac{\partial \alpha}{\partial t}+w\,\frac{\partial \alpha}{\partial z}=\dot{R}_\alpha \qquad [12.2]$$

where α, the degree of cure (DOC), is defined as the ratio of the heat released to the total heat of cure reaction. $\dot{R}_\alpha$ is the rate of cure reaction. The relationship between $d\alpha/dt$, T and α can be approximated using an Arrehenius equation as (Joshi *et al.*, 1999):

$$\frac{d\alpha}{dt} = B_0 \exp\left(\frac{-\Delta E}{RT}\right)(1-\alpha)^n \qquad [12.3]$$

where B_0 is the pre-exponential constant, ΔE the activation energy, R the universal gas constant and n the order of reaction. The dH/dt for a resin saturated preform is directly proportional to $\dot{R}_\alpha$, and these two terms can be related by an equation as:

$$\frac{dH}{dt} = \rho_r v_r H_t \dot{R}_\alpha \qquad [12.4]$$

where H_t is the total heat of reaction per unit mass of resin, v the volume fraction and the subscript r denotes the matrix (resin).

Changes in the material properties

Thermal expansion of resin depends on its temperature and volumetric shrinkage of resin is a function of DOC (Batch and Macosko, 1993; Hackett and Zhu, 1992). The change in a unit dimension of a volume due to these effects, δ_r^j, may be calculated as:

$$\delta_r = \left[1 + \varepsilon_r (T_T - T_0)\right] \times \left(1 - \frac{\gamma_r \alpha_T}{100}\right)^{\frac{1}{3}} \qquad [12.5]$$

where, ε_r is the coefficient of thermal expansion, and γ_r is the percentage volumetric shrinkage of resin.

It may be observed from Eq. [12.5] that when a composite part is being cured in a die, its dimensions change continuously. It is reasonable to assume that during pultrusion only the cross-sectional (i.e., X and Y) dimensions of a pultrudate change and the axis of gravity of the component (i.e., the Z-axis along which the pultrudate is moving) acts as a zero-shrinkage axis because the continuous lengths of the fibre reinforcement and the applied pulling force allow very little dimensional changes due to thermal expansion and shrinkage of resin in the longitudinal direction.

Therefore, ignoring the effects of thermal expansion and shrinkage of resin in the longitudinal direction and applying a rule of mixture based on volume fractions (v), the density of material content of pultrudate may be calculated as:

$$\rho_c = v_r \left[\frac{\rho_r}{(\delta_r)^2}\right] + v_f \rho_f \qquad [12.6]$$

where, subscript f refers to fibre reinforcement.

The thermal properties, specific heat (C_{pc}) and directional thermal conductivity ($\bar{k}_c$), of composites can also be calculated in real-time using rules of mixture based on final DOC and mass fractions (m) as:

$$C_{pc} = m_r\left[(1-\alpha_T)C^u_{pr} + \alpha_T C^s_{pr}\right] + m_f C_{pf} \quad [12.7]$$

$$\frac{1}{\bar{k}_c} = \frac{m_f}{\bar{k}_f} + \frac{m_r}{\left[(1-\alpha_T)k^u_r + \alpha_T k^s_r\right]} \quad [12.8]$$

where, $m_f = v_f(\rho_f/\rho_c)$, $m_r = 1 - m_f$, and superscripts u and s represent uncured and fully cured resin, respectively.

12.3 Improvements in pultrusion

Some disadvantages exist in the traditional thermosetting and thermoplastic pultrusion processes due to the limitation in different major aspects of the processes. However, studies conducted in various key features and problem areas, as listed in Table 12.1, of pultrusion process have resulted in better understanding and major advancements in the process.

As a result, a number of pultrusion variants that combine conventional pultrusion and other processes such as compression moulding, injection, braiding, etc., have gained attention due to the ability to integrate the best qualities of the basic processes (Strong, 2002; Sumerak and Martin, 1987).

12.3.1 Traditional pultrusion

Published investigations on the operations, equipment and materials used in thermosetting and thermoplastic pultrusion provide a base for understanding the current and prospective developments in these areas.

Table 12.1 Aspects of pultrusion that contributed to the improvement of the process

Areas related to equipment	Areas related to process
• Fibre drying facility	• Resin viscosity and fibre wet-out
• Resin preheating	• Additives used and their fraction
• Die design with multiple heating zones	• Pulling force
• Number of heaters, heating power and heater locations	• Fibre content distribution
• Post-curing facility	• Line speed

In thermosetting pultrusion, overall mechanical properties of a pultrudate depend largely upon the degree and uniformity of cross-linking or cure. Non-uniform curing of the composites during pultrusion leads to an anisotropic and inhomogeneous product. It has been shown that the extent and the uniformity of the curing achieved during pultrusion can be improved by optimizing some of the parameters listed in Table 12.1 (Lam *et al.*, 2003).

In thermoplastic pultrusion, there is no need for the wet resin stage for fibre impregnation. The plastic is added in one of its solid forms such as films, strips, powders or fibres. The function of the die, as in thermosetting pultrusion, is to heat the material above the melting temperature of the polymer as the composite moves through it to allow complete consolidation and forming of the part. As such, there is no curing of the resin and the related volume change. However, it is necessary to have a second die (see Fig. 12.4) placed immediately after the heating/consolidation die to control the cool-down rate for the part and facilitate its cooling. It is important that the resin be fully crystallized so as to achieve the best mechanical properties (Ahmed *et al.*, 2004).

Experimental studies and optimization

This section reviews some key experimental studies leading to process optimization.

Ma *et al.* (1986) designed unique thermosetting pultrusion equipment that consisted of a fibre drying chamber, a preheated resin tank, a die with multiple heating zones and a post-cure chamber. The equipment reportedly improved the pultrusion rate and the properties of the pultruded part.

A systematic study on complete wet-out of fibre reinforcement by resin was conducted by Dharia and Schott (1986). They considered two aspects: (a) outer coating of resin and (b) resin penetration into fibre roving. As expected, with the increasing pull speed and the resin viscosity, the amount of resin that coats the rovings from outside increased while the resin penetration into the fibre bundles decreased. It was determined that the penetration of resin at lower pull speeds is mainly governed by capillary forces

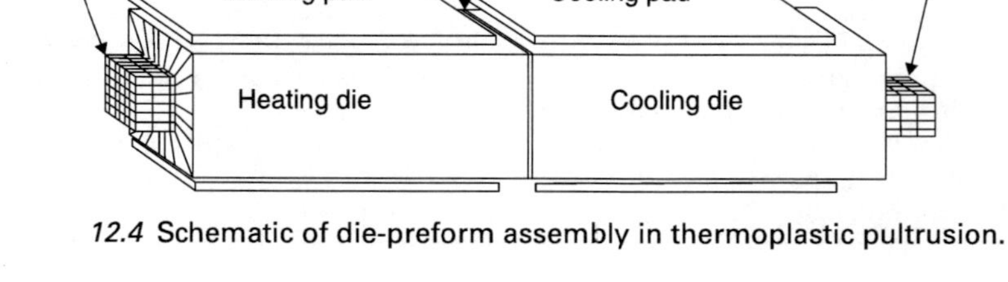

12.4 Schematic of die-preform assembly in thermoplastic pultrusion.

while at higher pull speeds the resin is forced in by the squeezing force. It was noted that the line speed and viscosity of resin can be optimized for a proper wet-out and optimum product quality.

A statistical design experimentation technique was presented by Vaughan (1988). The technique was based on a three-stage experimentation method that provides for the determination, characterization and optimization of various processing variables in pultrusion. The first stage induces the identification of all process variables. A set of screening experiments subsequently follow in the second stage in order to determine the level of influence of each process variable. The optimization of the process forms the third stage. This optimization method finally provides a mathematical description relating product quality or process efficiency value to any number of important process control variables.

A study on the design of the heating configuration for an existing pultrusion die has been carried out by Awa *et al.* (1992). The design optimization technique was used to synthesize the heating configuration in terms of the number of heaters, the power input as well as the locations of the heaters for a laboratory-scale die to produce the required temperature profile for a pultrusion product.

An analysis of the pultrusion machine pulling force, including its importance in optimizing quality and productivity was done by Sumerak and Martin (1993). Factors considered that could cause changes in the pulling force included loss or excess of the reinforcement, poor fibre wet-out, change in resin reactivity or viscosity, loss of temperature or change in the line speed.

Kowsika and Mantena (1996) proposed a statistical test pattern in manufacturing unidirectional graphite–epoxy composite beams using pultrusion. The influences of significant variables of the pultrusion process together with their interactive effects on the dynamic mechanical properties were investigated. Mathematical models were subsequently derived to determine the optimal pultrusion process conditions for improved dynamic mechanical properties of the finished product.

An experimental procedure was developed by Squires *et al.* (1996) to simulate the thermoplastic pultrusion process. They compared an intentionally designed compression moulding die with the same cross-sectional geometry pultrusion machine. The results of the experiments helped in determining the range of acceptable processing parameters. This knowledge can be blended with theoretical models of the thermoplastic pultrusion process to arrive at an optimized die configuration with necessary heating-cooling systems.

A novel process was developed by Chen and Wang (1998) for manufacturing glass/polystyrene composite pultrudate. They studied effects of the processing parameters such as the die temperature, the pulling rate, the post-cure temperature and time, the filler type and their contents, as well as the glass fibre content, etc., on the mechanical properties of the fibre composites. It was identified that the mechanical properties increased with increasing die temperature and glass fibre content, and with decreasing pulling rate. The post-curing also improved the properties. However, increasing talc content caused the properties to deteriorate.

The mechanism of blister formation and the effect of process variables such as die temperature, pulling speed, die length and composite thickness on blister formation was investigated by Li *et al.* (2002a). Dies with different length with a stop-start method were used to investigate blister formation. The results showed that for a given die and resin system, low temperature and higher pulling speed lead to blister formation. Longer dies on the other hand can prevent blister formation. The study suggests that heater power input should be optimized in high-speed pultrusion to avoid blisters.

In summary, it may be concluded that:

- most process parameters such as pull speed, die temperatures, pulling force, resin viscosity, fibre wet-out, etc., are interdependent and have coupled effect and
- post curing can be effectively used for improving pultrudate properties.

Numerical simulation and optimization

A numerical model has been developed by Viola *et al.* (1990) to optimize the operating parameters of a pultrusion line. Their model incorporates a one-dimensional finite difference simulation of the pultrusion process. Based on given target parameters, the model primarily optimizes the temperature set points along the die.

The use of a numerical heat transfer model and a design optimization procedure to simulate and synthesize the heater configuration in a laboratory-scale pultrusion die was developed and studied by Awa and West (1992). A two-dimensional steady-state conduction heat transfer model was developed to compute the temperature profile within the laboratory-scale die.

The study for predicting the velocity profiles in pultrusion by Gorthala *et al.* (1994) used a variable viscosity model. A comprehensive two-dimensional mathematical model in cylindrical coordinates was developed for resin flow, cure and heat transfer. A control-volume-based finite difference method (FDM) (Patankar method) was used for solving the governing equations. The use of artificial neural networks (ANNs) for pultrusion modelling in terms of the real process data and their potential for intelligent machine control was proposed by Wilcox and Wright (1998). Liu *et al.* (2000; Liu, 2001; Liu and Hillier, 1999) implemented a finite element/control volume technique for the simulation of the heat transfer and curing processes during the pultrusion of a fibreglass-vinyl ester I beam. Numerical simulations were conducted to obtain the temperature and curing profiles for different temperature settings on the die, pull speeds as well as for different fibre volume fractions. It was noted that increasing the control temperature had similar effect on the temperature and curing profiles as those incurred by the reducing pull speed.

A mathematical relationship between the DOC profile across the cross-section of a pultruded part and the die-heater temperatures was established by Li *et al.* (2002b). The relationship was employed to optimize the die-heating profile such that a near-uniformly cured component could be obtained finally. The algorithm was implemented for simulation of pultrusion process using the three-dimensional finite element/nodal control volume (FE/NCV) approach developed by Joshi and Lam (2001).

An objective function representing the effects of die-heater temperatures and pull speed on the distribution of DOC across the cross-section of a pultruded part at the die exit was used by Lam *et al.* (2003) to develop an algorithm for calculating the required changes to achieve a desirable DOC with maximum possible uniformity. The algorithm was implemented in the three-dimensional FE/NCV approach for process simulation, in which a general-purpose FE package was used to perform heat transfer analysis, together with other routines developed to perform cure modelling and the cure optimization.

The FE/NCV method was employed by Joshi *et al.* (2003) on optimization studies to achieve the desired DOC with minimum local variations across the pultrudate cross-section. The die-heating environment was optimized for a few cases, with different initial temperatures for a glass/epoxy wet preform and for the cooler installed within the pultrusion die near its entrance. The role of these temperature parameters in moderating the optimization constraints was examined.

The plots of maximum temperature experienced within the composite in the die, the mean DOC as a function of the number of optimization iterations for these parameters are shown in Fig. 12.5 (Joshi *et al.*, 2003).

Various optimization studies on thermosetting as well as thermoplastic pultrusion parameters were conducted (Chew, 2004; Joshi *et al.*, 2003; Kyaw, 2003; Li, 2001; Tat, 2006; Zaw, 2003) using the same FE/NCV method developed by Joshi and Lam (2001). Chew (2004) worked on cooling optimization in thermoplastic pultrusion whereas Tat (2006) studied the impact of mounting locations for the feedback thermocouples controlling heater temperatures. It was observed that if the thermocouples sense either the maximum or the minimum temperature on a hot die surface, the power input to the heater will be rather uneven and the process may not completely optimize. Joshi *et al.* (2003) looked for improved pultudate quality with the adjustments in the preheating and die-cooler temperatures. Li (2001) managed to optimize the combined effect of pull speed and the heater temperatures whereas Kyaw (2003) went on to study the process optimization with immediate post-curing of the pultrudate. Zaw (2003) focused on the number and size of die-heating pads. His studies determined that the 4-heater die configuration (see Fig. 12.6) works better than the 6-heater die configuration. His further optimization of the curing process revealed that after optimization the percentage improvement in the standard deviation of the DOC was 76.36% whereas the degree of cure itself improved by 23.79%. Figure 12.7 depicts the cure profile of the C-section at die exit.

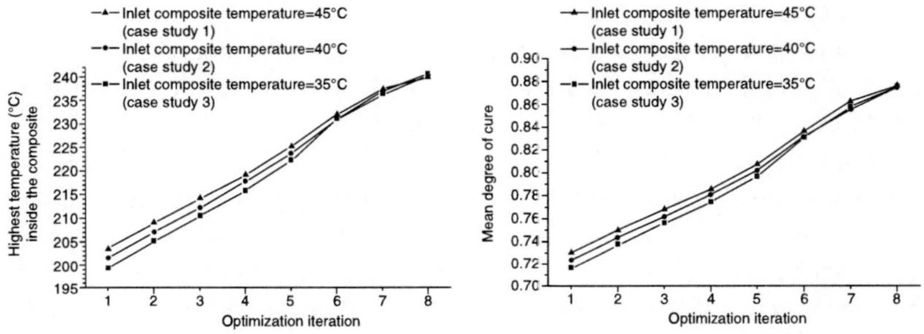

12.5 Variations in the highest temperature within the composite and the mean DOC as a function of composite pre-die temperature (Joshi *et al.*, 2003).

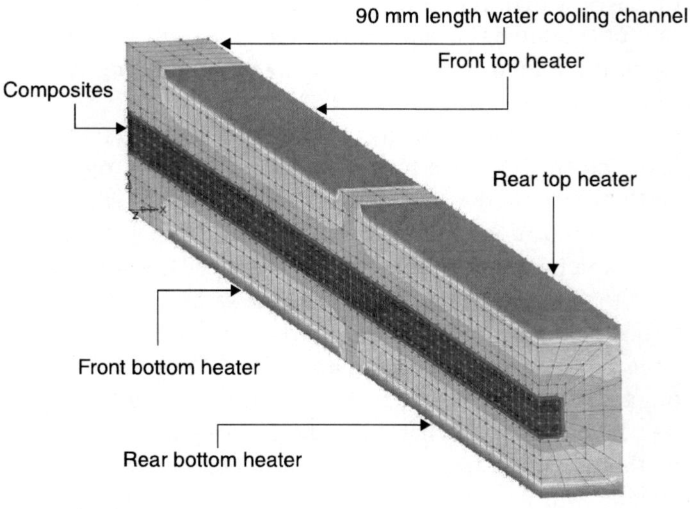

12.6 Four-heater pull-die configuration.

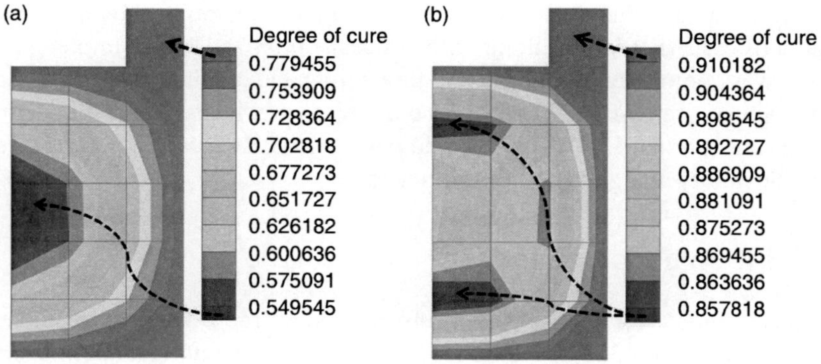

12.7 State of cure (a) before optimization and (b) after optimization.

Table 12.2 Correlation between FEM and FDM pultrusion analyses

Data	Temperature peak (°C)	Mean DOC	Standard deviation of DOC
Joshi *et al.* (2003)	217.72	0.892	0.0045
Finite difference model	215.85	0.8897	0.0042
Finite element model	217.27	0.8913	0.0049

Two different computational methods, finite differences and elements, were developed and analysed by Carlone *et al.* (2006). A comparison between the peak temperature of the pultrudate, the mean DOC at the exit section, and the standard deviation of the DOC, in the same section, are compared in Table 12.2.

The same authors (Carlone *et al.*, 2007) further developed a hybrid approach, based on genetic algorithms and simplex method, to optimize the accuracy of pultrusion process using a finite difference scheme.

According to the curing kinetics and the heat transfer theories, Chen *et al.* (2010) established the models for unsteady field of temperature and curing. The finite element method (FEM) and FDM were combined, and the indirect decoupling method based on ANSYS was adopted to simulate the temperature profile in pultrudate composites during pultrusion.

A mathematical model was established for the variation of temperature distribution and state of curing by Lu *et al.* (2008) on the basis of curing kinetic and unstable thermal conduction theories. A numerical model was established for the temperature and curing degree by combining the FEM with FDM and the coupled question was solved using the Euler-Cauch step-by-step iterative method. The effects of processing parameters including the pulling speed, die wall temperature and the initial temperature on the profile of curing degree and temperature in the die were analysed and discussed.

In summary, the FE/NCV approach seemed versatile allowing three-dimensional simulation along with process optimization. Its implementation was demonstrated in terms of die-heating and cooling, resin shrinkage, pull speed, state and uniformity of degree of cure, heater feedback operations and post-curing. It also allowed partial use FE software thereby offering flexibility and good computing speed.

12.3.2 Injection pultrusion

Injected pultrusion (IP) is a hybrid technique of traditional pultrusion and resin transfer moulding (RTM) used in manufacturing polymer composites. The process is similar to RTM in that the thermoset resin is injected through the top and/or bottom injection gates into the dry reinforcement after it is pulled into the injection-pultrusion die (see Fig. 12.8). The rest of the process is similar to traditional pultrusion (Liu, 2003a, 2003b; Srinivasagupta and Kardos, 2004; Srinivasagupta *et al.*, 2003).

This process not only allows a higher pull speed but also significantly reduces the volatile emissions due to the elimination of the open resin bath in the process.

Experimental studies and optimization

Kim *et al.* (1991) presented a study on controlling the injection pressure and optimizing the die design for IP process. Experimental results revealed that injection pressure depends upon resin viscosity, material permeability, fibre volume fraction and pulling speed.

Dubé *et al.* (1995) addressed both the issues of volatiles and fibre wetting with reaction injection pultrusion (RIP) process used to produce thermoplastic polyurethane and thermoset polyisocyanurate matrix composites. The low viscosity constituents used in RIP help in improving fibre impregnation, while the small volume of the impregnation bath reduces emissions.

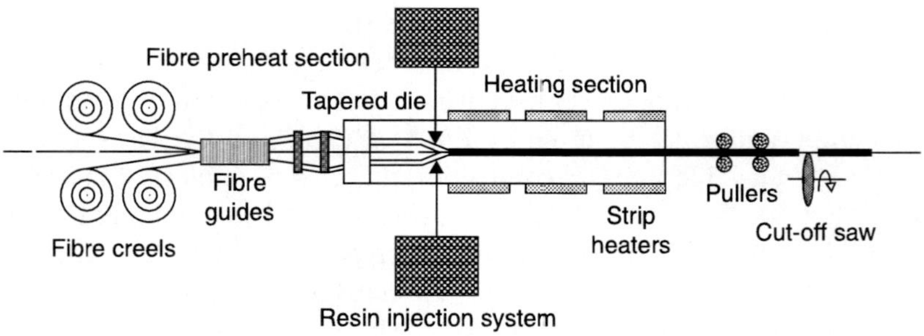

12.8 Schematic of injection pultrusion (Kim *et al.*, 1991).

Lackey and Vaughan (1997) presented basic information that can be used in the development of resin injection pultrusion systems. Experimental results and observations were presented to demonstrate the importance of factors such as injection chamber design, injection pressure, type of fibre and the elimination of voids for resin injection pultrusion systems.

It is clear from the above discussion that the resin injection process parameters have significant bearing on the quality of the pultruded product.

Numerical simulation and optimization

Kommu *et al.* (1998) worked on a computer simulation model for the IP process. First, the governing equations for conservation of mass, momentum and energy were developed using a local volume averaging approach. Subsequently, a simulation model of the IP process was developed using finite element/control volume (FE/CV) and FD techniques. It was shown that the model could be effectively used in die geometry design as well as optimization of the operating conditions for a given pultrudate.

Voorakaranam *et al.* (1998) presented a study on maximizing the production rates while maintaining IP quality. Based on statistical significance tests, streamlined regression models were generated by identifying processing variables and parameters that have a crucial bearing on the part quality as well as by eliminating superfluous variables.

Optimization and numerical simulations were carried out by Ding *et al.* (2000; Ding and Lee, 1999). Models for predicting compressibility and permeability of dry fibre rovings and random mats were developed and implemented in the process simulation code. The Control Volume based Finite Element Method (CV/FEM) was used to solve the resin flow problem and simulation results were verified by experiments.

An FE/NCV technique was developed by Liu (2003a, 2003b) to simulate the resin flow during IP process. In particular, an algorithm was developed to advance the flow front by taking into account both the resin flow relative to the reinforcement and the movement of the pultruded part as a whole.

Srinivasagupta and Kardos (2004) developed an algorithm for economical design of the IP process, subject to controllability considerations. The multi-objective approach was able to determine the optimal values of the processing parameters such as heating zone temperatures and resin injection pressure, as well as the equipment specifications (die dimensions, heater, puller and pump ratings) that satisfy the various objectives in a weighted sense.

It may be seen that hybrid numerical methods based on FE and CV concepts are able to provide necessary process analysis capability including the process optimization. Phenomena such as resin injection, resin flow, state and degree of cure as function of different process variables can be simulated.

12.3.3 Microwave pultrusion

The microwave assisted pultrusion process (Methven *et al.*, 2000) is an innovative variant of the pultrusion process. Microwave heating (Metaxas and Meredith, 1983) characterized by a high frequency electromagnetic energy source is a fast, instantaneous, non-contact and volumetric heating, and therefore, very interesting for materials processing.

Rapid microwave heating allows shorter die lengths, higher line speeds and much smaller pulling forces. It also allows the manufacture of large section profiles (Carlone and Palazzo, 2008).

Experimental studies and optimization

Metaxas and Meredith (1983) wrote a book on the theory and practice of industrial microwave heating, which also holds good for the pultrusion as well.

Lin and Hawley (1993) were the ones to conduct preliminary tests on the continuous microwave pultrusion on the prepregs. A microwave transparent ceramic die was designed to fit into the microwave cavity. The feasibility of the process was proved by the heating and curing results.

Smith *et al.* (1998; Smith and Hawley, 1998) conducted experiment studies wherein it was observed the extent of cure increased with microwave input power, yet decreased with pulling speed.

Methven *et al.* (2000; Methven and Ghaffariyan, 1992) presented reviews of the potential for microwave pultrusion. It was suggested that the extent of cross-linking in the die may be diffusion limited. Methven (1999) also observed that fibre microwave offers faster line speeds, reduced pulling forces, greater cure uniformity and reduced floor area.

Alternate heating using microwave has proved an energy and time saving scheme for resin curing in many composites manufacturing processes including pultrusion.

Numerical simulation and optimization

Carlone and Palazzo (2008) developed a computational modelling of microwave assisted pultrusion. This model is based on an electromagnetic submodel, meant to evaluate the electric field distribution and the heat generation rate due to the microwave source and on a thermochemical submodel, used to determine the temperature and degree of cure distributions. The performed simulations revealed the relevance of design of the microwave cavity, the curing die, and the importance of the dielectric properties of the materials in microwave pultrusion process.

12.3.4 Other varieties of pultrusion

Braiding pultrusion

Braiding pultrusion (see Fig. 12.9) is another hybrid process that combines braiding and pultrusion processes. Braiding creates the required fibrous preform strong enough to withstand pultrusion forces, while pultrusion provides the usual resin impregnation, curing and line speed (Freger *et al.*, 2005). Byun *et al.* (1996) developed a braided-pultrusion process by utilizing a novel resin impregnation device. Tubular composites were produced using Kevler 49 fibre and polyester.

Michaeli and Juerss (1996) presented results indicating that thermoplastic pull-braiding (reinforcing and polymer fibres are mingled together in dry form) of complex-shaped profiles is also possible. Bechtold *et al.* (2002) presented a pull-braiding process combined with a contact preheating system for commingled glass fibre/polypropylene yarns. In the study by Hidekuma *et al.* (2008), unidirectional fibres were covered with braided yarns to reduce the anisotropy. It was found that the braiding layer constrained development of cracks.

Pulforming

Pultrusion generally can only produce straight, constant-volume products or profiles using continuous reinforcements. Pulforming does not have this type of limitation (Ewald, 1981; Goldsworthy, 1979).

This process uses the fibre spools, alignment bushings, resin bath and shaping die to produce a continuous strip of uncured moulding material. In its most advanced form, moulds are incorporated in line with the impregnating equipment, producing parts continuously and automatically (Lewis, 1991).

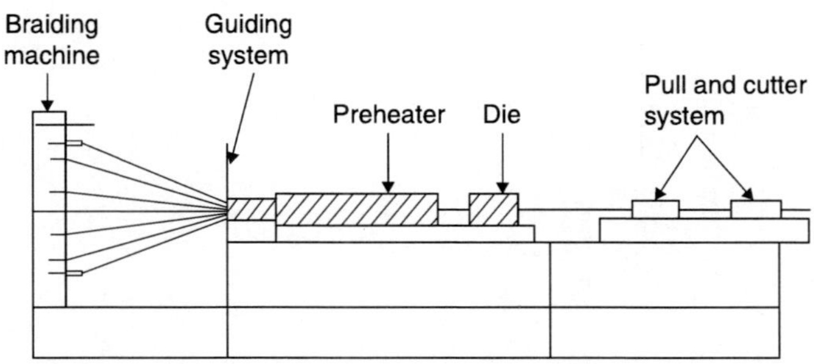

12.9 Schematic of pull-braiding process (Michaeli and Juerss, 1996).

UV assisted bent pultrusion

In this process, cure takes place outside the die resulting in very low pulling forces (see Fig. 12.10) thereby lowering the capital cost of the plant (Britnell *et al.*, 2003; Hay and O'Gara, 2006). The curing relies primarily on irradiating the resin-impregnated fibres with ultraviolet (UV) light, as the profile exits the shaping die.

A process was developed by Kennedy and Kusy (1995) to adapt radiation polymerization instead of thermal curing in pultrusion. Faster rates resulted in partial curing of methacrylate-based quartz-fibre-reinforced-composite profiles at a rate of 1.25 mm/s.

Britnell *et al.* (2003) studied the relation between pull speed and pull force in UV-assisted bent pultrusion process. As expected, pull force increased and the degree of cure decreased with increase in pull speed.

Sandwich pultrusion

Yun and Lee (2004) investigated pultrusion of phenolic foam composites. The effect of process variables on the foaming characteristics of phenolic resin during pultrusion was studied experimentally with the consideration of variables such as the heating temperature, the pulling speed and the mass fraction of the blowing agent.

Ben and Shoji (2005) designed a technique to mould a sandwich beam in which the phenolic foam as a core and a thin fibre-reinforced phenolic composites as a facesheet are used.

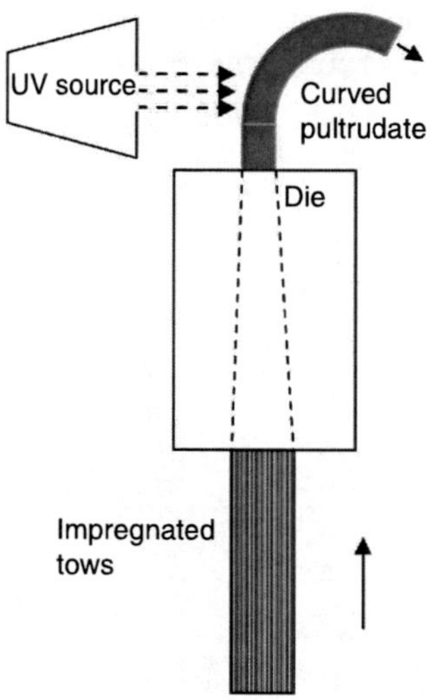

12.10 Schematic of bent pultrusion (Britnell *et al.*, 2003).

Liang *et al.* (2005) worked with pultruded composite sandwich panels that were designed, manufactured, tested and evaluated for their mechanical performance against the vacuum assisted resin transfer moulding (VARTM) products. The results have demonstrated that pultruded panels present significant advantages over VARTM panels, including 36% increase in tensile strength and stiffness, 87–97% increase in bending stiffness at panel level, 8–20% reduction in area weight, and estimated 46% reduction in material and production costs.

12.4 Innovation, industrial and future trends

The pultrusion industry has been growing and the growth is driven by newer applications for pultruded composites. Advances in materials and innovations in the process have enabled manufacturers to design new or improved products (Busel, 2008).

12.4.1 Innovations

Various patents have been granted to manufacturers and operators of pultrusion equipment that have come out with innovations so as to improve upon the conventional pultrusion – such as pultrusion with variants for multi-layer, multi-directional, multi-material products of different forms.

A patent granted in 1989 to Murakami *et al.* (1994) describes a unique method of manufacturing bi-directionally fibre-reinforced composite pultrudate. The process takes in reinforcing fibres arranged axially in a usual manner. However, a helical fibre layer of reinforcing fibres is directed normally to the axis. Thus, there may be axially oriented and helical fibre layers alternately formed one over another. Henrik *et al.* (2009) patented a novel pultrusion method in 2009. They allowed a fibre composite profile to enter the die at a non-zero so that the pultrudate is pulled through the die with higher force before its complete curing. This results in improved fibre alignment, better consolidation, reduced void content and overall increase in fibre-dependent properties. Michael (2011) invented a pultrusion process for a continuous profile with at least one prepreg strip along with other usual reinforcement. Schlueter and Bond (2001) offered an improved pultrusion process similar to filament winding. Fibres are drawn from the creels and taken through a resin tub to soak them. Subsequently the wet fibres are wound around a mandrel such that they form a wet layer gradually covering the mandrel (see Fig. 12.11). A die is then passed through the hollow mandrel so as to apply pressure and temperature from the inner side to cure the wound layer. A multi-layered form can also be made by repeating the described process. Goichi's (2003) novel pultrusion method offered continuous form of glass fibre-reinforced phenol foam composites.

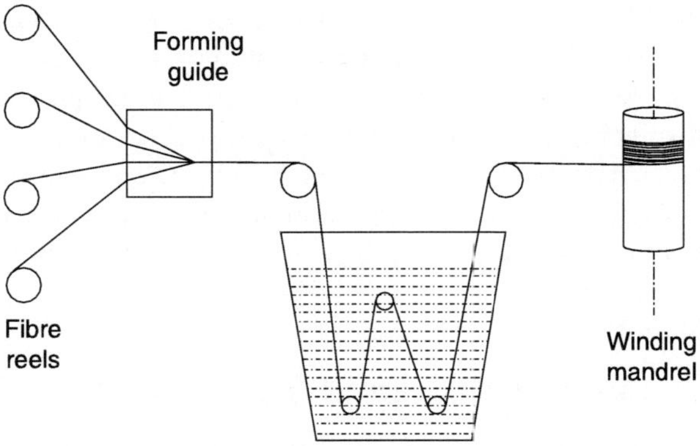

12.11 Schematic of pultrusion process by Schlueter and Bond (2001).

In this process, firstly, a foaming agent containing phenol resin is mixed with the curing agent. The mixed solution is then used to impregnate continuous glass fibres at temperature lower than the foaming temperature. Subsequently, the impregnated glass fibres are drawn into a heated die at the foaming temperature of the resin so as to form foam and cure the resin thereby producing a foam–fibre sandwich section. A unique method for performing resin injection pultrusion with multiple resins was patented by Jay and James (1996). In this process, the first layer of a fibrous reinforcement is injected with a first resin and then allowed to combine with second layers of the reinforcement pre-impregnated with a second resin. Thus, a pultruded part may be formed having different layers with different properties. Johnson *et al.* (1998) presented a high shear strength pultrusion by using woven fabric. Unidirectional fibres are stitched to the bi-directional woven fabric in such a way that the stitch lines run longitudinally in the pultrusion direction. Also, the unidirectional fibres pass through the terminated ends of the transverse tows of the bi-directional woven cloth. Nuno *et al.* (2001) invented a pultrusion head to produce composite profiles from thermoplastic powder coated fibre tows (towpregs). The designed head is composed of three main components: a preheating furnace, a consolidation die and a cooling die.

12.4.2 Developments

Pultrusion is one of the fastest growing processes within the industry for manufacturing composite products. The developments such as standards available to date, global market scenario, novel applications and future trends of the pultrusion industry are of interest to composite manufacturers and researchers.

Standards

Standards play an important role in the acceptance of new materials. The following standards are available regarding the manufacture, quality determination and quality assurance of pultruded products:

- EN 13706 General European specifications for pultruded profiles
- DIN 18820 Reinforced laminates for load-bearing building components
- DIN 2768 Free dimensional tolerances
- ISO 178/ISO 527/ISO 604/ISO 1183/ISO R62
- IEC 93 and 112 Formal test specifications for material testing
- ASTM E84 – US fire standard
- ASTM D-3917-84 – dimensional tolerances
- BS 3396:P3-87 – woven fabrics
- BS 3691-69 – rovings
- NF B 38–301-78 – mats

In 2007, ACMA (American Composites Manufacturers Association) and ASCE (American Society of Civil Engineers) jointly developed an LRFD (Load Resistance Factor Design) standard for pultruded composites. LRFD refers to a design methodology that makes use of load and resistance factors based on the known variability of applied loads and material parameters. The development of the LRFD standard is expected to significantly influence the universal acceptance of pultruded composites (Busel, 2008).

Global market

The global market for pultruded composites is 250–300 metric tons (Owens Corning, 2009), which account to about $1.2 billion (year 2004 estimate). The market spread amounts to approximately 40% in the Asia Pacific region, 40% in North America and 20% in Europe. As of 2010, there were more than 300 pultruders in the world, but many of them operate on a smaller scale (Jacob, 2006). In terms of the revenue generated by pultrusion products, Exel, Strongwell and Werner make the top three pultruders in the world currently.

Novel applications

Through employing wide panel pultrusion technology, KaZaK Composites Inc., USA, has developed a high expansion ration shelter (HERS) (see Fig. 12.12). The company has successfully produced single-piece foam-cored sandwich panels with internal stiffeners, which are further used to build HERS useful as easily deployable sheds for military troops, field hospitals and emergency housing. The company also demonstrated that components like hollow wing sections for unmanned aerial vehicles can be produced using pultrusion. One such demonstrator fabricated using stitched unidirectional carbon fabric reinforcement is shown in Fig. 12.13 (Black, 2009).

Another novel pultruded panel has been developed by Mastercore System Ltd., Canada. This could be used in applications such as truck walls, roofs, floors and doors, boat decks, freight containers and sound barriers. The mastercore panel consists of an outer shell of a polyurethane-impregnated fibre carbon/glass fibre hybrid composites with an inner core of a structural foam (Jacob, 2006).

12.12 KaZaK Composites Inc.'s high expansion ratio shelter (Jacob, 2006).

12.13 A pultruded hollow wing without internal ribs. (*Source*: KaZaK Composites [Black, 2009].)

Some new applications were showcased at the American Composites Manufacturers Association annual conference, Composites & Polycon 2007, in Tampa, Florida. These included scaffolding for the construction industry and a new steel-free concrete deck (see Fig. 12.14). Creative Pultrusions Inc. has developed a tough, light and more corrosion resistant composite sheet piling system (Fig. 12.15) named SuperLoc profiles. These profiles consist of polyester resin matrix with more than 100 fibreglass rovings in the core, plus two plies of continuous strand mat. SuperLoc so far has been utilized in more than 20 shorelines in the United States (Black, 2004).

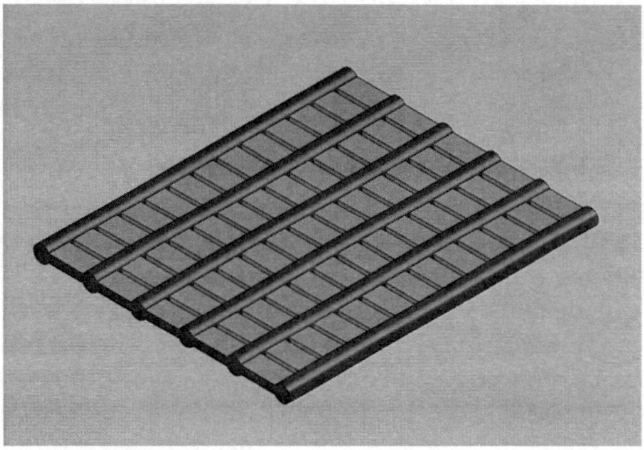

12.14 Pultruded planks used as scaffolding for bridge construction repair applications (Busel, 2008).

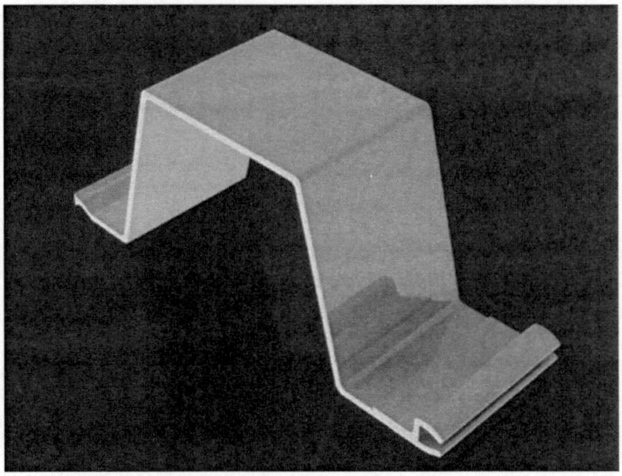

12.15 Cross-section of sheet piling showing the interlocking attachments (Busel, 2008).

12.4.3 Future trends

It appears that the pultrusion industry in North America and Western Europe is looking for cheaper production, and in future, is expected to relocate to Central and Eastern Europe and to Asian countries such as China and India. These options not only are economical but also provide a much wider and more open market with a scope for innovation.

The materials used in pultrusion are also expected to change with the change in developments on the materials front. Pultrusion machinery has so far evolved in terms of handling complex shapes of very large sizes. The capabilities to optimize product quality and to monitor the process *in situ* are also available. Further advances are still expected until a truly real-time processing product quality feedback system is achieved.

The continued quest for improvement and innovation are expected to bring pultrusion process science to a level whereby newer products can be achieved with cleaner and more efficient energy options.

On numerical fronts, the use of fully integrated design-analysis-optimization softwares with FEA and solid modelling are expected to become norms in the pultrusion industry.

In conclusion, the future of pultrusion is very promising as one of the most cost-effective continuous composites manufacturing methods available to produce composite solutions to design problems (Sumerak and Martin, 1987). Pultrusion will not only remain a primary production process for secondary applications but also will flourish with the development of more and more novel products.

12.5 Acknowledgement

The author would like to take this opportunity to express his heartfelt gratitude and appreciation to the people who have either worked with him on different aspects of pultrusion or supported the work that guided the design of this handbook chapter.

12.6 References

Ahmed, F., Joshi, S. C. and Lam, Y. C. (2004), 'Three-dimensional FE/NCV modelling of thermoplastic composites pultrusion', *Journal of Thermoplastic Composite Materials*, **17**, 447–462.

Awa, T. W. and West, R. L. (1992), 'Pultrusion die design through the coupling of a boundary element heat transfer model and a design optimization technique', *2nd International Conference on Advanced Computational Methods in Heat Transfer*, Milan, Italy.

Awa, T. W., West, R. L. and Price, H. L. (1992), 'Heater configuration design of a pultrusion die using design optimization techniques', *Proceedings of the ASME International Computers in Engineering Conference and Exposition*, San Francisco, USA.

Batch, G. L. and Macosko, C. W. (1993), 'Heat transfer and cure in pultrusion: model and experimental verification', *American Institute of Chemical Engineers Journal*, **39**(7), 1228–1241.

Bechtold, G., Wiedmer, S. and Friedrich, K. (2002), 'Pultrusion of thermoplastic composites: New developments and modelling studies', *Journal of Thermoplastic Composite Materials*, **15**, 443–465.

Ben, G. and Shoji, A. (2005), 'Pultrusion techniques and evaluations of sandwich beam using phenolic foam composite', *Advanced Composite Materials*, **14**, 277–288.

Black, S. (2004), 'Pultruded composite sheet piling replace degraded wood pilings', *Composites Technology*, February.

Black, S. (2009), 'Pultruding cost out of aerospace parts', *High-Performance Composites*, March.

Britnell, D. J., Tucker, N., Smith, G. F. and Wong, S. S. F. (2003), 'Bent pultrusion-a method for the manufacture of pultrudate with controlled variation in curvature', *Journal of Materials Processing Technology*, **138**(1–3), 311–315.

Busel, J. B. (2008), 'State of the North American pultrusion industry', *Composites Manufacturing*, April 2008, 28–54.

Byun, J. H., Lee, S. K. and Kim, B. S. (1996), 'Development of braided-pultrusion process and structure-property relationships for tubular composites', *Acta Metallurgica Sinica*, **9**(6), 555–564.

Carlone, P. and Palazzo G. S. (2008), 'Numerical modelling of the microwave assisted pultrusion process', *International Journal of Material Forming*, **1**, 1323–1326.

Carlone, P., Palazzo, G. S. and Pasquino, R. (2006), 'Pultrusion manufacturing process development by computational modelling and methods', *Mathematical and Computer Modelling*, **44**(7–8), 701–709.

Carlone, P., Palazzo, G. S. and Pasquino, R. (2007), 'Pultrusion manufacturing process development: Cure optimization by hybrid computational methods', *Computers & Mathematics with Applications*, **53**(9), 1464–1471.

Chen, C. H. and Wang, W. S. (1998), 'Development of glass fibre reinforced polystyrene for pultrusion. II: Effect of processing parameters for optimizing the process', *Polymer Composites*, **19**, 423–430.

Chen, X., Xie, H., Chen, H. and Zhang, F. (2009), 'Optimization for CFRP pultrusion process based on genetic algorithm-neural network', *International Journal of Material Forming*, **3** (Suppl 2), S1391–S1399.

Chew, K. S. P. (2004), 'Optimization of thermoplastic pultrusion', M.Sc. Dissertation, School of Mechanical and Aerospace Engineering, Nanyang Technological University, Singapore.

Dharia, A. N. and Schott, N. R. (1986), 'Resin pick-up and fibre wet-out associated with coating and pultrusion processes', *Proceedings of the Society of Plastics Engineers 44th Annual Technical Conference*; Boston, USA.

Ding, Z. and Lee, L. J. (1999), 'Process analysis and simulation of resin injection pultrusion', *Proceedings of the 31st International SAMPE Technical Conference*, Chicago, USA.

Ding, Z., Li, S., Yang, H., Lee, L. J., Engelen, H. and Puckett, P. M. (2000), 'Numerical and experimental analysis of resin flow and cure in resin injection pultrusion (RIP)', *Polymer Composites*, **21**, 762–778.

Dube, M. G., Batch, G. L., Vogel, J. L. and Macosko, C. W. (1995), 'Reaction injection pultrusion of thermoplastic and thermoset composites', *Polymer Composites*, **16**, 378–385.

Ewald, G. W. (1981), 'Details on a new way to make curved RP shapes', *Modern Plastics*, **58**, 74–76.

Freger, G. E., Kestelman, V. N. and Freger, D.G. (2005), *Braiding pultrusion technology*, New York: McGraw-Hill.

Goichi, H. (Nihon University, Japan) (2003), Pultrusion Method for Glass Fiber-Reinforced Phenol Foam Composite Material, Patent abstracts of Japan 2003–145570. 20 May 2003.

Goldsworthy, W. B. (1979), 'New technology for continuous reinforced plastics processing: It's called "Pulforming", and it permits extrusion of variable cross-section parts, curves, or both', *Modern Plastics*, **56**, 82–83.

Gorthala, R., Roux, J. A. and Vaughan, J. G. (1994), 'Resin flow, cure and heat transfer analysis for pultrusion process', *Journal of Composite Materials*, **28**, 486–506.

Hackett, R. M. and Zhu, S. Z. (1992), 'Two-dimensional finite element model of the pultrusion process', *Journal of Reinforced Plastics and Composites*, **11**, 1322–1351.

Hay, J. N. and O'Gara, P. (2006), 'Recent developments in thermoset curing methods', *Proceedings of the Institution of Mechanical Engineers, Part G: Journal of Aerospace Engineering*, **220**(3), 187–195.

Henrik, D., Holger, P., Christoph, H. and Axel, H. (University of Bremen) (2009), Pultrusion device and method for manufacturing profiles from composite fibre materials, EP 2105286, 30 September 2009.

Hidekuma, Y., Ohtani, A., Nakai, A. and Hamada, H. (2008), 'Processing and mechanical properties of pultruded moldings with braiding technique', *9th International Conference on Textile Composites: Recent Advances in Textile Composites*, Newark, USA.

Hota, V. S., GangaRao, P. V. V. and Taly, N. (2009), 'Manufacturing of composite components'. In *Reinforced Concrete Design with FRP Composites*, Tailor & Francis Group, LLC, CRC Press, USA, pp. 63–77.

Jacob, A. (2006), 'Globalisation of the pultrusion industry', *Reinforced Plastics*, **50**(5), 38–41.

Jay, J. B. and James, V. G. (Owens Corning Fiberglass Technology Inc.) (1996), Method for Performing Resin Injected Pultrusion Employing Multiple Resins, WIPO Patent WO/1996/040489, 19 December 1996.

Johnson, D. W., Goldsworthy, W. B. and Korzeniowski, G. (Ebert Composites Corporation) (1998), High Shear Strength Pultrusion, WIPO Patent WO/1998/032591, 30 July 1998.

Joshi, S. C. and Lam, Y. C. (2001), 'Three-dimensional finite-element/nodal-control-volume simulation of the pultrusion process with temperature-dependent material properties including resin shrinkage', *Composites Science and Technology*, **61**, 1539–1547.

Joshi, S. C., Lam, Y. C. and Tun, U. W. (2003), 'Improved cure optimization in pultrusion with pre-heating and die-cooler temperature', *Composites Part A: Applied Science and Manufacturing*, **34**(12), 1151–1159.

Joshi, S. C., Liu, X. L. and Lam, Y. C. (1999), 'A numerical approach for modelling of polymer curing in fibre reinforced composites', *Composites Science and Technology*, **59**(7), 1003–1013.

Kennedy II, K. C. and Kusy, R. P. (1995), 'UV-cured pultrusion processing of glass-reinforced polymer composites', *Proceedings of the 53rd Annual Technical Conference*, Part 1 (of 3), Boston, US.

Kim, Y. R., McCarthy, S. P. and Fanucci, J. P. (1991), 'Study of resin flow during injection-pultrusion process', *49th Annual Technical Conference – ANTEC '91*, Montreal, Canada.

Kommu, S., Khomami, B. and Kardos, J. L. (1998), 'Modeling of injected pultrusion processes: A numerical approach', *Polymer Composites*, **19**, 335–346.

Kowsika, M. V. S. L. N. and Mantena, P. R. (1996), 'Optimal pultrusion process conditions for improving the dynamic properties of graphite-epoxy composite beams', *Materials Evaluation*, **54**, 386–392.

Krolewski, S. and Gutowski, T. (1986), 'Effect of the automation of advanced composite fabrication process on part cost', *Sampe Quarterly*, **18**(1), 43–51.

Kyaw, S. M. (2003), 'Optimization of pultrusion of composites with post-curing', M.Sc. Dissertation, School of Mechanical and Aerospace Engineering, Nanyang Technological University, Singapore.

Lackey, E. and Vaughan, J. G. (1997), 'Experimental development and evaluation of a resin injection system for pultrusion', *Proceedings of the 42nd International SAMPE Symposium and Exhibition*, Anaheim, USA.

Lam, Y. C., Joshi, S. C. and Liu, X. L. (1998), 'Numerical simulation of mould-filling process in resin transfer moulding', *Composites Science and Technology*, **60**(6), 845–855.

Lam, Y. C., Li, J. and Joshi, S. C. (2003), 'Simultaneous optimization of die-heating and pull-speed in pultrusion of thermosetting composites', *Polymer Composites*, **24**, 199–209.

Lewis, C. F. (1991), 'Giving pultrusion an added dimension', *Materials engineering*, **108**, 24–25.

Li, J. (2001), 'Optimization of die-heating and pull speed for pultrusion', M.Eng. Dissertation, School of Mechanical and Aerospace Engineering, Nanyang Technological University, Singapore.

Li, S., Ding, Z., Xu, L., Lee, J. and Engelen, H. (2002a), 'Influence of heat transfer and curing on the quality of pultruded composites. I: Experimental', *Polymer Composites*, **23**, 947–956.

Li, J., Joshi, S. C. and Lam, Y. C. (2002b), 'Curing optimization for pultruded composite sections', *Composites Science and Technology*, **62**, 457–467.

Liang, R., Mutnuri, B. and Ganga Rao, H. (2005), 'Pultrusion and mechanical characterization of GFRP composite sandwich panels', *Society of Plastics Engineers Annual Technical Conference 2005*, Boston, USA.

Lin, M. and Hawley M. C. (1993), 'Preliminary tests of the application of continuous microwave technique to pultrusion', *Proceedings of the 38th International SAMPE Symposium and Exhibition*, Anaheim, USA.

Liu, X. L. (2001), 'Numerical modeling on pultrusion of composite I beam', *Composites Part A: Applied Science and Manufacturing*, **32**, 663–681.

Liu, X. L. (2003a), 'A finite element/nodal volume technique for flow simulation of injection pultrusion', *Composites Part A: Applied Science and Manufacturing*, **34**(7), 649–661.

Liu, X. L. (2003b), 'Iterative and transient numerical models for flow simulation of injection pultrusion', *12th International Conference on Composite Structures*, UK.

Liu, X. L. and Hillier, W. (1999), 'Heat transfer and cure analysis for the pultrusion of a fiberglass-vinyl ester I beam', *10th International Conference on Composite Structures*, Melbourne, Australia.

Liu, X. L., Crouch, I. G. and Lam, Y. C. (2000), 'Simulation of heat transfer and cure in pultrusion with a general-purpose finite element package', *Composites Science and Technology*, **60**, 857–864.

Lu, S., Xie, H. and Chen, P. (2008), 'Simulating and optimizing of GFRP pultrusion process', *Fuhe Cailiao Xuebao/Acta Materiae Compositae Sinica*, **25**, 46–51.

Ma, C. C. M., Lee, K. Y., Lee, Y. D. and Hwang, J. S. (1986), 'Correlations of processing variables for optimizing the pultrusion process', *SAMPLE Journal*, **22**(5), 42–48.

Mazumdar, S. K. (2009), 'Manufacturing techniques'. In *Composites manufacturing: Materials, product, and process engineering,* , CRC Press, USA, Chapter 6, pp. 1–135.

Metaxas, A. C. and Meredith, R. J. (1983), 'Industrial microwave heating', *IEE Power Engineering Series*, The Institution of Engineering and Technology, UK.

Methven, J. (1999), 'Microwave Assisted Pultrusion (MAP) continuous manufacture of composite profiles', *Materials Technology*, **14**, 183–186.

Methven, J. M. and Ghaffariyan, S. R. (1992), 'Microwave assisted pultrusion', *Proceedings of the American Chemical Society Spring Meeting*, San Francisco, USA.

Methven, J. M., Ghaffariyan, S. R. and Abidin, A. Z. (2000), 'Manufacture of fiber-reinforced composites by microwave assisted pultrusion', *Polymer Composites*, **21**, 586–594.

Meyer, R. W. (1985), *Handbook of pultrusion technology*. London: Chapman & Hall.

Michael, B. (Airbus operations GMBH) (2011), Pultrusion process for production of a continuous profile, Patent Application Publication, US2011/0049750A1, 3 March 2011.

Michaeli, W. and Juerss, D. (1996), 'Thermoplastic pull-braiding: Pultrusion of profiles with braided fibre lay-up and thermoplastic matrix system (PP)', *Composites Part A: Applied Science and Manufacturing*, **27**, 3–7.

Murakami, S., Manabe, K., Miyao, M., Enomoto, M., Ishida, Y. and Inoue, H. (Toa Nenryo Kogyo Kabushiki Kaisha) (1994), Carbon fibre-reinforced composite resin pultrusion products and method of manufacturing the same, EP0308237 B1, 10 August 1994.

Nuno Ferreira Da Mota, J., Pedro Lourenço Gil Nunes, J. and Sérgio Duarte Pouzada, A. (Universidade Do Minho) (2001), Pultrusion head to produce long fibre reinforced thermoplastic matrix profiles, WIPO Patent WO 02/06037 A1, 13 July 2001.

Owens Corning Corporation (2003), *Pultrusion of Glass Fibre Composites – A Technical Manual*, June 2003.

Owens Corning Composite Materials LLC (2009), *Pultrusion*, March 2009, France.

Schlueter, J. E. L. and Bond, W. E. (Xerox Corp, Japan) (2001), Pultrusion Method, Patent abstracts of Japan 2001–219474, 14 August 2001.

Sharma, D., McCarty, T. A., Roux, J. A. and Vaughan, J. G. (1998), 'Investigation of dynamic pressure behaviour in a pultrusion die', *Journal of Composite Materials*, **32**(1), 929–950.

Smith, A. C. and Hawley, M. C. (1998), 'Microwave pultrusion processing', *Proceedings of the 33rd Microwave Power Symposium*, Manassas, USA.

Smith, A. C., McMillan, A. M. and Hawley, M. C. (1998), 'Microwave pultrusion of graphite/epoxy composites using single frequency and variable frequency processing techniques', *Proceedings of the 30th International SAMPE Technical Conference*, San Antonio, USA.

Squires, C., Almaraz, J. and O'Toole, B. (1996), 'Experimental simulation of the thermoplastic pultrusion process', *International SAMPE Symposium and Exhibition*, **41**, 1667–1677.

Srinivasagupta, D., Kardos, J. L. and Joseph, B. (2003), 'Rigorous dynamic model-based economic design of the injected pultrusion process with controllability considerations', *Journal of Composite Materials*, **37**, 1851–1880.

Srinivasagupta, D. and Kardos, J. L. (2004), 'Ecologically and economically conscious design of the injected pultrusion process via multi-objective optimization', *Modelling and Simulation in Materials Science and Engineering*, **12**, S205–S219.

Strong, A. B. (2002), 'Pultrusion – High Productivity Now, Getting Even Better', personal note, Brigham Young University, USA.

Sumerak, J. E. and Martin, J. D. (1993), 'Pulse of pultrusion: Pull force trending for quality and productivity management', *Polymers and Polymer Composites*, **1**, 199–210.

Sumerak, J. E. and Martin, J. D. (1987), 'Pultrusion'. In *Engineered materials handbook*, ASM International, USA, pp. 533–543.

Tat, T. L. K. (2006), 'Optimization of thermoset composites pultrusion with feedback sensors', M.Sc. Dissertation, School of Mechanical and Aerospace Engineering, Nanyang Technological University, Singapore.

Vaughan, J. G. (1988), 'Use of statistical design experimentation to characterize and optimize the pultrusion process', *Proceedings of the Technical Sessions of Composites Institute 43rd Annual Conference*, Cincinnati, USA.

Viola, G., Portwood, T., Ubrich, P. and DeGroot, H. R. (1990), 'Numerical optimization of pultrusion line operating parameters', *35th International SAMPE Symposium and Exhibition – Advanced Materials: the Challenge for the Next Decade*, Part 2, Anaheim, USA.

Voorakaranam, S., Kardos, J. L. and Joseph, B. (1998), 'Model-based control of injection pultrusion process', *Proceedings of the American Control Conference*, Evanston, USA.

Wilcox, J. A. D. and Wright, D. T. (1998), 'Towards pultrusion process optimisation using artificial neural networks', *Journal of Materials Processing Technology*, **83**(1–3), 131–141.

Yun, M. S. and Lee, W. I. (2004), 'Process optimization of pultrusion process of phenolic foam composites', *SAMPE 2004*, Long Beach, USA.

Zaw, K. (2003), 'Optimization of die heater parameters for pultrusion of polymer composites', M.Sc. Dissertation, School of Mechanical and Aerospace Engineering, Nanyang Technological University, Singapore.

13
Autoclave processing for composites

P. HUBERT, McGill University, Canada
and G. FERNLUND and A. POURSARTIP,
The University of British Columbia, Canada

Abstract: This chapter discusses the state-of-the art of autoclave processing of polymer composites. First, the architecture and typical formulation of a process model applied to the autoclave process is presented. Then, the steps for autoclave process development using the proposed modelling approach are discussed. In particular, the thermal management during the process and the dimensional control of the final part are reviewed in more detail.

Key words: autoclave processing, process modelling, flow, compaction, heat transfer, cure, dimensional control, residual stresses, cure cycle.

13.1 Introduction

Autoclave processing is usually the method of choice to manufacture fibre reinforced plastic composite components for high-performance applications, especially involving large complex structures. Using this technique, thin layers of high modulus fibre impregnated with partially cured resin (prepreg) are cut and stacked to form a component of desired shape. In addition to uncured prepreg, typical laminated composite structures also include such materials as honeycomb core, pre-cured composite stiffeners and structural adhesives to bond together the various parts. After assembly, the structure is covered with various layers of cloth (bleeder and breather) and sealed inside a vacuum bag, as illustrated in Fig. 13.1. Bleeders assist in achieving an optimal fibre volume fraction by absorbing excess resin. Breathers provide a path for removal of air and volatile gasses from the part during cure. The tool–laminate assembly is placed in an autoclave, essentially a large temperature and pressure controlled vessel, and the bag is connected to the vacuum system (Fig. 13.2). In order to cure the part, pressure and temperature are applied to the laminate in a predetermined cure cycle (Fig. 13.3). The temperature cycle is necessary to trigger the resin polymerization reaction. The pressure is applied to the laminate to conform the laminate to the tool surface, and to compact the laminate at the desired fibre volume fraction and collapse any voids that may develop during the resin cure. The

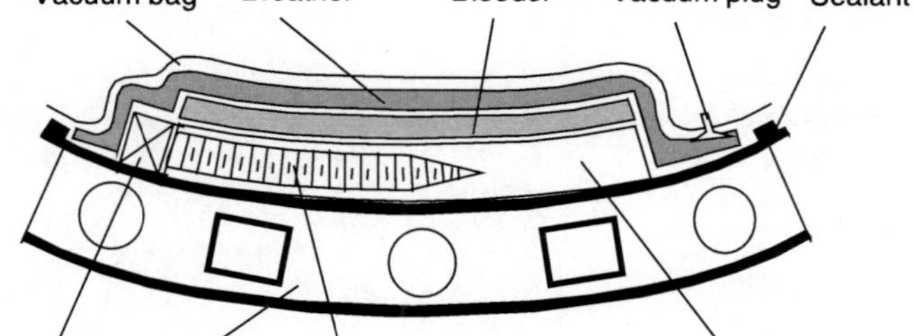

13.1 Component and tooling prepared for autoclave processing.

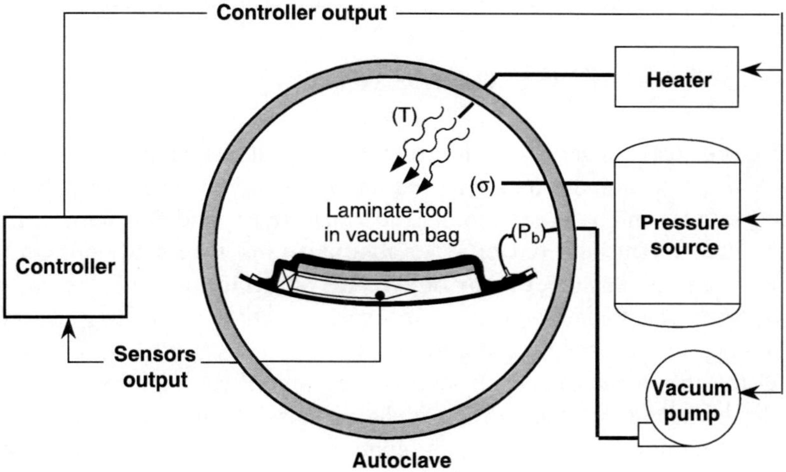

13.2 Typical autoclave processing system.

bag pressure is controlled to initially remove any entrapped air during layup and volatiles during cure. The bag pressure must be less than the autoclave pressure to ensure compaction of the laminate. Finally, the cured part is debagged and ready for secondary and finishing processes.

The main challenge of the curing process is to determine the cure cycle, design the tool and define a bagging procedure that will produce a fully cured, void free and undistorted part in the shortest time and the most economical fashion. Residual deformations or warpage are affected by residual stress development during cure. Resin-rich and resin-poor regions are a consequence of resin movements. Flow and resin pressure distributions in the laminate play an important role in void formation and migration. For sandwich panels, surface dimpling and core buckling also depend on the pressure distribution. At the same time, the dimensions of the produced

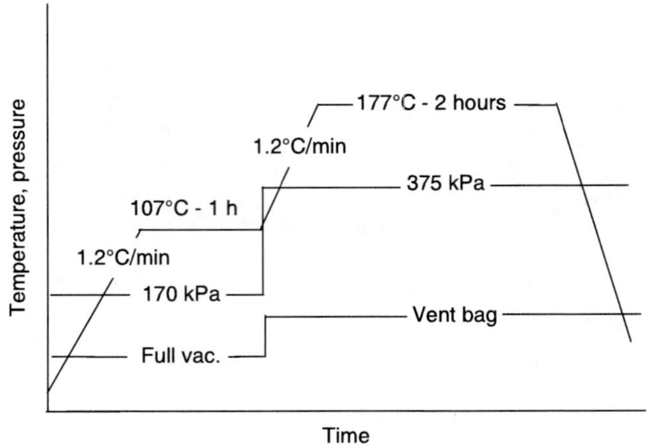

13.3 Typical autoclave process cycle.

structure must not vary beyond preset tolerance limits. Meeting these objectives has traditionally entailed trial-and-error modification of a baseline process cycle and design of tooling and structure design.[1] This same iterative procedure might be repeated when modifications were made in materials, component dimensions, tooling or even the autoclave in which the part was processed. This technique is both expensive and time consuming, especially when applied to large complex structures.

Using this technique, even after an 'optimum' combination of process cycle, tooling and structure design is chosen, it is difficult to gauge the robustness of the developed process, let alone forecast potential problems which might arise due to random variability in process variables such as material properties, part layup or even autoclave loading. One way to help minimize process variation is through use of 'intelligent process control'.[2–4] Using this approach, various types of instruments such as thermocouples or dielectric sensors are typically embedded in the part and their measurements used for real-time autoclave process cycle control. These systems often employ expert systems, sometimes in conjunction with simple process models[5–8] to meet certain predefined objectives such as minimum process cycle time or minimum temperature gradients within a part. While potentially very useful, this approach is not a panacea for the composites processor. One deficiency of such systems is that they neither allow *a priori* prediction of processing outcomes (such as process-induced deformation) nor provide any insight into robust process design. Also, the rules on which such systems are based are usually very material dependent as well as inflexible and difficult to modify.

Many of the shortcomings of the intelligent process control approach can be addressed through the use of autoclave process modelling. Using this approach, computational models are developed for the various phenomena involved in autoclave processing such as heat transfer, resin cure, resin flow and the development of residual stress and deformation. These models are then used to create 'virtual parts' which can be subjected to a simulated autoclave process cycle. Using these models, a very large number of potential processing scenarios, part and tooling designs can be examined for such things as uniformity of resin cure and structure warpage before a single real part is constructed. In combination with processing trials, this can greatly reduce the number of process and structural design iterations and consequently both cycle time and development costs. By exploiting the inherent suitability of such models for sensitivity analyses, 'sweet spots' can be identified in which the process is relatively insensitive to material and manufacturing variability. The major remaining variability drivers can then be targeted for tight monitoring and control.

13.2 Autoclave processing model

The overall structure of the autoclave process model is similar to the 'integrated sub-model' approach, first applied to composites process modelling by Loos and Springer.[9] Using this approach, a complex coupled problem such as the current one is divided into a series of simpler problems that are tackled independently in a series of 'sub-models', or 'modules' (to emphasize the modularity of the approach). Coupling between modules is maintained by solving each in sequence as the solution marches forward in time. In this way, the predictions of one module can be used by another. In COMPRO,[10] modules are created not only for the analysis of the structure of interest, but also for simulation of the autoclave response. As shown in Fig. 13.4, six different modules are currently employed including:

1. *Autoclave simulation* module: simulates the variation of autoclave internal temperature and pressure and the vacuum bag pressure with time. Also simulates other parameters associated with structure boundary conditions such as local heat transfer coefficient (HTC).
2. *Thermochemical* module: predicts internal temperature in the structure and tooling and the resin degree of cure in composite structural components.
3. *Flow* module: uses a percolation model to predict resin flow and fibre bed compaction in composite components.
4. *Stress* module: models the development of residual strain and deformation in the structure and any process tooling.
5. *Input* and *output* modules: control acquisition of all input parameters required for a program run and output of all parameters to external devices.

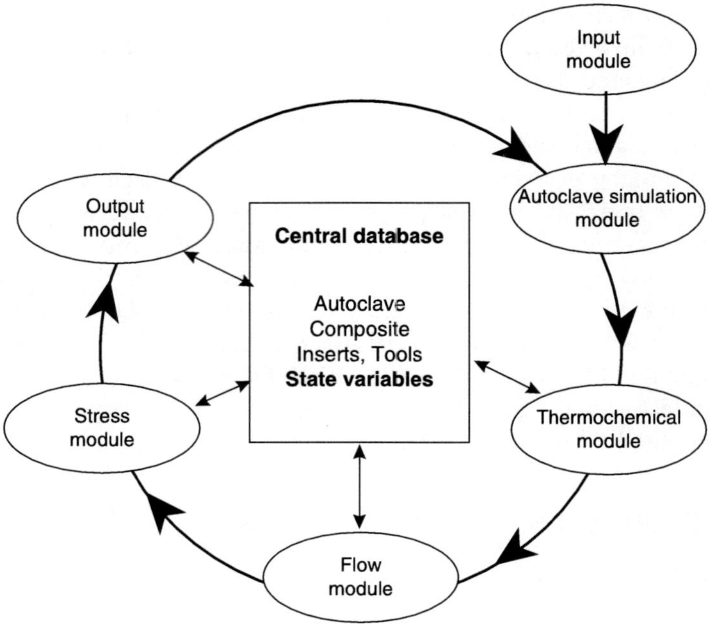

13.4 Schematic of COMPRO structure and program flow.

13.2.1 Heat transfer and cure

The governing equation of the heat transfer in autoclave processing is the transient Fourier anisotropic heat conduction equation, with a heat generation term from the exothermic resin cure reaction:

$$\frac{\partial}{\partial t}(\rho C_P T) = \nabla(k\nabla T) + \dot{Q} \qquad [13.1]$$

where ρ is the composite density, C_P is the specific heat, k is the anisotropic thermal conductivity and $\dot{Q}$ is the resin reaction heat generation rate. The composite thermophysical properties used in Eq. [13.1] (ρ, C_P, k), can be calculated from the local fibre volume fraction (V_f), and measured resin and fibre properties. Although they are assumed to be constant in the majority of models, these properties may actually change considerably during processing. As shown by Mijovic and Wijaya[11] and Twardowski *et al.*,[12] changes in these properties during processing may be large enough in some cases to have a significant influence on thermochemical model predictions.

The rate of heat generation in Eq. [13.1] can be determined from:

$$\dot{Q} = \frac{d\alpha}{dt} H_R \qquad [13.2]$$

where $d\alpha/dt$ is designated as the cure rate and H_R is the total amount of heat generated during a complete resin reaction. A large number of different mechanistic, empirical and semi-empirical models have been proposed for predicting the cure rate of various resin systems.[5,9,11]

The boundary conditions that may be used with the thermochemical module include: specified boundary temperature, convective heat transfer or no heat transfer (adiabatic). Different conditions (e.g., different HTCs) can be applied to each element as desired. Either explicit or implicit techniques may be chosen to solve the heat transfer (Eq. [13.1]) and cure rate equations. Using either technique, these two equations are uncoupled during each solution time-step. This approach facilitates a simplified and modular solution procedure and is sufficiently accurate if small time steps are used.

13.2.2 Flow compaction

The prepreg fibre bed is typically assumed to be an elastic porous medium with incompressible and inextensible fibres and fully saturated with the resin. The resin is assumed to flow in the pores between the fibres, and the fibre mass in the laminate remains constant during cure. The governing equations of the system must describe the behaviour of the composite constituents: the fibre bed and the resin. Firstly, the equilibrium of forces on the representative element is considered. Secondly, the mass conservation for the representative element must be satisfied. For a porous medium saturated with a single phase fluid, the total stress tensor σ_{ij} is separated into two parts as (tensile stresses are considered positive):

$$\sigma_{ij} = \bar{\sigma}_{ij} - \delta_{ij} P \qquad [13.3]$$

where $\bar{\sigma}_{ij}$ is the fibre bed effective stress, δ_{ij} is the Kronecker delta ($\delta_{ij} = 1$ for $i = j$ and $\delta_{ij} = 0$ for $i \neq j$) and P is the resin pressure. The resin flows through the fibre bed according to Darcy's law. Accordingly, the resin velocity v_i relative to the fibre bed is expressed as:

$$v_i = -\frac{K_{ij}}{\mu}\left(P + \rho_R g h\right)_{,j} \qquad [13.4]$$

where K_{ij} is the fibre bed permeability tensor, μ is the resin viscosity, ρ_R is the resin density, g is the acceleration due to gravity and h is the height above a reference point. The resin viscosity is a function of the temperature and degree of cure and can be obtained from any number of empirical models.[5,9] The fibre bed compaction behaviour and the fibre bed permeability are functions of the fibre volume fraction, V_f. Both are calculated using empirical or analytical models available from the literature.[13–16]

The boundary conditions that may be simulated with the flow-compaction module are: autoclave pressure, impermeability or permeability with prescribed bag pressure, no displacement or no normal displacement (tangent sliding condition). The governing equations (Eqs [13.3] and [13.4]) are coupled during individual time-steps of the transient solution. A Newton–Raphson iterative procedure is used to solve the resulting nonlinear system of equations. Details in the solution of the flow-compaction for autoclave processing can be found in reference 17.

13.2.3 Dimensional stability and residual stresses

During autoclave processing, a number of different processes lead ultimately to the development of residual stress and deformation. A convenient way of categorizing and analysing these processes is to distinguish individual residual stress and deformation 'sources' and 'mechanisms'. In the current context, stress *sources* may be thought of as the 'driving forces' behind residual stress generation. Stress generation *mechanisms* can then be thought of as the *combination* of these stress sources with *material behaviour*. For example, a commonly identified source of residual stress is anisotropic thermal strain. The mechanism for stress development in this case is the combination of these thermal strains with changing resin modulus. Unless *both* are present, no residual stress and deformation will be generated.

From the mechanisms for the development of process-induced stress and deformation described in the literature, four main sources have been identified:

- Thermal strains.
- Resin cure shrinkage strains.
- Gradients in component temperature and resin degree of cure.
- Tooling mechanical constraints.

Thermal strains

Changes in component and tooling temperatures during processing are accompanied by thermal strains. There are a number of mechanisms by which these thermal strains contribute to the development of residual stress and deformation including: thermal strains of both unidirectional and woven prepreg plies are highly anisotropic, typically being much lower in fibre-dominated directions than transverse to the fibres. As discussed in reference 18, the majority of residual stress analyses are based on the assumption that this is the *only* source of process-induced stress and deformation.[19] In these analyses, stress is generally assumed to develop only during the cool-down process as the component temperature is decreased to room temperature from some assumed 'stress-free' temperature, usually chosen as the maximum processing temperature.[20,21] For flat laminates, thermal strain anisotropy results in the generation of residual stress but residual deformation can be prevented simply by using a symmetric layup. For curved sections, however, differences in strains in the through-thickness and in-plane directions will result in a change in part shape. Another important source of process-induced stress is strain mismatch between components of the composite structure. The mechanism for stress development in this case is a combination of this strain mismatch with changes in resin modulus and stress relaxation.

Resin cure shrinkage strains

During polymerization, thermosetting resins undergo a significant increase in density and a corresponding reduction in volume, commonly referred to as 'cure shrinkage'. One effect of resin cure shrinkage is both in-plane and through-thickness reductions in part dimensions. Shrinkage strains will be much larger transverse to the fibre direction than in the fibre direction in which they will be largely constrained. Thus, the effect of cure shrinkage on residual stress will be very similar to that of a *decrease* in part temperature and may be analysed in much the same way. The relative significance of cure shrinkage strains to the final residual stress varies from case-to-case. Clearly, therefore, the effect of cure shrinkage strains depends greatly on the material and process cycle employed.

Cure and temperature gradients

In thin components, through-thickness variations in temperature and resin degree of cure are usually very small and their contributions to residual stress and deformation may be neglected. In thicker parts, however, the low composite transverse thermal conductivity coupled with the rapid heat generation of the resin reaction may result in significant temperature and cure gradients. The stress development mechanism in this case is a combination of spatial and temporal variation in resin modulus with gradients in thermal and cure shrinkage strains. The residual stresses resulting from temperature and cure gradients develop regardless of part layup and can be generated even in unidirectional components which are normally assumed to be stress-free after processing. Also, if component boundary conditions during processing are not symmetric (due, for example, to the presence of process tooling) an asymmetric stress state may develop, resulting in component warpage.

Process tooling

Process tooling affects component stress development in two basic ways: by influencing the component internal temperature,[22] and via mechanical loads and constraints applied at tool/part interfaces. The first of these effects is a result of both the often large heat capacity of the tool and the insulating effects of the bagging and bleeder/breather layers.[23] Even at relatively low rates of heating or cooling, component toolside and bagside temperatures may be quite different. This results in through-thickness variations in the rate of resin cure and the potential development of an asymmetric residual stress state. In addition to altering component internal temperature during processing, tools also apply boundary loads to the part, both in shear and normal to the tool/part interface. For simply shaped components, shear loads result mainly from differences in tool and component thermal and cure shrinkage strains.[24] These interface stresses have long been identified as sources of process-induced warpage, and are at least partially responsible for the extensive use of expensive low-*CTE* substrate tooling materials. The mechanism leading to warpage has usually been identified as the preferential relief of stresses induced in plies *away* from the tool/part interface by resin flow.[24–26]

Resin mechanical behaviour during processing

One of the primary objectives of processing thermosetting composite materials is to transform the matrix resin from a low molecular weight semi-solid material to a highly cross-linked macromolecular structure. As the resin undergoes this transition its effective elastic modulus increases by several orders of magnitude. Understanding and accurately modelling resin behaviour during processing is critical to understanding the development of residual stress and the roles played by the identified stress sources.

As part temperature is increased during the early stages of processing, the resin transforms from a semi-solid to a viscous state. As cure progresses, resin viscosity increases rapidly until reaching an effectively infinite value at gelation. It is at this point that the resin also first attains a non-zero equilibrium modulus and thus becomes capable of supporting non-hydrostatic stress.[18] Even after gelation, however, resin temperature is still generally high and material response highly viscoelastic. Thus stresses generated soon after gelation will still quickly decay.

Continued cross-linking leads to an increase in both resin equilibrium modulus and viscoelastic relaxation times. Thus, material behaviour becomes increasingly elastic and generated stresses begin to decay more slowly. Finally, when the resin instantaneous glass transition temperature (T_g) exceeds its local temperature (i.e., vitrification occurs), relaxation times jump sharply and mechanical behaviour becomes highly elastic.

It is convenient to divide the continuum of resin mechanical property development during processing into three stages as follows:

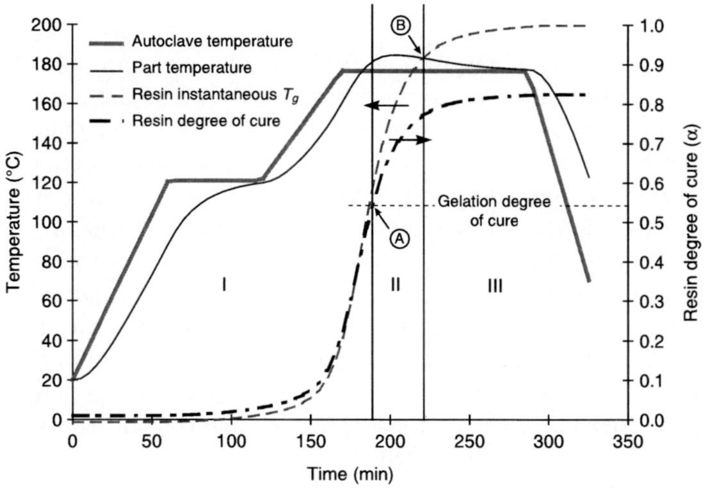

13.5 Stages in resin property development for a resin with a glass transition temperature greater than the maximum curing temperature. 'A' indicates the point of gelation ($\alpha = \alpha_{gel}$), 'B' indicates vitrification ($T_g > T$).

I. *Purely viscous behaviour.* During this stage, the low molecular weight resin is unable to support shear loads and thus develops no internal stress.
II. *Viscoelastic behaviour.* After resin gelation, but prior to vitrification, resin behaviour will be highly viscoelastic. Any stresses generated during this stage will decay to some degree, although they may still be important to the final stress state.
III. *Elastic behaviour.* At some point during processing, the resin will undergo vitrification and mechanical response will become effectively elastic. The time at which this transition occurs depends on the process cycle and the development of resin T_g with the progression of cure.

An example of when each of these stages might be encountered in a typical autoclave process cycle is shown in Fig. 13.5. In this example, the viscous-viscoelastic transition (point A) occurs when the resin undergoes gelation at a degree of cure of about 0.55. The viscoelastic-elastic transition (point B) follows quickly in this case, occurring about 30 min later. Thus, stresses developed in the early stages of processing will probably not have time to relax during the remainder of the process cycle. However, if vitrification were delayed until much later in the process (e.g., during cool-down), most of the stress generated in Stage II would probably decay. The time of the viscoelastic-elastic transition, and thus the length of Stage II, is dependent on both the process cycle and the material employed.

Stress analysis solution method

The stress module is responsible for prediction of process-induced strain and deformation throughout the structure of interest and the process tooling. Instantaneously linear elastic constitutive models assuming orthotropic material behaviour are employed for all types of materials. For non-composite materials, mechanical properties at any time during processing are calculated as linear functions of temperature. For composites, the current industrial state of the art is that the isotropic matrix resin is modelled as a 'cure-hardening/instantaneously linear elastic' (CHILE) material. Calculated elastic constants are used in micromechanics models to determine transversely isotropic ply properties. Numerous techniques are employed to minimize required computational effort including use of an adaptive 'time-stepping' algorithm, using an efficient 'skyline' matrix solver, and allowing incorporation of multiple composite plies within each element. Coupled with recent advances in available computing power, these methods allow examination of structures of intermediate size and complexity using only a mid-range personal computer. Details on the solution method for residual stresses and deformations in autoclave processing can be found in reference 27.

The CHILE approach has been known, from early days, to be an approximation to what is a full viscoelastic material response. More recent work[28] has shown that there are approximations that allow the full viscoelastic approach to be simplified progressively. The most simple approximation, namely the choice of a constant time or frequency to evaluate the modulus, reduces to the CHILE approach. The advantage of positioning the CHILE approach in the viscoelastic framework, apart from providing a proper justification, is that the choice of testing protocols can be made using validated and systematic decisions. Thus, in general, one can consider a continuum of trade-off of investment versus accuracy as one goes from a full viscoelastic approach to the simplest CHILE approach.

There are increasingly cases where a CHILE approach may not be adequate. A good example is the off-tool post-cure of a structure. This will be done in cases where the tool or perhaps the autoclave cannot handle the required final hold temperatures, or where there is a desire to maximize use of expensive tooling and equipment. In this situation, the CHILE approach will be unable to properly predict the relaxation of stresses after the first hold, and a full viscoelastic approach is required. Whereas most full viscoelastic approaches are integral in formulation, and present significant issues in terms of implementation and use for industrial use, recent work[29] has provided a differential viscoelastic formulation that is backwardly compatible with the CHILE approach, thus allowing the user to reasonably seamlessly move back and forth between the current industrially validated CHILE approach and the required viscoelastic capability for the more complex problems of tomorrow.

Tool removal simulation

At the end of a process cycle, the manufactured structure will typically be removed from the autoclave, still affixed to any process tooling, and allowed to cool to room temperature. At this point, depending on the case, large residual stresses might still exist between part and tooling, sometimes making separation quite difficult. Consequently, the stress state in the processed structure, and thus the residual deformation, will be quite different before and after tool removal.

The situation in the simulated process cycle is quite similar: at the end of the process cycle simulation, the tooling and processed structure remains attached with residual stresses remaining between them. Thus, to predict the 'final' post-processed component shape and residual strain state, a simulation of the tool removal process is undertaken.

The tool removal simulation is performed in a single elastic step and the calculated change in displacements added to the pre-tool-removal displacements to give the final component shape. The procedure used is as follows:

1. Create a finite element description of the tool removal problem. The new F.E. mesh is identical to that used in the main stress module simulation, except that the elements identified by the user as 'tooling' are not included. A new set of displacement boundary conditions is also required since the part will see different boundary conditions during tool removal than during the process cycle.
2. Assemble the tool removal global stiffness matrix, $[\mathbf{K}_{\mathrm{TR}}]$, and set the tool removal global force vector, $\{\mathbf{F}_{\mathrm{TR}}\}$, to zero.
3. Add the *tooling element* nodal forces, $\{\mathbf{f}_{\sigma}\}_e$, at all nodes along the tool/part interface to the tool removal global force vector. Since these forces will be the *negative* of the forces which the tool applied to the part (from equilibrium), the effect of adding these forces is to force a stress-free interface.
4. Solve for the part displacements during tool removal ($\{\delta_{\mathrm{TR}}\} = [\mathbf{K}_{\mathrm{TR}}]^{-1}\{\mathbf{F}_{\mathrm{TR}}\}$).
5. Add the calculated displacements to those calculated prior to the tool removal simulation to determine the post-tool-removal part shape.

This procedure is illustrated schematically in Fig. 13.6.

13.3 Process development

Three common concerns when developing an autoclave process for a specific component in the aerospace industry are: (1) to meet the cure cycle specification, that is, to ensure that all regions of the component are experiencing

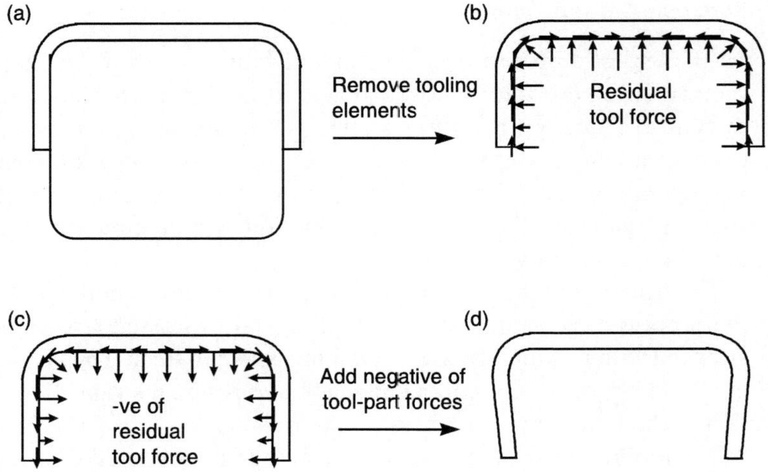

13.6 Schematic of the tool removal process: (a) prior to tool removal, part and tool in equilibrium, part conformed to tool shape; (b) tooling removed, residual tool/part interface forces remain; (c) add negative of interface loads to obtain stress free interface; (d) predicted part shape after tool removal.

the prescribed cure cycle with respect to minimum and maximum heat-up and cool-down rates, and length of temperature holds; (2) to ensure that the part quality is acceptable, for example acceptable porosity levels, fibre orientation and part thickness; (3) to ensure dimensionality fidelity, that is, that the cured part meets geometric tolerances. All of these concerns can be addressed through process modelling. Although current process models are not able to capture all interacting chemical, physical and mechanical phenomena to the same level of detail and accuracy, they can still be very useful in guiding process development and to reduce risk in composites processing. A useful and effective process model has the right balance of scientific rigour and engineering pragmatism. Two case studies, one on thermal management, and the other on dimensional control are presented as examples.

13.3.1 Thermal management

As discussed previously, accurate prediction of the thermal history of the part is critical. Not only is the management of the thermal history important in its own right, for example to avoid exotherms, but it determines the degree of cure evolution. In turn, the coupled temperature and degree of cure history, which in a real part will vary with position throughout the structure, will determine flow behaviour and residual stress/deformation behaviour. Further to Section 13.2.1, determination of the system boundary conditions

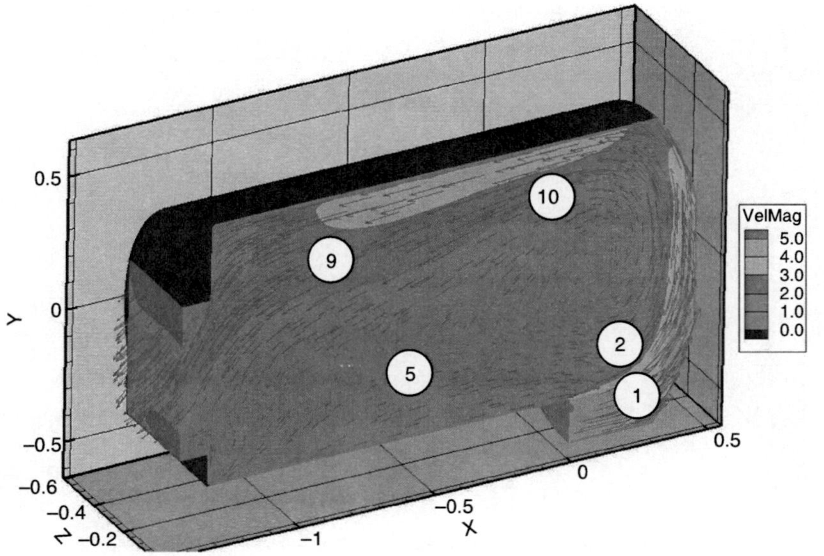

13.7 CFD simulation of airflow within a small research autoclave (1.5 m long, 1.15 m diameter). The circled numbers indicate the location of calorimeters used to measure the local HTC. (From reference 30.)

is a very important requirement for accurate prediction of the part response. In autoclave processing, the appropriate heat transfer boundary condition is to use a convective HTC. In general, the convective heat transfer value will be a function of position in the autoclave, with an example of the airflow in a small research autoclave shown in Fig. 13.7,[30] with additional complexity when airflow under the toolskin, through the tooling substructure, is considered. The combination of autoclave and tooling complexity typically forces the user to simplify the problem to a representation of the part, the tooling skin, and any other significant components such as a bagside caul-plate and/or internal inserts such as mandrels.

Typical values for HTC range from 10 to 200 W/m^{2}K, with bagside values typically being higher due to unimpeded airflow, and toolside values typically lower. It is important to emphasize that toolside behaviour is tooling material dependent. For aluminum tooling, where the material diffusivity is very high, the aluminum tooling substructure can act as a significant heat fin, and compensate significantly for any loss in airflow velocity. The range of HTC values is such that it can play an important role in the temperature history, in both thermal lag and exotherm, as shown in Fig. 13.8 taken from reference 30. As can be seen, it is not necessarily true that a high HTC is a desirable attribute, as in this case a lagging Invar tool will provide a significant heat sink to manage the exotherm. Process simulation then becomes a

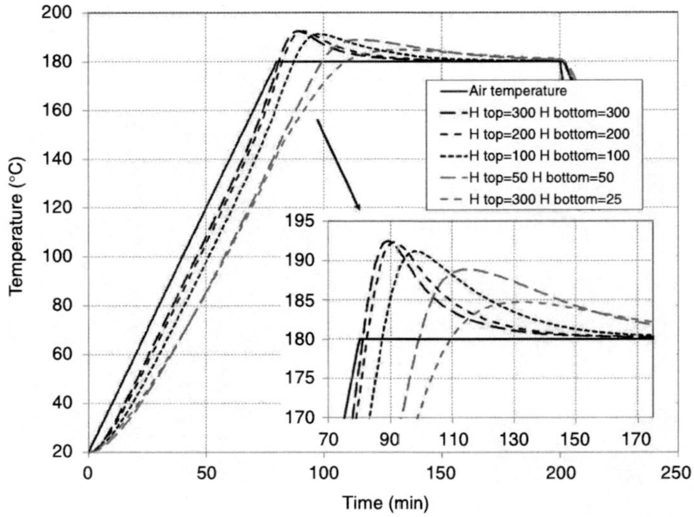

13.8 Evaluation of thermal histories for a 15 mm thick AS4/8552 part on a 15 mm thick Invar tool using a 1D thermal simulation in COMPRO. (Re-created from reference 30.)

very effective tool for parametrically evaluating the design space in terms of top (bagside) and bottom (toolside) HTC values, as well as tooling material and thickness choices.

Inspection of experimental and numerical results for the temperature evolution through the thickness of a part shows that the response can be explained in terms of an initial transient response in the system, followed by a quasi-steady state condition, as shown schematically in Fig. 13.9, and described in greater detail in reference 31.

As the temperature increases, the cure kicks off, and the heat generated during cure increases the local temperature, which further accelerates the cure, and consequently the temperature history of a curing part rapidly diverges from that of an equivalent inert part. An example of this is shown in Fig. 13.10, where the part was subjected to the same autoclave cycle twice.[32]

Experience to date indicates that if there is adequate characterization of material properties (in particular the cure kinetics response) and the HTC boundary conditions, then the prediction of thermal history and consequent degree of cure evolution is sufficient for most practical purposes. Currently, the biggest impediments to further adoption of process simulation for thermal management are materials characterization costs (which are one-off) and cost-effective boundary condition characterization, particularly for the case of process variations and deviations such as random positioning of tools in autoclaves and similar uncontrolled practical issues.

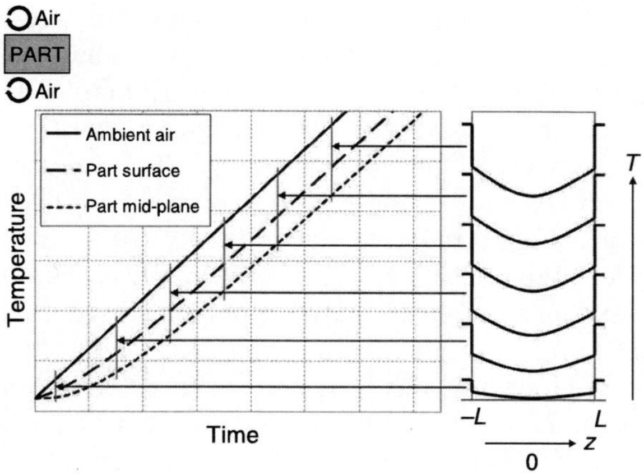

13.9 Evolution of the temperature profile in an inert part as a function of time. The response is described simply as an initial transient response until a steady state trough-thickness profile is developed and which then increases in step with the air temperature increase. (Re-created from reference 31.)

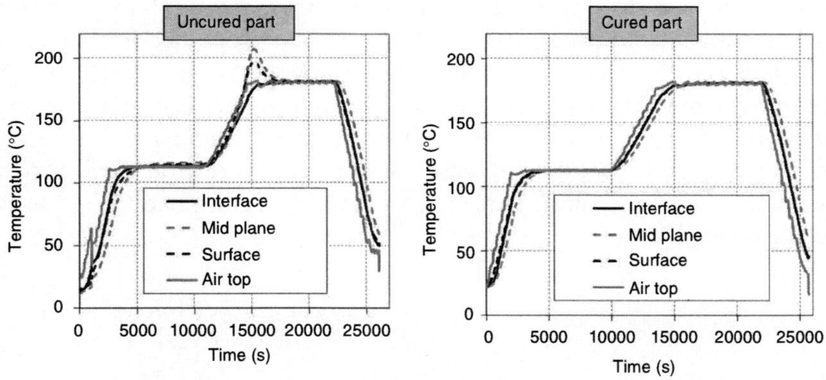

13.10 Experimental demonstration of the effect of cure on temperature history on a 37 mm thick part made from AS4/8552 on a 25 mm thick aluminum tool. The thermal lags are the same pre- and post-cure, and the effect of cure is the addition of a significant exotherm to the base response. (Re-created from reference 32.)

13.3.2 Dimensional control

The following case study illustrates the use of process models to ensure dimensional fidelity of large aerospace honeycomb structures.[33] The concern in the current case was whether geometric compensation of the process tool was necessary in order to get the part within dimensional tolerances. The part under consideration is a thick honeycomb structure with carbon/epoxy composites skins (Fig. 13.11). The component is approximately 3.5 m long (fore-aft) and 2.5 m wide (inboard-outboard), with a maximum thickness of about 0.1 m. It has a distinct curvature in the inboard-outboard plane and a slight curvature in the fore-aft plane. The honeycomb core is chamfered down and the panel is a solid laminate at the edges. The component is processed at 177°C on a steel tool and the desired geometric tolerances are ±0.5 mm of nominal engineering dimensions. In this case all the sources of residual stress identified in Section 13.2.3 – thermal strain, cure shrinkage, cure gradients and tool constraints – have the potential to cause the dimensions to be outside tolerances.

Without process modelling, this problem has to be solved empirically using trial-and-error. This approach can be very time consuming and costly for large structures. Furthermore, it gives little or no insight to the main drivers of the problem and last-minute changes in part geometry, layup, tool material or process conditions cannot be addressed. An alternative approach is to develop a validated process model. For the current structure, cure kinetics and resin mechanical behaviour sub-models had already been developed for the skin material. The thermal expansion/contraction behaviour of the core was measured using a thermal-mechanical-analyser from TA instruments, which together with mechanical properties from the core supplier and thermal/mechanical properties of the tool material gave the required input for the process models. To develop confidence in model predictions, a series of models with varying complexity and detail were developed using the COMPRO software.[33] By developing several models of the same problem, the main drivers of process-induced deformation in this case could be identified.

13.11 Cured test article placed on process tooling to measure dimensional fidelity.

The predicted deformations from a full 3D process model of the part are shown in Fig. 13.12 (insert). The figure shows small process-induced deformations of the flatter section but significant 'spring-in' of the curved section. To validate the process model, a test article was manufactured and the gap (deformation) between the cured part and the process tool was measured with a feeler gauge while the cured part was clamped back on the process tool it was made on as shown in Fig. 13.11. A comparison of predicted and measured process-induced deformations along the perimeter of the part is shown in Fig. 13.12. In contrast to an experiment, a process model does not only give information about what is going to happen but also why. In the current case it is seen that the expected process-induced deformation is significantly greater than the desired tolerances of ±0.5 mm. Action needs to be taken to ensure dimensional fidelity. The deformations are dependent on tool and part material properties, part layup, tool and part geometries and the process conditions. To not alter the performance of the part, it was decided to modify the process tool. There are two options when modifying the process tool in the current case. The first one is to compensate the tool geometry such that when the part springs in, it will spring in to the correct geometry. The tool should then be compensated with the opposite of the predicted deformation shown in Fig. 13.12. The second option is to change the tool material to a material with the same coefficient of thermal

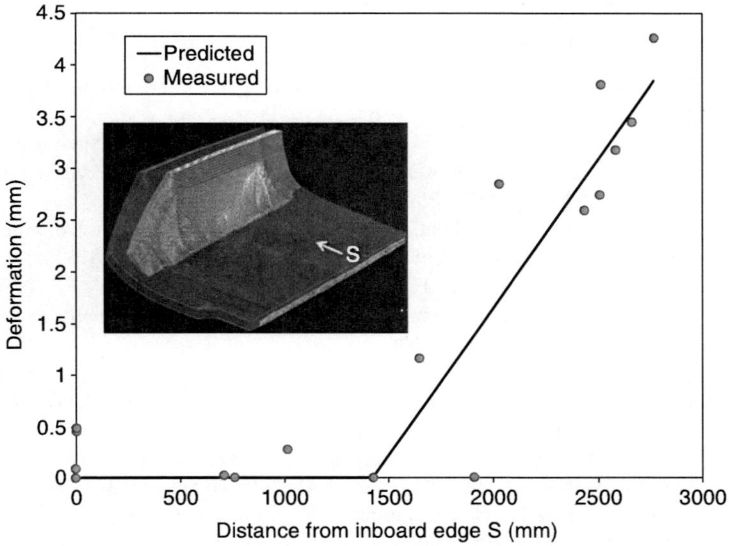

13.12 Comparison of predicted and measured out-of-plane deformation for the full scale test article. Insert: undeformed and deformed finite element mesh of the panel. Out-of-plane deformations magnified 10 times.

expansion as the core has through-thickness. When heated up, the tool would then expand such as the part sprung back into the required shape. Finding a tool material with the appropriate coefficient of thermal expansion was not practical and it was decided to compensate the tool with the opposite of the predicted spring-in shown in Fig. 13.12.

13.4 Conclusions and future trends

In this chapter, the state-of-the-art modelling approach applied to the processing of composite with the autoclave process was summarized. This approach enables the reduction of the risks associated with the manufacturing of complex integrated composite structures by reducing the number of expensive experiments in order to develop the cure cycle and tooling. The autoclave processing technology can be considered a mature technology as it has not dramatically changed for the last four decades. However, the complete effort to develop the process is still very long and complex. Therefore, the advances in process modelling and simulation of the last two decades can significantly reduce the development cost of the process. This will lead to development of more cost-effective tooling, more thermally efficient autoclaves and more robust autoclave control systems. In the future, autoclave processing will remain the composite manufacturing process of choice for high quality, high performance and low part count applications.

13.5 References

1. Purslow, D. and Childs, R., 'Autoclave moulding of carbon fibre-reinforced epoxies', *Composites*, **17**(2), 1986, 127–136.
2. Holl, M. W. and Rehfield, L. W., 'An evaluation of the current status of automated process control for thermosetting composites', *Composites Materials Testing and Design*, ASTM STP 1120, Philadelphia, 1992, 308–319.
3. Kalra, L., Perry, M. J. and Lee, L. J., 'Automation of autoclave cure of graphite/epoxy composites', *Journal of Composite Materials*, **26**(17), 1992, 2567–2585.
4. LeClair, S. R. and Abrams, F., 'Qualitative process automation', *Proceedings of the IEEE Conference on Decision and Control Including the Symposium on Adaptive Processes*, **1**(3), 1988, 558–563.
5. Kenny, J. M., 'Integration of process models with control and optimization of polymer composites fabrication', *Proceedings of the Third Conference on Computer Aided Design in Composite Materials Technology*, 1992, 530–544.
6. Kline, R. A. and Altan, M. C., 'Curing of composite materials using extended heat transfer models', United States Patent 5,453,226, 26 September 1995.
7. Pillai, V. K., Beris, A. N. and Dhurjati, P. S., 'Implementation of model-based optimal temperature profiles for autoclave curing of composites using a knowledge-based system', *Industrial and Engineering Chemistry Research*, **33**, 1994, 2443–2452.
8. Pillai, V. K., Beris, A. N. and Dhurjati, P., 'Intelligent curing of thick section composites using a knowledge-based system', *Journal of Composite Materials*, **31**(1), 1997, 22–51.

9. Loos, A. C. and Springer, G. S., 'Curing of epoxy matrix composites', *Journal of Composite Materials*, **17**(2), 1983, 135–169.
10. COMPRO, 'Virtual Autoclave' simulation tool, http://www.convergent.ca/products/index.html, Convergent Manufacturing Technologies, Vancouver, BC.
11. Mijovic, J. and Wijaya, J., 'Effects of graphite fiber and epoxy matrix physical properties on the temperature profile inside their composite during cure,' *SAMPE Journal*, **25**(2), 1989, 35–39.
12. Twardowski, T. E., Lin, S. E. and Geil, P. H., 'Curing in thick composite laminates: Experiments and simulation', *Journal of Composite Materials*, **27**(3), 1993, 216–250.
13. Davé, R., Kardos, J. L. and Dudukovic, M. P., 'Process modeling of thermosetting matrix composites: A guide for autoclave cure cycle selection', *Proceedings of the American Society for Composites, 1st Technical Conference*, 1986, 137–153.
14. Gutowski, T. G. and Dillon, G., 'The elastic deformation of lubricated carbon fiber bundles: Comparison of theory and experiments', *Journal of Composite Materials*, **26**(16), 1992, 2330–2347.
15. Astrom, B. T., Pipes, R. B. and Advani, S. G., 'On flow through aligned fiber beds and its application to composites processing', *Journal of Composite Materials*, **26**(9), 1992, 1351–1373.
16. Hubert, P. and Poursartip, A., 'A method for the direct measurement of the fibre bed compaction curve of composite prepregs', *Composites: Part A: Applied Science and Manufacturing*, **32**, 2001, 179–187.
17. Hubert, P., Vaziri, R. and Poursartip, A., 'A two-dimensional flow model for the process simulation of complex shape composite laminates', *IJNME*, **44**(1), 1999, 1–26.
18. Adolf, D. and Martin, J. E., 'Calculation of stresses in crosslinking polymers', *Journal of Composite Materials*, **30**(1), 1996, 13–34.
19. Griffin Jr., O. H., 'Three-dimensional curing stresses in symmetric cross-ply laminates with temperature-dependent properties', *Journal of Composite Materials*, **17**, 1983, 449–463.
20. Hahn, H. T., 'Effects of residual stresses in polymer matrix composites', *Journal of the Astronautical Sciences*, **32**(3), 1984, 253–267.
21. Hahn, H. T. and Pagano, N. J., 'Curing stresses in composite laminates', *Journal of Composite Materials*, **9**, 1975, 91–108.
22. Bogetti, T. A., 'Process-induced stress and deformation in thick-section thermosetting composites', Technical Report CCM-89-32, Center for Composite Materials, University of Delaware, Newark, Delaware, 1989.
23. Roberts, R. W., 'Cure quality control'. In Dostal, C. A., Woods, M. S. and Ponke, A. W. (eds.), *Engineered materials handbook, Vol. 1: Composites*. Ohio: ASM International, 1987, pp. 745–760.
24. Ridgard, C., 'Accuracy and distortion of composite parts and tools: Causes and solutions', SME Technical Paper EM93-113, 1993.
25. Nelson, R. H. and Cairns, D. S., 'Prediction of dimensional changes in composite laminates during cure', 34th International SAMPE Symposium, 8–11 May 1989, pp. 2397–2410.
26. Shi, J. and Flannagan, R., 'A simple spring-damper-slider model for laminate slippage'. In Poursartip, A. and Street, K. (eds.), *Proceedings of the 10th International Conference on Composite Materials (ICCM – 10)*. Whistler, B.C., Canada, August, 1995, pp. III-197–III-204.
27. Johnston, A., Vaziri, R. and Poursartip, A., 'A plane strain model for process-induced deformation of laminated composite structures', *Journal of Composite Materials*, **35**(16), 2001, 1435–1469.

28. Zobeiry, N., Vaziri, R. and Poursartip, A., 'Computationally efficient pseudo-viscoelastic models for evaluation of residual stresses in thermoset polymer composites during cure', *Composites Part A: Applied Science and Manufacturing*, **41**(2), 2010, 247–256.
29. Zobeiry, N., Vaziri, R. and Poursartip, A., 'Differential implementation of the viscoelastic response of a curing thermoset matrix for composites processing', *Journal of Engineering Materials and Technology*, **128**(1), 2006, 90.
30. Slesinger, N., Shimizu, T., Arafath, A. R. A. and Poursartip, A., 'Heat transfer coefficient distribution inside an autoclave', Proceedings of ICCM-17, Edinburgh, Scotland, 2009.
31. Rasekh, A., Vaziri, R. and Poursartip, A., 'Simple techniques for thermal analysis of the processing of composite structures', 36th International SAMPE Technical Conference, 15–18 November 2004, San Diego, USA.
32. Shimizu, T., Kotlik, J. C., Arafath, A. R. and Poursartip, A., 'Evaluation of temperature profiles in thick composite parts during autoclave processing', Proceedings of 23rd ASC Technical Conference, Memphis, TN, September 2008.
33. Fernlund, G. and Floyd, A., 'Process analysis and tool compensation for a complex composite panel', ASC 22nd Annual Technical Conference. Seattle, WA, September 2007, 17–20.

14

Out-of-autoclave curing process in polymer matrix composites

J. SCHLIMBACH and A. OGALE,
Schlimb@ch Business Solutions, Germany

Abstract: Traditional autoclave curing technology has its own limitations which has motivated many researchers and industries to consider an out-of-autoclave (OoA) alternative. Since the last couple of decades number of autoclave substitutes have been developed and demonstrated by the technologists around the globe. This chapter provides information on most of the OoA technologies that are used on the industrial and lab scale. These technologies are compared with the conventional autoclave process and their merits and demerits are also illustrated. Expectations of the composite industry from such technologies have been presented at the end of this chapter.

Key words: out-of-autoclave (OoA), liquid composite molding, Quickstep, electron beam, microwave, vacuum bag only (VBO).

14.1 Introduction

This chapter explains the need for out-of-autoclave (OoA) processing of composites and the means that have been developed to achieve this concept. Various aspects of composite curing without an autoclave are explained in detail. Curing techniques based on fluid heating, electromagnetic heating and cure kinetics are thoroughly discussed with examples and the latest available information. Although the OoA philosophy has been established for more than two decades, aspects like qualification of vacuum bag only (VBO) prepregs, OoA processes and automation have not yet completely been adopted by the composite manufacturing industry. Partial acceptance of the techniques and related materials is reflected in techno-economic discussions. At the end of this chapter, the authors' recommendations on the future trends in terms of materials, processes and component size are summarized.

14.2 Reasons for using the out-of-autoclave (OoA) process

Most of the high performing composite components are currently produced by using epoxy based prepreg laminates cured in an autoclave at high

Published by Woodhead Publishing Limited, 2012

temperature and pressure. Most major prepreg materials are designed for autoclave curing and therefore, optimal material properties can be achieved with high reliability. As a result, the manufacturing process is optimized to its physical limits. In the last few decades when it comes to stability, reliability and experience, autoclave curing has taken its place as the reference in high performance composite manufacturing.

Apart from reliable performance and maturity, autoclave curing has numerous disadvantages that make it inapplicable in a variety of fields of application. These disadvantages include:

- Long cycle time/turnaround time.
- High pressure.
- Massive investment.
- Costly tooling.
- Low temperature control (high latency).
- Substantial cost for 'lost' nitrogen.
- Huge footprint.
- Excessive energy consumption.

On their own, some of these disadvantages are self-explanatory but it is worth examining some of the specifics in detail. The long cycle time itself (normally ~8 h but easily higher) would not be a huge problem, but as a batch curing process, when combined with cost of the multiple tools for each batch, costs are often so massive the process becomes unaffordable for most parts and markets. The tooling usually costs a minimum of US $0.5 M, and the total investment cost is one of the strongest drivers. An example of the world's largest autoclave is shown in Fig. 14.1 (9 m diameter and 23 m length). This autoclave was built for production of fuselage sections for the Boeing Dreamliner.

Of course the high variable cost is also something to look at, which is by and large driven by the physics of the process. To heat up a large quantity of inert gas and transfer heat to a massive tool to cure a composite prepreg laminate, the smallest portion of the whole setup, is obviously time consuming and therefore quite inefficient.

In the end, all these factors result in very high part cost and restrict the volume to relatively low numbers per day. This has driven continuing development to try and reduce the manufacturing costs, starting in the segments of lower performance applications where the extreme cost of autoclaves could not be warranted. The development made in these low end performance sectors has fundamentally changed the entire composites manufacturing market.

The achievements of the past 20 years have seen the composite manufacturing industry mature greatly and FRP has become a material group

14.1 Example of an autoclave. (From reference 1.)

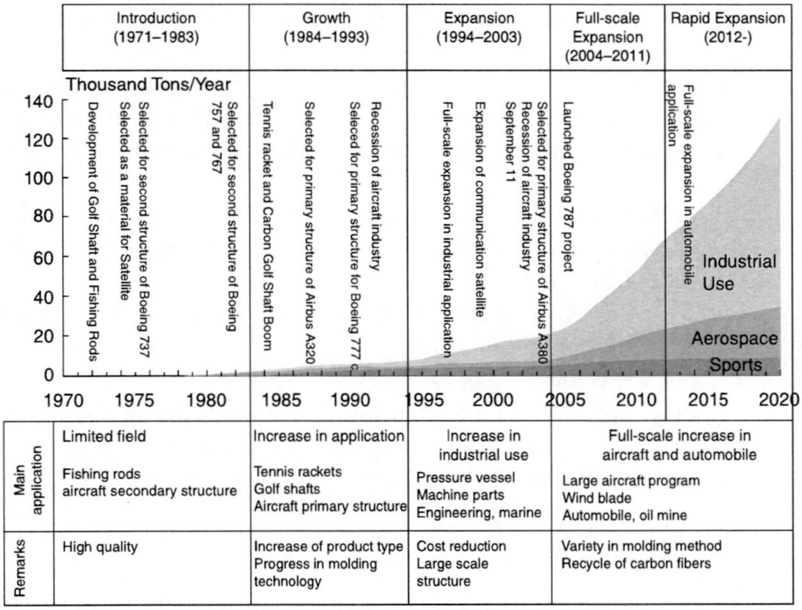

14.2 Growth for different markets in terms of carbon fibers. (*Source*: Toray.)

that is seen as very interesting to many different markets for many different applications. The reinforcing material market, especially carbon fibers, has grown exponentially in the last 40 years and forecast use for 2020 is double that of 2010 (Fig. 14.2). At the same time, it has become obvious to aerospace suppliers that the industry will sooner or later be non-sufficient and face more or less a single customer or at least single market problem. Consequently, the interest in using more economical alternatives than autoclave curing is evolving and becoming more and more a necessity for a lot of companies. This would allow diversification strategies for the suppliers to become a reality and on the other hand to fulfill future cost reduction targets as required by aerospace manufacturers.

But the trend is not only related to cost and market, it is also driven by government regulations. Carbon emissions are in particular a topic with a high priority both politically and macroeconomically. This topic has a strong influence on all industries, especially for the composites manufacturers. In addition to the previously mentioned disadvantages of autoclave manufacturing, the high-energy consumption of the process and the numerous pre- and post-processes that are required, a large number of boundary conditions exist pushing the composites industry to develop OoA solutions. These boundary conditions include, but are not limited to:

- The production of carbon fibers on their own in particular needs a lot of energy.
- Autoclave tools are massive and produce often a very high carbon emission during the production of their source materials (e.g., aluminum).
- Most resins are oil based at the moment and therefore use fossils twice over.
- Repair of CFRP is still challenging or near to impossible (e.g., automotive applications where 'look' is highly important or crash elements, structural defects, etc.).
- Recycling of CFRP is a real problem (only down-cycling or thermal recycling seems possible).

By taking these implications into consideration as well, the significant need for this industry to develop different composite manufacturing solutions for different markets, preferably OoA, is clear.

14.3 Strategies

To overcome the challenges involved in autoclave curing and the pre- and post-processes, alternative composite manufacturing technologies with the aim of producing cost effective aerospace grade composites in a shorter processing time are developing rapidly. These include resin transfer molding (RTM), resin film infusion (RFI) and vacuum-assisted resin infusion (VARI). But how can those approaches be categorized?

There are main 'steps' or 'physical treatments' a composite material needs to go through before it ends up as a part (Fig. 14.3).

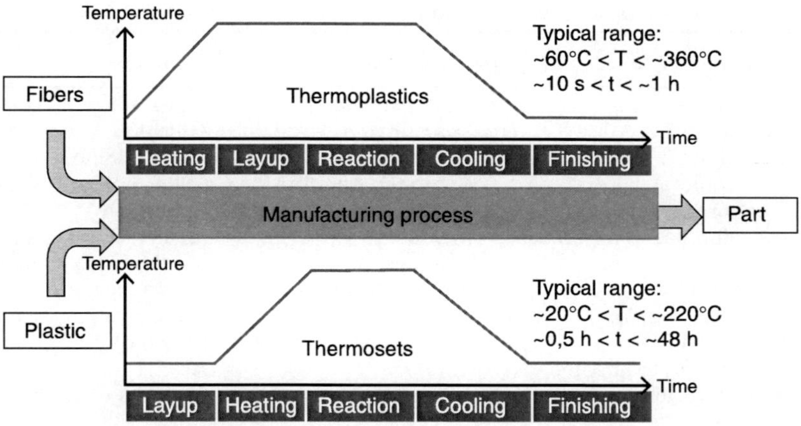

14.3 Composite manufacturing.

All these steps can be influenced by their efficiency. Some of the alternative manufacturing technologies not only try to change a single step, but also perhaps multiple steps, often similar ones. Excluding material modifications, by following this rationale there are only three (and a half) different strategies to make composite manufacturing processing more economical:

Heating and cooling steps	→	change the energy transfer method
Reaction steps	→	change resin formulation or resin kinetics
Layup and finishing	→	automation and '*net-shape*' as half of a strategy

Category based approaches are: means of energy transfer, cure kinetics and possibilities for automation. Table 14.1 shows list of various approaches.

14.4 Technical description of different OoA processes

This chapter provides a quick technical overview of different OoA processes including recent results and applications. The Achilles' heel in any strategy to overcome autoclave curing is often the resin systems themselves. Most currently qualified systems are specifically designed for high pressure and slow temperature ramp up and cool down rates. In fact, in many cases, resin chemistry has been modified and inhibitors are added to make the resin suitable to be cured very slowly. On the other hand, many OoA manufacturing processes work in a low pressure environment. When manufacturing autoclave materials with low pressure processes, a lot of development has to be done using various strategies to achieve the required material performance.

Table 14.1 OoA approaches by category

Energy transfer	Resin cure kinetics	Automation
• Induction heating • Gas nozzles • Vibration • Ultrasonic • Electromagnetic radiation (laser, infrared, microwave, radio frequency) • Oven heating, for example, VBO • Fluid curing (oil heated tooling, Quickstep and others)	• LCM also called infusion and injection • Room temperature Resin systems • *In situ* Polymerization • E-beam curing	• Preforming (stitching, binder, AFP, braiding) • Roll forming • Pultrusion • Press technologies (double-belt, semi-continuous and static press, injection molding) • ATP • Filament winding

As a logical step, new materials, designed specifically for OoA processing with the same performance should be developed. Especially in aerospace where high qualification requirements and costs are required to put processes into production and very long service lifetimes exist for aircraft and long-term production contracts exist, it is often commercially not viable and sometimes impossible to change materials. The underlying reason for this is that aircraft parts are designed with a specific material and qualified to particular manufacturing processes. This lengthy and laborious process makes for a very robust and stable process but also makes it extremely restrictive to enter the market. This often results in less than optimal manufacturing solutions in the field. For the first generation of OoA parts in production, optimization has been limited. Only with further penetration of the market will the demand for new and adopted resin formulations be developed and used for the subsequent generation of parts.

14.4.1 Oven/vacuum bag only (VBO) processes

The principal idea of OoA is to get away from high pressure curing. This will allow for more economical equipment acquisition and maintenance, and eliminate the need for compressed nitrogen. However, usually VBO processes can only be used for thermosets not for thermoplastics due to this characteristic low pressure. There are two main areas where oven curing is used:

- VBO which is related to pre-impregnated materials (Prepreg)
- Liquid composite molding (LCM) processes where the resin is infused into a fiber structure or preform

14.4 Convection oven.

The heat transfer of oven processes is based on the same principle as the autoclave: convection heating. The heat transfer medium, normally air, is different than for an autoclave, where it is usually compressed nitrogen. However, the specific heat capacity of air is lower than compressed nitrogen which makes it slower to react. This means that the effect and risk of exothermic reaction can only be restricted by ramping up the temperature very slowly and with very accurate control. Figure 14.4 shows an example of convection oven used for composite curing.

In comparison to an autoclave, ovens are much cheaper but are still batch production processes. Therefore, they are also a bottleneck and if the part does not fit in the oven a completely new piece of equipment needs to be acquired to produce the part. A summary of the pros and cons concerning convection oven curing are shown in Table 14.2.

Vacuum bag only (VBO) → OoA prepregs

OoA prepregs are primarily designed for VBO processing, ensuring even resin distribution and avoiding the dry spots and resin-rich pockets common with infusion processes. Furthermore, OoA prepregs can be cured at lower pressures and temperatures (atmospheric pressure and cure at approximately 120°C compared to 7–10 bar and 180°C autoclave cure). Nevertheless, after curing a part at 120°C, a free-standing post-cure at elevated temperature (e.g., 180°C) is recommended to achieve a final higher T_g and improved

Table 14.2 Pros and cons for oven curing

Pros	Cons
• Low investment • Large number of parts per cure possible (good for high variety of parts but low volume each) • Easy to use • Long history of experience in industry • Applicable for perform and infusion processing	• Lower heat transfer rates than an autoclave • Long process time (heating but especially cooling!) • Relatively high energy consumption (a lot of futile thermal mass, air and tooling, is heated) • Space limited (like autoclave) • Batch processing (like autoclave)

mechanical performance. As far as manufacturing is concerned, two common methods have been employed to produce VBO prepregs. One method involves the use of a perforated resin film that results in a prepreg with gaps in the resin film. The second method involves partial impregnation of fibers/fabric with a resin film. The perforated resin film method allows gas to flow out of the laminate in directions parallel and perpendicular to the plane of the prepreg. However, the partial impregnation in practice leaves dry spaces or channels in the prepregs. During subsequent compaction, these channels facilitate airflow in a direction parallel to the laminate plane. Partial impregnation of dry fibers and utilizing an edge-breathing consumable stack during the processing is the more common method of producing VBO laminates.[2]

Since the conception of OoA, great efforts have been made by the composites industry, especially aerospace, to develop optimum prepreg for OoA processing. Figure 14.5 shows a consolidated view of the actual and targeted specifications for three generations of OoA prepregs. Here, prepregs available for autoclave curing and laminates made from them are taken as a base line. First and second generation OoA prepregs were not at all up to the current autoclave specifications and therefore, a new specification of third generation prepregs was set. Based on this and other research, there are number of third generation OoA prepregs available in the market which are listed in Table 14.3.

Material qualification

Some OoA prepregs are already undergoing qualification for aerospace production and have comparable mechanical properties to autoclave-cure systems for primary aircraft structures. Nevertheless, material qualification and the availability of a suite of OoA prepregs are still not at a mature level. To get Federal aviation authority (FAA) approval, a database of performance data from qualification testing is mandatory. Qualification of the

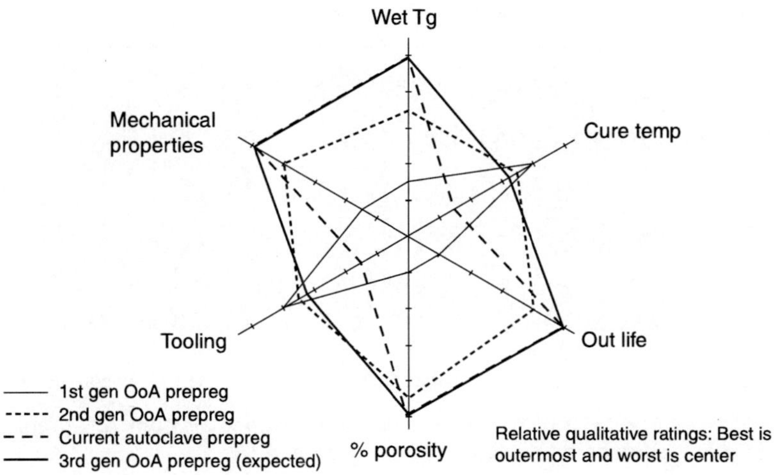

14.5 Boeing's target for 3rd generation OoA prepregs. (From reference 3.)

Table 14.3 Partial list of available OoA prepregs

Manufacturer	Cytec	Hexcel	ACG	Toray
Available prepregs	Cycom 5320 Cycom 5215 Cycom 5320-1	Hexply M56	MTM44-1 MTM45-1 MTM46 XMTM47	2510

materials based on testing is extensive, requiring thousands of test samples to be produced and tested to meet the minimum dataset requirement for various applications. To generate such an extensive database can cost between $3 million and $5 million per material.[4]

Regardless, some programs are undertaken by commercial aircraft manufacturers. For example, at Airbus some trials are underway to develop a primary structure using OoA prepreg to show potential use of the material. However, there are still no specifications for the prepreg product that is being used for the development program.

Long evacuation for low void

Out-of-autoclave prepreg systems do not mean faster production rates. As previously explained, unlike an autoclave, in VBO processing of OoA prepreg, removal of volatiles, including air entrapped and absorbed moisture, is generally achieved by using an edge-breathing strategy (Fig. 14.6). The laminate edges must be in contact with the breather and materials must

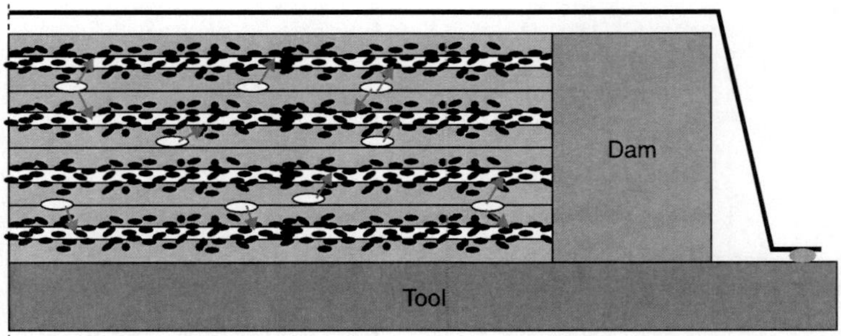

14.6 Schematic of evacuation of entrapped air from VBO prepreg via edge dam principle. (From reference 5.)

be arranged in a way that it maintains air escape paths. The vacuum bag materials and sequence are the same as with autoclave processing, but vacuum quality is vital. As entrapped air extraction is a time-dependent process, OoA cure cycles are typically longer. After debulking, vacuum must be held for an extended period before initiating cure; this hold time period depends on the part size and complexity, ranging from as low as 4 h for 0.4 m^2 to 16 h for a 72 m^2.[6]

Furthermore, slow ramp rates are also recommended for the OoA prepreg because fast ramp rates reduce the resin viscosity, allowing the resin to penetrate through the fibers very quickly, which hinders the removal of air from the laminate. However, by utilizing slower ramp ups, additional evacuation time is achieved allowing for the maximum extraction of air and generated volatiles.

External parameters such as relative humidity also play a very important role in void content in the laminate. Generally, epoxy resin tends to absorb moisture in the air and the trapped moisture is very difficult to remove under VBO processing. Influence of relative humidity on void content of VBO processed laminate has been systematically studied by Nutt *et al.*[7]

Cure cycle recommendations from the manufacturer for the curing of Cycom 5320 are[6]:

- Prior to heating a vacuum hold at full vacuum (minimum 28 in Hg or 711 torr at sea level) is required.
- Full vacuum should be within 2 in Hg (50 torr) of absolute vacuum for the given altitude.
- Vacuum hold times will depend on the part size and complexity, but general recommendations are 4 h minimum hold for any uniform thickness

parts smaller than 2 ft × 2 ft (0.6 m × 0.6 m) and 16 h minimum hold for larger or more complex parts.

- Fastest cure cycle: 121°C (1 h) dwell + 177°C (2 h dwell); ramp rates 0.6–2.8 K/min.
- Total cycle time could be between 8 and 23 h.

These recommendations give an idea regarding the long cure cycles for void-free laminates.

Possibility of automation

Automated tape laying (ATL) of OoA prepreg is still a while away from the production environment. The original equipment manufacturers' (OEM) are investigating key process parameters and layup time for manufacture of large aircraft structures using various automated technologies. For ATL, prepreg should be highly impregnated; in contrast, standard OoA prepregs are deliberately under-impregnated (at least at the edges). To use similar a prepreg system as VBO, the right combination of tape width, impregnation level at the centre of the tape and edge trimming attachment within the machine are important. Nevertheless, there is an issue of entrapped air that needs to be solved, quite the opposite to VBO, where the air path can be blocked during consolidation by the roller pressure and create voids in the laminate.

However, significant progress has been made to address these issues. Based on prepreg manufacturer and OEM studies, the optimal combination of parameters such as roller pressure, heat and evacuation strategies are solved to large extent and therefore, some of the OoA prepregs can be used for automated layup. The concept of ATL is now being studied for a number of OoA prepreg systems from ACG, Cytec and Hexcel.

As far as resin is concerned, Hexcel's thermoplastically toughened HexPly M56 is now also rated for primary parts, with a number of Tier 1 suppliers cited as being in the process of qualification. From these trials Hexcel reports very positive feedback regarding product handling and processing. In particular, the resin appears well adapted for ATL and has demonstrated less than 0.5% porosity.[8] Figure 14.7 shows an ATL machine in action using Hexply M56 prepreg.

Small aircraft manufacturers are taking steps towards manufacturing composite structures entirely OoA. However, the manufacturers of large commercial aircraft structures are still evaluating OoA prepregs for various sub-structures (Fig. 14.8) and this will take some more time before they are implemented in production. The military aircraft sector is moving faster than the commercial one in accepting OoA philosophy due to the longer development time and higher research funds.

14.7 Example of ATL of Hexcel's M56 layers.[9]

14.8 Cargo Aircraft's fuselage comprises eight pieces molded from MTM-45 prepreg (From reference 10.) and launch vehicle fairing laid up via automated fiber placement, using Cycom 5320. (From reference 11.)

Despite the numerous advantages of OoA prepreg for VBO (Table 14.4) and its claimed cost effectiveness, the work based on OoA is primarily R&D oriented. In fact, the OoA materials available right now costs same as the autoclave primary structure materials in use. But considering the reduced tooling costs and manufacturing costs of these OoA laminates, OoA prepregs could win over the existing standard prepregs.

14.4.2 Fluid heating

The basic idea behind this approach is the use of liquid fluids as heat transfer media, so-called heat transfer fluid (HTF). There are a variety of ways

Table 14.4 Pros and cons for VBO

Pros	Cons
• Low investment • Large number of parts per cure possible (good for high variety on parts but low volume each) • Easy to use	• Lower heat transfer than autoclave • Long process time (heating but especially cooling!) • Relatively high energy consumption (a lot of futile thermal mass, air and tooling, is heated) • Space limited (like autoclave) • Batch processing (like autoclave) • Extremely long debulking time • Often special (expensive) bagging needed • Automation seems challenging • Material cost and availability • Qualification needed (new materials)

to implement this basic idea and use the higher heat capacity of liquid fluids which can be categorized into two main concepts: closed mold and open mold. In principle, heated tools will be described for closed mold and for single side mold the Quickstep technology.

Liquid heated tools

The basis of this approach goes back to the early days of composite manufacturing and is widely used in a broad range of applications. In most cases, matched metal tooling was used to build the geometry on both sides of the part containing a manifold system or pipe construction inside each mold where the liquid was pumped through to heat the tool itself.

Since autoclave or oven cure could, alone, consume a huge amount of time to heat up a tool, in one of the examples shown in Fig. 14.9 a more cost- and time-effective direct-heating process was investigated. This uses a simple pressurized water system to rapidly heat and cool the thin nickel tool. Hot water circulates through copper piping attached to the backside of the tool. As the cure cycle finishes, heat exchangers replace the hot water with cold water to expedite cooling. Water pressure of 6 bar (87 psi) prevents the hot water from turning to steam at 130°C (266°F). The directly heated metallic tool promotes a fast manufacturing cycle through rapid heating and cooling, which reduces the need to make multiple tool sets to meet throughput requirements.

14.9 The nickel tool connected to the sources for hot/cold water circulation through copper piping on the backside of the tool. (*Source*: Gurit.)

The same principle of water heated tools can be applied for oil heated tools. Instead of pressurized water, temperature stable and high heat-transfer coefficient oil can be circulated through the tubes. Here, unlike the water heated system, high pressure is not mandatory; nevertheless, sufficient pressure to circulate fluid through the narrow pipes is necessary. An attachment of an oil heating bath can be incorporated in the loop. Oil heated tools can be used for more than 180°C (350°F) curable resin systems.

The fluid is mostly heated and cooled instantly during the process by a heating/cooling unit which also supplies the fluid flow. However, some approaches are trying to separate the heating from the cooling and are therefore isothermal for each station. The isothermal approach is mostly used in high volume production such as the automotive industry, often in combination with the use of thermoplastic CRFP and/or other thermal control units.

As outlined before, this approach is quite well established and used in different production setups, also in combination with other OoA techniques. Some examples are as follows:

- Press process with oil heated tooling (e.g., rotor blades).
- Water heated standalone tools (e.g., windmill blade production).
- Combination with GMT, BMC, SMC, RFI, RTM and other processes.
- Prepreg presses.
- Heated tools for example for RTM, vacuum-assisted resin transfer molding (VARTM) or same qualified resin transfer molding (SQRTM) (same qualified RTM).
- In combination with other heating technologies (e.g., induction heating (RocTool), etc.).

As it is a closed environment containing the HTF, it is possible to pump the fluid with high velocity and pressure. Therefore, the process can be very efficient and allows increasing the HTF performance above the fluid's gas transition temperature. This can be very useful as the performance of the system is dependent on the HTF values for heat capacity and heat transfer plus viscosity. For example, with water having one of the best sets of values but being limited to 100°C under atmospheric conditions, the operating limits of the fluid can be increased using higher pressure. Further pros and cons of fluid heated tools are stated in Table 14.5.

However, by increasing the pressure these fluid heated tools verge on the classification of pressure vessels which raises a large number of regulations concerning (re-)certification and also the heating unit where they are plugged into. This in itself implies higher acquisition costs of the whole system as well as for maintenance and safety. Moreover, higher pressure will require very rigid, heavy tooling and also inserts to meet the requirements on low or zero deflection, etc. Another cost factor is related to the high design efforts on the tooling. In contrast to gas based heating, fluid heated systems are likely to have a large degree of non-uniformity over the tooling surface geometry. Therefore if not modeled accurately before the manifold system is designed and built temperature distribution can be greatly affected.

The potential variability in performance of these systems with regards to temperature control and ramp speed is enormous but mostly restricted by the very heavy tooling that it needs to meet all the requirements. Therefore, higher production volumes are required to build up an economic business case.

In terms of product quality, matched metal tools usually achieve consistently and reproducibly the highest surface quality combined with predictable mechanical properties. Furthermore, the fiber volume fraction can

Table 14.5 Pros and cons for fluid heated tools

Pros	Cons
• Precise temperature control • Heating *and* cooling possible • Higher heat transfer rates • No need for an oven • Lower energy compared to autoclave (for larger parts) • Line production vs. batch processing • Combination with other heating methods possible (mainly used as cooling then)	• High investment for larger tools and for each part • Complex prediction of heat distribution • Higher maintenance costs for metallic tools • Tool insulation required (heat radiation) • High thermal mass • Mostly pressure vessels (certification) • Limited to geometry and complexity

reach theoretical limits due to the predefined volume of the matched metal tool.

Although each geometry requires always two tool halves, the variety of parts producible using liquid heated tools is high. It can be implemented for, but not limited to, thermoplastic and thermoset CFRP production, low tech and high performance parts, small to oversized structures, prepreg systems and liquid composite molding technologies, etc.

Quickstep

The Quickstep™ process is one of the OoA technologies with the potential to be qualified for fiber reinforced polymer composite (FRPC) manufacturing. This process utilizes liquid to transfer heat to and from the part, as fluids have a much greater volumetric heat energy capacity than gases, and the heat transfer rate between the fluid and laminate is much higher than that achievable in an autoclave. This is the central idea behind the Quickstep process developed by an Australian company (Quickstep Technologies Pty Ltd.). It is a balanced pressure, heated mold process that can be utilized for the OoA curing of advanced composite materials. Tight temperature control is maintained by circulating the fluid through pressure chambers (Fig. 14.10). This provides rapid and precise heating/cooling of the laminate and allows excellent consolidation to be obtained at low application pressures.

The heat transfer fluid can also act as a thermal sink, removing any excess heat generated in an exothermic reaction, so that a constant cure temperature may more easily be maintained, even for thick laminates. The process benefits from versatile production facilities, fast cure cycles and reduced capital, tooling and operational costs.[12,13]

The high heat-up rates allow for rapid cure of parts and a consequential highly flexible manufacturing route through high-rate production principles. This would allow composites manufacturers to move away from a

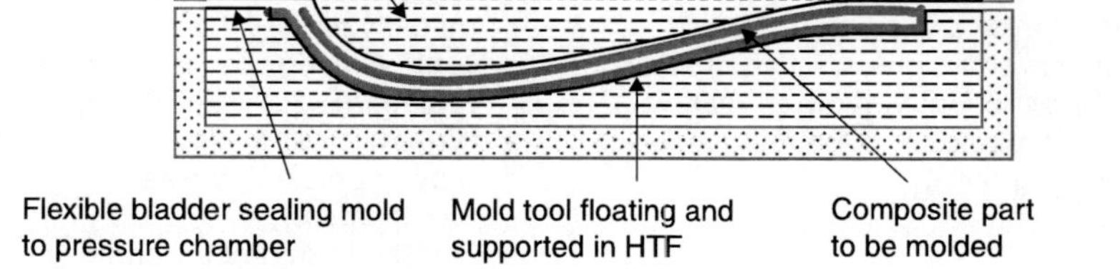

14.10 Principle of Quickstep fluid curing.

batch manufacturing philosophy, where a large number of uncured parts are installed in the processing plant and cured together, to a more flexible, one-piece flow where individual parts are prepared and cured in an assembly line process.

In an autoclave or oven, while manufacturing thick laminates, many of the epoxy resins have a danger of exothermic reaction due to accumulation of heat inside the laminate. The Quickstep process does have a capability to remove the excess heat generated by the reaction by effectively transferring it into the HTF. During the extremely critical exothermic reaction, immediate cooling down of the laminate is possible with Quickstep. The rate of cooling down can be up to 15 K/min for 50 mm thick laminate.[14] Figure 14.11 shows a Quickstep single tank machine and curing chambers.

Substitution of an oven for manufacturing of dry fiber preforms (based on binder activation) and curing of resin infused preforms are other aspects of the Quickstep process. Preforming and resin infusion techniques in combination with the Quickstep process help to accelerate the composite manufacturing process. Due to the additional pressure applied on the infused fiber (atm. + up to 0.8 bar), a significant increase in fiber volume has been observed (typically 56–58%). Furthermore, by using this technology, there are possibilities of partial curing of resin infused net-shape dry preforms. These pre-impregnated partially cured preforms can be utilized for manufacturing complex parts and various onsite repairs of composite structures.

By combining the effect of heating and cooling on the same laminate at specific desired locations, it is possible to engineer laminate curing in such a way that enables the elimination of adhesive bonding steps. The process called melding (combination of melting, welding and/or bonding of

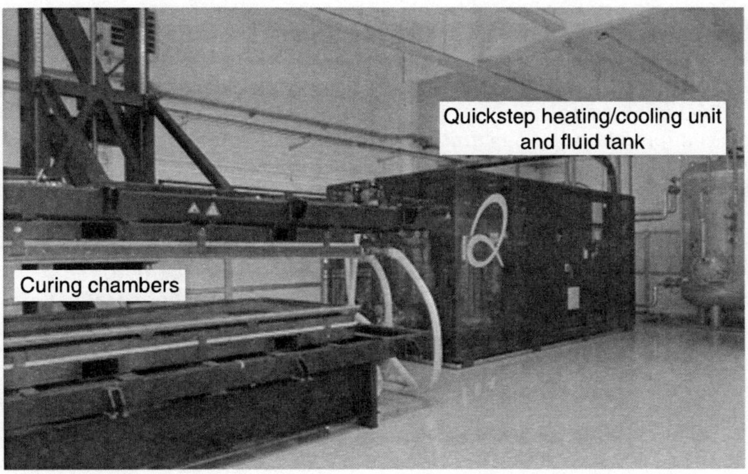

14.11 QSE250 curing machine. (*Source*: Quickstep.)

uncured components) allows this.[15] Melding can overcome the limitations of the adhesives by retaining a chemically active surface to react and form a bonded joint. Surfaces are left partially cured so that chemical cross-linking can occur between joined surfaces to create a structure without secondary adhesive bonds that are prone to failure.

This process is cost competitive with generic production methodologies like the vacuum bag technique. It requires lower investment and non-recurring costs are lower compared to conventional autoclaves.[16] Utilization of flexible tools and lightweight composite tools are additional advantages. Further advantages and limitations of the Quickstep process are listed in Table 14.6.

Although this process offers many potential benefits it is not yet used in serial production. Therefore, a lot of questions related to stability and reliability cannot be answered. It can be expected that the lifetime of a flexible membrane is limited and replacement could take some time. It is also possible that the flexible membrane, when aged, will fail during a cure cycle, which could potentially compromise the parts being cured.

14.4.3 Electromagnetic heating

This section will give an overview of different heating methods using electromagnetic (EM) radiation. The concepts differ mainly between the used radiation ranges or spectrums defined by the wavelength and/or frequency.

Table 14.6 Pros and cons for Quickstep

Pros	Cons
• Higher heat transfer rates • Lower energy consumption (especially 24/7 processing) • Multiple parts can be cured in a single tray • Low pressure process • Elimination of standard process steps (e.g., core stabilization) • Lower NRC and RC compared to autoclave • Use of autoclave materials • Useable for prepreg *and* infusion processing • Flexible membrane processing (different geometries in the same pressure chamber system) • Accurate temperature control • Restriction of exotherm • Heating *and* cooling possible	• Heat transfer rates are dependent on fluid domain • Part specific trays required for complex 3D components • Restricted to medium complexity parts (low pressure process) • Cure cycle optimization maybe required • No serial production data available • Restricted life time of flexible membrane • Extensive CFD for complex trays

This defines the theoretical enthalpy that can be inserted by radiation on and within the materials.

Generally, EM radiation is classified by wavelengths in the radio wave, microwave, infrared, the visible region, ultraviolet, X-rays and gamma rays. The behavior of EM radiation depends on its wavelength. When EM radiation interacts with single atoms and molecules, its behavior also depends on the amount of energy per quantum it carries. Tables 14.7 and 14.8 give an overview of these different ranges by wavelength, frequency and enthalpy. For a better understanding, some examples of applications are included.

The following sections will describe the most commonly known techniques as to how this energy is transferred. These are:

- Induction heating.
- Infrared.
- Microwave and
- Radio frequency curing.

Table 14.7 Electromagnetic spectrum 1

Range	Y-X-radiation	UV		NIR/VIS
Wavelength	10^{-9}	10^{-8}	10^{-7}	10^{-6}
Frequency	3×10^{17} s^{-1}	3×10^{16} s^{-1}	3×10^{15} s^{-1}	3×10^{14} s^{-1}
Enthalpy	119630 kJ/mol	11963 kJ/mol	1196 kJ/mol	119,6 kJ/mol
Application			Hg Vapor discharge lamps Photopolymerization Excimer laser	HL diode laser Argon ion-Laser 2 × Nd:YAG laser

Table 14.8 Electromagnetic spectrum 2

Range	Infrared		Microwave			Radio
Wavelength	10^{-5}	10^{-4}	10^{-3}	10^{-2}	10^{-1}	10^{-0}
Frequency	3×10^{13} s^{-1}	3×10^{12} s^{-1}	3×10^{11} s^{-1}	3×10^{10} s^{-1}	3×10^{9} s^{-1}	3 × 108 s^{-1}
Enthalpy	12 kJ/mol	1.2 kJ/mol	0.12 kJ/mol	0.012 kJ/mol	0.0012 kJ/mol	0.00012 kJ/mol
Application	IR spectroscopy CO_2 laser		Gyrotron	Q-band ESR	Magnetron UMTS GMS1800	GSM900 Radio Television

Induction heating

Induction heating is the process of heating conducting tools through electromagnetic induction. In this process eddy currents are generated within the conducting metal tools and their inherent resistance to these currents leads to heating of the metal. An induction heater consists of an electromagnet, through which a high-frequency AC current is passed. Heat may also be generated by magnetic hysteresis losses in materials that have significant relative permeability. The frequency of AC used depends on the object size, material type, coupling (between the work coil and the object to be heated) and the penetration depth.

Though not widely popular, induction heating has had long time advocates as an efficient method of heating molds. RocTool's is one of the example with an approach of heating and cooling of only the mold surface for much faster cycles, rather than conventionally heated molds. In this technique, turning on electric power to the inductors for only a few minutes heats only a 0.2 mm deep section of the tool surface while the rest of the mold stays cold. The mold surface heats rapidly when the power is on, and when power shuts off, the tool cools rapidly with chilled-water channels under the skin. The conductive skins, however, have a limited operational life and do not last long, needing replacement after only about 100 cycles. Also, a cast ceramic tool cannot be modified and has specific process and design limitations, so any design changes mean an entirely new tool.

Induction could be used with conventional tools built by any tool maker. Induction coils can be mounted with quick disconnects, so that they can be reconfigured for a tool change, which can be done by two people in half a day. Nevertheless, the mold block must be made of a magnetic steel alloy containing chromium or nickel for efficient electromagnetic conduction. Aluminum and P20 tool steel are not conductive enough. Some alloys heat fast by induction and reach temperatures up to 400°C (750°F). Others heat more slowly and peak at temperatures as low as 250°C (480°F).

The inductors use a frequency of approximately 100 kHz and are located on opposite sides of the tool. Current flows in opposite directions around each mold half separated by the composite laminate molded in between (Fig. 14.12). Electromagnetic flows around the mold halves (a), helping to choose optimum mold alloys and inductor and cooling-line positions. As shown in (b), only the surface of the mold gets hot.[17] With this, however, if the two mold halves make contact without an insulator between them, a short-circuit will result.

The main advantage of this process is by far the very fast heating (up to 250 K/min). As the majority of the tool is kept cold with the additional cooling circuit, a lot of the energy is lost during the complete cycle. These two complex circuits result in high cost for the whole equipment to support

14.12 (a, b) Concept of induction heating employed at the RocTool systems. (*Source*: RocTool.)

Table 14.9 Pros and cons for induction heating

Pros	Cons
• Very high volumetric heating rates • Inverters of different frequencies can be used (depending on heating need, localized heating) • Rapid ramp-up rates (up to 250 K/min) • Extremely short cycle time	• Heat uniformity hard to predict • Part specific tools required • Precision tooling is mandatory • Tool size is restricted by induction capacity • High to very high energy consumption • Automation and handling is complex • Heating rate depends on electrical conductivity of the material • High equipment cost • Health implications • Limited serial production data available

this. It is also a thermal challenge which results in high requirements on the tooling materials: different materials combined with high temperature differences result usually in tremendous deflection issues. In addition, the concept necessitates double sided closed molds. Therefore, the engineering of such tools is quite demanding to achieve highly accurate parts. Furthermore, it is restricted in regard to part size. The overall cost and other technical restrictions make this process best suitable for small to medium size parts in the section of low to medium performance at high volume. Table 14.9 gives further information on the pros and cons of this curing technology.

A flow control for the VARTM process based on localized induction heating is one of the applications of induction heating. Localized heating elements can be used at the low permeability preform region which assists lowering localized resin viscosity and increased resin penetration through the preform. There are promising results in terms of improved flow patterns and reduced air entrapment leading to less dry spots within the preform. In addition to improving flow uniformity, this helps to reduce fill times significantly. Overall the localized induction heating flow control scheme was shown to provide practical improvements to a standard VARTM process.[18]

Furthermore, the same localized heating technology can be used for the following applications:

- Preform binder activation and manufacture of 3D net shape preforms.
- Preform tow placement.
- Prepreg tape laying.
- Welding of thermoplastic laminates.

Infrared

Infrared (IR) energy is a form of radiation which falls between visible light and microwaves in the electromagnetic spectrum. Like other forms of electromagnetic energy, IR travels in waves and there is a known relationship between the wavelength, frequency and energy level. That is, the energy (temperature) increases as the wavelength decreases.

Unlike convection, which first heats air to transmit energy to the part, infrared radiation is directly incident on the top surface of the composite laminate to be cured and the inner layers are heated by conduction of heat from the top layers. Since the laminate receives heat from all sides, as it is a volumetric heating process, the curing process is more uniform. Infrared radiation is transparent to the medium of air in between the IR source and laminate. It is ideally suited for flat surfaces.

Infrared cure is gaining increased attention from coaters as a result of shorter cure cycles and the possibility of smaller floor space requirements when compared to convection oven curing. Figure 14.13 shows an example of an IR oven set up for composite curing. IR curing is limited in its use for composite curing as detailed in Table 14.10; however, it has potential for fast and efficient heating for various applications such as preform binder activation.

Infrared systems are relatively compact heating units, which can heat large components on a conveyor belt, without the need for a very large oven for the complete part. Infrared radiation can be precisely matched to the product and process and modern numerical methods such as ray tracing and computational fluid dynamics can also help to heat large surfaces homogeneously.[19]

Microwave

Microwave heating is a widespread technology within the food industry, but is also used in various other applications, such as communication systems (WiFi, satellite communication, mobile phones, Bluetooth), weapons

14.13 Medium wave IR oven for the curing of carbon composites. (*Source*: Voetsch.)

Table 14.10 Pros and cons for infrared heating

Pros	Cons
• Localized heating to exact amount (tape laying) • Fast heating rate • Energy efficient (flat and thin laminates) • Easy to install	• Uneven heating of complex component • Heat penetration through thick part is difficult (surface to thickness heating) • Very high energy consumption for large components • Extra cooling strategy needed • Issues with exotherm control

(ADS), sensor technology (microwave resonators) and for lighting (sulfur lamps). In the composites industry, research has been ongoing for the last three decades with limited results. However, even with limited results, the beauty of this technology is the direct use of the energy on a molecular level. Therefore, it is well positioned for ienergy efficiency. Microwaves have a frequency between 300 MHz and 200 GHz, which is similar to a wavelength of between 1 mm and 1 m (Tables 14.7 and 14.8). At the moment, free to use frequencies are 433 MHz, 2.54 GHz and 5.8 GHz. According to international agreement, industrial microwaves operate at a frequency of 2.54 GHz, which is powered by a variable power generator up to 1.26 kW.[20] An example of a typical microwave oven used in the industry is shown in Fig. 14.14.

14.14 GKN uses microwave curing to reduce OoA cure cycles.

Microwave heating of materials occurs due to dielectric loss mechanisms. When an external electric field is applied to a dielectric material, three types of polarization may occur: (1) electronic, (2) ionic or atomic and (3) orientational or dipolar polarization. The principal mechanism of coupling microwave radiation to polymer dielectrics is through dipole orientation by the electric field. The efficiency of coupling microwave energy into a material is dependent on a number of factors, including the dipole strength, the mobility of the dipole and the mass of the dipole. The amount of microwave energy absorbed by a material is given by the following relation:

$$P_{\text{absorbed}} = 2\pi f E_{\text{rms}}^2 \varepsilon_0 \varepsilon' \qquad [14.1]$$

where $P_{absorbed}$ is the dissipated or absorbed power per unit volume, E_{rms} is the root mean square electric field strength, ε_o is the permittivity of free space, ε' is the relative loss factor and f is the frequency.[21] In most cases it is the water molecule H_2O that reacts to the microwaves (such as in conventional microwave ovens). This is achieved by generating a force on the molecules, inducted by the electrical field, so that these will rotate and move in one direction (Fig. 14.15). Due to a steady change of polarity, this movement is reverse revealing, and therefore creates oscillation of the molecules which in itself causes the material to heat up; usually from the inside outwards.

In the case of heating composite materials, in particular fiber reinforced plastics, only the resin can be heated efficiently due to the molecular structure and the relating dielectric values of most used fibers (Fig. 14.15). As the free for use range of microwaves is limited, the process is only applicable to suitable resin systems; for example a lot of thermoplastic resins are permeable.

In fixed frequency microwave heating, areas with higher electric field strength are heated more, resulting in hotspots, which will generally stay in the same position over the heating period and will finally lead to thermal degradation. Hotspots are caused due to an inhomogeneous field distribution in the process area and because the materials have different dielectric properties (Fig. 14.16).

In a carbon fiber reinforced polymer, the polymer matrix is heated more quickly in the microwave field than the fiber. This is the reason why accumulation of resin or resin-rich volumes should be avoided when designing a component to be processed in microwave heating, since these areas cause heating momentum that can damage the component. Due to the possible leakage of the microwave system, safety measures need to be taken while processing composite components. Furthermore, due to the risk of

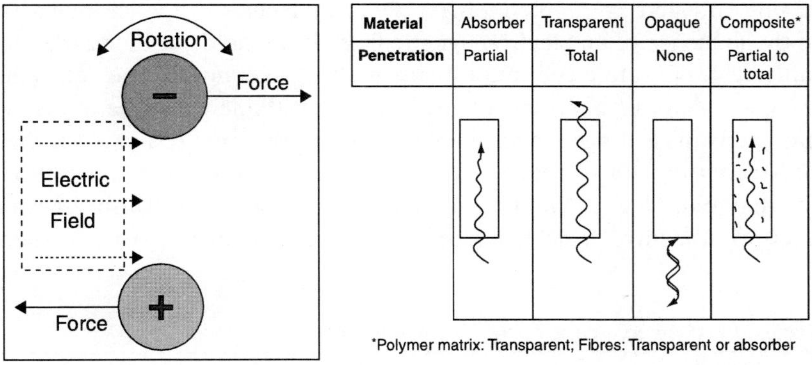

14.15 Microscopic principle of microwave heating and interaction of microwaves with materials.

14.16 Hotspots during microwave cure. (*Source*: Schmuhl GmbH.)

overheating and burning, during microwave processing the door of the microwave oven should face a wall to reduce damage in case of explosion. Microwave oven doors should also be closed until the curing process ends completely and equipment is switched off due to the health hazard posed by the radiation. This technology is therefore also restricted to a certain part size. Furthermore, each geometry requires a different setup to guarantee uniform heating.

To overcome some of these problems (listed in Table 14.11) there are concepts currently being trialed where only the tooling is heated via microwave radiation. The limitation with this approach, however, is that another spectrum (induction) is far better suited to provide this (see section 'Induction heating').

Radio frequency curing (RFC)

Radio frequency (RF) technology, as used in resin and composite curing applications, has existed since the early 1900s. It is a type of dielectric heating which causes molecular rotation in materials containing polar molecules having an electrical dipole moment. Polar molecules will align themselves in the presence of an electromagnetic field. If the field oscillates, the polar molecules will rotate continuously in an effort to maintain magnetic alignment. Oscillating molecules will push, pull and collide with other molecules increasing the average kinetic energy of the atoms or molecules in the material. Increasing the average kinetic energy in turn raises the temperature of the material. Thus, dipole rotation (orientation) is a mechanism by which energy in the form of electromagnetic radiation is converted to heat energy

Table 14.11 Pros and cons for microwave heating

Pros	Cons
• Very well known technique • Very fast heating rates • Direct heating of the resin • No material modification is mandatory	• Medium to high equipment cost • Uneven heating and hot spots (in case of fixed frequency system) • Specific tooling material is mandatory • Safety risk involved; health issues • Processing of thick laminate is very difficult • No cooling • Not suitable for all resin systems (sometimes additives needed) • Heating and burning of consumable stack • No serial production data available

in matter. The range of electromagnetic spectrum for RF curing is shown in Fig. 14.17.

From year 1920 to 1940, RF was an exciting heating technology. Manufacturing industry giants of the period (GE, Westinghouse, Bell and RCA) invested in the development of RF generators and industrial heating equipment. Although RF generators were readily available, difficulties arose as molders and other end-use fabricators were left to themselves to develop their own electrode configurations.

Three elements are necessary for an RFC system to function properly: an RF generator, a proper electrode configuration and a material sensitizer. Each element must be present and properly applied in order for the entire system to function efficiently.

At the core of the RF generator is a standard oscillator circuit which makes use of triodes to generate electromagnetic energy. This energy is then applied by way of various electrode configurations. Upper output levels for an RF generator typically reside somewhere around 100 kW with lower levels falling in the 3 kW range. All of this depends on the size and dielectric loss factor of the part to be manufactured. Triode tubes will typically come guaranteed up to 2000 h though most will run in excess of this value.

According to a study based on radio frequency (27.12 MHz) activation of the curing reaction for DGEB-A resin, there is no specific effect of the electromagnetic field on the structure of the cross-linked epoxy materials.[23] Although there are lots of advantages (Table 14.12) for this technology to be used in commercial composite manufacturing, there is very limited data on the curing on polymer composites using the technique. The main reason for this is the disadvantages of the system, in particular the cost of the system.

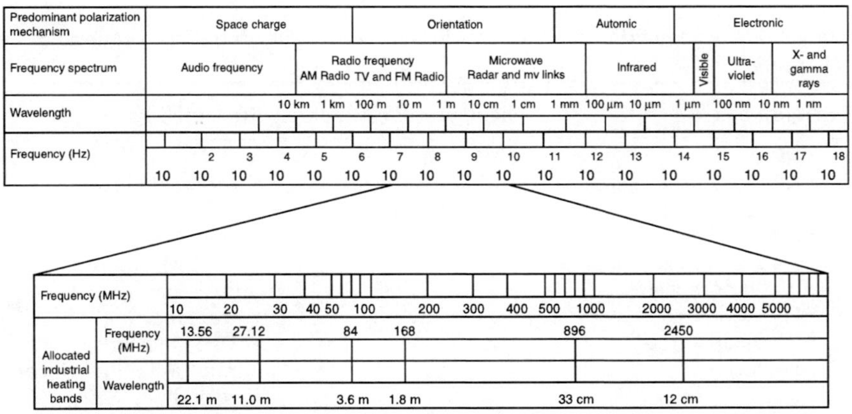

14.17 The electromagnetic spectrum for radio frequency. (From reference 22.)

14.4.4 Cure kinetics

Liquid composite molding (LCM)

The basic idea behind this group of processes is the integration of formally two production processes in one step. Fibers are no longer pre-impregnated and later cut and applied before curing but rather, the fiber fabrics are cut or manufactured to fiber preforms and within the curing process infused or injected and cured to the final part. Most scrap is therefore produced only by the dry fiber fabrics and not pre-processed fibers and resin. In principle there are three physical approaches or basic categories for how the resin is actually transported through the fiber structure: applying pressure, using vacuum, or both, by over- and under-pressurizing the fiber structure. From these basic principles many different process types have originated, with various names all over the world and differing from industry to industry (Table 14.13).

Table 14.12 Pros and cons for RFC

Pros	Cons
• Rapid heating / short production time • Can be used for variety of resin systems • Compact instrument • Low maintenance cost • Rapid start up and shut down • Flexibility inherent in an IR heating • Quiet and emission-free	• Higher capital cost • Higher installation cost • Higher energy cost • Lack of uniform heating • Limited to thin parts (RF wave penetration) • No cooling circuit • No serial production data available

Table 14.13 Selection of different liquid composite molding processes by category

Vacuum based	Vacuum and pressure	Pressure based
• VARI • VAP (vacuum-assisted process) • SCRIMP (Seeman composite resin infusion molding process) • RFI • RST (resin spray transfer) • ...	• VARTM • LRTM (light resin transfer molding) • ...	• RTM • Inflatable tube process • GAP impregnation • TERTM (thermal expansion resin transfer molding) • URTRI (ultimately, reinforced thermoset resin injection) • ...

These processes are considered to be very cheap for several reasons. The main cost advantage is quite often based on the use of preforming processes which are on the one hand the basis of one-shot solutions for complex structures and on the other hand designed with a high level of automation. Other benefits can be achieved by using vacuum based methodology and single sided tools (if possible in means of part requirements), integral and large structures (less bonding and/or assembly), faster material layup, no debulking steps and of course a lower material cost. This not only originates from lower scrap costs (mostly only fibers without resin and no outlife restrictions that exist for prepregs) and cheaper direct material cost (single components vs. semi-finished products), but also lower storage cost (most prepreg materials need a controlled cold storage at –18°C and less).

Resin transfer molding (RTM)

The RTM process begins with a dry fiber preform. The preform is placed into a matched metal mold and the mold is closed resulting in the compaction of the preform to the specified fiber volume fraction. A liquid thermosetting resin is then injected into the mold (typically at high pressure, such as 5–7 bar). The mold and resin can be preheated before injection, or the mold can be heated after injection to cure the resin. Due to the high injection pressures and often high temperatures involved, RTM tools are bulky and costly to manufacture and to process. To aid filling of the mold, vacuum may also be applied to remove trapped air. In addition, vacuum can be used to optimize the resin infusion/injection. This leads to VARTM. More information on RTM can be found in Chapters 9–11.

Vacuum-assisted resin infusion (VARI)

Vacuum-assisted resin infusion is one process variation of vacuum-driven LCM techniques, using a flexible vacuum bag as the upper mold. During the VARI process, dry reinforcements are placed onto an open rigid tool. The layup is sealed by a flexible vacuum bag and vacuum is applied, resin is drawn into the vacuum bag by the negative pressure and distributed throughout the dry layup and the fabric is infiltrated in x-y and through-thickness direction. Curing takes place from room temperature up to 180°C, depending on the application and resin system.

Resin film infusion (RFI)

In RTM, one of the requirements of the resin is that its viscosity must be low enough throughout the entire processing temperature window to fully wet out the preform. Since many of the resins currently in use were initially designed for prepreg systems, their viscosity tends to be too high for traditional RTM. To compensate for this, a new process called RFI was developed.[24] Instead of injecting the resin into the mold, a sheet of neat resin is cast as a thin film and placed directly under the preform. The thickness of the film depends on the fiber volume fraction desired and the permeability of the preform. Tool inserts are then placed around the preform, and the whole assembly is vacuum bagged and placed in an oven. When the tool is heated and pressurized, the resin melts, flows into the preform, and is cured. With this fabrication method it is possible to manufacture monolithic or sandwich-type structures. Compared with RTM, RFI reaches higher fiber volume fraction and molds are not so expensive.

While processing with RFI technology, various aspects need to be considered, such as preform characteristics, resin flow through the preform, heating rates, resin characteristics, cure kinetics, etc. Pros and cons of RFI are listed in Table 14.14 and more information on RFI can be seen in Chapter 14.

Same qualified resin transfer molding (SQRTM)

SQRTM has been developed and is now in the process of commercialization by Radius Engineering Inc. (Salt Lake City, Utah). SQRTM is a closed molding method that combines prepreg processing and liquid molding to produce net-shape, highly unitized composite parts.

The difference between SQRTM and standard RTM is that, in place of a dry fiber preform, it substitutes a prepreg layup. Prepreg plies are arranged within the mold, the mold is closed, and then, somewhat counterintuitively, liquid resin is injected into the tool. The injected resin is the same as that used in the prepreg, and, therefore, those who adopt the process need not re-qualify materials.

There are other advantages compared to conventional RTM, for example part thickness is controlled by matched tooling, avoiding the potential thickness variation inherent in the vacuum bagging process. Starting with fully impregnated, qualified, toughened prepreg eliminates the risk of dry spots

Table 14.14 Pros and cons of RFI

Pros	Cons
• Repeated debulking operation for thick laminates not required • Resins easily toughened • Good quality laminates produced • No need of special storage of dry fibers	• High material costs (resin film) • Difficult to process complex shaped parts • Resin film placement inside the mold is tricky and require high labor skill

during injection and the need to introduce toughening agents to the part via the liquid resin. An example of an SQRTM tool used for manufacture of the composite structure is shown in Fig. 14.18.

These processes are not all directly linked to a curing method but can be combined with a lot of different methods. Usually, the vacuum based methods are combined with low pressure processes as follows:

- Quickstep.
- Microwave.
- Radio frequency curing.

Principally the pressure based methods could also use conductive oven heating but would usually be used with a press to withstand the high pressure (usually between 5 and 8 bar). The mandatory use of closed mold arrangements leads to the most economic and feasible processes as follows:

- Liquid heated tools and other direct tool heating mechanisms.
- Induction heating.

Table 14.15 summarizes the pros and cons of these manufacturing techniques.

As a closing remark it should be mentioned that technically, LCM processing can be combined with autoclave curing. This is not only theory; there are some manufacturers that use autoclave cure with infusion technology (e.g., Bombardier in Ireland for wing skins). The economics of such a process chain is often based on written-off and otherwise unused equipment

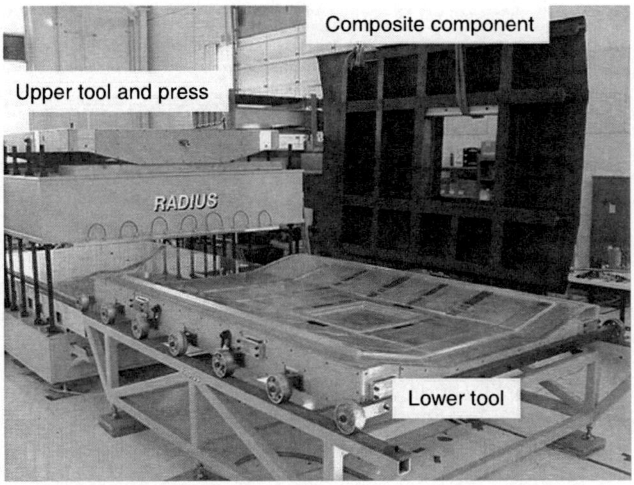

14.18 Example of SQRTM. (*Source*: Radius Engineering.)

compared to a complete new process chain acquisition. As a result, and with the focus on OoA, it can be neglected.

In situ *polymerization*

Use of low viscosity pre-polymers, which polymerize (after adding a catalyst) during or after impregnation into a polymer, is called *in situ* polymerization. For this process different conditions have to be fulfilled:

- Low viscosity < 1Pa.s.
- Relatively short polymerization time.
- Polymer with good mechanical properties.
- No by-products.

There are various systems that fulfill the different conditions, namely: thermoplastic PU, anionic ring-opening polymerization of polyamide, entropy driven ring-opening polymerization of cyclic oligo-esters.

In situ polymerization of PBT from cyclic oligomers (CBT) has been very well studied in the recent past. The cyclic oligomers are readily polymerized to high molecular weight polyesters in the presence of suitable catalysts within a few minutes. Polymerization of CBT was reported over the range 160–200°C.[25] Before that temperature was reached, the CBT began to melt at about 140°C, and it was completely molten at 160–190°C. CBT need a suitable catalyst to perform *in situ* polymerization in order to produce the final polymer (Fig. 14.19).

Table 14.15 Pros and cons for LCM processing

Pros	Cons
• Cheaper materials (compared to prepreg) • High level of automation possible • Discrete fiber orientations possible • Large, oversized, complex and integral structures possible • Reduced bonding and assembly efforts • One-Shot and near net-shape possible • Less scrap cost • Can be combined with a high variety of curing / heating processes • Versatile fields of application • Different fiber volume fractions possible	• Resin flow prediction critical and very difficult • Preform process most quality defining and still immature • For pressurized systems: - extremely high invest on tooling and presses - Race tracking effects - Extensive tooling rework (tolerance limits and tool degradation) - Often very long cycles (thermal mass of the closed pressure toolings) • For vacuum based systems: - Single sided tooling results in only one tool face on the part - Lower fiber volume fraction achievable

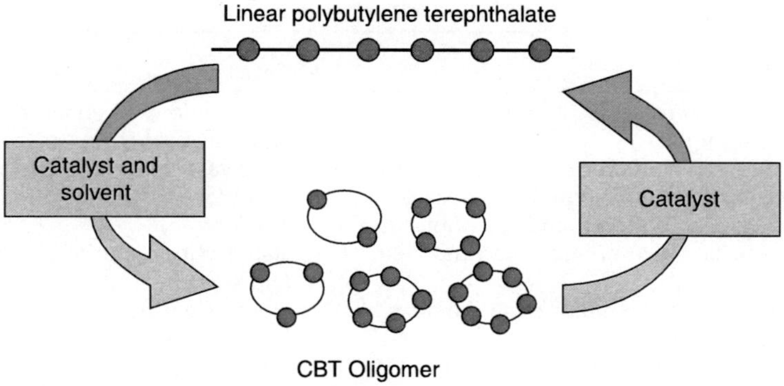

14.19 Schematic of CBT to PBT polymerization. (*Source*: Fraunhofer ICT; From reference 26.)

In situ polymerization of PA12 is also very common. Here the monomer viscosity can be reduced to 3 Pa·s before polymerization. Another well known material for *in situ* polymerization is PA6. Furthermore, to increase the toughness of the laminate, *in situ* polymerization can be combined with nanoparticles or carbon nanotubes.

Although this concept is very interesting as it allows for the possibility of 'infused thermoplastics', a lot of drawbacks have hindered development. Besides the fact that there exists very low market relevance and experience, the process itself is very hard to control over the whole chain:

- The material needs to be dried and kept very dry, which is sometimes time critical.
- The window of reaction is very narrow, which implies the need for very accurate temperature control combined with fast ramping (heating method could be costly).
- Even with the relatively low viscosity of the monomer (compared to thermoplastic materials) fiber impregnation is still challenging.
- The high pressure needed for impregnation and injection will not allow using single sided molds but requires closed mold processing (high cost, accuracy, deflection, etc.).
- On top of these, all deficits of closed mold resin injection processes apply as well (racetracking, preform quality issues).

Pros and cons of the *in situ* polymerization are listed in Table 14.16.

Electron-beam curing

Another interesting approach to avoid autoclave curing is electron-beam (EB) or E-beam curing, often combined with automated fiber placement (AFP). EB curing activates the cross-linking process of the resin system by

Table 14.16 Pros and cons for *in situ* polymerization

Pros	Cons
• Cheaper materials (compared to prepreg) • High level of automation possible • Discrete fiber orientations possible • Reduced bonding and assembly efforts (thermoplastic material after reaction) • One-shot and near net-shape possible • Less scrap cost • Can be combined with a high variety of curing/heating processes	• High equipment cost for accurate and ultra fast cure • Only very little materials available (therefore restricted fields of application) • Issues with material shrinkage and creeping • Fiber print through • Drying of material quality and time critical • Fiber impregnation difficult • Resin flow prediction critical and very difficult • Very narrow process windows • Preform process most quality defining and still immature • Extremely high investment on tooling and presses • Race tracking effects • Extensive tooling rework (at least every 80 parts) • No relevant serial production data available (immature process)

ionizing the monomer chains. This eliminates the need for any additional heating. The ionizing or activation is initiated by shooting electrons through a gun on the resin prepreg (Fig. 14.20).

This process uses a beam of high-energy electrons from a device called an accelerator to initiate resin cure in place of the usual autoclave that applies heat and pressure. EB curing is suitable for composite parts as thick as 5 cm. Parts are cured by spraying these electrons onto the composite part. The cure is controlled through the application rate of the electrons and the total number of electrons or dose applied to the part. It is conducted at room temperature so that normal thermal stresses associated with heat curing are minimized[27] across the part; however, local residual stresses are induced due to the localized curing that occurs only within the exposure area.

All reinforcing fibers that are traditionally used for advanced composites can be employed for EB cured composites. Carbon fibers have shown to be particularly suited for EB processing, since they exhibit excellent radiation stability, tolerating about 1000 times the typical curing dose, and as good electrical conductors they ensure that no charge is built up in thicker composite structures. Aramid fibers are also unaffected by EB treatment up to the typical curing doses, a consequence of its highly aromatic structure.

The effect of EB treatment on the mechanical properties, density and chemical durability of glass fibers was shown to be negligible up to a dose of 3 MGy. The only visible effect was discoloration. The use of organic fibers, such as polyethylene or nylon fibers, for reinforcing EB cured composites must be considered with caution, since the properties of the fibers are affected by the EB irradiation.[28] An industrial EB accelerator for 1.2 m working width is shown in Fig. 14.21.

Electron beam irradiation is linked to the chemical incompatibility and physical phenomenon. This affects the interfacial properties of the composites. Specific surface treatment for the reinforcing fibers and suitable sizing is a part of preparations required specifically for EB cured composites.

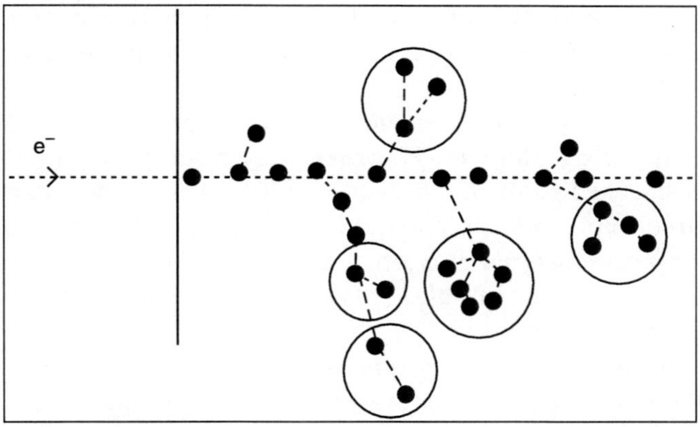

14.20 Principle of electron beam curing.

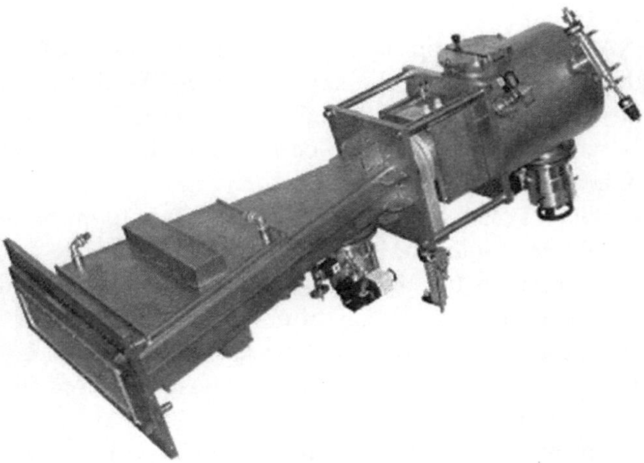

14.21 Electron beam accelerator, 250 kV accelerating voltage and 1.2 m working width.[29]

Concerning the irradiation doses, it was shown that a lower dose per pass leads to higher cure levels and lower residual stresses at a given dose level. In addition, cure shrinkage induces residual stresses that can be appreciable in EB cured multi-directional laminates. Therefore, there are much fewer residual stresses induced during EB curing as observed with thermally cured composite. However, during cure, due to the exothermic polymerization reaction in the core of the material the process is limited in the thickness of parts that can be produced. This is due to conduction and convection boundary effects, as the process is performed in ambient conditions. High thermal conductivity of the mold is preferred for distribution of exothermic natured resin systems. Additionally, a balance between irradiation dose, number of passes, amount of initiator, initiator dispersion and mold properties must be examined case by case in order to minimize the effect of internal stresses in EB cured composite materials and structures.[28]

There are two groups of monomers that can be activated by EB: acrylate (radical) and epoxy (cationic). However, some resin systems would require additives, so-called photo initiators, to enable the resin system to respond in the required way. These photo initiators are special 'salts' and could potentially change the material microstructure and also the mechanical performance. Therefore re-qualification of the material is required if it is to be used in primary structures.

The major advantage of this approach is that it requires extremely low energy, theoretically only the activation energy for the chemical cross-linking. EB curing is also a very fast process, as the chemical reaction is forced to happen immediately. The possibility to cure instantaneously makes it predestined for use in continuous processes, such as automated fiber placement (AFP).

The US NASA took the approach of *in situ* curing of thermosets with AFP and developed specific EB curable resin prepregs. The first real approach was presented by Goodman, Byrne and Yen who used EB curing for large integrated structures. As a result, EB curing would be used, often directly on the AFP head, next to the roller unit, to *in situ* cure layer-by-layer. The studies undertaken so far showed no uniform result in terms of mechanical performance. While Goodman, Byrne and Yen state that the *in situ* cured composites performed better than post-cured composites and that residual stresses were reduced, other studies from Airbus, University of Beyreuth *et al.* show the opposite. This was mainly based on the technology maturity level and availability of appropriate EB guns. An EB cured fiber placed prepreg is shown in Fig. 14.22.

Another thorough study shows that E-beam exposure cannot propagate the polymerization of BMI systems until the temperature goes up to 100°C. Higher intensity E-beam can give high reaction conversion of BMI

14.22 E-beam cured composite laminate. (From reference 27.)

Table 14.17 Pros and cons for E-beam curing

Pros	Cons
• High level of automation possible (AFP) • Very low energy consumption (theoretical) • Direct initiating the reaction on molecular level (no heat losses through tools)	• Deadly radiation • Extremely high cost for infrastructure (safety measures) • Limited availability • Limited materials tested • Often additives (photo initiators) needed • Immature process • Material quality issues still unsolved • No serial production data available

above 75%. However, the temperature of the BMI reached up to 250°C.[29,30] Further pros and cons of EB curing are mentioned in Table 14.17.

Despite the enormous potential of this approach it is highly questionable whether the process will see full-scale production. The first reason for this is related to the irradiation itself. As a serious health risk to humans, EB systems need to be enclosed in a huge concrete shielding building. Access to the EB area could require furthermore up to seven opposite walls at the entrance area to ensure that the radiation is fully stopped depending on the radiation strength that is used. This could potentially destroy the economic advantage. Another drawback is the size limitations that the current EB systems still have. At the moment it is unknown whether development in this area can provide smaller EB heads in the future.

14.5 Process comparison and classification

14.5.1 Techno-economic portfolios

The main driver for OoA manufacturing is manufacturing cost per part. Each process can offer different advantages when used for the right case. The main factors affecting the cost per part are volume in comparison to capital expenditure and cycle time, while capital expenditure and cycle time determine which kinds of parts can be produced most efficiently. The portfolio shown in Fig. 14.23 categorizes parts in three different economic areas: low, medium and high gross margin. It is obvious that parts with low gross margin demand low investment and low cycle time. Whereas parts with high gross margin will allow for having both high cycle times and high investment costs and still result in a viable business case.

There would be no objection for high gross margin parts to have lower investment costs and also low cycle time. However, often the technical restrictions of the processes will not allow for production of these higher value parts. As it is only a very limited view for initial categorization, more indicators are needed to define the economic range of products for each process.

It is therefore necessary to look deeper into the factors that determine the gross margin of such parts. In terms of high market prices, the main indicators for composite parts are related to part size, thickness and complexity. Another criterion, especially for FRP, is the form accuracy required for

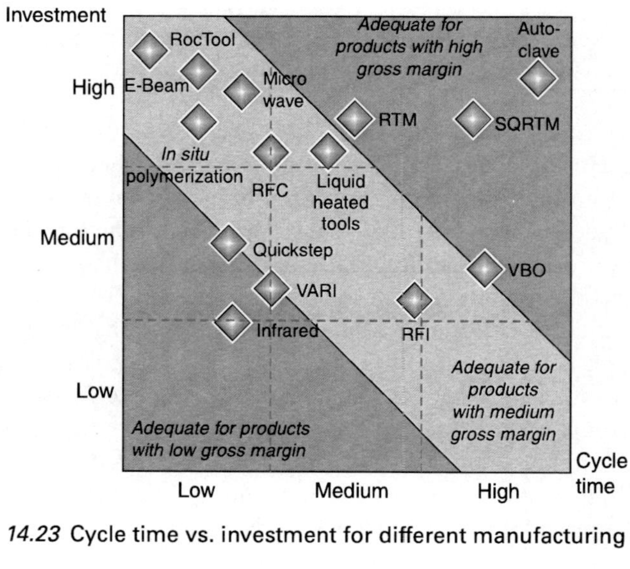

14.23 Cycle time vs. investment for different manufacturing processes.

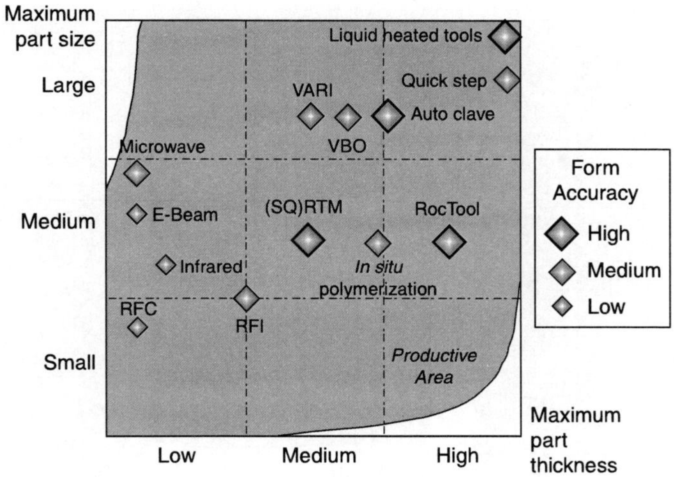

14.24 OoA processes by performance criteria.

the part to allow maximum performance. Figure 14.24 categorizes the OoA processes by these three elements.

If those two portfolios are combined to establish a full picture of the 'best fit' for each process it needs another variable to make the final decision. One of the main advantages of fiber reinforced composites is the inhomogeneous material behavior compared to isotropic materials like metal. It allows designing and producing very lightweight and structurally optimized parts. The dedicated fiber orientation, therefore, is the final major element to define the product and its price. The more efficiently the material can be used, the higher the functionality. Figure 14.25 shows for different OoA processes divided in categories the achievable range in terms of part size, accuracy, complexity and dedicated fiber orientation. For a better understanding, the portfolio also shows characteristic groups of parts within these areas as well.

All three portfolios show that each process can operate best in a certain area or better in a range of different applications. According to the specific parts and requirement, the manufacturing process has to be chosen wisely. This also links back to the design process. However, the maturity of the processes also plays a role in making the decision. A lot of approaches have been taken in categorizing maturity. The aerospace industry has established 'technology readiness levels' (TRL) from 1 to 9, where in simple words 1 is basic research level and 9 is serial production readiness. Even with the differences between the Boeing (USA) TRLs and those used by EADS (Europe), the main message remains the same.

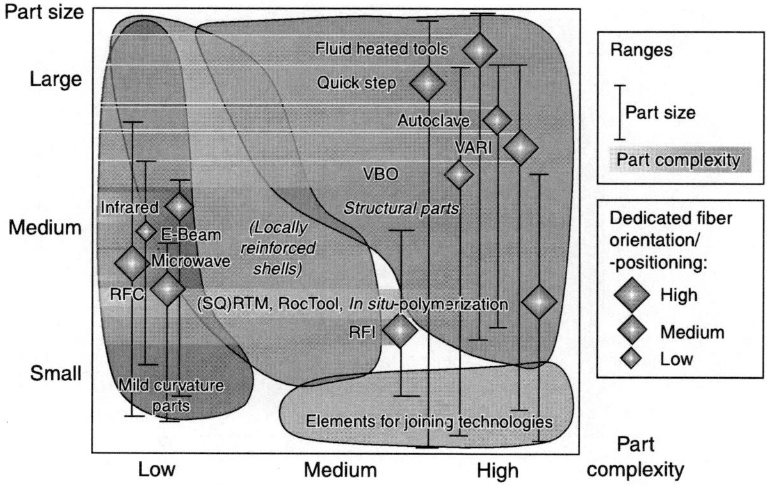

14.25 OoA processes categorized by part size, complexity and dedicated fiber orientation.

This TRL can also be transferred to other industries. Figure 14.26 categorizes the processes regarding their maturity or use in the market. This is a high level study and covers a lot of changes over time. Therefore, this can only be a snapshot at a certain point in time. The authors tried to reflect the current situation in a very low level of detail for the year 2011.

It is obvious that newer technologies are lower in terms of their maturity. Sometimes, this is because major issues or problems have not yet been solved (such as inhomogeneous heating (e.g., microwave or RFC) or safety issues (e.g., E-beam)). In other instances, it could be due to difficulty in finding the right business case. However, very often it is simply based on the fact that the process development has started in the aerospace industry first which has very high requirements and regulations with regards to qualification. These developments usually have a minimum time to market of 10–15 years, often even longer. As an example, the development for RTM to be implemented in aerospace production took more than 20 years. If the first market is not aerospace but any other with lower barriers to entry, it can make a large difference for aerospace and other branches regarding TRL (e.g., RocTool (consumer and automotive), infrared (automotive), fluid heated tools (wind)).

Additionally, it has to be mentioned that there are a lot of 'unproven' claims for each process, which creates some anomalies for the evaluation and therefore can be just an indicative rating. Furthermore, the whole selection for a suitable process is a 'chicken and egg problem'. The less the new processes are chosen, the less information or proof will be available.

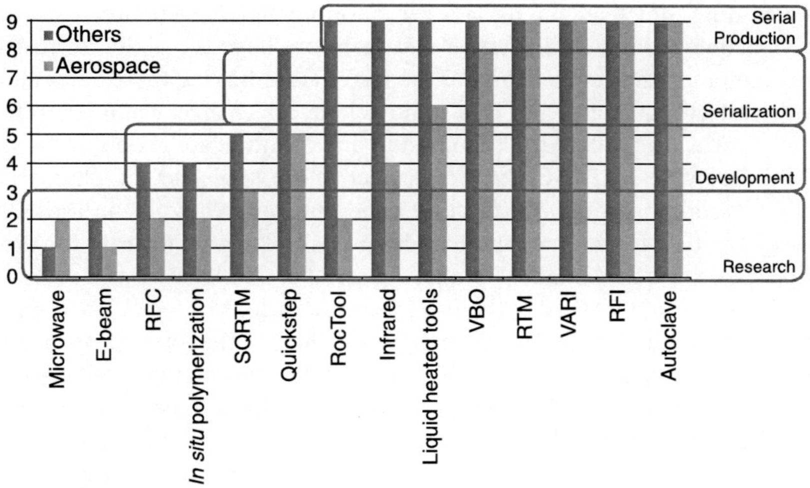

14.26 Maturity of different OoA processes in year 2011.

14.6 Future trends

It can be stated that OoA will play a major role in future manufacturing of CFRP where market growth is the main driver. In the area of high performance composite parts production this in the end means cost and time, but this trend also involves a lot of other factors such as, but not limited to:

- government regulations; mainly carbon emissions
- extreme growth of CFRP parts in all industries
- new materials
- maturity of CFRP.

All the described methods of OoA processing seem to have their own validity in their claims if implemented for the right range of products regarding size, complexity, geometry, quality and production volume. However, this is still a pretty immature area compared to other industries but it is growing and maturing very fast.

14.6.1 Materials

The demand will certainly drive the trend to have more suitably adapted resin systems for OoA developed. The VBO approach could just be an intermediate step until resins are developed that can be cured at much higher ramp rates combined with shorter dwell times and at the same quality as autoclave designed materials. This applies for all processes that can increase the heating effectiveness. It can even be more suitable for processes that have active cooling implemented and therefore have the ability to restrict the higher exothermic reaction of the faster cross-linking that comes along with fast curing materials. This minor chemical change would therefore apply perfectly for RocTool, Quickstep and liquid heated tools. Processes like microwave and E-Beam would need more sophisticated changes to enable good absorption on the one hand but low exotherm on the other hand. The development of these additives will play a major role for those processes to be implemented broadly in serial production.

Another general trend is the increased use of LCM processes to reduce waste and material cost. The main drawback here is the lower performance in impact strength which is solved with (mostly) thermoplastic fillers integrated into prepregs. Most trials to infuse such thermoplastic fillers have failed because of the filtering effect of the fabric materials. Therefore, the fiber materials will have to incorporate these tougheners within the weave. There are a lot of materials already available that try to fulfill this requirement, such as PriForm from Cytec or PA/PES veils. These fillers can also act as a binder and support the perform process as well.

The use of composites opens huge possibilities for new products and greater performance to be achieved. At the same time this versatility in combination with less experience compared to mature materials such as metal generates a lot of questions. It has been a trend in metal manufacturing for a long time to use materials according to the specific requirements. This can also be foreseen as a major trend in composites in general and is particularly important to OoA. Hybrid material systems will be one of the future challenges OoA technologies will face. Processes will have to be adapted rapidly to accommodate these new material systems. As a quick example: if thermoplastic and thermoset FRP have to be combined in one part in the same production technology, it will be difficult for all 'low pressure processes' to accommodate the material curing requirements. Therefore, even for hybrid material systems, a lot of development must be undertaken to better the autoclave.

14.6.2 Combination

Taking all this into account OoA will find its way to real cost benefits only by combining more than one approach. The savings of just one little piece in the full manufacturing chain is often too small to justify an investment in new technology. For example, does it make sense to change just the curing system from batch to line production when the whole pre-production, such as layup, etc., is not converted accordingly?

As described previously, there are already a lot of developments taking place to automate processes. One of the major trends to achieve a fully economical process chain is to use automated pre-processes combined with OoA curing systems. This applies for both prepreg (ATL, ATP) and LCM (stitch and/or binder performing, AFP). There are also developments to use different heating methods for these processes as well. It could well be that, for example, heating techniques like infrared, induction, laser or even the Quickstep process can be used to activate binder preforms.

Another example is the use of ultrasonic or E-Beam for pre-compaction or pre-consolidation in layup processes. In the second case several trials were undertaken to fully consolidate during the layup process; unfortunately with limited success so far.

14.6.3 Large integral parts

In the last 20 years it has been shown that FRP parts are not only becoming larger and larger but also more and more integral structures. The effect is that for such parts a restricted curing chamber size is a problem. As a result, approaches that need a dedicated area to process will be less favorable than those that can increase their production area easily without an investment for the main aggregates. This development will also lead to more single sided tooling processing to avoid the high cost for closed molds.

As a result, closed mold processes and closed chamber designs, such as RTM, RocTool, microwave and ovens tend to be less adaptable. For these structures, infrared, RFC, liquid heated tools or Quickstep would be a better approach. Examples are the windmill blades used in the wind energy industry which uses liquid as well as induction heated tools. E-Beam is a special case as it has high health implication restrictions but it could be very adaptable if used with robot and/or gantry systems.

Another implication for these structures will be the use of cheap tooling. This will not only be achieved by single sided tools but also by implementation of cheaper tooling technologies and materials. Examples are CFRP tools, spray tools and other plastic tooling.

These large and integral, sometimes even oversized parts will also lead to more automation in general as already mentioned in this chapter. Furthermore, preforming and layup techniques have to be applied to produce these parts more efficiently.

14.6.4 High volume

Two major events have changed the automotive industry very recently: the financial crisis and global warming. Both resulted in a complete paradigm shift. Lightweighting is no longer a niche application for sports cars but a future core competence for car builders. This is not only caused by the short-term reduction of emissions the OEM have to guarantee by secured penalty fees. Another reason is the long-term goal of 'electro mobility'. To increase operating distances, the energy consumption has to be lowered dramatically. The best way to achieve this is of course lightweight construction. In addition to the structure of the car body the battery weight is another issue that seems hard to solve. Therefore, the logical rationale is to lower the car weight in the structure and body panels. FRP are considered to be a main contributor to achieve this from a material and performance point of view. The main drawback in automotive is as always cost (and also surface quality for body panels in particular). Another automotive specific requirement is the range of production volume per part. In civil, military and general aviation aircraft manufacture, a typical volume is between 30 up to a maximum of 500 aircraft per year. The typical production volume for cars is between 300 per year for very special vehicles up to 3000 per day. There is at the moment no composite process in use that can deliver endless fiber reinforced composite parts at this volume with the required reliability. A lot of effort is being made to achieve this for future car manufacture.

Another area of growth is the aircraft update or upgrade packages such as wingtips or winglets. These are also at very much higher rates than the original volume for new aircraft deliveries. If this is done solely with autoclave manufacture, a huge number of tools per part will be needed.

Therefore, it is obvious that OoA technologies will be the only way to deliver high performance and affordable composite parts in higher volume. Another major trend resulting from these higher volumes is the definite need for automation, as mentioned earlier. By getting away from hand layup to robotized manufacturing environment, the production will increase both flexibility and quality.

Table 14.18 Books for further reference

Sr. No.	Title	Authors	Publisher	Year
1	Electron-Beam (E-Beam) processing as a means of achieving high performance composite structures	J. M. Sands, S. H. McKnight, B. K. Fink, A. Yen, G. R. Palmese	Storming Media	2001
2	Microwave-enhanced polymer chemistry and technology	D. Bogdal, A. Prociak	Blackwell Publishing	2007
3	Liquid composite molding	R. S. Parnas	Hanser	2000
4	Principles of the manufacturing of composite material	S. V. Hoa	DEStech Publications	2009

14.7 Sources of further information and advice

Recent OoA Sessions/Special Issues:

- SAMPE Long Beach 2011 Proceedings.
- SAMPE Seattle 2010 Proceedings.

Table 14.18 lists some of the books on specific technologies and composite manufacturing methods.

14.8 References

1. http://www.aschome.com/ASC30x75_Picture1.htm (Accessed 25 May 2011).
2. Thomas, S., 'Vacuum bag-only processing of composites', dissertation presented to the Faculty of the Graduate School, University of South California. May 2009.
3. Bond, G. G., Griffith, J. M., Haahn, G. L., Bongiovanni, C. and Boyd, J., 'Non-autoclave (prepreg) manufacturing technology'. http://www.dtic.mil/cgi-bin/GetTRDoc?Location=U2&doc=GetTRDoc.pdf&AD=ADA510683 (Accessed 15 June 2011).
4. 'Out-of-autoclave prepregs: Hype or revolution?' http://www.compositesworld.com/articles/out-of-autoclave-prepregs-hype-or-revolution (Accessed 3 June 2011).
5. Nutt, S. and Boyd, J., 'Out-of-autoclave processing with vacuum-bag-only (VBO) prepregs'. http://www.ltt.de/uploads/pdfs/Events/ILA2006/ILA06_Praesentat_OOA_Processing_USC_Nutt.pdf (Accessed 15 June 2011).
6. https://www.cytec.com/engineered-materials/products/Datasheets/CYCOM%205320Rev1-3.pdf (Accessed 8 August 2011).
7. Gruenenfelder, L. K. and Nutt, S. R., 'Void formation in composite prepregs: Effect of dissolved moisture', *Composites Science and Technology*, **70**(16), 2010, 2304–2309.
8. Calder, N., 'Who needs the pressure?' 10 Aerospace Manufacturing, May 2009. www.cytec.com (Accessed 16 June 2011).
9. Mortimer, S. and Smith, M. J., 'Product development for Out-of-Autoclave (OOA) manufacture of aerospace structure'. http://www.hexcel.com/Innovation/Documents/Product-Process%20Development%20For%20Out-of-Autoclave.pdf (Accessed 15 June 2011).
10. www.compositesworld.com/ (Accessed 10 June 2011).
11. www.timecompression.com (Accessed 17 June 2011).
12. Bader, M. G., 'Selection of composite materials and manufacturing routes for cost effective performance', *Composites Part A: Applied Science and Manufacturing*, **33**, 2002, 913–934.
13. Griffiths, B. and Noble, N., 'Process and tooling for low cost, rapid curing of composite structures', *SAMPE Journal*, **40**(1), 2004, 41–46.
14. Schlimbach, J. and Ogale, A., 'Presentation on: Quickstep – A cost effective composite manufacturing solution for large scale and thick composite structures', 3rd International Conference: Supply on Wings, 2008, Frankfurt, Germany.
15. Corbett, T., Forrest, M., Law, H. and Fox, B. L., 'Melding: A new alternative to adhesive bonding', SAMPE Conference Proceedings 2005.
16. Schlimbach, J., 'Presentation on: Quickstep – Potentials and applications', Materialica exhibiter's forum, 16–18 October 2007, Munich, Germany.

17. Shut, J. H., 'Induction heated molds produce class A thermoplastic composites', *Plastic Technology*, November 2009.
18. Johnson, R. J. and Pitchumani, R., 'Active flow control in a VARTM process using localized induction heating'. http://www.me.vt.edu/amtl/index_htm_files/c0404.pdf (Accessed 14 June 2011).
19. 'Infrared heat for state-of-the-art composites'. Cover story. www.developmentscout.com (Accessed 24 June 2011).
20. Yarlagadda, P., 'Experimental studies on the curing of alternate un-reinforced mold materials using microwave heating'. http://eprints.qut.edu.au/6158/1/6158_1.pdf (Accessed 7 June 2011).
21. Tanikella, R. V., Allen, S. A. B. and Kohl, P. A., 'Variable frequency microwave curing of benzocyclobutene'. http://www.microcure.com/pdf/WaferProcessing.pdf (Accessed 7 June 2011).
22. Love, C. K., 'Investigation of RF curing parameters in resin infusion molding', Master's Thesis, Brigham Young University, 2010.
23. Alazard, P. and Gourdenne, A., 'Radiofrequency (27.12 MHz) activation of the curing reaction of epoxy resins of DGEBA type', *Macromolecular Symposia*, **122**, 1997, 185–190.
24. Chen, V., Hawley, A., Klotzsche, M., Markus, A. and Palmer, R., 'Composite technology for transport primary structure', 1st NASA Advanced Composite Technology Conference, 1990.
25. Steeg, M., 'Prozesstechnologie für Cyclic Butylene Terephthalate im Faser-Kunststoff-Verbund'. Dissertation, IVW Publication, Vol. 90, Kaiserslautern 2010.
26. http://www.ict.fraunhofer.de/kernko/PE/Faserverbundwerkstoffe_FVW/RTM/index.jsp (Accessed 10 June 2011).
27. http://www.acsion.com/ (Accessed 7 June 2011).
28. Wolff-Fabris, F., Altstädt, V., Arnold, U. and Döring, M., *Electron beam curing of composites*. Munich: Hanser.
29. Laurell, B. and Föll, E., 'Electron beam technology and coatings'. http://www.crosslinking.com (Accessed 7 June 2011).
30. Yuntao, L., 'Synthesis and cure characterization of high temperature polymers for aerospace applications'. Dissertation, Graduate Studies of Texas A&M University 2004.

Index

Adaptive Response Surface Method (ARSM), 372
additives, 388
advanced composites, 255
aramid fibres, 254
Arrhenius equation, 390
autoclave, 10–11, 414–32
  component and tooling prepared, 415
  dimensional stability and residual stresses, 420–5
    cure and temperature gradients, 421
    process tooling, 421–2
    resin cure shrinkage strains, 421
    resin mechanical behaviour during processing, 422–3
    stages in resin property development, 423
    stress analysis solution method, 424
    thermal strains, 420–1
    tool removal simulation, 425
  future trends, 432
  model, 417–25
    COMPRO structure and program flow, 418
    flow compaction, 419–20
    heat transfer and cure, 418–19
  process development, 425–32
    CFD simulation of airflow, 427
    cured test article placed on process tooling, 430
    dimensional control, 429–32
    effect of cure on temperature history, 429
    predicted vs measured out-of-plane deformation, 431
    temperature profile evolution, 429
    thermal histories evaluation, 428
    thermal management, 426–9
    tool removal process schematic, 426
  typical cycle, 416
  typical system, 415
automated tape laying (ATL), 445

B-factor, 235
ball milling, 109
Barone–Caulk model, 74, 76
Bentonite, 98
bias extension test, 161–3
  actual direction of the yarns from the DIC, 163
  setup, 162
  shear angle contours, 163
biocomponent fibres, 197
bipolar plate, 52
blisters, 62
braiding pultrusion, 401
bulk moulding compound (BMC), 4, 56–7
  continuous compounding system, 57

cantilever beam, 159–60
carbene, 114
carbon fibres, 254
carbon nanofibre (CNF), 5, 95
carbon nanotube
  alignment, 110–12
    ex situ, 110–11
    induced, 111–12
  dispersion, 108–10
  functionalisation, 112–15

carbon nanotube (*Cont.*)
covalent, 112–13
non-covalent, 114–15
SWCNT sidewall methods, 113
paper-based nanocomposites, 103–5
selection, 108
Carmen-Kozeny model, 355
chopped strand mats (CSM), 255
co-injection moulding, 17
illustration, 18
co-injection resin transfer moulding (CIRTM), 250
coefficient of friction, 152
commingled yarns, 196–7
illustration, 196
thermoplastic, 213
Composite Design and Simulation Software (CDS), 201
composite manufacturing
polymer matrix composites, 1–11
overview, 3–11
processing, 1–3
compression moulding, 3, 4
future trends, 90–2
materials, 53–61
bulk moulding compound (BMC), 56–7
directionally reinforced moulding compound (XMC), 58
glass mat thermoplastics (GMT), 58–61
sheet moulding compound (SMC), 53–6
thick moulding compound (TMC), 57–8
modelling and simulation, 69–90
dimensions, loading and supporting conditions of plate, 90
fibre orientation, 81–5
fibre separation, 85–8
material flow and mould filling, 74–81
maximum displacements under the load for three cases, 91
numerical analysis issues, 70
precharge location and dimensions, 91
structural performance and process parameters, 88–90
temperature and cure cycle, 71–4
overview, 47–53
automotive FEM (front end module) carrier, 51
industrial application, 50–3
particles forming percolation networks, 52
polymer-graphite bipolar plates for direct methanol fuel cell, 52
process description, 47–50
temperature and pressure variation, 49
time for each operation cycle, 49
polymer matrix composites, 47–92
press design and process optimisation, 67–9
principal parameters, 68
process window, 70
process defects and remedies, 61–7
blisters, 62
compression moulded parts, 61
delamination, 62
fibre separation, 65–7
knit lines, 63–4
preferential fibre orientation, 64–5
sink marks, 62–3
voids, 61–2
warpage or residual stress, 64
compression resin transfer moulding (CRTM), 8, 250–1, 348–76
CRTM-1 process modelling and analysis, 357–64, 365, 366, 367, 368
4 mm thick truncated pyramid part geometry, 366
normal stress distributions comparison, 362, 363
solution methods, 360
total clamping force trace comparisons, 364, 365, 367
total force trace comparisons, 368
validation, 360–4
viscoelastic compaction model application, 358
CRTM-2 process modelling and analysis, 364–9

future trends, 375–6
material properties and characterisation, 355–7
compress and hold compaction tests, 356
optimisation, 370–5
final result for helmet geometry optimisation, 375
fireman's helmet, 374
Genetic Algorithm, 373
output from GA for helmet geometry, 374
process description, 351–4
CRTM-1 and CRTM-2 process variants, 354
injection flow rate, cavity thickness and clamping force, 353
key process parameters, 352
schematic description, 350
COMPRO, 417, 430
conjugated polymers, 115
consolidation, 125
degree, 202
force, 189–90
compactor roller, 190
filament break, 189
continuous approach, 159–60
continuous compression moulding, 7, 225–38
application areas, 238–9
circumference of different profiles contacting the die, 239
specific production of pultrusion and CCM process, 239
material in use, 225–7
machine, 226
powder prepreg surface, 228
powder prepreg unit, 227
mechanical properties, 237–8
crash test behaviour of closed cap profile, 238
modelling, 235–6
approach for B-Factor determination, 236
B-factor calculation, 236
process, 227–31
creel of CCM press and strap brake, 228
different release films made of paper, polyetherimd and steel, 229
heating and cooling section of a CCM machine, 230
material supply/input, 228–9
pressing unit, 229–31
sketch of a pressing unit, 230
profile production, 231–5
capable profile geometries, 234
CCM profiles, 232
closed cap profiles with different reinforcements and matrix materials, 234
fibre drawn in profile curves, 235
pre-forming unit in front of the CCM machine, 233
pull-off unit, 232
continuous fibre composites, 129
continuous fibre reinforced profiles, 209–39
application areas, 238–9
circumference of different profiles contacting the die, 239
specific production of pultrusion and CCM process, 239
continuous compression moulding, 225–38
approach for B-Factor determination, 236
B-factor calculation, 236
capable profile geometries, 234
CCM profiles, 232
closed cap profiles with different reinforcements and matrix materials, 234
crash test behaviour of closed cap profile, 238
creel of CCM press and strap brake, 228
different release films made of paper, polyetherimd, and steel, 229
fibre drawn in profile curves, 235
heating and cooling section of a CCM machine, 230
machine, 226
material in use, 225–7

continuous fibre reinforced profiles (*Cont.*)
mechanical properties, 237–8
powder prepreg surface, 228
pre-forming unit in front of the CCM machine, 233
pressing unit sketch, 230
process, 227–31
process modelling, 235–6
profile production, 231–5
pull-off unit, 232
schematic drawing of powder prepreg unit, 227
pultrusion, 210–24
commercial industrial and consumer applications, 225
commercial products and applications, 224, 225
common polymer materials properties for the pultrusion process, 212
creel assembled with glass fibre, 215
different pull off units, 220
elements of resistance in the die, 219
horizontal pultrusion unit, 211
material in use, 210–14
on-line forming of profiles, 224
PAZ process, 223
pin impregnation of the reinforcement material, 217
prepreg types in use, 214
process, 214–21
process chart, 214
process combinations and variations, 221–4
pull-extrusion process, 223
pull-winding technology, 222
pultrusion die, 216
reinforcement fibres tension control, 215
reinforcement materials properties, 212
special profile geometries, 218
standard profile geometries, 218
viscosity of reactive and non-reactive thermoplastic materials, 213
continuous fibres, 1, 2
continuous filament mat (CFM) reinforcement, 374
continuous filament random mats (CFRM), 256
control volume, 297
cooling, 125
Creative Pultrusions Inc., 407
cross winding, 184–5
cure shrinkage, 421
cure-hardening/instantaneously linear elastic (CHILE) material, 424
curing, 48
cycle, 71–4
model parameters for SMC, 73
model parameters for unsaturated polyester and vinyl ester, 72
Cycom 5320, 444

Darcy's law, 279, 280, 315, 419
velocity, 281, 359
Darcy, H., 279
delamination, 62
Delesseian Law, 360
diaphragm, 132–4
precursor between extendable sheets, 133
diazonium, 113–14
dielectric sensors, 302–3
digital image correlation (DIC), 149, 162
dip bath, 385
dipole rotation, 460
direct impregnation, 193–5
air pressure vessels for truck application, 196
multifeed configuration, 195
process chains, 194
direct synthesis, 105
directionally reinforced moulding compound (XMC), 4, 55, 58
production, 59
discrete approach, 159–60
dissolvable cores, 191
domino pushing, 104–5
illustration, 105
Double Dome, 144
double-walled carbon nanotube (DWCNT), 95
dough moulding compound (DMC), 57

E-beam curing *see* electron-beam curing
E-glass fibres, 254
electric time-domain reflectometry (E-TDR) wire sensors, 302
electron-beam curing, 467–71
 E-beam cured composite laminate, 471
 electron beam accelerator, 469
 principle, 469
 pros and cons, 471
electrophoretic deposition (EPD), 106–7
 process, 107
epoxy resin, 97–8
expanded graphite (EG), 97–8

fabric, 254–5
 architecture, 145–7
  harness satin-weave, 146
  stitched double-bias, 147
  twill-weave, 145
  unidirectional, 146
 bending stiffness, 155–9
 deformation, 155–9
  bending stiffness test, 157
  cantilever experimental setup, 156
  comparison of experimental data to FE model, 158
  FE model of a fabric sample compressed from each side, 158
  post-processing of experimental data, 157
 friction-test, 154
fabric thermostamping, 6
 future trends, 175
 material characterisation, 147–59
  deformation modes, 155–9
  friction, 151–5
  shear frame testing, 147–50
  tensile testing of yarns, 150–1
 methods for product quality improvement, 169–74
 modelling, 159–69
  bias extension, 161–3
  discrete mesoscopic model principle, 161
  forming process, 164–9
 overview, 139–42
  in-plane yarn waviness and out-of-plane wave defects, 141
  plain-weave fabric, 141
  process, 142
  shearing of a deformed woven-fabric, 142
 polymer matrix composites, 139–78
 process description, 142–7
  fabric architecture, 145–7
  finite element model of the double dome tooling, 143
  formed parts resulting from the process using Twintex, 144
  Twintex commingled glass/polypropylene yarns, 144
fast remotely actuated channelling (FASTRAC), 249–50
FiberSIM, 274
fibre
 buckling, 273
 deformation, 135–6
  changes in orientation during forming, 135
 exposure, 38–9
  phenomenon in moulded composite parts, 39
 extension, 274
 orientation, 81–5
  distribution function and tensor, 82
  experimental and simulation results in compression moulding, 84
  single representation on spherical coordinates, 81
 placement, 135–6
fibre motion model, 201
fibre segregation *see* fibre separation
fibre separation, 65–7, 85–8
 content distribution in compression moulded part, 89
 drag force, 86
 fibre interaction and acting forces, 85
 fibre orientation dependence, 66
 homogeneous and heterogeneous suspension, 66
 orientation distribution function with respect to angle and flow directions, 87

fibre separation (*Cont.*)
resin rich zone at the end of the rib, 67
rib and flange dimension, 67
fibre-reinforced polymer (FRP) composites, 381
fibreoptic sensors, 302
filament winding, 6
basics, 184–6
comparison of hoop, helical and polar winding, 185
filament winding *vs.* lay-up process, 185
process chain, 184
component quality, 202–3
consolidation degree, 202
inhomogeneous fibre matrix distribution, 203
oxidation/degradation, 203
residual stress, 203
void content, 202
future trends, 203–4
component quality, 204
machinery and materials, 203–4
simulation, 204
overview, 182–4
possible forms, 183
process, 186–99
comparison different supply forms of semi-finished materials, 200
configurations, 191–2
different techniques comparison, 199
head technologies, 192–9
steps, 186–91
simulation tools, 199–202
path generation, 199–200
process simulation, 200–2
thermoplastics, 182–204
filled polymers, 20–1
fillers, 388
finite element, 296
flat bed, 191–2
explanation of the different axis, 193
flow under pressure, 125
fluid control panel, 19
fluid-assisted injection moulding, 17–19
process setup, 19
fluorination, 113
fluorocarbon single-walled carbon nanotubes (F-SWCNT), 113
Folgar–Tucker model, 35–6
forming process, 164–9
90/0 fabric orientation, 164
90/0 fabric orientation final stage experimental-simulation comparison, 165
90/0 part showing large spacing of the weft yarn, 167
detail 90/0 fabric orientation final stage experimental-simulation comparison, 166
finite element mesh of the floorpan punch, 168
finite element mesh of the tub punch, 168
plan view of floorpan demonstration part, 172
shear angle contour 15/105 and 60/-3, 170
shear angle contour in degrees for the 45/-45 fabric orientation, 169, 171
shear angle contour in degrees for the 90/0 fabric orientation, 170, 171
yarn tension contour for 45/-45 fabric orientation, 174
yarn tension contour for 90/0 fabric orientation, 173
yarn tension contour for the 22.5/112.5 fabric orientation, 172
friction, 151–5
coefficient of friction *vs.* displacement curves, 156
fabric friction-test apparatus, 154
fabric/fabric setup, 155
flow curves for polypropylene at different temperatures, 153
Stribeck curve, 153
front end module (FEM), 51
fully consolidated fibres, 198–9
illustration, 198
thermoplastic tape wound bearing, 199

Galerkin finite element method, 297
gantry winding systems, 191–2
  explanation of the different axis, 193
  illustration, 193
gap-wise averaged velocities, 75
generalised Hele-Shaw model, 74–6
Genetic Algorithm (GA) approach, 373
glass fibre reinforced polypropylene, 30
glass fibre reinforced polystyrene, 29–30
glass fibres, 254
glass filament reinforced pellets, 29
glass filament un-reinforced pellets, 29
glass mat thermoplastics (GMT), 4, 58–61
  compression moulding process, 61
  production, 60
Goldsworthy, W.B., 210
grapheme sheets (GS), 5
graphene paper-based nanocomposites, 106–7
graphite nanoplatelet (GNP), 95
green strength, 247

heat transfer fluid (HTF), 446
heated mandrels, 190
heating, 125, 129
heating zone, 187–9
  heating methods and their thermal expansion, 188
  illustration, 189
helical winding, 184
HexPly M56, 445
high expansion ration shelter (HERS), 405
high viscosity thermoplastic, 5
hoop winding, 185
hydraulic presses, 68

I-DEAS, 279
in situ consolidation, 183
in situ polymerisation, 96, 100–1, 466–7
  CBT to PBT polymerisation, 467
  process, 100
  pros and cons, 468
in-line braiding process, 222
in-mould coating, 69
in-mould decoration (IMD) injection moulding, 19–20
  process, 20
induction heating, 454–6
  concept employed at RocTool systems, 455
  pros and cons, 455
infrared, 456, 457
  medium wave IR oven for curing carbon composites, 457
  pros and cons, 457
inhomogeneous fibre matrix distribution, 203
injection and double vacuum assisted resin transfer moulding (IDVARTM), 336
injection compression moulding, 250
injection moulding, 3, 4
  characterisation and prediction of fibre orientation, 34–7
    definition of the orientation angles, 35
    fibre orientation in injection moulded plates, 36
    long-fibre orientation in injection moulded plates, 37
  compounds, 21–33
  defects, 37–41
    fibre exposure, 38–9
    jetting, 37–8
    warpage, 341
    weldlines, 39–40
  future trends, 41–2
  machines, 18–19
  overview, 15–21
    co-injection (sandwich), 17
    fluid-assisted injection, 17–19
    illustration, 16
    in-mold decoration (IMD) injection, 19–20
    microcellular injection of filled polymers, 20–1
  polymer matrix composites, 15–42
injection unit, 18–19
inter-bundle microvoid, 276
inter-fibre (intraply) shear, 273
inter-fibre slip, 273
internal fibre motion, 125
intra-bundle microvoid, 276

jetting, 37–8
  tab gate, 38

KaZaK Composites Inc., USA, 405
Kevlar, 254
knit lines, 63–4
  L-shaped and circular opening mould, 63
Kozeny-Carman equation, 320
Kronecker delta, 419
light resin transfer moulding (LRTM), 250
LIMS, 295
liners, 191
liquid composite moulding (LCM), 7, 11, 246, 348, 440, 462–3
  different processes by category, 462
  pros and cons, 466
Load Resistance Factor Design (LRFD), 405
long fibre composites, 128
long fibre reinforced thermoplastics (LFT), 51, 59
long glass fibre reinforced injection moulding, 32–3
  Izod impact strength and tensile strength of long glass fibre reinforced parts, 33
  length distribution of glass fibre moulded by conventional plastication unit, 34
low equilibrium model, 292–3

macrovoids, 274
mandrels
  different components, 191
    disassembled, 192
  lay-on, 190
Mastercore System Ltd., Canada, 406
mat, 254
matched die, 130–2
  two tool surfaces to drive the composite sheet into final shape, 131
material flow, 74–81
  illustration, 75
material rack, 186–7
matrix flow, 134
matrix systems, 3
melding, 451–2
melt-mixing, 96, 102
melting, 101–3
  illustration, 102
membrane-based infusion processing, 339–40
  process description and infusion behaviour, 339–40
  VAP process lay-up, 340
  VARTM and VAP processing thickness and pressure behaviour, 341–2
microwave, 456–60
  GKN uses microwave curing to reduce OoA cure cycles, 458
  hotspots during microwave cure, 459
  microscopic principle, 459
  pros and cons, 460
mirocellular injection moulding, 20–1
moldability diagram, 30–1
montmorillonite (MMT), 5, 95
mould closure, 48
mould filling, 74–81
  closing force *vs.* cavity height for different closing speeds, 80
  geometry, mesh, initial charge position and boundary condition, 79
  patterns *vs.* fibre orientations, 81
mould temperature, 68
moulding compounds, 21–33
  fibre orientation, 31–2
    injection and fluid-assisted injection moulded parts, 32
  glass fibre filled materials, 29–31
    fibre length distribution of injection moulded short and long glass fibre, 30
    moldability diagram for the injection moulding process, 31
  long glass fibre reinforced injection moulding, 32–3
  mechanical properties of injection moulded plaques, 26–8
  processing properties, 23–5
  rheological properties, 22, 29
multi-walled carbon nanotube (MWCNT), 5, 95, 97, 111

multistage curing (MSC) technique, 329
mutation, 373

nanoclay and polyhedral oligomeric silsesquioxanes (POSS), 5
nanoparticle/polymer/solvent solution, 96–7
nanoparticles, 3
nitrene, 114
non-crimp fabrics (NCF), 257
non-heated mandrels, 190–1
NSGA-II (Non-dominated Sorting Genetic Algorithm), 373

on-line forming process, 224
open bath, 385
out-of-autoclave curing process, 11, 435–78
  books for further reference, 478
  cure kinetics, 462–72
    CBT to PBT polymerisation, 467
    different liquid composite moulding processes by category, 462
    E-beam cured composite laminate, 471
    E-beam curing pros and cons, 471
    electron beam accelerator, 469
    electron-beam curing, 467–71
    example of SQRTM, 465
    in situ polymerisation, 466–7
    in situ polymerisation pros and cons, 468
    LCM processing pros and cons, 466
    liquid composite moulding, 462–3
    principle of electron beam curing, 469
    resin film infusion, 463–4
    resin transfer moulding, 463
    RFI pros and cons, 464
    same qualified resin transfer moulding, 464–6
    vacuum assisted resin infusion, 463
  electromagnetic heating, 452–61, 462
    concept of induction heating employed at RocTool systems, 455
    electromagnetic spectrum 2, 453
    electromagnetic spectrum for radio frequency, 461
    GKN uses microwave curing to reduce OoA cure cycles, 458
    hotspots during microwave cure, 459
    induction heating, 454–6
    infrared, 456, 457
    medium wave IR oven for curing carbon composites, 457
    microscopic principle of microwave heating, 459
    microwave, 456–60
    pros and cons of induction heating, 455
    pros and cons of infrared heating, 457
    pros and cons of microwave heating, 460
    radio frequency curing, 460
    RFC pros and cons, 462
  fluid heating, 446–52
    fluid heated tools pros and cons, 449
    liquid heated tools, 447–50
    principle of Quickstep fluid curing, 450
    pros and cons of Quickstep, 452
    QSE250 curing machine, 451
    Quickstep, 450–2
    simple pressurised water system to rapidly heat and cool the thin nickel tool, 448
  future trends, 475–8
    combination, 476–7
    high volume, 477–8
    large integral parts, 477
    materials, 475–6
  oven/vacuum bag only processes, 440–6
    ATL of Hexcel's M56 layers, 446
    available OoA prepregs, 443
    Boeing's target for 3rd generation OoA prepregs, 443
    cargo aircraft's fuselage, 446
    convection oven, 441
    evacuation of entrapped air from VBO prepreg, 444

out-of-autoclave curing process (*Cont.*)
long evacuation for low void, 443–5
material qualification, 442–3
oven curing pros and cons, 442
possibility of automation, 445–6
vacuum bag only - OoA prepregs, 441–2
VBO pros and cons, 447
process comparison and classification, 472–5
cycle time vs investment for different manufacturing processes, 472
different OoA processes divided in categories, 474
maturity of different OoA processes in 2011, 475
OoA processes by performance criteria, 473
techno-economic portfolios, 472–
reasons for using out-of-autoclave process, 435–8
example of an autoclave, 437
growth for different markets in terms of carbon fibres, 437
strategies, 438–9
composite manufacturing, 439
technical description of different processes, 439–71
convection oven, 441
OoA approaches by category, 440
oxidation/degradation, 203

PAM-RTM, 295
part release, 48
path generation, 199–200
PATRAN, 279
PAZ, 222
photo initiators, 470
plain-weave fabric, 146
plastic deformation, 123
polar winding, 184
polyhedral oligomeric silsesquioxanes (POSS), 95
polymer matrix, 107–8
polymer matrix composites
autoclave processing, 414–32
future trends, 432
model, 417–25
process development, 425–32
composite manufacturing, 1–11
compression moulding, 47–92
future trends, 90–2
materials, 53–61
modelling and simulation, 69–90
overview, 47–53
press design and process optimisation, 67–9
process defects and remedies, 61–7
compression resin transfer moulding, 348–76
CRTM-1 process modelling and analysis, 357–64
CRTM-2 process modelling and analysis, 364–9
future trends, 375–6
material properties and characterisation, 355–7
optimisation, 370–5
process description, 351–4
continuous fibre reinforced profiles, 209–39
application areas, 238–9
continuous compression moulding, 225–38
pultrusion, 210–24
fabric thermostamping, 139–78
future trends, 175
material characterisation, 147–59
methods for product quality improvement, 169–74
modelling, 159–69
overview, 139–42
process description, 142–7
injection moulding, 15–42
characterisation and prediction of fibre orientation, 34–7
compounds, 21–33
defects, 37–41
future trends, 41–2
overview, 15–21
out-of-autoclave curing process, 435–78
future trends, 475–8
process comparison and classification, 472–5

reasons for using out-of-autoclave process, 435–8
strategies, 438–9
technical description of different processes, 439–71
overview, 3–11
processing, 1–3
pultrusion process, 381–408
improvements, 391–403
innovation, industrial and future trends, 403–8
process description, 384–91
resin transfer moulding, 245–303
fibres, fabrics and preform manufacturing, 253–9
future trends, 303
heat transfer and cure model, 292–5
issues that influence manufacturing, 270–7
need for process modelling, 277–8
numerical simulation of resin flow, 295–300
process control, 300–3
process steps, 252–3
resin flow models, 278–92
resin injection equipment, 267–70
resin system, 259–61
RTM mould, 261–6
sheet forming, 123–37
future trends, 136–7
key objectives, 123–4
matrix flow and fibre deformation, 134–6
process changes for product quality improvement, 136
process description, 124–34
vacuum assisted resin transfer moulding (VARTM), 310–45
defects and challenges, 323–35
fundamentals, 315–23
future trends, 337–9
membrane evaluation, 340–2
membrane-based infusion processing, 339–40, 344–5
process and material property improvement, 342–4
recent advances, 335–7
VARTM processing, 310–15
polymer matrix, polyhydroxyaminoether (PHAE), 97
polymer nanocomposites
methods to improve process, 107–15
carbon nanotubes alignment, 110–12
carbon nanotubes dispersion, 108–10
carbon nanotubes functionalisation, 112–15
carbon nanotubes selection, 108
polymer matrix selection, 107–8
process description, 96–107
carbon nanotube paper, 103–5
graphene paper, 106–7
in situ polymerisation process, 100–1
melting process, 101–3
solution processing, 96–100
processing, 95–115
powder impregnated fibres, 197–8
illustration, 198
precharge placement, 48
precharge preheating, 69
precharge preparation, 48
preferential fibre orientation, 64–5
changes during compression moulding process, 64
distribution with respect to the initial coverage of the mould by SMC, 65
preform manufacturing, 257–9
common processes, 258–9
3D weaving, 259
braiding, 259
filtration of chopped fibre slurry on perforated screen, 259
knitting, 259
spray-up of chopped fibres and binder, 258
stamping of mats and fabrics in two-sided mould, 259
fibre and fabric types and fibre volume fraction selection, 257–8
cost, 258
fibre wet out, 258

preform manufacturing (*Cont.*)
formability, 258
mechanical performance, 257–8
permeability to resin flow, 258
pressure filtration, 103
pressure sensors, 301
ProSimFRT, 202
pulforming, 401
pull-braiding process schematic, 401
pull-extrusion process, 222–4
pull-winding process, 221
pultrus, 7
pultrusion process, 210–24, 381–408
application areas, 238–9
circumference of different profiles contacting the die, 239
specific production of pultrusion and CCM process, 239
applications, 383–4
advantages, 383
limitations, 384
commercial products and applications, 224, 225
industrial and consumer applications, 225
components of pultrusion process operation, 384–7
cut-off saw, 387
forming guides, 386
hollow section pultrusion requiring mandrels, 385
puller and clamp system, 386–7
reinforcement dispenser, 384–5
resin impregnator, 385–6
temperature controlled die, 386
type of die design, 387
controlling equations and formulae, 389–91
changes in the material properties, 390–1
governing equations, 389–90
developments, 404–7
global market, 405
KaZaK composites Inc.'s high expansion ratio shelter, 406
novel applications, 405–7
pultruded hollow wing without internal ribs, 406
pultruded planks used as scaffolding for bridge construction repair, 407
sheet piling cross-section, 407
standards, 405
future trends, 408
historical background, 382
improvements, 391–403
studies conducted in various key features and problem areas, 391
injection pultrusion improvements, 398–9
experimental studies and optimisation, 398–9
numerical simulation and optimisation, 399
schematic, 398
innovation, 403–4
innovation, industrial and future trends, 403–8
material in use, 210–14
common polymer materials properties, 212
horizontal pultrusion unit, 211
prepreg types in use, 214
reinforcement materials properties, 212
viscosity of reactive and non-reactive thermoplastic materials, 213
materials, 387–8
microwave pultrusion improvements, 400
experimental studies and optimisation, 400
numerical simulation and optimisation, 400
other varieties pultrusion improvements, 401–2
bent pultrusion schematic, 402
braiding pultrusion, 401
pulforming, 401
pull-braiding process schematic, 401
sandwich pultrusion, 402–3
UV assisted bent pultrusion, 402
process, 214–21
advantages and disadvantages, 221

creel assembled with glass fibre, 215
different pull-off units, 220
elements of resistance in the die, 219
impregnation, consolidation and solidification, 216–18
material supply, 214–15
pin impregnation of the reinforcement material, 217
process chart, 214
pull-off and cut-off unit, 220–1
pulling force, 218–19
pultrusion die, 215–16
pultrusion die illustration, 216
special profile geometries, 218
standard profile geometries, 218
process combinations and variations, 221–4
on-line forming of profiles, 224
PAZ process, 223
pull-extrusion process, 223
pull-winding technology, 222
process description, 384–91
schematic, 382
schematic by Schlueter and Bond, 404
technological challenges, 388–9
heat transfer and pull speed, 389
material composition, 388
pultrudate shaping, 388–9
traditional pultrusion, 382, 391–8
correlation between FEM and FDM pultrusion analyses, 397
die-preform assembly in thermoplastic pultrusion, 392
experimental studies and optimisation, 392–4
four-heater pull-die configuration, 396
numerical simulation and optimisation, 394–8
state of cure before and after optimisation, 397
variations in highest temperature within the composite and mean DOC, 396

Quickstep, 11, 450–2
principle, 450
pros and cons, 452
QSE250 curing machine, 451

racetracking, 271
racetracking channels, 271–3
radio frequency curing (RFC), 460
electromagnetic spectrum for radio frequency, 461
pros and cons, 462
Reactive Injection Pultrusion (RIP), 217
reduced-strain closure (RSC), 36
reproduction, 373
residual stress, 64, 203
resin film infusion, 9, 124, 463–4
pros and cons, 464
resin flow phenomenon, 315–19
resin infusion between double flexible tooling (RIDFT), 251
resin infusion under flexible tooling (RIFT), 9
resin injection equipment, 267–70
resin transfer moulding, 245–303
compression RTM process, 251
FASTRAC process, 250
fibres, fabrics and preform manufacturing, 253–9
common fibre mats and fabrics used in LCM, 255
fibres, 253–4
mats and fabrics, 254–7
periodic unit cell of plain weave fabric ply, 256
preform manufacturing, 257–9
flow of liquid thermoset polymer through porous fabric preform, 246
future trends, 303
heat transfer and cure model, 292–5
resin cure models, 294
temperature and cure dependent viscosity, 294–5
issues that influence manufacturing, 270–7, 278
dual scale fibre structure in a preform, 276
effect of racetracking channel on flow patterns, 273

resin transfer moulding (*Cont.*)
- fibre structure deformation during draping, 273–4
- formation of intra-bundle microvoids, 278
- macrovoid formation, 274
- macrovoids trapped in composite part, 275
- material and process variations that can alter flow in RTM, 271
- microvoid formation, 276–7
- microvoids trapped in composite part, 277
- mould cavity edge with and without racetracking channel, 272
- racetracking channels, 271–3
- transverse flow in the thickness direction, 274–5

need for process modelling, 277–8
numerical simulation of resin flow, 295–300
- mould filling simulations, 299–300
- properties of polyester resin system with inhibitor and glass fibre, 296

permeability measurement experimental set-up
- steady 1D constant-flow rate injection, 286
- transient filling 1D constant-flow rate injection, 285
- transient filling 1D constant-pressure injection, 286
- transient filling radial constant-flow rate injection, 287

process control, 300–3
- sensor systems for flow and cure monitoring, 301–3

process steps, 252–3
- demoulding and final processing, 253
- fibre preform manufacturing, 252
- lay-up and draping, 253
- mould closure, 253
- resin injection and cure, 253

resin flow models, 278–92
- anisotropic Darcy flow through porous medium, 280–3
- boundary conditions for resin flow, 283, 284
- Darcy flow through a porous medium, 279–80
- experimental set-up used by Darcy in 1856, 280
- permeability experimental characterisation, 283–92
- permeability measurement experiments, 288–9
- samples of experimentally measured permeability values for reinforcing materials, 291
- volume-averaged Darcy's velocity through a porous fibre preform, 281

resin injection equipment, 267–70
- classification, 267
- flow-rate controlled injector and pressure-controlled injector, 269
- inlet pressure and flow rate, 268

resin system, 259–61
RTM mould, 261–6
- different silicone seals and metal-to-metal joint, 266
- lower plate of an RTM mould, 264
- matched (two-sided) mould to manufacture complex composite structure, 263
- matched mould made from smaller pieces, 263
- with two ejector pins, 266

SCRIMP process, 248
VARTM process, 248
VIPR process, 249

resin transfer moulding (RTM), 8, 124, 140, 463
resistive sensors, 301–2
Reynolds number, 280
robotised winding, 192
RocTool, 454
Rohacell, 227
RTM-Worx, 295
rubber pad, 132
rule of mixtures, 257

S-glass fibres, 254
same qualified resin transfer moulding (SQTRM), 464–6
  example, 465
sandwich moulding *see* co-injection moulding
sandwich pultrusion, 402–3
satin-weave fabric, 145
seeman's composite resin infusion moulding process (SCRIMP), 248–9
semi-discrete approach, 159–60
semi-finished materials, 195
shear frame test, 147–50
  before and after deformation, 149
  digital image correlation contour, 150
  illustration, 148
  woven-fabric shear behaviour curve, 149
shear mixing, extrusion and milling, 109–10
sheet forming
  future trends, 136–7
  key objectives, 123–4
    turning of flat metal disk into a three-dimensional structure by plastic deformation, 124
  matrix flow and fibre deformation, 134–6
  polymer matrix composites, 123–37
  process changes for product quality improvement, 136
  process description, 124–34
    design, 125–8
    diaphragm method, 132–4
    fundamentals, 128–30
    heating of flat laminated materials by conveying oven, 130
    interactive design and development, 126
    matched die method, 130–2
    physical, 125
    picture frame mould, 127
    rubber pad method, 132
sheet moulding compound (SMC), 4, 48, 50–1, 53–6
  additives for resin paste formulation, 56
  composition of SMC-R in weight fraction, 55
  designation, 55
  manufacturing process, 54
short fibre composites, 1, 3
  illustration, 128
SimLCM, 361, 373
single-walled carbon nanotube (SWCNT), 5, 95, 109, 111, 112, 113–14, 114–15
sink marks, 62–3
  rib corner designs and prevention, 63
skin–core–skin fl ow sequence, 17
SMARTweave, 302
SMC-CR, 55
SMC-R, 55
SolidWorks, 279
solution processing, 96–100
  clay/polymer composites structures, 99
  illustration, 97
  solvents properties, 98
sonication, 108–9
spin-coating, 104
spring-in, 330–1
strain tensor, 77
Stress module, 417
stress tensor, 77
structural reaction injection moulding (S-RIM), 250
SuperLoc profiles, 407

tape materials, 213
technology readiness levels (TRL), 473
Tekscan products, 301
temperature, 71–4
  changes and degree of cure profiles during SMC moulding, 74
  profile change with time in thickness direction during SMC moulding, 73
tensile test, 150–1
  fabric tensile behaviour determination, 151
  load-displacement curve from yarn, 151
tension controller, 187
Terzaghi's Law, 359

Thermochemical module, 417
thermocouples, 301
thermoplastic matrix materials, 213
thermoplastic prelaminate, 5
thermoplastics, 1
  filament winding, 182–204
    basics, 184–6
    component quality, 202–3
    future trends, 203–4
    overview, 182–4
    process, 186–99
    simulation tools, 199–202
thermoplastics resins, 388
thermosets, 1
thermosetting resins, 388
thick moulding compound (TMC), 4, 57–8
  production, 58
three-roll mills *see* shear mixing, extrusion and milling
trellis-frame test *see* shear frame test
twill-weave fabric, 145
Twintex, 144–5, 237

ultrasonic sensors, 302
unidirectional fabrics (UDF), 257
UV assisted bent pultrusion, 402

vacuum assisted resin infusion (VARI), 463
vacuum assisted resin transfer moulding (VARTM), 8–9, 247–8, 310–45
  defects and challenges, 323–35
    air entrapment, 323–8
    comparison of experimental spring-in angles and modelling results, 334
    curing and thermal management, 329
    feasibility of automated VARTM design process, 326
    illustration of spring-in problem, 330
    material properties of polyester resin, E-glass fibre and CNFs, 333
    microvoid in a random E-glass fibre/unsaturated polyester FRP sample, 327
    spring-in, 329–35
    spring-in angle data comparison, 333
    thickness and fibre volume fraction uniformity, 328–9
  fundamentals, 315–23, 324, 325
    analytical VARTM flow model results, 319
    composite curing behaviour, 321–3
    critical elements of VARTM process design, 323
    cure kinetic parameters for epoxy resin and unsaturated polyester resin, 322
    fibre preform compaction, 319–21
    geometry and parameters of analytical VARTM flow model, 317
    resin flow phenomenon, 315–19
    resin viscosity, 321
    simulated temperature and curing history of VARTM experiment, 324–5
    thermal properties of random E-glass fibre/polyester composites, 322
  future trends, 337–9
  membrane evaluation, 340–2
    Albatros membrane, 343
    permeability behaviour for Albatros membrane/fluid system, 344
  membrane-based infusion processing, 339–40, 344–5
    process description and infusion behaviour, 339–40
    VAP process lay-up, 340
    VARTM and VAP processing thickness and pressure behaviour, 341–2
  process and material property improvement, 342–4
    fibre volume fraction and void content for panels, 344
  recent advances, 335–7
    IDVARTM process set-up, 336
  VARTM processing, 310–15

major advantages and disadvantages, 313–15
set-up and procedure, 311–13
typical process, 311
vacuum bag only (VBO), 435, 440, 441–2
vacuum bag only out-of-autoclave (VBO-OOA), 11
vacuum filtration, 103, 106
vacuum induced preform relaxation (VIPR), 249
vacuum injection preform relaxation (VIPR)
vacuum-assisted moulding, 69
vacuum-assisted process (VAP) process, 339
void content, 202
voids, 61–2
SMC flow showing defect, 62

warpage, 64, 341
weldlines, 39–40
formation, 40
whiskers, 1
winding angle, 184
woven fabrics, 256–7